Edited by
Tran Quoc Khanh, Peter Bodrogi,
Quang Trinh Vinh, and Holger Winkler

LED Lighting

Related Titles

De La Rue, R., Yu, S., Lourtioz, J. (eds.)

Compact Semiconductor Lasers

2014
ISBN: 978-3-527-41093-4

Bodrogi, P., Khanh, T.Q.

Illumination, Color and Imaging

Evaluation and Optimization of Visual Displays

2012
ISBN: 978-3-527-41040-8
Also available in digital formats

Okhotnikov, O.G. (ed.)

Fiber Lasers

2012
ISBN: 978-3-527-41114-6
Also available in digital formats

Osten, W., Reingand, N. (eds.)

Optical Imaging and Metrology

Advanced Technologies

2012
ISBN: 978-3-527-41064-4
Also available in digital formats

Tsujimura, T.

OLED displays fundamentals and applications

2012
ISBN: 978-1-118-17306-0

Cristobal, G., Schelkens, P., Thienpont, H. (eds.)

Optical and Digital Image Processing

Fundamentals and Applications

2011
ISBN: 978-3-527-40956-3
Also available in digital formats

Okhotnikov, O.G. (ed.)

Semiconductor Disk Lasers

Physics and Technology

2010
ISBN: 978-3-527-40933-4
Also available in digital formats

*Edited by
Tran Quoc Khanh, Peter Bodrogi, Quang Trinh Vinh, and
Holger Winkler*

LED Lighting

Technology and Perception

Verlag GmbH & Co. KGaA

Editors

Prof. Tran Quoc Khanh
Technische Universität Darmstadt
Laboratory of Lighting Technology
Darmstadt, Germany

Dr. Peter Bodrogi
Technische Universität Darmstadt
Laboratory of Lighting Technology
Darmstadt, Germany

Dr. Quang Trinh Vinh
Technische Universität Darmstadt
Laboratory of Lighting Technology
Darmstadt, Germany

Dr. Holger Winkler
Merck KGaA
Darmstadt, Germany

All books published by **Wiley-VCH** are carefully produced. Nevertheless, authors, editors, and publisher do not warrant the information contained in these books, including this book, to be free of errors. Readers are advised to keep in mind that statements, data, illustrations, procedural details or other items may inadvertently be inaccurate.

Library of Congress Card No.: applied for

British Library Cataloguing-in-Publication Data
A catalogue record for this book is available from the British Library.

Bibliographic information published by the Deutsche Nationalbibliothek
The Deutsche Nationalbibliothek lists this publication in the Deutsche Nationalbibliografie; detailed bibliographic data are available on the Internet at <http://dnb.d-nb.de>.

© 2015 Wiley-VCH Verlag GmbH & Co. KGaA, Boschstr. 12, 69469 Weinheim, Germany

All rights reserved (including those of translation into other languages). No part of this book may be reproduced in any form – by photoprinting, microfilm, or any other means – nor transmitted or translated into a machine language without written permission from the publishers. Registered names, trademarks, etc. used in this book, even when not specifically marked as such, are not to be considered unprotected by law.

Print ISBN: 978-3-527-41212-9
ePDF ISBN: 978-3-527-67017-8
ePub ISBN: 978-3-527-67016-1
Mobi ISBN: 978-3-527-67015-4
oBook ISBN: 978-3-527-67014-7

Cover Design Grafik-Design Schulz
Typesetting Laserwords Private Limited, Chennai, India
Printing and Binding Markono Print Media Pte Ltd, Singapore

Printed on acid-free paper

Foreword

The lighting industry has been going through a revolution with the phasing out of the tungsten source and the advance of LED technology. The latter is well known for its greater energy efficiency, longer life, and its ability to provide a more adjustable spectral power distribution, illuminance, and beam shape. These features lead to large changes in lighting applications: LED is now not just meeting the basic need of illumination, but also moving into the improvement of work performance, the provision of optimum atmosphere environment, and the achievement of better health and wellbeing. However, its characteristic of providing intensive illuminance in small fields leads to some large performance discrepancies when compared with conventional light sources such as fluorescent and tungsten. Many of the earlier quality measures cannot be applied to LED sources and for this reason new standards and guidelines have recently been developed. So far, there has been no good textbook in illumination engineering to focus mainly on LED technologies.

The book covers a comprehensive range of topics on LEDs from the fundamental sciences of vision, radiometry and photometry, colorimetry, and the circadian rhythm; followed by the manufacturing techniques of radiation generation and packaging, the design, and modeling of light sources; then lighting quality measures under photopic and mesopic regions; and finally, the optimization and characterization of indoor and outdoor lighting.

A reader who is unfamiliar with color science should have no problem to go through the book, as it gives sufficient background information. Thus the book will be suitable for beginners as well as for more experienced readers. A number of topics are focused on the development of new standards by the International Commission on Illumination (CIE), which is based on recent results of LED lighting research and applications. The book can be considered as the most comprehensive and an up-to-date textbook available, which will inject new knowledge for both color engineers and academic researchers.

My congratulations go to the authors for the great efforts in compiling this text. They include a large number of examples of their research results, and have demonstrated that the research and teaching go hand in hand.

I am strongly recommending this textbook as a valuable research and development tool.

Ronnier Luo
Professor of Zhejiang University (China),
Leeds University (UK),
National Taiwan University of Science and Technology (ROC),
Director CIE Division 1

Contents

Foreword *V*
Table of the Coauthors *XIX*
Preface *XXI*

1   **Introduction** *1*
       Peter Bodrogi and Tran Quoc Khanh
       Reference *5*

2   **The Human Visual System and Its Modeling for Lighting Engineering** *7*
       Peter Bodrogi and Tran Quoc Khanh
2.1   Visual System Basics *7*
2.1.1   The Way of Visual Information *7*
2.1.2   Perception *8*
2.1.3   Structure of the Human Eye *8*
2.1.4   The Pupil *9*
2.1.5   Accommodation *10*
2.1.6   The Retina *10*
2.1.7   Cone Mosaic and Spectral Sensitivities *12*
2.1.8   Receptive Fields and Spatial Vision *14*
2.2   Radiometry and Photometry *16*
2.2.1   Radiant Power (Radiant Flux) and Luminous Flux *18*
2.2.2   Irradiance and Illuminance *19*
2.2.3   Radiant Intensity and Luminous Intensity *19*
2.2.4   Radiance and Luminance *20*
2.2.5   Degrees of Efficiency for Electric Light Sources *21*
2.3   Colorimetry and Color Science *22*
2.3.1   Color Matching Functions and Tristimulus Values *24*
2.3.2   Color Appearance, Chromatic Adaptation, Color Spaces, and Color Appearance Models *27*
2.3.2.1 Perceived Attributes of Color Perception *27*
2.3.2.2 Chromatic Adaptation *27*

2.3.2.3	CIELAB Color Space *29*
2.3.2.4	The CIECAM02 Color Appearance Model *31*
2.3.3	Modeling of Color Difference Perception *36*
2.3.3.1	MacAdam Ellipses *36*
2.3.3.2	u', v' Chromaticity Diagram *36*
2.3.3.3	CIELAB Color Difference *37*
2.3.3.4	CAM02-UCS Uniform Color Space and Color Difference *38*
2.3.4	Blackbody Radiators and Phases of Daylight in the x, y Chromaticity Diagram *39*
2.4	LED Specific Spectral and Colorimetric Quantities *41*
2.4.1	Peak Wavelength (λ_p) *41*
2.4.2	Spectral Bandwidth at Half Intensity Level ($\Delta\lambda_{0.5}$) *42*
2.4.3	Centroid Wavelength (λ_C) *43*
2.4.4	Colorimetric Quantities Derived from the Spectral Radiance Distribution of the LED Light Source *43*
2.4.4.1	Dominant Wavelength (λ_D) *43*
2.4.4.2	Colorimetric Purity (p_C) *43*
2.5	Circadian Effect of Electromagnetic Radiation *44*
2.5.1	The Human Circadian Clock *44*
	References *47*

3	**LED Components – Principles of Radiation Generation and Packaging** *49*
	Holger Winkler, Quang Trinh Vinh, Tran Quoc Khanh, Andreas Benker, Charlotte Bois, Ralf Petry, and Aleksander Zych
3.1	Introduction to LED Technology *49*
3.2	Basic Knowledge on Color Semiconductor LEDs *50*
3.2.1	Injection Luminescence *50*
3.2.2	Homo-Junction, Hetero-Junction, and Quantum Well *52*
3.2.2.1	Homo-Junction *52*
3.2.2.2	Hetero-Junction *53*
3.2.2.3	Quantum Well *54*
3.2.3	Recombination *56*
3.2.3.1	Direct and Indirect Recombination *56*
3.2.3.2	Radiative and Nonradiative Recombinations and Their Simple Theoretical Quantification *57*
3.2.4	Efficiency *60*
3.2.4.1	Internal Quantum Efficiency (η_i) *60*
3.2.4.2	Injection Efficiency (η_inj) *60*
3.2.4.3	Light Extraction Efficiency ($\eta_\text{extraction}$) *60*
3.2.4.4	External Quantum Efficiency (η_ext) *62*
3.2.4.5	Radiant Efficiency (η_e, See Section 2.2.5, Eq. (2.13)) *62*
3.2.4.6	Luminous Efficacy (η_v) *62*
3.2.5	Semiconductor Material Systems – Efficiency, Possibilities, and Limits *63*

3.2.5.1	Possible Semiconductor Systems	63
3.2.5.2	Semiconductor Systems for Amber–Red Semiconductor LEDs	64
3.2.5.3	Semiconductor Systems for UV–Blue–Green Semiconductor LEDs	66
3.2.5.4	The Green Efficiency Gap of Color Semiconductor LEDs	66
3.3	Color Semiconductor LEDs	67
3.3.1	Concepts of Matter Waves of de Broglie	68
3.3.2	The Physical Mechanism of Photon Emission	68
3.3.3	Theoretical Absolute Spectral Power Distribution of a Color Semiconductor LED	70
3.3.4	Characteristic Parameters of the LEDs Absolute Spectral Power Distribution	70
3.3.5	Role of the Input Forward Current	71
3.3.6	Summary	71
3.4	Phosphor Systems and White Phosphor-Converted LEDs	72
3.4.1	Introduction to Phosphors	72
3.4.2	Luminescence Mechanisms	79
3.4.3	Aluminum Garnets	82
3.4.4	Alkaline Earth Sulfides	86
3.4.5	Alkaline Earth *Ortho*-Silicates	89
3.4.6	Alkaline Earth Oxy-*Ortho*-Silicates	92
3.4.7	Nitride Phosphors	93
3.4.7.1	CASN	94
3.4.7.2	2-5-8-Nitrides	97
3.4.7.3	1-2-2-2 Oxynitrides	99
3.4.7.4	β-SiAlON	103
3.4.8	Phosphor-Coating Methods	104
3.4.9	Challenges of Volumetric Dispensing Methods	106
3.4.10	Influence of Phosphor Concentration and Thickness on LED Spectra	107
3.5	Green and Red Phosphor-Converted LEDs	109
3.5.1	The Phosphor-Converted System	109
3.5.2	Chromaticity Considerations	111
3.5.3	Phosphor Mixtures for the White Phosphor-Converted LEDs	112
3.5.4	Colorimetric Characteristics of the Phosphor-Converted LEDs	115
3.6	Optimization of LED Chip-Packaging Technology	118
3.6.1	Efficiency Improvement for the LED Chip	120
3.6.2	Molding and Positioning of the Phosphor System	122
3.6.3	Substrate Technology – Integration Degree	127
	References	130

4 Measurement and Modeling of the LED Light Source 133
Quang Trinh Vinh, Tran Quoc Khanh, Hristo Ganev, and Max Wagner

4.1	LED Radiometry, Photometry, and Colorimetry	133

4.1.1	Spatially Resolved Luminance and Color Measurement of LED Components	*134*
4.1.2	Integrating Sphere Based Spectral Radiant Flux and Luminous Flux Measurement	*139*
4.2	Thermal and Electric Behavior of Color Semiconductor LEDs	*143*
4.2.1	Temperature and Current Dependence of Color Semiconductor LED Spectra	*143*
4.2.1.1	Temperature Dependence of Color Semiconductor LED Spectra	*143*
4.2.1.2	Current Dependence of Color Semiconductor LED Spectra	*144*
4.2.2	Temperature and Current Dependence of Radiant Flux and Radiant Efficiency of Color Semiconductor LEDs	*145*
4.2.2.1	Temperature Dependence of Radiant Flux and Radiant Efficiency of Color Semiconductor LEDs	*145*
4.2.2.2	Current Dependence of Radiant Flux and Radiant Efficiency of Color Semiconductor LEDs	*146*
4.2.2.3	Conclusion	*147*
4.2.3	Temperature and Current Dependence of the Chromaticity Difference of Color Semiconductor LEDs	*147*
4.2.3.1	Temperature Dependence of the Chromaticity Difference of the Color Semiconductor LEDs	*147*
4.2.3.2	Current Dependence of Chromaticity Difference of the Color Semiconductor LEDs	*148*
4.3	Thermal and Electric Behavior of White Phosphor-Converted LEDs	*149*
4.3.1	Temperature and Current Dependence of Warm White PC-LED Spectra	*149*
4.3.1.1	Temperature Dependence of Warm White PC-LED Spectra	*150*
4.3.1.2	Current Dependence of Warm White PC-LED Spectra	*151*
4.3.2	Current Limits for the Color Rendering Index, Luminous Efficacy, and White Point for Warm White PC-LEDs	*152*
4.3.2.1	General Considerations	*152*
4.3.2.2	Comparison of Color Rendering Index and Luminous Efficacy	*153*
4.3.2.3	White Point of the Warm White PC-LEDs	*153*
4.3.3	Temperature and Current Dependence of the Luminous Flux and Luminous Efficacy of Warm White PC-LEDs	*153*
4.3.3.1	Temperature Dependence of the Luminous Flux and Luminous Efficacy of Warm White PC-LEDs	*154*
4.3.3.2	Current Dependence of Luminous Flux and Luminous Efficacy of Warm White PC-LEDs	*155*
4.3.4	Temperature and Current Dependence of the Chromaticity Difference of Warm White PC-LEDs	*156*
4.3.4.1	Temperature Dependence of the Chromaticity Difference of Warm White PC-LEDs	*156*

4.3.4.2	Current Dependence of the Chromaticity Difference of Warm White PC-LEDs *157*	
4.4	Consequences for LED Selection Under Real Operation Conditions *157*	
4.4.1	Chromaticity Differences Between the Operating Point and the Cold Binning Point *157*	
4.4.2	Chromaticity Difference Between the Operating Point and the Hot Binning Point *158*	
4.5	LED Electrical Model *160*	
4.5.1	Theoretical Approach for an Ideal Diode *160*	
4.5.2	A LED Experimental Electrical Model Based on the Circuit Technology *163*	
4.5.3	An Example for a Limited Electrical Model for LEDs *163*	
4.5.3.1	Limited Operating Range *163*	
4.5.3.2	Mathematical Description of the LEDs Forward Current in the Limited Operating Range *164*	
4.5.3.3	An Example for the Application of the Limited Electrical Model *165*	
4.5.3.4	Evaluation and Improvement of the Electrical Model *167*	
4.6	LED Spectral Model *167*	
4.6.1	Spectral Models of Color Semiconductor LEDs and White PC-LEDs *167*	
4.6.1.1	Mathematical Approach *168*	
4.6.2	An Example for a Color Semiconductor LED Spectral Model *174*	
4.6.2.1	Experiments on Spectral Models for Color Semiconductor LEDs *174*	
4.6.3	An Example for a PC-LED Spectral Model *178*	
4.6.3.1	Experiments for the Spectral Models of White PC-LEDs *178*	
4.7	Thermal Relationships and Thermal LED Models *181*	
4.7.1	Thermal Relationships in LEDs *181*	
4.7.1.1	Thermal Structure of a Typical LED *182*	
4.7.1.2	A Typical Equivalent Thermal Circuit *182*	
4.7.1.3	External Thermal Resistance *183*	
4.7.2	One-Dimensional Thermal Models *185*	
4.7.2.1	The First Order Thermal Circuit *185*	
4.7.2.2	Second Order Thermal Circuit *187*	
4.7.2.3	The nth Order Thermal Circuit *188*	
4.7.2.4	The Transient Function and Its Weighting Function *189*	
4.7.2.5	Conclusions *190*	
4.8	Measurement Methods to Determine the Thermal Characteristics of LED Devices *190*	
4.8.1	Measurement Methods and Procedures *190*	
4.8.1.1	Selection of an Available Measurement Method *190*	
4.8.1.2	Description of the Cooling Measurement Procedure *191*	

4.8.2	Description of a Typical Measurement System and Its Calibration	*192*
4.8.2.1	Components and Structure of the Measurement System	*192*
4.8.2.2	Determination of Thermal Power and Calibration Factor for Several LEDs	*193*
4.8.3	Methods of Thermal Map Decoding	*194*
4.8.3.1	Decoding of the Thermal Map by the Method of the Structure Function	*194*
4.8.3.2	Thermal Map Decoding by the Euclidean Algorithm	*195*
4.9	Thermal and Optical Behavior of Blue LEDs, Silicon Systems, and Phosphor Systems	*197*
4.9.1	Selection of LEDs and Their Optical Behavior	*197*
4.9.2	Efficiency of the LEDs	*197*
4.9.3	Results of Thermal Decoding by the Structure Function Method	*198*
4.9.4	Results of Thermal Decoding by the Method of the Euclidean Algorithm	*200*
4.10	Aging Behavior of High-Power LED Components	*201*
4.10.1	Degradation and Failure Mechanisms of LED Components	*202*
4.10.2	Research on the Aging Behavior of High-Power-LEDs	*204*
4.10.2.1	Change of Spectral Distribution and Chromaticity Coordinates	*205*
4.10.2.2	Change of Electrical and Thermal Behavior	*209*
4.10.2.3	Change of Luminous Flux – Lumen Maintenance	*212*
4.11	Lifetime Extrapolation	*214*
4.11.1	TM 21-Method	*215*
4.11.2	Border Function (BF)	*215*
4.11.3	Vector Acceleration (Temperature Acceleration – Vector Method – Denoted by TA–V)	*217*
4.11.4	Arrhenius Behavior	*217*
4.11.5	Groups of LEDs	*218*
4.11.6	Exponential Function (Belonging to the Definition "Other Mathematical Fit Functions, Flexible (MFF-FLEX)")	*218*
4.11.7	Root Function (Belonging to the Definition "Other Mathematical Fit Functions, Flexible (MFF-FLEX)")	*218*
4.11.8	Quadratic Function (Belonging to the Definition "Other Mathematical Fit Functions, Flexible (MFF-FLEX)")	*220*
4.11.9	Limits of the Extrapolation Procedure	*221*
4.11.10	Conclusions	*221*
4.12	LED Dimming Behavior	*222*
4.12.1	Overview on the Dimming Methods	*223*
4.12.2	Experiments: Setup and Results	*224*
4.12.2.1	Experimental Setup	*224*
4.12.2.2	Test Results for White LEDs	*225*
4.12.2.3	Test Results for Red LEDs	*227*
	References	*229*

5	**Photopic Perceptual Aspects of LED Lighting** *233*	

Peter Bodrogi, Tran Quoc Khanh, and Dmitrij Polin

5.1	Introduction to the Different Aspects of Light and Color Quality *233*	
5.2	Color Rendering Indices: CRI, CRI2012 *242*	
5.2.1	CIE CRI Color Rendering Index *243*	
5.2.2	Deficiencies of the CIE CRI Color Rendering Method *246*	
5.2.3	CRI2012 Color Rendering Index *247*	
5.2.3.1	Test Color Samples in the CRI2012 Method *249*	
5.2.3.2	Root Mean Square and Nonlinear Scaling *251*	
5.3	Semantic Interpretation of Color Differences and Color Rendering Indices *253*	
5.3.1	Experimental Method of the Semantic Interpretation of Color Differences *254*	
5.3.2	Semantic Interpretation Function $R(\Delta E')$ for CAM02-UCS Color Differences *256*	
5.3.3	Semantic Interpretation of the CRI2012 Color Rendering Indices *257*	
5.3.4	Semantic Interpretation of the CIE CRI Color Rendering Indices *258*	
5.4	Object Specific Color Rendering Indices of Current White LED Light Sources *261*	
5.4.1	Spectral Reflectances of Real Colored Objects *261*	
5.4.2	Color Rendering Analysis of a Sample Set of White LEDs *266*	
5.4.2.1	Definition of the Sample Set of White LEDs *266*	
5.4.2.2	Definition of the Sample Set of Object Reflectances *268*	
5.4.2.3	Examples for the Relationship among the Color Rendering Indices in Terms of the Semantic Interpretation Scale (R) *269*	
5.4.2.4	Object Specific Color Rendering Bar Charts with Semantic Interpretation Scales (R) *270*	
5.4.3	Summary *272*	
5.5	Color Preference Assessment: Comparisons Between CRI, CRI2012, and CQS *273*	
5.5.1	CQS: The Color Quality Scale *275*	
5.5.1.1	Components of the CQS Method *275*	
5.5.1.2	Discussion of the CQS Method *279*	
5.5.2	Relationship between the Color Quality Scale (CQS Q_a, Q_p) and the Color Rendering Indices (CRI, CRI2012) *280*	
5.6	Brightness, Chromatic Lightness, and Color Rendering of White LEDs *285*	
5.6.1	Modeling the Chromaticity Dependence of Brightness and Lightness *287*	
5.6.1.1	CIE Brightness Model *287*	

5.6.1.2	Ware–Cowan Brightness Model 288
5.6.1.3	A Chromatic Lightness Model (L**) and Its Correlation with Color Rendering 290
5.7	White Point Characteristics of LED Lighting 292
5.7.1	Whiteness Perception, Correlated Color Temperature, and Target White Chromaticities for the Spectral Optimization of White LEDs 293
5.7.2	Analysis of the White Points of the Sample Set of 34 White LEDs 296
5.7.3	Summary and Outlook 297
5.8	Chromaticity Binning of White LEDs 298
5.8.1	The ANSI Binning Standard 299
5.8.2	A Visually Relevant Semantic Binning Strategy 301
5.8.3	Comparison of the Semantic Binning Method with a Visual Binning Experiment 304
5.8.4	Evaluation of the White Points of the Sample set of 34 White LEDs in Terms of the Visually Relevant Semantic Binning Strategy 306
5.8.5	Summary 307
5.9	Visual Experiments (Real Field Tests) on the Color Quality of White LEDs 309
5.9.1	Color Quality Experiments in the Three-Chamber Viewing Booth 309
5.9.1.1	The First Color Quality Experiment 310
5.9.1.2	The Second Color Quality Experiment 312
5.9.2	Concluding Remarks 315
5.10	Circadian Stimulus, Color Temperature, and Color Rendering of White LEDs 315
5.10.1	The Rea *et al.* Model 316
5.10.2	Application of the Rea *et al.* Model to White LED Light Sources 317
5.10.3	Relationship between the Color Rendering Index $R_{a,2012}$ and the Circadian Stimulus for the White LED Light Sources 320
5.10.4	Summary 321
5.11	Flicker and Stroboscopic Perception of White LEDs under Photopic Conditions 321
5.11.1	Flicker Research Results in the Past 323
5.11.2	The Bullough *et al.* Study on Flicker and the Stroboscopic Effect with a LED-Luminaire 323
5.11.3	The Study of the Present Authors on Flicker Perception and the Stroboscopic Effect with an LED Luminaire 325
5.11.3.1	Experimental Setup 325
5.11.3.2	The Subjects' Tasks 325
5.11.3.3	Experimental Results 329
5.11.4	Conclusions 331
	References 331

6 Mesopic Perceptual Aspects of LED Lighting *337*
Tran Quoc Khanh, Peter Bodrogi, Stefan Brückner, Nils Haferkemper, and Christoph Schiller

6.1 Foundations and Models of Mesopic Brightness and Visual Performance *337*
6.1.1 Visual Tasks in the Mesopic Range *337*
6.1.2 Mesopic Vision and Its Modeling *338*
6.1.3 The Mesopic Visual Performance Model of the CIE *344*
6.2 Mesopic Brightness under LED Based and Conventional Automotive Front Lighting Light Sources *347*
6.2.1 Experimental Method *348*
6.2.2 Mean Results of Brightness Matching *349*
6.2.3 Interobserver Variability of Mesopic Brightness Matching *350*
6.2.4 Conclusion *353*
6.3 Mesopic Visual Performance under LED Lighting Conditions *353*
6.4 Visual Acuity in the Mesopic Range with Conventional Light Sources and White LEDs *357*
6.4.1 Introduction *357*
6.4.2 Test Method *358*
6.4.2.1 Test Chart *359*
6.4.3 Letter Contrast Acuity Results *361*
6.5 Detection and Conspicuity of Road Markings in the Mesopic Range *362*
6.5.1 Introduction *362*
6.5.2 Spectral Characterization of Light Sources and Road Markings *363*
6.5.2.1 Automotive Light Sources *364*
6.5.2.2 Light Sources for Street Lighting *364*
6.5.2.3 Spectral Characteristics of Road Markings *365*
6.5.3 Contrast Calculations *366*
6.5.4 Results and Evaluation *367*
6.6 Glare under Mesopic Conditions *368*
6.6.1 Introduction: Categories of Glare Effects *368*
6.6.2 Causes and Models of the Glare Phenomenon *370*
6.6.2.1 Disability Glare *370*
6.6.2.2 Discomfort Glare *375*
6.6.3 Discomfort Glare under Mesopic Conditions – Spectral Behavior and Mechanisms *376*
6.6.3.1 Introduction *376*
6.6.3.2 Experimental Method of a Discomfort Glare Experiment *377*
6.6.3.3 Results and Discussion of the Discomfort Glare Experiment *378*
6.6.4 Experiments to Determine the Spectral Sensitivity of Disability Glare in the Mesopic Range under Traffic Lighting Conditions *379*
6.6.5 Glare in the Street Lighting with White LED and Conventional Light Sources *383*

6.7	Bead String Artifact of PWM Controlled LED Rear Lights at Different Frequencies *388*	
6.7.1	Introduction *389*	
6.7.2	A visual experiment on the bead string artifact *390*	
6.7.2.1	Key Question *390*	
6.7.2.2	Experimental Setup *391*	
6.7.2.3	Peripheral Observation *391*	
6.7.2.4	Foveal Observation *392*	
6.7.3	Results of the Visual Experiment on the Bead String Artifact *392*	
6.7.3.1	Effect of Viewing Condition (Peripheral vs Foveal) *393*	
6.7.3.2	Effect of the Observer's Age *393*	
6.7.4	Conclusion *394*	
6.8	Summarizing Remarks to Chapter 6 *394*	
	References *395*	

7 Optimization and Characterization of LED Luminaires for Indoor Lighting *399*

Quang Trinh Vinh and Tran Quoc Khanh

7.1	Indoor Lighting – Application Fields and Requirements *399*	
7.2	Basic Aspects of LED-Indoor Luminaire Design *403*	
7.2.1	LED, Printed Circuit Board, Electronics *403*	
7.2.2	Optical Systems *405*	
7.2.3	Controller-Regulation Electronics *406*	
7.2.4	Thermal Management *407*	
7.3	Selection Criteria for LED Components and Units *409*	
7.3.1	Geometry *410*	
7.3.2	Spectral and Colorimetric Angular Distribution *411*	
7.3.3	Warm/Cold White LED *412*	
7.3.4	Color Shift *412*	
7.3.5	Forward Voltage *413*	
7.3.6	Choice of the Optimal Current for LEDs *413*	
7.3.7	Thermal Resistance *413*	
7.4	Application Fields with Higher Color and Lighting Requirements *414*	
7.4.1	Museum and Gallery Lighting *414*	
7.4.2	Film and TV Lighting *416*	
7.4.3	Shop Lighting *419*	
7.4.4	Requirements for LED Luminaires with High Color Quality *420*	
7.5	Principles of LED Radiation Generation with Higher Color Quality and One Correlated Color Temperature *421*	
7.6	Optimization and Stabilization of Hybrid LED Luminaires with High Color Rendering Index and Variable Correlated Color Temperature *426*	
7.6.1	Motivation and General Consideration *426*	
7.6.1.1	Hybrid LED lamp and spectral LED models *427*	

7.6.1.2	Lighting Quality Parameters, Their Limits and Proposals for the Most Appropriate LED-Combination for Hybrid LED Lamps	*427*
7.6.1.3	Two Methodical Demonstration Examples and Their Tasks	*427*
7.6.2	Spectral Reflectance of Color Objects in Museum, Shop and Film Lighting	*428*
7.6.2.1	Analysis of the Spectral Reflectance Curves of the Color Objects	*429*
7.6.2.2	Color Objects and Their Spectral Reflectance Curves in the Museum (Oil Paintings)	*430*
7.6.2.3	Qualification and Prioritization of the Color Objects for the Optimization of Museum Lighting	*431*
7.6.3	Optimization Process for the Hybrid LED Combination with High Color Quality	*433*
7.6.3.1	Role of LED Components and Primary Proposals for LED Selection	*433*
7.6.3.2	LED Selection for the Most Available LED Combination of the Hybrid LED Lamp	*433*
7.6.3.3	Optimization of the Hybrid LED Lamp for Oil Color Paintings	*434*
7.6.3.4	Optimized Spectral Power Distributions of the Hybrid LED-Lamps	*436*
7.6.4	Stabilization of the Lighting Quality Aspects of the Hybrid LED Lamp	*438*
7.6.4.1	Control System Structure for the Stabilization of Lighting Quality	*439*
7.6.4.2	Results of Optimization and Stabilization	*440*
	References	*442*
8	**Optimization and Characterization of LED Luminaires for Outdoor Lighting**	*443*
	Tran Quoc Khanh, Quang Trinh Vinh, and Hristo Ganev	
8.1	Introduction	*443*
8.2	Construction Principles of LED Luminaire Units	*445*
8.2.1	Mechanical Unit	*445*
8.2.2	Electronic Unit	*446*
8.2.3	Optical System	*449*
8.3	Systematic Approach of LED Luminaire Design for Street Lighting	*451*
8.3.1	General Aspects	*451*
8.3.2	Definition of Specifications – Collection of Ideas	*451*
8.3.3	LED Characterization and Selection	*456*
8.3.4	Thermal and Electronic Dimensioning	*458*
8.4	Degradation Behavior of LED Street Luminaires	*460*
8.5	Maintenance Factor for LED Luminaires	*463*
8.5.1	Basic Aspects of Maintenance Factor	*463*

8.5.2	Basic Aspects of the Maintenance Factor of LED Street Luminaires *467*	
8.6	Planning and Realization Principles for New LED Installations *471*	
8.6.1	Technical Approach *473*	
8.6.1.1	Phase 1: Coordination *473*	
8.6.1.2	Phase 2: Measurement and Evaluation of the Old System *473*	
8.6.1.3	Phase 3: Lighting Measurement and Evaluation of the LED System *474*	
8.6.2	Qualification of the LED Luminaires and Selection of the Luminaires *476*	
8.6.3	Installation and Measurement of the LED System *477*	
8.6.3.1	Phase 4: Documentation by the Scientific Partner and Public Opinion Poll *478*	
	References *478*	
9	**Summary** *479*	
	Peter Bodrogi and Tran Quoc Khanh	
	Index *483*	

Table of the Coauthors

Coauthors	Chapter
Andreas Benker, Company Merck KGaA (Darmstadt)	3.4
Dr. Ralf Petry, Company Merck KGaA (Darmstadt)	3.4
Aleksander Zych, Company Merck KGaA (Darmstadt)	3.4
Charlotte Bois, Technische Universität Darmstadt & Company Merck KGaA (Darmstadt)	3.5
Hristo Ganev, Technische Universität Darmstadt	4.10 and 8.4
Max Wagner, Technische Universität Darmstadt	4.11
Dmitrij Polin, Technische Universität Darmstadt	5.11
Nils Haferkemper, Technische Universität Darmstadt	6.4
Christoph Schiller, Technische Universität Darmstadt	6.5
Stefan Brückner, Technische Universität Darmstadt	6.7

Preface

LED technology is a dynamically developing technology. It has a substantial impact on worldwide technological development and the way of social thinking. With the rapid growth of LED technology, new materials, industrial value chains, manufacturing methods, and optimizations processes have been established in order to improve energy efficiency, light, and color quality and to reduce the amount of material, environmental, and human resources utilized for lighting applications, and, consequently, to contribute to today's worldwide energy saving efforts. Recognizing the huge potential of LED components and LED systems to generate smart and flexible spectral power distributions, correlated color temperatures, angular luminous intensity distributions and absolute luminous flux levels, the disciplines of lighting engineering, and vision science should improve their current photometric and colorimetric quantities and attributes and calculation methods and research deeply the new aspects of human eye physiology and psychophysics to achieve a new quality for the fundamental description of human visual information processing including light, color, and spatial structure.

In this context, there have been numerous LED related research and development projects in Europe, North America, and Asia that characterized the technological and human physiological aspects of LED components and LED luminaires for indoor and outdoor applications. *LED Lighting – Technology and Perception* is based on the results of several research and development projects as well as engineering projects in Europe. These projects were conducted under the leadership of the Laboratory of Lighting Technology of the Technische Universität Darmstadt (Germany). This book also summarizes current international research and development outcomes and shows the development tendencies in the field of LED technology and research.

The authors would like to acknowledge the German Federal Ministry for Education and Research (BMBF) and Federal Ministry for Economy (BMWi) for funding our LED research and advancing our LED projects. The authors also would like to thank the German companies Merck (Darmstadt), Arnold &

Richter Cine Technik (Munich), Trilux (Ansberg), Bäro (Leichlingen), TechnoTeam (Ilmenau), and Ilexa (Ilmenau) for allowing us the use of their picture materials.

Darmstadt,
September 2014

Tran Quoc Khanh
Peter Bodrogi
Trinh Quang Vinh
Holger Winkler

1
Introduction

Peter Bodrogi and Tran Quoc Khanh

The technology of LED lighting (the illumination with light emitting diodes), especially with modern high-power LEDs has developed very rapidly in the past decade. According to the LEDs' tendency to achieve high energy efficiency with superior lighting quality and accordingly, a high level of user acceptance, LEDs have acquired a substantial economical relevance worldwide. The market share of LED light sources and LED luminaires is increasing rapidly at the time of writing. Nevertheless, to ensure good lighting quality for the human light source user, the user's perceptual characteristics (e.g., the way they perceive colors) are considered as important optimization criteria during the design and development of high-tech LED illumination systems. This book presents optimization guidelines for LED technology in the view of human perceptual features within the interdisciplinary framework of lighting engineering.

Lighting engineering deals with the energy efficient and application dependent production, characterization, transmission, and effects of optical radiation on human users taking the aspects of visual perception and light and health aspects into account [1]. The four main subject areas of lighting engineering are shown in Figure 1.1.

As can be seen from Figure 1.1, the principles of light production techniques are important to manufacture light sources (lamps, lamp modules, luminaires, headlights) with certain desirable spectral, energetic, geometric properties, and spatial light distributions of the light they generate. Using advanced illumination techniques, this light is projected onto the object arranged in different room geometries (in interior lighting) or street geometries (in exterior lighting). Light measurement techniques, in turn, have the task to physically measure the descriptor quantities of the visual (e.g., luminance, chromaticity) and light and health (e.g., circadian) characteristics of the illuminating system.

Colorimetry and color science (together with eye psychology and visual psychometry) analyze the answers of the human visual system to the spectral (and spatial) properties of the visual stimulus (i.e., the optical radiation reaching the human eye) in detail. In colorimetry and in color science, numeric descriptor quantities are defined to quantify human perceptions together with the circumstances under which a human visual model is valid, for example, different models

---

LED Lighting: Technology and Perception, First Edition.
Edited by Tran Quoc Khanh, Peter Bodrogi, Quang Trinh Vinh and Holger Winkler.
© 2015 Wiley-VCH Verlag GmbH & Co. KGaA. Published 2015 by Wiley-VCH Verlag GmbH & Co. KGaA.

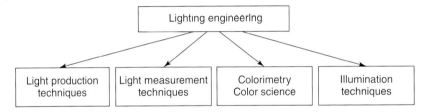

Figure 1.1 Subject areas of lighting engineering [1].

for different adaptation levels – ranging between daytime vision down to nighttime vision, or different models for different sizes of the visual stimulus (e.g., 2° vs 10°).

This lighting engineering framework is complemented by theoretical knowledge on the chemistry and material science of phosphors and semiconductors as well as relevant aspects of electronics and thermodynamics. The necessary deep theoretical knowledge is applied to practical problems. Principles of LED design are illustrated by real-life examples so that, at the end, real know-how is conveyed in terms of easy-to-understand and easy-to-use numeric criterion values. The aim is to help apply this knowledge in the everyday practice of engineers involved in the design, development and manufacturing of LED components, light sources, lamps, luminaires, and LED lighting systems. Engineers can use this book as a theoretical reference and a practical guide to solve problems related to these types of questions:

- How can the technology of LED light sources and luminaires be optimized for the human user to increase user acceptance? How can the LED user's visual performance and light and health aspects be enhanced?
- In this respect, what are the most important components (that generate and distribute electromagnetic radiation and dissipate heat) of LED light sources and luminaires?
- What materials can be used to manufacture them and how can these materials be improved and their arrangement improved?
- How can the radiation of LED chips be converted by today's diverse variety of phosphors? How can phosphor converted LEDs be combined with pure (colored) chip LEDs to achieve high lighting quality in a high-end interior lighting system?
- What are the most important chemical, physical, and technological parameters of these LED devices exhibiting previously unknown design flexibility hence huge optimization potentials for human users?
- How can the input (e.g., current, temperature) and output (e.g., spectral radiance) parameters of the LED devices be measured physically?
- How can LED devices be modeled and controlled (e.g., by pulse width modulation) – including their aging phenomena?
- How does the visual stimulus of the scene illuminated by the LED light source come to existence?

- How can LED illumination systems exploit the human visual system's properties to provide an excellent (interior or exterior) environment lit by LEDs for excellent visual performance (excluding glare, flicker, and stroboscopic artifacts) and high lighting quality?
- What interactions are there between the light from the LED light source and the colored objects that reflect it and how does this interaction influence the perceptions or the light and health aspects of human users?
- How can LED lighting systems be optimized for lower luminance levels in typical nighttime automotive lighting and street/road lighting applications from the point of view of the human visual system in order to optimize brightness perception, visual performance, and visual acuity?
- How can objective (numeric) criteria be derived from human measurements that can be used to optimize a LED lighting system by engineering methods? How can we formulate such criteria in terms of really usable numbers in engineering practice?

To answer the above questions and to cope with related practical subjects in real-world application effectively, Figure 1.2 shows the interdisciplinary workflow concept of the present book for the technological and perceptual co-optimization of LEDs.

As can be seen from Figure 1.2 (going from the upper box in the middle toward the right), the LED lighting system illuminates an indoor scene or an outdoor scene with an arrangement of colored objects with certain spectral reflectance properties at a certain luminance level (daytime, twilight, or nighttime). The light

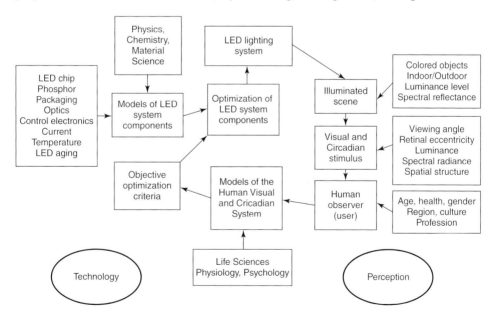

Figure 1.2 Interdisciplinary workflow concept for the technological and perceptual optimization of LED lighting systems.

coming from the light source reaches the human eye sometimes directly but is often reflected from the objects of the scene (see also Figure 2.14, left). Reflection changes the spectral composition of the light and a plethora of different color stimuli arises. Parallel to this, the light also generates nonvisual brain signals, for example, via the circadian stimulus which is responsible for the timing of the daily rhythm of the human user.

The stimulus (i.e., the light reaching the eye) has some characteristic properties that strongly influence the perception (see also Figure 2.14, right) it evokes in the human visual system: its viewing angle, its retinal eccentricity after being imaged onto the retina (i.e., the photoreceptive layer in the eye), its luminance, spectral radiance, and spatial structure (e.g., a homogeneous disk or a complex letter structure). Perceptions also depend on the characteristics of the human observer (age, health, and gender). The cognitive interpretations of the perceptions and the decisions based on them (e.g., to purchase or not to purchase a lighting system) also depend on the profession and culture of the user and his or her region of origin he or she happens to live in or has grown up in.

To provide numeric optimization criteria for engineers, usable (i.e., practice oriented and not too complex) models of the human visual system (and nonvisual systems like the Circadian system) can be described systematically and they must be well understood. These models compute the objective optimization criteria for the LED lighting system from the physically measured characteristics of the stimulus (e.g., its spectral radiance distribution). These criteria can be used, in turn, for the optimization process of the different components of the LED illumination system. Now, going back to the upper box in the center of Figure 1.2 again, these optimized components constitute an advanced LED lighting system whose enhanced visual properties can be validated by measuring it physically, computing its improved visual criterion numbers (e.g., a higher color rendering index) and, finally, validating it in a dedicated field study.

As can be seen from Figure 1.2, the objective optimization criteria derived from the human models belong to the left (technological) side of the workflow concept as these criteria represent technological optimization targets. Anyway, technological optimization can be carried out only if usable models of the LED system components are available for the engineer. Models based on the knowledge from physics, chemistry, and material science can be formulated for the LED chip (semiconductor structure), the phosphor, the packaging of the LED light source, its optics, control electronics, the temperature, and current dependence of its light output as well as for the aging of LEDs after thousands or ten thousands of operating hours. If such models and their optimization criteria are known then LED technology will be able to achieve an important one of its ultimate goals: better-quality human perception. This book is intended to help the engineer achieve this ambitious objective by the systematic concept of its chapter structure shown in Figure 1.3.

As can be seen from Figure 1.3, Chapters 3 and 4 deal with technological aspects (left-hand side) while Chapters 2, 5, and 6 are related to human perceptual issues (right-hand side). Chapter 3 describes the principles of how LEDs

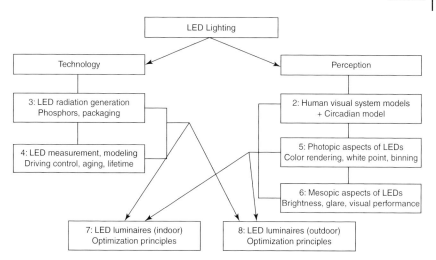

Figure 1.3 Chapter structure of the book.

generate electromagnetic radiation – including semiconductors, phosphors and packaging. Chapter 4 presents LED specific measurement procedures from the lighting engineer's point of view and applies them to collect data as inputs for usable LED models, both for short-term modeling and for modeling LED aging and to predict their lifetime. Chapter 2 introduces the basics of the human visual system and its models – extended to a Circadian model (timing human daily rhythms by light).

Chapter 5 deals with the photopic perceptual aspects relevant to the design of interior LED lighting, for example, color rendering, white tone quality, and the chromaticity binning of LEDs. Chapter 6 communicates the mesopic (twilight) perceptual aspects of LED illuminating systems relevant to exterior applications requiring lower light levels, for example, brightness perception, glare, and visual performance in nighttime driving with LED car headlamps or LED based street/road lighting. Finally, Chapters 7 and 8 combine and apply the knowledge accumulated in the earlier chapters according to the workflow concept of Figure 1.2 to deduce optimization procedures and principles for LED luminaire design with lots of demonstrative practical and numeric examples. Finally, Chapter 9 recapitulates the most important lessons and findings of the book.

Reference

1. Khanh, T.Q. (2013) Licht- und Farbforschung: Augenphysiologie, Psychophysik, Technologie und Lichtgestaltung. Z. Licht, **4** (2013), 60–67.

2
The Human Visual System and Its Modeling for Lighting Engineering

Peter Bodrogi and Tran Quoc Khanh

Chapter 2 defines the most important concepts necessary to understand the perceptual aspects of LED (light emitting diode) technology throughout this book including the basics of the human visual system, radiometry and photometry, colorimetry and color science as well as the human circadian system. According to its relevance to the subject, a short section on LED specific colorimetric quantities is added.

Chapter 2 incorporates issues like the photoreceptor structure of the human retina (density and spectral sensitivity of rods and cones), spatial and temporal contrast sensitivity (CS) of the human visual system, color appearance (related and unrelated colors, lightness and brightness, hue, colorfulness, and saturation), color difference perception, chromatic adaptation, blackbody radiators and phases of daylight, color matching functions, and the use of color spaces (CIELAB, CIECAM02). Chapter 2 is intended to provide a short introduction while the interested reader is encouraged to refer to literature (e.g., [1, 2]).

2.1
Visual System Basics

2.1.1
The Way of Visual Information

Figure 2.1 shows the way of visual information to and through the human brain [3].

The left side of Figure 2.1 shows the two overlapping parts of the visual fields of the two eyes. The optics of the eye images these two parts of the visual field onto the two retinae (in the left eye and the right eye) that contain photoreceptor mosaics and neural preprocessing cells. In the photoreceptors of the retina, light signals are converted into neural action potentials. The neuronal layers of the retina preprocess these signals and forward them through the optic nerves toward the so-called LGN (lateral geniculate nucleus). Nerve fibers from the two eyes cross (partially) in the so-called chiasma (opticum). About 90% of the visual nerves (also called optic nerves) reach the visual cortex (containing the processing units of the different visual sensations and interpretations like motion, color, spatial

LED Lighting: Technology and Perception, First Edition.
Edited by Tran Quoc Khanh, Peter Bodrogi, Quang Trinh Vinh and Holger Winkler.
© 2015 Wiley-VCH Verlag GmbH & Co. KGaA. Published 2015 by Wiley-VCH Verlag GmbH & Co. KGaA.

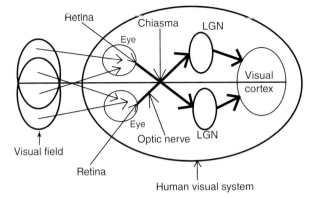

Figure 2.1 Schematic view of the way of visual information to and in the human visual system (see text).

structure, object segmentation, and recognition) over the LGN. About 10% of the retinal signals reaches other regions of the brain (in the parietal and temporal lobes) responsible, for example, for the control of the release of hormones (see Section 2.5)

2.1.2
Perception

Perception is a psychophysical process. The human being receives physical information (the so-called stimulus) about the state and changes of the environment via sensory organs and processes this information in the brain to obtain perceptions and to take decisions on the basis of the quality and magnitude of these perceptions [3]. The perceptual process is flexible: it depends on the context of the stimulus being perceived and also on previous experience (knowledge) of the human subject. It should be mentioned that not all stimuli result in a perception: some of the stimuli are not perceived at all (e.g., a light signal of very low contrast or electromagnetic radiation with 2 µm wavelength).

2.1.3
Structure of the Human Eye

Figure 2.2 illustrates the structure of the human eye. As can be seen from Figure 2.2, the human eye is an ellipsoid with an average length of about 26 mm and a diameter of about 24 mm. The eye is rotated in all directions by the aid of eye muscles. The outer layer is called *sclera*. The sclera is continued as the transparent *cornea* at the front. The *choroidea* supplies the *retina* with oxygen and nutrition. The retina is the photoreceptive (interior) layer of the eye also containing the visual preprocessing cells (see Sections 2.1.6–2.1.8). The *vitreous body* is responsible for maintaining the ellipsoid form of the eye. It consists of a suspension of water (98%) and hyaluronic acid (2%).

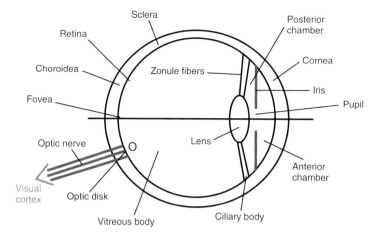

Figure 2.2 Structure of the human eye (the optic nerve is also called visual nerve). O, optic disk – the point at which the optic nerve passes through the eye and transmits the preprocessed neural signals of the retina toward the visual cortex.

The optical system of the human eye is a complex, slightly decentered lens system projecting an inverted and downsized image of the environment onto the retina. The cornea, the anterior chamber, and the iris constitute the front part of this optical system and then, the posterior chamber and the biconvex eye lens follow. The lens is held by the zonule fibers. By contracting the ciliary muscles, the focal length of the lens can be changed. The visual angle intersects the retina at the *fovea* (centralis), the location of sharpest vision.

The most important optical parameters of the components of the eye media include refractive indices (ranging typically between 1.33 and 1.43) and spectral transmission factors. All parameters vary among different persons considerably and are subject to significant changes with aging. Especially, accommodation, visual acuity, and pupil reactions are impaired with increasing age. The spectral transmission of the eye media decreases with age significantly, especially for short wavelengths.

After having reached the retina, light rays have to travel through the retinal layers and in the central retina also the so-called macula lutea (a yellow pigment layer that protects the central retina) *before* reaching the photoreceptors placed at the *rear* side of the retina. The *optic disk* (also called optic nerve head or blind spot) is the point (designated by O in Figures 2.2 and 2.4) at which the optic nerve passes through the eye. The retina is blind at the location O (as the density of rods and cones equals zero there).

2.1.4
The Pupil

A hole located at the center of the iris constitutes the pupil, see Figure 2.2. Its size depends on the amount of irradiance the retina receives, generally varying

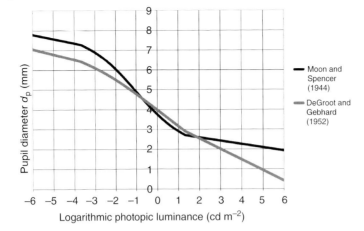

Figure 2.3 Pupil diameter d_P (in millimeter units) as a function of adaptation luminance in the viewing field of the observer (L_A in cd m^{-2} units), see Eqs. (2.1) and (2.2).

between 2 (in a bright environment) and 8 mm (in darkness but 8 mm is seldom unless a drug is administered). This range corresponds to a dynamic luminous flux range of 1 : 16 and represents one component of the adaptation mechanisms of the human visual system to changing light levels. The pupil diameter d_P (in millimeter units) can be computed (approximately) by the so-called Moon and Spencer equation (Eq. (2.1)) or the so-called DeGroot and Gebhard equation (Eq. (2.2)) from the adaptation luminance in the viewing field of the observer (L_A in cd m^{-2} units).

$$d_P = 4.9 - 3 \cdot \tanh(0.4 \cdot \log L_A + 1) \tag{2.1}$$

$$d_P = 0.8558 - 0.000401 \cdot (\log(L_A + 8.6))^6 \tag{2.2}$$

Both equations are visualized in Figure 2.3.

2.1.5
Accommodation

Accommodation is a property of the eye that enables to focus on an object located at an arbitrary distance in front of the eye so that the image of this object on the retina becomes sharp. As the eye lens becomes less elastic with increasing age, the old eye is unable to accommodate to near objects. The range of possible accommodation distances diminishes with decreasing adaptation luminance level.

2.1.6
The Retina

The retina (a layer of a typical thickness of 250 μm in average) is part of the optical system of the eye. The so-called Müller cells (or Müller glia, the support cells for the neurons of the retina) funnel light (like optical fibers) to the photoreceptors that are situated at the opposite side of the retina (viewed from the eye lens). The retina contains a complex cell layer with two types of photoreceptors, rods, and cones. Both the rod receptors and the cone receptors are connected to the nerve fibers of the optic nerve via a complex network that computes neural signals from receptor signals. This network is part of the retina: it is a mesh of different types of processing cells, so-called horizontal, bipolar, amacrine, and ganglion cells.

The retina contains about 6.5 millions of cones and 110–125 millions of rods while the number of nerve fibers is about 1 million. The density of rods and cones is different and depends on retinal location. Recently, a third type of photosensitive cell has been discovered, the so-called ipRGC (intrinsically photosensitive retinal ganglion cell containing the pigment melanopsin) responsible for regulating the circadian rhythm (Section 2.5). Figure 2.4 shows rod density and cone density as a function of retinal location.

As can be seen from Figure 2.4, there are no receptors at the position of the optic disk or blind spot as the optic nerve exits the eye at this place (designated by O, compare with Figure 2.2). The fovea is located in the center of the macula lutea region. A characteristic value to represent the diameter of the fovea is 1.5 mm corresponding to about 5° of visual angle. The fovea is responsible for best visual acuity according to the high cone receptor density; see the cone density maximum in Figure 2.4.

The center of the fovea is the foveola (or central pit) which has a diameter of about 0.2 mm (0.7°). The foveola has the highest cone density hence highest visual acuity. In its central part (in the angular range of about 20′), every cone (with a

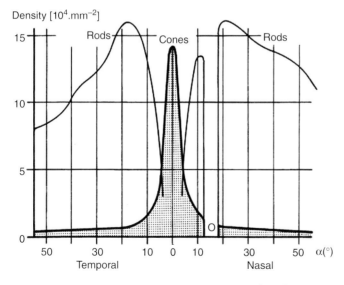

Figure 2.4 Rod density and cone density (ordinate, $10^4/mm^2$) as a function of retinal location (abscissa: α in degrees). (Drawn after Oesterberg [4].) O, optic disk (blind spot), compare with Figure 2.2.

cone diameter of 1.5 µm) is connected to its individual nerve fiber via a single bipolar cell and ganglion cell. There are only cone photoreceptors and no rods in a central region of about 0.350 mm (1.25°) horizontal diameter within the fovea [5] (see the rod density minimum in Figure 2.4). Outside the fovea, cone diameter increases up to about 4.5 µm, cone density decreases, and rod (diameter of rods: 2 µm) density increases to reach a rod maximum at about 20° (temporally rather at 18°), according to Figure 2.4.

Rods are responsible for nighttime vision (also called scotopic vision without the ability of accurate fixation and high visual acuity). Rods are more sensitive than cones (cones are responsible for daytime or photopic vision). Besides pupil contraction, the transition between rod vision and cone vision (in the so-called twilight or mesopic range, see Chapter 6) constitutes a second important adaptation mechanism of the human visual system to changing light levels. (There is a third adaptation mechanism, the *gain control* of the receptor signals.)

Photoreceptors contain pigments (opsins, certain types of proteins) that change their structure when they absorb photons and generate neural signals that are preprocessed by the horizontal, amacrine, bipolar, and ganglion cells of the retina to yield neural signals for later processing via the different visual (and nonvisual) pathways. Above a luminance of about $100\,\text{cd}\,\text{m}^{-2}$, rods become saturated (they do not produce any signal). Vision is then mediated purely by cone signals.

2.1.7
Cone Mosaic and Spectral Sensitivities

There are three types of cone with pigments of different spectral sensitivities, the so-called L (long-wavelength sensitive), M (middle-wavelength sensitive), and S (short-wavelength sensitive) cones that yield the so-called L, M, and S signals for the perception of homogeneous color patches and colored spatial structures (e.g., a red–purplish rose with fine color shadings). The number of different cone signals (three) has an important psychophysical implication for color science that the designer of a LED light source should be aware of: color vision is trichromatic, color spaces have three dimensions (Section 2.3), and there are three *independent* psychological attributes of color perception (hue, saturation, brightness).

L, M, and S cones build a retinal cone mosaic. The central (rod-free) part of the cone mosaic is illustrated in Figure 2.5.

As can be seen from Figure 2.5, the central part of the foveola (subtending about 20′ of visual angle or about 100 µm) is free from S cones. This fact results in the so-called small-field tritanopia, that is, the insensitivity to bluish light for very small central viewing fields. There are *in average* (among different observers) 1.5 times as many L cones as M cones in this region of the retina [1]. L and M cones represent 93% of all cones while S cones represent 7%.

The relative spectral sensitivities of the L, M, and S cones are depicted in Figure 2.6 together with some other important functions are discussed forward. A database of all characteristic functions of the human visual system (including these functions) can be found in the web [6].

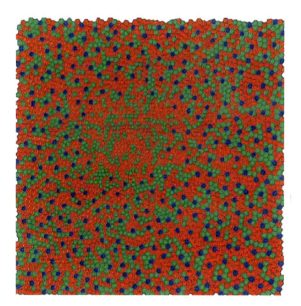

Figure 2.5 Cone mosaic of the rod-free inner fovea subtending about 1.25°, that is, about 350 μm. Red dots: long-wavelength sensitive cone photoreceptors (L cones). Green dots: middle-wavelength sensitive cones (M cones). Blue dots: short-wavelength sensitive cones (S cones). (Figure 1.1 from Ref. [1]. Reproduced with permission from Cambridge University Press.)

The spectral sensitivities in Figure 2.6 were measured at the cornea of the eye. They incorporate the average spectral transmission of the ocular media and the macular pigment at a retinal eccentricity of 2°. As can be seen from Figure 2.6, the spectral bands of the L, M, and S cones [6, 7, 8] yield three receptor signals for further processing. From these signals, the retina derives two chromatic signals: (i). L−M (the red–green opponent signal or its mediating neural channel) and (ii) S − (L + M) (the yellow–blue opponent signal or channel), and (iii) one achromatic signal, L + M. The L + M channel is usually considered as a *luminance* channel. The most important role of the luminance channel is that it enables the vision of fine image details (Section 2.1.8). In these signals, the "+" and "−"characters are only symbolic. In vision models, the L, M, and S signals must be weighted, for example, $\alpha S - (\beta L + \gamma M)$.

As can be seen from Figure 2.6, the L, M, and S cone sensitivity curves have their maxima at 566, 541, and 441, respectively [1]. In photometry (Section 2.2), for stimuli subtending 1°–4° of visual angle, the spectral sensitivity of the L + M channel is approximated by a standardized function, the so-called luminous efficiency function ($V(\lambda)$) (the basis of photometry, see Section 2.2) while for spatially more extended (e.g., 10°) stimuli, the so-called $V_{10}(\lambda)$ function (the CIE - International Commission on Illumination) 10° photopic photometric observer [9] is used.

For practical applications (e.g., the prediction of brightness perception under mesopic, that is, twilight conditions, see Figure 5.41; or to model mesopic

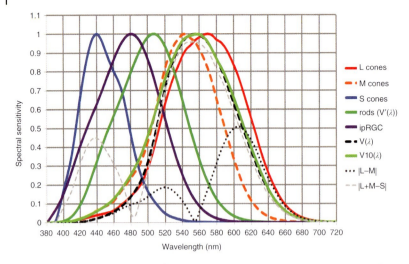

Figure 2.6 Relative spectral sensitivities of the L, M, and S cones (for 2°) [6–8] as well as other visual mechanisms that use the LMS cone signals as input. The sensitivity of the ipRGC mechanism (Section 2.5) is also shown. The spectral sensitivity of the rods (dark green curve) is approximated by the $V'(\lambda)$ function. $V(\lambda)$: luminous efficiency function (the basis of photometry, for stimuli of standard viewing angle, about 1°–4°); $V_{10}(\lambda)$: its alternative version for stimuli of greater viewing angle (about 10°).

spectral detection sensitivity, see Figure 6.2), it is important to compare the spectral sensitivity of the rod (R) mechanism (approximated by the so-called $V'(\lambda)$ function), with the spectral sensitivities of the L, M, S cones, the spectral sensitivity of the two chromatic mechanisms (L − M and S − (L + M)), with $V(\lambda)$ and $V_{10}(\lambda)$ (that roughly represent the L + M signal as mentioned before) and also with the already mentioned ipRGC mechanism (Section 2.5), see Figure 2.6.

2.1.8
Receptive Fields and Spatial Vision

Objects with characteristic spatial details of different spatial frequencies (e.g., a pedestrian on the roadside for exterior lighting or a red rose on the table for interior lighting) illuminated by the LED light source can be discerned by the human observer from their background [10]. The LED light source (and the illuminated objects) can be designed so that the coarse or fine spatial structures are perceived. For correct light source design, in order to consider the properties of this so-called spatial (contrast) vision correctly, the *spatial frequency characteristics* of the earlier-mentioned channels (L + M, L − M, S − (L + M)) of the human visual system can be understood.

To do so, it is important to study how the human visual system analyzes a spatial structure in a retinal image: ganglion cells gather and process the signals from several cones (at least outside the very central part of the retina where there is a one-to-one correspondence) located inside their so-called *receptive fields*. Receptive fields of ganglion cells are able to amplify spatial contrasts inside

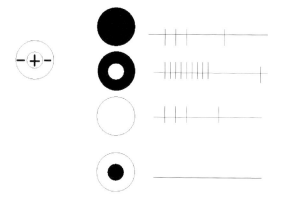

Figure 2.7 Left: Schematic representation of the receptive field of an "on-center" ganglion cell, +: center, −: surround. Middle column: black – no light, white – light stimulus, from top to bottom: 1. no light over the whole receptive field; 2. contrast – light on the center, no light on the surround; 3. light over the whole receptive field; 4. light on the surround. Right column: firing rate, from top to bottom: weak, strong, weak, no response [10]. (Reproduced with permission from Wiley VCH.)

the image: every receptive field has a circular center and a concentric circular surround, see Figure 2.7.

A stimulation of the center and the surround by a light signal leads to opposite firing reactions of the ganglion cell. The ganglion cell is firing when the stimulus is in the center ("on-center cell") while it is inhibited when the stimulus is in the surround. The other type of ganglion cell ("off-center cell") is inhibited when the stimulus is in the center and firing when the stimulus is in the surround. This way, spatially changing stimuli (contrasts or edges) increase firing while spatially homogenous stimuli generate only a low response level, see Figure 2.7.

On the human retina, achromatic contrast (i.e., spatial changes of the L + M signal) is detected according to the principle of Figure 2.7. Similar receptive field structures produce the chromatic signals for chromatic contrast, that is, spatial changes of the L − M or S − (L + M) signals. But in the latter case, the spectral sensitivity of the center differs from the spectral sensitivity of the surround because of the different combinations of the L, M, and S cones in the center and in the surround. Such a receptive field structure is called *double opponent* because there is spatial opponency (center/surround) *and* spectral (cone) opponency (L/M or S/(L + M)) [10].

It is the size and sensitivity of the receptive fields and the spatial aberrations of the eye media (cornea, lens, vitreous humor) that determine the spatial frequency characteristics of the achromatic and chromatic channels [11]. In practical applications including LED lighting, the question is how much achromatic (or chromatic) contrast is needed to detect or recognize a visual object of a given size corresponding to a given spatial frequency while size can be expressed in degrees of visual angle and spatial frequency is expressed in cycles per degree (cpd) units (e.g., 5 cpd means that there are five pairs of thin black and white lines within one degree of visual angle).

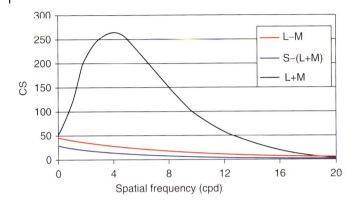

Figure 2.8 Chromatic contrast sensitivity functions of the L − M and S − (L + M) channels compared with the achromatic contrast sensitivity function (at a high retinal illuminance level and in the fovea). Abscissa: spatial frequency in cpd units, ordinate: contrast sensitivity (relative units) [10, 11]. (Reproduced with permission from Wiley VCH.)

Contrast (C) can be measured either by the contrast ratio, that is, the signal value (L + M, L − M or L + M − S) of the object (S_O) divided by the signal value of its background (S_B), that is, S_O/S_B or by the so-called Michelson contrast ($S_O - S_B)/(S_O + S_B$). Contrast sensitivity is defined as the reciprocal value of the threshold value of contrast needed to detect or recognize an object at a given spatial frequency. Chromatic CS functions (depending on spatial frequency) of the L − M and S − (L + M) channels are compared with the achromatic contrast (L + M) sensitivity function at a high retinal illuminance level in Figure 2.8.

As can be seen from Figure 2.8, achromatic or L + M (luminance) CS is a band-pass function of spatial frequency increasing up to about 3–5 cpd and then decreasing toward high spatial frequencies. For about 40 cpd (corresponding to a visual object of about 1 min of arc) or above, achromatic CS equals zero. This means that it is no use increasing the contrast (even up to infinity, i.e., black on white) if the object is smaller than about 1 min of arc. This is the absolute limit of (foveal) visual acuity. Contrary to the achromatic CS function with bandpass nature, chromatic CS functions are low pass functions of spatial frequency [11]. The most important point is that it is the L + M (luminance) channel (its spectral sensitivity is approximated by the $V(\lambda)$ function) that enables high visual acuity, that is, the vision of high spatial content but only in the fovea where the density of L cones and M cones is very high, see Figure 2.4.

2.2
Radiometry and Photometry

As mentioned in Chapter 1, lighting engineering deals with the energy efficient and application dependent production, characterization, transmission, and effects of optical radiation on human users taking the aspects of visual perception and light and health aspects into account. Accordingly, after the introduction of the basic properties of the human visual system in Section 2.1, the concepts of radiometry and photometry are summarized as a further important basic knowledge (characterization, transmission) for lighting engineering. Production of LED radiation is dealt with in Chapter 3 while LED specific issues of radiometry, photometry, and colorimetry are sketched in Section 2.4 (about basic concepts) and Section 4.1 (about advanced characterization and measurement).

The subject of radiometry and photometry is more complex and more extensive than what would be possible to present in this short section and its literature is abundant. Hence, the purpose of this section is just to define and illustrate the basic concepts to get an overview and a feeling for a better understanding of this book. The interested reader is advised to study literature (e.g., [12]) for more detail.

The concept of radiometry can be defined as follows: "Radiometry is the measurement of energy content of electromagnetic radiation fields and the determination of how this energy is transferred from a source, through a medium, and to a detector"[12]. The concept of photometry can be defined as follows: "The radiation transfer concepts … of photometry are the same as those for radiometry. The exception is that the spectral responsivity of the *detector*, the *human eye*, is specially defined. Photometric quantities are related to radiometric quantities via the spectral efficiency functions defined for the photopic and scotopic CIE Standard Observer" [12].

These functions are the two standard luminous efficiency functions, the CIE (1924) photopic $V(\lambda)$ function (for daytime vision) and the CIE (1951) scotopic $V'(\lambda)$ function (for nighttime rod vision), see Figure 2.6. Therefore, photometry can be considered as a special case of radiometry which is applied to the spectral sensitivity of the human eye as a detector of radiation. To be illustrative, the most important photometric concepts (based on the $V(\lambda)$ function) can be defined parallel to their radiometric counterparts. Figure 2.9 illustrates the transition from radiometric quantities to photometric quantities.

As can be seen from Figure 2.9 (left), if the spectral dependence $X_{e\lambda}$ of the radiometric quantity X_e is known then the radiometric quantity itself can be

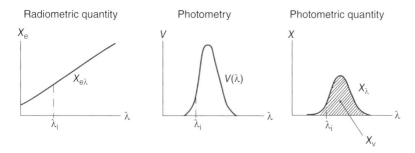

Figure 2.9 Photometric quantities are $V(\lambda)$-weighted radiometric quantities.

obtained by integration in the visible wavelength range between 380 and 780 nm, see Eq. (2.3).

$$X_e = \int_{380 \text{ nm}}^{780 \text{ nm}} X_{e\lambda} \, d\lambda \tag{2.3}$$

As can be seen from Figure 2.9 (right), to obtain the corresponding photometric quantity, X_V, the function $X_{e\lambda}$ can be weighted by the CIE (1924) photopic luminous efficiency function ($V(\lambda)$ in Figure 2.9, middle) across the visible spectrum (between 380 and 780 nm). Then, this weighted function (X_λ, right) can be integrated in the visible range (between 380 and 780 nm) to obtain the corresponding photometric quantity, X_V. This is expressed by Eq. (2.4).

$$X_V = K_m \int X_{e\lambda} \cdot V(\lambda) \cdot d\lambda \tag{2.4}$$

In Eq. (2.4), X_V is a photometric (luminous) quantity for photopic vision, $X_{e\lambda}$ is a radiant quantity, K_m is the luminous efficacy of radiation (LER) for photopic vision, $K_m = 683 \text{ lm W}^{-1}$, and $V(\lambda)$ is the CIE (1924) spectral luminous efficiency function for photopic vision. The value of K_m converts the power of electromagnetic radiation to a corresponding photometric unit, lumen (lm). Similarly, scotopic quantities can also be defined but then the $V'(\lambda)$ function (see the dark green curve in Figure 2.6) shall be used in Eq. (2.4) instead of $V(\lambda)$ and the value of K_m shall be changed to $K_m = 1699 \text{ lm W}^{-1}$ (scotopic value) instead of $K_m = 683 \text{ lm W}^{-1}$ (photopic value).

2.2.1
Radiant Power (Radiant Flux) and Luminous Flux

"Radiant power or radiant flux is the power (energy per unit time t) emitted, transferred or received in the form of electromagnetic radiation [12]." It is designated by Φ_e (unit: W), see Eq. (2.5).

$$\Phi_e = \int_{380 \text{ nm}}^{780 \text{ nm}} \Phi_{e\lambda} \cdot d\lambda \tag{2.5}$$

The corresponding photometric quantity is luminous flux (Φ_V, unit: lumen, lm) as defined by Eq. (2.6).

$$\Phi_V = K_m \int_{380 \text{ nm}}^{780 \text{ nm}} \Phi_{e\lambda} \cdot V(\lambda) \cdot d\lambda \tag{2.6}$$

Figure 2.10 illustrates the concept of radiant flux Φ_e.

As can be seen from Figure 2.10, radiant flux contains all rays emitted by the light source in all possible directions and the power of all rays shall be integrated to measure it.

Figure 2.10 Illustration of the concept of radiant flux: the total power of all rays emanating from the light source in all directions is considered.

2.2.2
Irradiance and Illuminance

Irradiance (E_e) is the ratio of the radiant power ($d\Phi_e$) incident on an element of a surface (dA) to the area of that element (unit: W m^{-2}), see Eq. (2.7).

$$E_e = \frac{d\Phi_e}{dA} \qquad (2.7)$$

The corresponding photometric quantity is illuminance (E_v, unit: lux, lx = lm m^{-2}) as defined by Eq. (2.8).

$$E_v = \frac{d\Phi_v}{dA} \qquad (2.8)$$

Figure 2.11 illustrates the concept of irradiance (E_e).

2.2.3
Radiant Intensity and Luminous Intensity

"Radiant intensity (I_e, unit: W sr^{-1}) is the ratio of the radiant power ($d\Phi_e$) leaving a source to an element of solid angle $d\Omega$ (unit: steradian, sr) propagated in the given

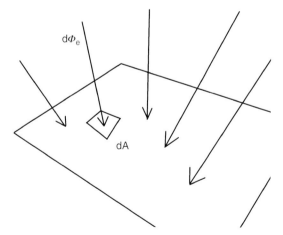

Figure 2.11 Illustration of the concept of irradiance: the radiant power $d\Phi_e$ reaches the surface element dA.

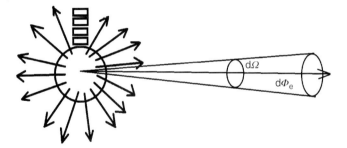

Figure 2.12 Illustration of the concept of radiant intensity: the radiant power $d\Phi_e$ leaves the light source to the element of solid angle $d\Omega$ in the indicated direction.

direction" [12] as defined by Eq. (2.9).

$$I_e = \frac{d\Phi_e}{d\Omega} \qquad (2.9)$$

The corresponding photometric quantity is luminous intensity (I_v, unit: candela, cd = lm sr^{-1}) as defined by Eq. (2.10).

$$I_v = \frac{d\Phi_v}{d\Omega} \qquad (2.10)$$

Figure 2.12 illustrates the concept of radiant intensity (I_e).

2.2.4
Radiance and Luminance

"Radiance (L_e, unit: W (sr m^2)$^{-1}$) is the ratio of the radiant power ($d\Phi_e$), at an angle α to the normal of the surface element, to the infinitesimal elements of both projected area ($dA \cos \alpha$) and solid angle ($d\Omega$)" [12] as defined by Eq. (2.11).

$$L_e = \frac{d\Phi_e}{d\Omega\, dA \cos \alpha} \qquad (2.11)$$

The corresponding photometric quantity is luminance (L_v, unit: cd m^{-2}) as defined by Eq. (2.12).

$$L_v = \frac{d\Phi_v}{d\Omega\, dA \cos \alpha} \qquad (2.12)$$

Figure 2.13 illustrates the concept of radiance (L_e).

As can be seen from Figure 2.13, radiance has relevance to a specific direction (2: direction of the detector) and a specific (small) surface (dA) on the light source or the luminaire that emits electromagnetic radiation. As light sources and luminaires are spatially extended objects that radiate nonuniformly, in a general case, their radiance depends on the point on their surface which is considered. Generally, radiance also depends on the direction from which it is detected (the direction 2 in Figure 2.13) except for the case of the so-called Lambertian radiators whose radiance does not depend on the direction where the detector is located. Note

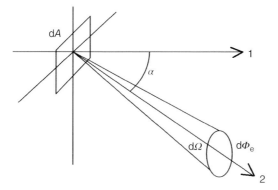

Figure 2.13 Illustration of the concept of radiance. 1, Direction which is normal to the surface element; 2, direction of the detector.

that "radiance plays a special role in radiometry because it is the propagation of the radiance that is conserved in a lossless optical system" [12].

From the point of view of the eye optics (see Figure 2.2), an interesting property of radiance (or luminance) is that, together with the area of the pupil which can be considered approximately constant for a moment at a fixed adaptation luminance level, this is the quantity that determines the values of retinal irradiance (or illuminance) at every point of the retinal image, that is, the stimulus for the retinal photoreceptor mosaic that acts like a digital camera, see Figure 2.5. Therefore, if the radiance (or luminance) distribution is measured at every point of a colored object by an imaging radiance (or luminance) meter (see Figure 4.4), then this dataset characterizes the image of that object projected to the retina by the optical system of the eye. This is the stimulus for the photoreceptor mosaic of the eye which is being captured by the L, M, and S cones and processed by the retina and later the visual brain.

A consequence is that it is the *luminance* of an extended light source, for example, a lamp with an opaque emitting surface, that determines its perceived brightness impression which is derived from the retinal LMS signals under photopic conditions (at least when the white tone is constant, e.g., when only warm white light sources are compared with each other; otherwise brightness comparisons also depend on chromatic signals and on chromaticity, see Figure 5.41) and *not* the total luminous flux of the light source.

2.2.5
Degrees of Efficiency for Electric Light Sources

From the point of view of electric power consumption and light output of the light source, the following three efficiency measures are defined:

1) Radiant efficiency: energy conversion efficiency with which the light source converts electrical power (P) into radiant flux (Φ_e), the ratio of the total output

radiant flux (Φ_e) to the input electrical power (P), see Eq. (2.13).

$$\eta_e = \frac{\Phi_e}{P} \qquad (2.13)$$

2) Luminous efficacy *of a source* (η_v or sometimes η, unit: lm W^{-1}) characterizes how efficiently a light source produces visible light (at least as predicted by spectral weighting by the $V(\lambda)$ function) from electric energy. It is the ratio of luminous flux (Φ_v) to electric power (P), see Eq. (2.14).

$$\eta_v = \frac{\Phi_v}{P} \qquad (2.14)$$

3) LER (unit: lm W^{-1}) characterizes how efficiently a given spectral power distribution is able to produce visible light (at least as predicted by spectral weighting by the $V(\lambda)$ function) from its total radiant flux (Φ_e). It is the ratio of luminous flux (Φ_v) to radiant flux (Φ_e), see Eq. (2.15).

$$\mathrm{LER} = \frac{\Phi_v}{\Phi_e} \qquad (2.15)$$

Note that, although widely used at the time of writing, the efficiency measures η_v and LER are only of limited relevance for the human visual system that the light sources are (or should be) designed for. The reason is that these efficiency measures take only the spectral sensitivity characteristics of a single channel, the L + M (luminance) channel, into account which is represented by the $V(\lambda)$ function (see Figure 5.6). One important aim of a major part of this book (Chapters 5–8) is to show novel ways for the spectral design of light sources that consider all relevant mechanisms of the human visual system.

2.3
Colorimetry and Color Science

As mentioned in Section 2.1.2, visual perception is the result of a psychophysical process: the human visual system converts electromagnetic radiation reaching the human eye (the stimulus) into neural signals and interprets these neural signals as different psychological dimensions of visual perception at later processing stages in the visual cortex (Figure 2.1). These perceptual dimensions (subjective aspects that are extracted from the processed neural signals in different processing centers of the brain) include shape, spatial structure (texture), motion, depth, and color. Understanding human color perception is essential for lighting engineering and produces huge economic interest: both the white tone of the light source and the colored objects illuminated by the light source should appear to be functional (for work performance) and/or aesthetic (to meet the user's demand). Otherwise the light source (and the luminaire) cannot be sold. To do so, the methodology of colorimetry and color science can be applied.

Colorimetry is the science and technology that quantifies human color perception and recommends methods to derive these so-called colorimetric quantities from the instrumentally measured spectral radiance distribution of the stimulus. Thus, it becomes possible to predict the psychological magnitude of a color perception and use it for the characterization, design, and optimization of a lit (interior or exterior) environment. In literature, the science of a more advanced modeling of sophisticated color perceptual phenomena is usually called color science (including e.g., the so-called color appearance models). A further aim of colorimetry is to define such quantities that are simple to measure instrumentally (without measuring spectral power distributions) yet enable lighting engineers to characterize a stimulus in such a way that its color perception or a property of its color perception can be easily described and understood (e.g., tristimulus values, chromaticity coordinates or the concept of correlated color temperature, CCT). Figure 2.14 illustrates how color perception (with its perceptual sub-dimensions brightness, colorfulness, and hue) comes to existence.

As can be seen from Figure 2.14, the electromagnetic radiation of the light source is reflected from the color sample which has a certain spectral reflectance $R(\lambda)$. After the reflection, the spectrally selective sample (the lilac rectangle in case of Figure 2.14) changes the spectral radiance distribution of the light source and this reflected light provides the color stimulus for the human observer who processes the color stimulus. The result of this processing is the color perception of the color sample with the perceived magnitudes of its brightness, colorfulness, and hue (the latter concepts are defined later).

The reflecting color stimulus in the example of Figure 2.14 is observed together with its immediate background (which is white in this example) and its surround (which is gray in this example). The visual system processes the color stimulus considered (the lilac rectangle in this example) together with the background and the surround, and the color perception of the color stimulus is influenced by (and related to) the background and the surround. Such a color stimulus is usually called a *related color (perception)*. If the light source itself is being observed directly (without reflection) then the light source itself represents the (so-called

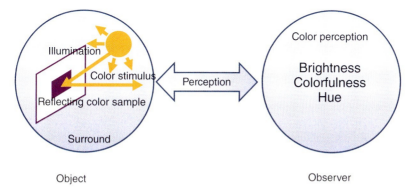

Figure 2.14 Illustration of how color perception (right) comes to existence from a color stimulus (left).

self luminous) color stimulus. If it is the only stimulus in the field of view then it is called an *unrelated color (perception)*.

In this section, a brief overview of the most important concepts and definitions of colorimetry and color science is given including the psychological dimensions of color perception (hue, saturation, and brightness), hue circle, color atlas, related and unrelated color stimuli, tristimulus values (XYZ), and chromaticity coordinates (x, y), chromaticity diagram with the loci of blackbody radiators and daylight phases, the concept of CCT, chromatic adaptation, CIELAB color space, color differences, and the CIECAM02 color appearance model as well as its advanced CAM02-UCS color difference metric. The aim of dealing with these issues is to provide an insight into modern colorimetry and color science to be able to understand the colorimetric concepts in this book.

2.3.1
Color Matching Functions and Tristimulus Values

Although it would be plausible to start from a quantity that is proportional to the signals of the three cone types (L, M, and S), in current standard colorimetric practice, LMS spectral sensitivities (see Figure 2.6) are *not* used to characterize electromagnetic radiation, to describe the color stimulus, and, in turn, to quantify color perception. Instead of LMS, for color stimuli subtending $1°-4°$ of visual angle, so-called *color matching functions* of the CIE 1931 standard colorimetric observer [13] are applied as the basis of standard colorimetry, denoted by $\bar{x}(\lambda), \bar{y}(\lambda), \bar{z}(\lambda)$. For visual angles greater than $4°$ (e.g., $10°$), the so-called CIE 1964 standard colorimetric observer is recommended [13] with the $\bar{x}_{10}(\lambda), \bar{y}_{10}(\lambda), \bar{z}_{10}(\lambda)$ color matching functions, see Figure 2.15.

The aim of the use of the color matching functions in Figure 2.15 is to predict which spectral radiance distributions result in the same color appearance (in other words, to predict *matching color stimuli*) provided that the viewing condition is the same, that is, both stimuli are imaged to the central retina for an average observer (the standard colorimetric observer). From the spectral radiance distribution of the color stimulus measured by a spectroradiometer, so-called *XYZ tristimulus values* are computed by the aid of the color matching functions via Eq. (2.16). If the *XYZ* tristimulus values of two color stimuli are the same, then they will result in the same color perception (in other words, there will be no *perceived color difference* between them). It should be noted that, from historical reasons, the $\bar{y}(\lambda)$ color matching function equals the $V(\lambda)$ function.

$$X = k \int_{360\,\text{nm}}^{830\,\text{nm}} L(\lambda)\bar{x}(\lambda)\,d\lambda$$
$$Y = k \int_{360\,\text{nm}}^{830\,\text{nm}} L(\lambda)\bar{y}(\lambda)\,d\lambda$$
$$Z = k \int_{360\,\text{nm}}^{830\,\text{nm}} L(\lambda)\bar{z}(\lambda)\,d\lambda \tag{2.16}$$

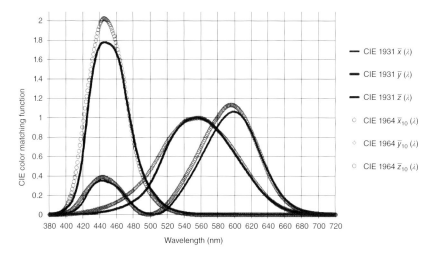

Figure 2.15 Black curves: color matching functions of the CIE 1931 standard colorimetric observer [13] denoted by $\bar{x}(\lambda)$, $\bar{y}(\lambda)$, $\bar{z}(\lambda)$ for color stimuli subtending 1°–4° of visual angle. Open gray circles: color matching functions of the CIE 1964 standard colorimetric observer [13] denoted by $\bar{x}_{10}(\lambda)$, $\bar{y}_{10}(\lambda)$, $\bar{z}_{10}(\lambda)$ for color stimuli subtending greater than 4°. The $\bar{y}(\lambda)$ color matching function was chosen to be equal $V(\lambda)$. (Reproduced with permission from Ref. [10].)

In Eq. (2.16), $L(\lambda)$ denotes the spectral radiance distribution of the color stimulus $L(\lambda)$ and k is a constant. According to the scheme of Figure 2.14, in case of reflecting color samples, the spectral radiance of the stimulus ($L(\lambda)$) equals the spectral reflectance ($R(\lambda)$) of the sample multiplied by the spectral irradiance from the light source illuminating the reflecting sample ($E(\lambda)$), see Eq. (2.17) (for diffusely reflecting materials).

$$L(\lambda) = \frac{R(\lambda)E(\lambda)}{\pi} \qquad (2.17)$$

In Eq. (2.17), the value of k is computed according to Eq. (2.18) [13].

$$k = 100 / \int_{360\,\text{nm}}^{830\,\text{nm}} L(\lambda)\bar{y}(\lambda)\mathrm{d}\lambda \qquad (2.18)$$

As can be seen from Eq. (2.18), for reflecting color samples, the constant k is chosen so that the tristimulus value $Y = 100$ for ideal white objects with $R(\lambda) \equiv 1$. For self-luminous stimuli (if the subject observes, e.g., a white LED light source directly), the value of k can be chosen to equal $683\,\text{lm}\,\text{W}^{-1}$ [13] and then, as $\bar{y}(\lambda) \equiv V(\lambda)$, the value of Y will be equal to the *luminance* of the self-luminous stimulus. For color stimuli with visual angles greater than 4°, the tristimulus values X_{10}, Y_{10}, and Z_{10} can be computed substituting $\bar{x}(\lambda)$, $\bar{y}(\lambda)$, $\bar{z}(\lambda)$ by $\bar{x}_{10}(\lambda)$, $\bar{y}_{10}(\lambda)$, $\bar{z}_{10}(\lambda)$ in Eq. (2.16). As can be seen from Figure 2.15, the two sets of color matching functions, that is, $\bar{x}(\lambda)$, $\bar{y}(\lambda)$, $\bar{z}(\lambda)$ and $\bar{x}_{10}(\lambda)$, $\bar{y}_{10}(\lambda)$, $\bar{z}_{10}(\lambda)$ differ significantly. The consequence is that two matching (homogeneous) color stimuli subtending, for example, 1° of visual angle will not match in a general case if their size is increased to, for example, 10°.

The so-called chromaticity coordinates (x, y, z) are defined by Eq. (2.19).

$$x = \frac{X}{X+Y+Z} \quad y = \frac{Y}{X+Y+Z} \quad z = \frac{Z}{X+Y+Z} \tag{2.19}$$

The diagram of the chromaticity coordinates x, y is called the CIE 1931 chromaticity diagram or the CIE (x, y) chromaticity diagram [13]. Figure 2.16 illustrates how color perception changes across the x, y diagram.

As can be seen from Figure 2.16, valid chromaticities of real color stimuli are located inside the so-called *spectral locus* which is the boundary of quasi-monochromatic radiations of different wavelengths and the so-called purple line in the bottom of the diagram. White tones can be found in the middle with increasing *saturation* (to be defined exactly below) toward the spectral locus. Perceived *hue* changes (purple, red, yellow, green, cyan, blue) when going around the region of white tones in the middle of the diagram.

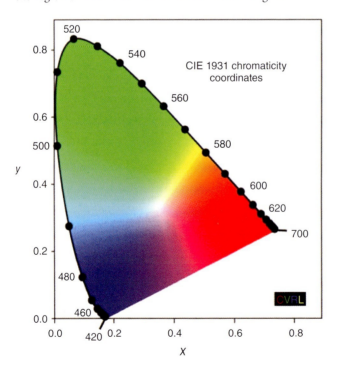

Figure 2.16 Illustration of how color perception changes across the CIE (x, y) chromaticity diagram [13]. The curved boundary of colors with three-digit numbers (wavelengths in nanometer units) represents the locus of monochromatic (i.e., most saturated) radiation. (Figure 7 from Ref. [14], Copyright Wiley-VCH Verlag GmbH & Co. KGaA, Berlin, 2004. Reproduced with permission from Wiley-VCH.)

2.3.2
Color Appearance, Chromatic Adaptation, Color Spaces, and Color Appearance Models

Although color stimuli can be fully described in the system of tristimulus values (X, Y, and Z), this description yields a nonuniform and nonsystematic representation of the color perceptions corresponding to these color stimuli. The psychological attributes of the perceived colors (lightness, brightness, redness–greenness, yellowness–blueness, hue, chroma, saturation, and colorfulness) cannot be expressed in terms of XYZ values directly. To derive a model that predicts the magnitude of these perceptual attributes (a so-called color space or color appearance model), mathematical descriptors can be defined to each one of these attributes (so-called numeric correlates). Numeric correlates can be computed from the XYZ tristimulus values of the stimulus and the XYZ values of their background and surround parameters. In the following, the psychological attributes are defined.

2.3.2.1 Perceived Attributes of Color Perception

Hue is the attribute of a visual sensation according to which a color stimulus appears to be similar to the perceived colors red, yellow, green, and blue, or a combination of two of them [15]. *Brightness* is the attribute of a color stimulus according to which it appears to emit more or less light [15]. *Lightness* is the brightness of a color stimulus judged relative to the brightness of a similarly illuminated reference white (appearing white or highly transmitting) [16]. Lightness is an attribute of *related colors*.

Colorfulness is the attribute of a color stimulus according to which the stimulus appears to exhibit more or less chromatic perceived color. For a given chromaticity, colorfulness generally increases with luminance [17]. In a lit interior (a built environment which is important for the application of white LED light sources), observers tend to assess the *chroma* of (related) surface colors or colored objects. The perceived attribute *chroma* refers to the colorfulness of the color stimulus judged in proportion to the brightness of the reference white [16].

Saturation is the colorfulness of a stimulus judged in proportion of its own brightness [15]. A perceived color can be very saturated without exhibiting a high level of chroma. For example, the color of the deep red sour cherry is saturated but it exhibits less chroma because the sour cherry is colorful compared to its (low) own brightness but it is not so colorful in comparison to the brightness of the reference white. Figure 2.17 illustrates the three perceived attributes – hue, chroma, and lightness.

In the mathematical construct of a *color space* (which has three dimensions), the three perpendicular axes and certain angles and distances carry psychologically relevant meanings related to the earlier-defined perceived color attributes. A schematic illustration of the structure of a color space can be seen in Figure 2.18.

2.3.2.2 Chromatic Adaptation

The perceived attributes of the color perception of a color stimulus depend not only on the color stimulus itself but also on the so-called adapted white point

2 The Human Visual System and Its Modeling for Lighting Engineering

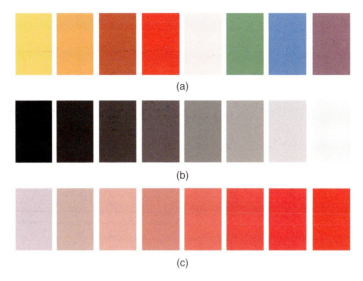

Figure 2.17 Illustration of three attributes of perceived color: (a) changing hue, (b) changing lightness, and (c) changing chroma. (Reproduction of Figure 1 [18] with permission from *Color Research and Application*.)

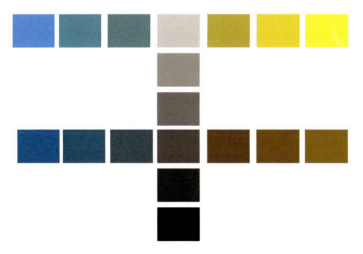

Figure 2.18 Schematic illustration of the general structure of a color space. Lightness increases from black to white from the bottom to the top along the gray lightness scale in the middle. Chroma increases from the gray scale toward the outer colors of high chroma. The perceptual attribute of hue varies when rotating the image plane around the gray axis in space (blue and yellow being opponent color perceptions). (Reproduced with permission from Ref. [10].)

(e.g., the large white wall surfaces in a room illuminated by a LED light source). This white point can be, for example, warm white (with a minor but perceptible yellowish shade in the white tone of the luminaire) or cool white (with a slight bluish shade). The chromaticity of the white tone that predominates in the visual environment influences the color perception of the object colors significantly. The reason is not just the spectral change of the stimulus triggered by the different light source emission spectra that illuminate the objects (e.g., warm white or cool white according to Figure 2.14).

There is another, visual effect (that *partially* compensates for these *spectrally induced changes*): *chromatic adaptation*. For example, if a cool white LED light source is switched on in the room instead of a warm white LED light source then the state of chromatic adaptation (the adapted white point in the scene) changes from "warm white adapted" to "cool white adapted". The mathematical description of chromatic adaptation is an important component of color spaces and color appearance models. The result of chromatic adaptation is that the perceived colors of the colored stimuli (colored objects) do *not* change *so much* (this is the tendency of the so-called *color constancy*) – despite the significant spectral changes caused by the change of the emission spectra (e.g., cool white vs warm white) that illuminate the objects. This is true as long as the light source that illuminates the objects has a shade of white (a white tone which can exhibit slight yellowish or bluish shades, see Section 5.7). If the objects are illuminated by non-white-light sources (e.g., a poor LED lamp with a greenish shade in its "white" tone) then color constancy breaks down and a "strange" greenish shade is perceived on every object.

To experience a feeling about chromatic adaptation, look at the still life in Figure 2.19 (top) illuminated by a warm white light source about 1 min and then look at the same still life illuminated by a cool white light source.

If the still life of Figure 2.19 is first illuminated by the warm white light source (top) and the illuminant is changed to cool white (bottom), then the process of chromatic adaptation starts immediately after the illuminant change: the bluish shade on the still life under the cool white light source diminishes continuously with elapsing time. At the end of the chromatic adaptation process, the state of adaptation of the observer stabilizes at the adapted white point of the cool white light source. The reader can imitate this process: cover the bottom still-life image of Figure 2.19 for 1 min and look at the warm white version only and then cover the top still-life image and look at the cool white version. Figure 2.20 offers another possibility to experience how chromatic adaptation works.

2.3.2.3 CIELAB Color Space

As can be seen from Figure 2.18, lightness increases from black to white from the bottom to the top along the gray lightness scale in the middle of color space. At every lightness level, chroma increases from the gray scale toward the most saturated outer colors. The perceptual attribute of hue varies when rotating the image plane around the gray axis of the color space. Standard CIE colorimetry

Figure 2.19 Illustration of chromatic adaptation. Example for two different states of chromatic adaptation: the same colorful still life of the same colored objects under a warm white (top) and a cool white (bottom) light source. (Image source: Technische Universität Darmstadt.)

Figure 2.20 Experience how chromatic adaptation of the human visual system works: cover the right circle and adapt to the left (yellowish) circle for 1 min then cover both circles and look at the middle (white). A bluish afterimage of the left circle should appear on the white background because of the fact that L and M cones have a lower gain (at the moment of looking at the middle) than the blue sensitive S cones because L and M cones were adapted to the relatively higher L and M signal content of the yellowish circle for 1 min. Repeat this procedure by adapting 1 min to the right (bluish) circle which reduces the gain of the S cones and try to explain the phenomenon.

recommends two such coordinate systems, CIELAB and CIELUV, the so-called CIE 1976 ("uniform") color spaces [13]. Here, only the more widely used CIELAB color space is described. The XYZ values of the color stimulus and the XYZ values of a specified reference white color stimulus (X_n, Y_n, Z_n) constitute its input and the CIELAB L^*, a^*, b^* value its output, see Eq. (2.20).

$$L^* = 116 f\left(\frac{Y}{Y_n}\right) - 16$$
$$a^* = 500 \left[f\left(\frac{X}{X_n}\right) - f\left(\frac{Y}{Y_n}\right) \right]$$
$$b^* = 200 \left[f\left(\frac{Y}{Y_n}\right) - f\left(\frac{Z}{Z_n}\right) \right] \qquad (2.20)$$

In Eq. (2.20), the function f is defined by Eq. (2.21).

$$f(u) = u^{1/3} \text{ if } u > \left(\frac{24}{116}\right)^3$$
$$f(u) = \left(\frac{841}{108}\right) u + \left(\frac{16}{116}\right) \text{ if } u \leq \left(\frac{24}{116}\right)^3 \qquad (2.21)$$

The CIELAB formulae of Eq. (2.20) try to account for chromatic adaptation by simply dividing the tristimulus value of the color stimulus by the corresponding value of the adopted reference white. In reality, chromatic adaptation is based on the gain control of the L, M, and S cones (see Figure 2.20). The modeling of photoreceptor gain control by the reference white (white point) control of the XYZ tristimulus values is a very rough approximation. This is one of the reasons why CIELAB is only of limited applicability for lighting practice.

The output quantities of the CIELAB system represent approximate correlates of the perceived attributes of color: L^* (CIE 1976 lightness) is intended to describe the perceived lightness of a color stimulus. Similarly, CIELAB chroma (C^*_{ab}) stands for perceived chroma and CIELAB hue angle (h_{ab}) for perceived hue, see Eq. (2.22). The quantities a^* and b^* in Eq. (2.20) can be considered as rough correlates of perceived redness–greenness (red for positive values of a^*) and perceived yellowness–blueness (yellow for positive values of b^*). L^*, a^*, and b^* constitute the three orthogonal axes of CIELAB color space (as illustrated in Figure 2.20). Eq. (2.22) shows how to calculate C^*_{ab} and h_{ab} from a^* and b^*.

$$C^*_{ab} = \sqrt{a^{*2} + b^{*2}}$$
$$h_{ab} = \arctan\left(\frac{b^*}{a^*}\right) \qquad (2.22)$$

Figure 2.21 shows the CIELAB a^*–b^* diagram with the numeric correlates of chroma (C^*_{ab}) and hue angle (h_{ab}). In Figure 2.21, the CIELAB hue angel (h_{ab}) changes between 0° (at $b^* = 0$, $a^* > 0$) and 360° around the L^* axis (the achromatic axis that stands perpendicular to the middle of the a^*–b^* plane, compare with Figure 2.18).

2.3.2.4 The CIECAM02 Color Appearance Model

To apply the CIELAB color space, it is important to read the notes of the CIE publication [13] carefully: CIELAB is "intended to apply to … object colors of the same size and shape, viewed in identical white to middle-gray surroundings by an observer photopically adapted to a field of chromaticity not too different

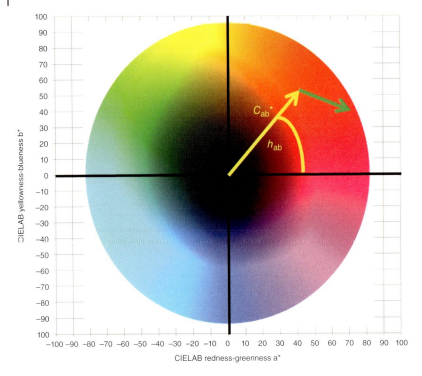

Figure 2.21 Illustration of a CIELAB a^*–b^* diagram with the correlates of chroma (C^*_{ab}) and hue angle (h_{ab}). The L^* axis is perpendicular to the middle of the a^*–b^* plane in which L^* is constant. At the end of the yellow arrow, there is a color stimulus with CIELAB chroma C^*_{ab} and CIELAB hue angle h_{ab}. The green arrow shows a color difference (between a red and an orange stimulus), here in the a^*–b^* plane only, that is, without a lightness difference (ΔL^*).

from that of average daylight." To describe the color appearance of color stimuli viewed under different viewing conditions from incandescent (tungsten) light to daylight, from dark or dim to average or bright surround luminance levels or color stimuli on different backgrounds, so-called color appearance models can be used. Figure 2.22 illustrates the effect of different viewing conditions on the color appearance of a color element considered.

As can be seen from Figure 2.22, the color perception of the color stimulus (color element whose color appearance is being assessed by the human visual system) depends not only on the tristimulus values of the stimulus but also on the viewing conditions (the same stimulus on the right appears to be lighter). Below, the most widely used the so-called CIECAM02 color appearance model [19] is described, which is able to account for the change in viewing conditions. The CIECAM02 model computes numeric correlates (mathematical descriptor quantities) for all of the earlier-defined perceived attributes of color (e.g., chroma, saturation, and

Figure 2.22 Two viewing conditions with the same color stimulus. Color stimulus (brownish color element considered): small filled circle in the middle, typically of a diameter of 2°; background: large gray filled circle, typically of a diameter of 10°; surround: the remainder of the viewing field outside the background; adapting field = background + surround. Left: color stimulus on a bright adapting field; right: the same stimulus on a dark adapting field (just an illustration, not intended to be colorimetrically correct).

brightness) for the color element considered. These CIECAM02 quantities correlate better with the perceived magnitude of all color attributes than the CIELAB quantities (L^*, C^*_{ab}, h_{ab}).

The CIECAM02 model uses six parameters to describe the viewing condition of the color stimulus: L_A (adapting field luminance), Y_b (relative background luminance between 0 and 100), F (degree of adaptation), D (degree of chromatic adaptation to reference white, e.g., the white walls of an illuminated room), N_c (chromatic induction factor), and c (impact of the surround). In the absence of a measured value, the value of L_A can be estimated by dividing the value of the adopted white luminance by 5. For typical arrangements of color objects illuminated by a light source in a room, the value of $Y_b = 20$ can be used.

The values of F, c, and N_c depend on the so-called surround ratio computed by dividing the average surround luminance by the luminance of the reference white (e.g., the white walls in an illuminated office) in the scene. If the surround ratio equals 0, then the surround is called *dark*; if it is less than 0.2 then the surround is *dim*; otherwise it can be considered an *average* surround. The values of F, c, and N_c are equal: 0.8, 0.525, and 0.8 for dark, 0.9, 0.59, and 0.95 for dim, and 1.0, 0.69, and 1.0 for average surrounds, respectively. For intermediate surround ratios, these values can be interpolated. In the CIECAM02 model, the value of D (degree of chromatic adaptation to reference white) is usually computed from the values of F and L_A by a dedicated equation. But the value of D can also be forced to a specific value instead of using that equation. For example, it can be forced to be equal to 1 to ensure complete chromatic adaptation to reference white for high illuminance levels (e.g., $E_v = 700$ lx) in a well-illuminated room.

Figure 2.23 shows the block diagram of the computational method of the CIECAM02 color appearance model. Because of their complexity, its defining

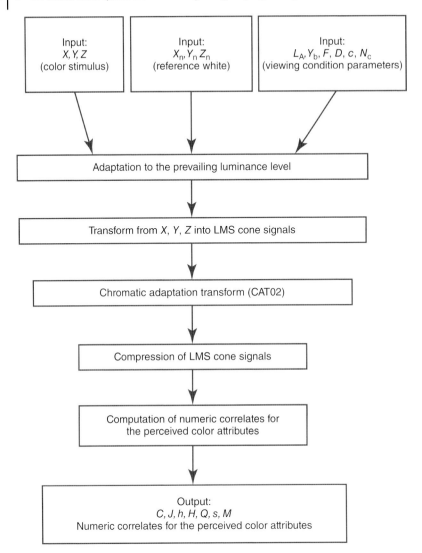

Figure 2.23 Block diagram of the computational method of the CIECAM02 color appearance model [19, 20, 21].

equations [19, 20] are not repeated here. A free worksheet can be found in the web [21].

As can be seen from Figure 2.23, the CIECAM02 model computes an adaptation factor to the prevailing luminance level, that is, an average luminance value of the adapting field. Then, the tristimulus values X, Y, Z (which can be measured and computed promptly by the most widely used standard instruments and their software including spectroradiometers and colorimeters) are transformed into LMS cone signals. Then, in turn, chromatic adaptation is

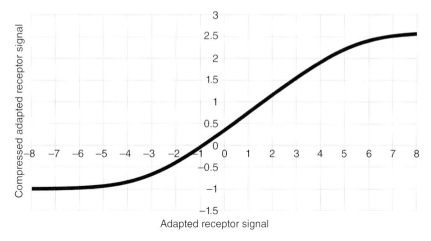

Figure 2.24 Illustration of the compression of the adapted receptor signal (L, M, or S) in the CIECAM02 color appearance model ($\log_{10}$–$\log_{10}$ diagram).

modeled by an advanced chromatic adaptation transform, the so-called CAT02. CAT02 represents a significant enhancement compared to the neurally incorrect "chromatic adaptation" transform of CIELAB that simply divides a tristimulus value by the corresponding value of the reference white. Adapted receptor signals are then, in turn, compressed. Figure 2.24 illustrates this important signal compression step.

As can be seen from the log–log diagram of Figure 2.24, the cone signal compression phase is nonlinear: this is an important feature of the human visual system. Note that there is noise for small adapted receptor signals while the compressed signal converges to an upper limit for high input signal values. After this compression phase, the numeric correlates of the perceived color attributes are computed as the output of the model (see Figure 2.23): C (CIECAM02 chroma), J (CIECAM02 lightness), h (CIECAM02 hue angle), H (CIECAM02 hue composition), Q (CIECAM02 brightness), s (CIECAM02 saturation), M (CIECAM02 colorfulness), a_M (CIECAM02 redness–greenness), and b_M (CIECAM02 yellowness–blueness).

To visualize the quantities C, J, h, a_M, and b_M, refer to Figure 2.21: they are analogous to CIELAB C^*_{ab}, L°, h_{ab}, a°, and b°, respectively. But the correlation between them and the magnitudes of the corresponding perceptual color attributes is significantly enhanced compared to CIELAB, as mentioned earlier. CIECAM02 H (hue composition) varies between 0 and 400 and not between 0° and 360°: 0 corresponds to so-called *unique red* (without any yellow and blue perception), 100 to *unique yellow* (without red and green), 200 to *unique green* (without yellow and blue), and 300 to *unique blue* (without red and green) while 400 is the same as 0 (red).

As a summary, it can be stated that CIECAM02 has the following advantages compared to CIELAB: (i) viewing condition parameters can be set to represent predominating viewing conditions in the environment illuminated by the light source; (ii) reference white can be changed in a wide range (from warm white to cool white) reliably; (iii) CIECAM02 provides more numeric correlates for all perceived attributes of color (colorfulness, chroma, saturation, hue, brightness, and lightness); (iv) numeric scales of these correlates correspond better to color perception than CIELAB correlates; and (v) a CIECAM02 based advanced color difference formula and a uniform color space was established [22].

2.3.3
Modeling of Color Difference Perception

2.3.3.1 MacAdam Ellipses

A disadvantage of the CIE x, y chromaticity diagram (Figure 2.16) is that, concerning color difference perceptions, it is perceptually not uniform in the following sense. In Figure 2.11, observe that a distance in the green region of the diagram represents a less change of perceived chromaticness than the same distance in the blue-purple region. The so-called MacAdam ellipses [23] quantify this effect around the different chromaticity centers, that is, the centers of the ellipses in Figure 2.25. Roughly speaking, perceived chromaticity differences are hardly noticeable between any two chromaticity points inside the ellipse (for a more precise definition of the MacAdam ellipses, see [23]). Note that the ellipses of Figure 2.25 are magnified ten times.

2.3.3.2 u', v' Chromaticity Diagram

As can be seen from Figure 2.25, MacAdam ellipses are large in the green region of the CIE x, y chromaticity diagram while they are small in blue–purple region, and the orientation of the ellipses also changes. To overcome these difficulties, the x and y axes were distorted so as to make identical circles from the MacAdam ellipses, and this resulted in the so-called CIE 1976 uniform chromaticity scale diagram (or simply u', v' diagram) defined by Eq. (2.23).

$$u' = \frac{4x}{(-2x + 12y + 3)}$$
$$v' = \frac{9y}{(-2x + 12y + 3)} \qquad (2.23)$$

The concept of u', v' chromaticity difference between two chromaticity points can be defined as the Euclidean distance on the plane of the u', v' diagram: $\Delta u'v' = \sqrt{(\Delta u')^2 + (\Delta v')^2}$. The u', v' diagram is perceptually *more uniform* than the x, y diagram. This means that equal $\Delta u'v'$ values *should* represent equal *perceived* chromaticity differences in *any* part of the u', v' diagram and in *any* direction at every chromaticity center provided that the relative luminance difference of the two color stimuli is small, for example, $\Delta Y < 0.5$. Therefore, the shape of the MacAdam ellipses in Figure 2.25 comes closer to a circular shape when they are depicted in the u', v' diagram. This means that the u', v' diagram is more useful (although far from perfect) than the x, y diagram to evaluate differences of perceived chromaticness without lightness differences.

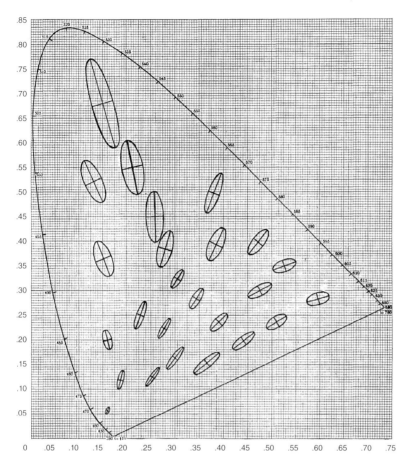

Figure 2.25 MacAdam ellipses [23] in the CIE x, y chromaticity diagram. Abscissa: chromaticity coordinate x, ordinate: chromaticity coordinate y. Roughly speaking, perceived chromaticity differences are not noticeable inside the ellipses. For a more precise definition of the MacAdam ellipses, see [23]. Ellipses are magnified ten times. (Reproduced from Ref. [23] with permission from the *Journal of the Optical Society of America*.)

2.3.3.3 CIELAB Color Difference

Perceived total (including lightness) color differences between two color stimuli can be modeled by the Euclidean distance between them in a rectangular color space, for example, in CIELAB, this quantity is denoted by ΔE^*_{ab}, see Eq. (2.24).

$$\Delta E^*_{ab} = \sqrt{(\Delta L^*)^2 + (\Delta a^*)^2 + (\Delta b^*)^2} \tag{2.24}$$

Lightness, chroma, and hue angle differences of two color stimuli (ΔL^*, ΔC^*_{ab}, and Δh_{ab}) can be computed by subtracting the lightness, chroma, and hue angle values of the two color stimuli. Hue differences (ΔH^*_{ab}) must not be confused with hue angle differences (Δh_{ab}). Hue differences include the fact that the same hue change results in a large color difference for large chroma and in a small color difference for small chroma (i.e., in the neighborhood of the CIELAB L^* axis), see Figure 2.21. CIELAB hue difference is defined by Eq. (2.25). In Eq. (2.25), ΔH^*_{ab} has the same sign as Δh_{ab}.

$$\Delta H^*_{ab} = \sqrt{(\Delta E^*_{ab})^2 - (\Delta L^*)^2 - (\Delta C^*_{ab})^2} \qquad (2.25)$$

2.3.3.4 CAM02-UCS Uniform Color Space and Color Difference

Unfortunately, the CIELAB color space and the color difference computed in it by using Eq. (2.24) exhibit perceptual nonuniformities depending on the region of color space (e.g., reddish or bluish colors) and on color difference magnitude (small, medium, or large color differences) [24]. To address these problems, a uniform color space (so-called CAM02-UCS) was introduced on the basis of the CIECAM02 color appearance model [22] to describe all types of color difference magnitudes. The superior performance of CAM02-UCS was corroborated in visual experiments on color rendering [25, 26].

The CAM02-UCS color space [22] is defined in the following two steps on the basis of the CIECAM02 color appearance model:

Step 1. The CIECAM02 correlates lightness (J) and colorfulness (M) are transformed according to Eqs. (2.26) and (2.27). The CIECAM02 hue angle h is not transformed.

$$J' = \frac{1.7 J}{0.007 J + 1} \qquad (2.26)$$

$$M' = \frac{1}{0.0228} \ln(1 + 0.0228 M) \qquad (2.27)$$

Step 2. The variables a' und b' are defined by Eq. (2.28).

$$\begin{aligned} a' &= M' \cos(h) \\ b' &= M' \sin(h) \end{aligned} \qquad (2.28)$$

The variables a' and b' can be imagined as new correlates of perceived redness–greenness and yellowness–blueness (similar to CIELAB a^* and b^* in Figure 2.21). But their real importance is to constitute – together with J' – three axes of a uniform color space (a', b', J') in which perceived color differences can be quantified reliably by the Euclidean distance in this space. The J' axis stands in the middle of the $a'-b'$ plane perpendicular to this plane. In order to describe the perceived color difference between two color stimuli in the CAM02-UCS color space, the so-called CAM02-UCS color difference ($\Delta E'$ or $\Delta E_{CAM02-UCS}$ or briefly ΔE_{UCS}) can be calculated by Eq. (2.29).

$$\Delta E' = \sqrt{(\Delta J')^2 + (\Delta a')^2 + (\Delta b')^2} \qquad (2.29)$$

The advantage of the CAM02-UCS color difference metric of Eq. (2.30) is that it is based on the CIECAM02 color appearance model hence it can be applied to all viewing conditions by adjusting the values of the CIECAM02 viewing parameters, the tristimulus values of the reference white as well as the parameters L_A, Y_b, F, D, c, and N_c.

2.3.4
Blackbody Radiators and Phases of Daylight in the x, y Chromaticity Diagram

For the practical applications of white LED lighting to be described in this book, it is necessary to further analyze the x, y chromaticity diagram and introduce some further essential concepts. As could be seen from Figure 2.16, the different white tones (e.g., warm white or cool white) are located in the middle region of the x, y chromaticity diagram. But this region deserves further attention, see Figure 2.26.

As can be seen from Figure 2.26 (compare with Figure 2.16), spectral color stimuli (of quasi-monochromatic electromagnetic radiation) constitute the left,

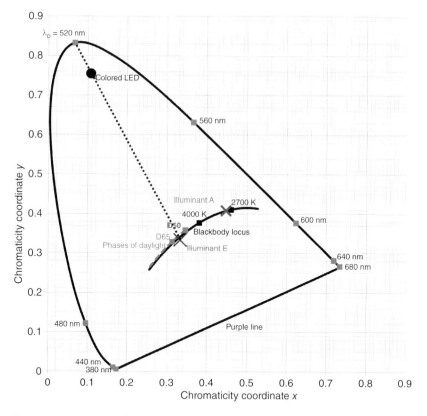

Figure 2.26 Important chromaticity points and curves in the x, y chromaticity diagram. The gray cross represents the chromaticity of CIE illuminant "A" ($T_{cp} = 2856$ K). Illuminant E: equi-energetic stimulus (its spectral radiance is constant at every wavelength, $x = 0.333$; $y = 0.333$); Filled black circle: a colored LED as an example to explain the concept of dominant wavelength ($\lambda_D = 520$ nm in this example) and colorimetric purity in Section 2.4.4.

upper, and right boundary of the region of the possible chromaticities of real color stimuli. In Figure 2.26, some spectral color stimuli (like 680, 640, and 600 nm) between 380 and 680 nm are marked. The purple line connects the chromaticities between the two extremes on the left (at the 380 nm point) and on the right (at the 680 nm point).

One of the important concepts that can be introduced is the curve of *blackbody radiators* (blackbody locus or Planckian curve), see the black line in the middle of Figure 2.26. This corresponds to the chromaticities (x, y) of blackbody radiators at different temperatures between 2000 (yellowish tones) and 20 000 K (bluish tones) in the figure. The figure contains two marked data points, 2700 and 4000 K, while 20 000 K is located at the left end of the black line. An important blackbody radiator is CIE illuminant "A" because it represents typical tungsten filament lighting (its chromaticity is well known among light source users). Its relative spectral power distribution is that of a Planckian radiator at a temperature of 2856 K.

Associated with the blackbody locus, the concept of CCT (correlated color temperature) is defined here according to its relevance to assess the color rendering of light sources. The CCT of a test light source (T_{cp}; unit: Kelvin, K) of a given chromaticity is equal to the temperature of the blackbody radiator of the nearest chromaticity. To compute the value of the CCT, the chromaticity point of the test light source can be projected onto the blackbody locus and, by the use of Planck's law, the temperature can be computed. The projection to the blackbody locus will occur in a uniform chromaticity diagram [13] and not in the x, y diagram as the latter diagram is nonuniform, see Figure 2.25. Standard colorimetry uses the so-called u, v diagram to this purpose, see Eq. (2.30) and compare with Eq. (2.23).

$$u = u' = \frac{4x}{(-2x + 12y + 3)}$$
$$v = (2/3)v' = \frac{6y}{(-2x + 12y + 3)} \qquad (2.30)$$

The chromaticities of average *phases of daylight* of different CCTs represent a further remarkable set which is located slightly above the blackbody locus (see the gray curve in Figure 2.26). Figure 2.26 shows them between $T_{cp} = 5000$ K and $T_{cp} = 20\,000$ K. In lighting engineering, daylight illuminants D65 ($T_{cp} = 6504$ K) and D50 ($T_{cp} = 5003$ K) are especially important as they are often used as target white points when optimizing the spectral radiance distribution of a light source. They are marked in the figure.

Figure 2.27 shows the relative spectral radiance distributions of two selected blackbody radiators (at 2700 and 4000 K) and the daylight illuminants D50 and D65.

As can be seen from Figure 2.27, the blackbody radiator at 2700 K (P2700) has a high amount of emission in the yellow and red wavelength range (this is why its white tone is yellowish, so-called warm white). This red–yellow amount decreases for P4000, D50, and D65, respectively. But the blue content of these light sources

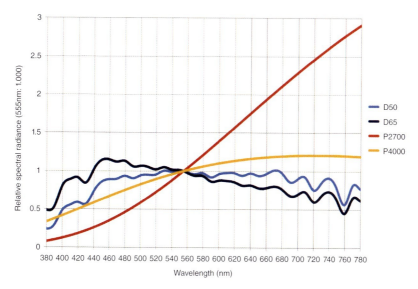

Figure 2.27 Relative spectral radiance distributions of two selected blackbody radiators (at 2700 and 4000 K) and the daylight illuminants D50 and D65; rescaled to ≡ 1 at 555 nm.

increases in this order. This is the reason why the daylight illuminant D65 has a bluish white tone (so-called cool white).

2.4
LED Specific Spectral and Colorimetric Quantities

In this section, LED specific spectral and colorimetric quantities are introduced to make understanding Chapter 4 on LED measurement and modeling easy. In this section, LED light sources without phosphor conversion (nonwhite or colored chip LEDs) are considered only while phosphor emission characteristics are dealt with in Section 3.4. Such colored LEDs emit characteristic spectral radiance distributions with a spectral bandwidth of "some tens of nanometers" [27]. In this sense, they differ from other light sources significantly. Some typical spectral radiance distributions are shown in Figure 2.28.

Characteristic wavelengths and wavelength intervals used to describe such spectral radiance distributions (Figure 2.28) are shown in Figure 2.29.

2.4.1
Peak Wavelength (λ_p)

This is the wavelength at the maximum of the spectral distribution at which the spectral power distribution is usually normalized (see Figure 2.29). Note that LEDs exhibit a characteristic shift of peak wavelengths with temperature, see Section 4.2.

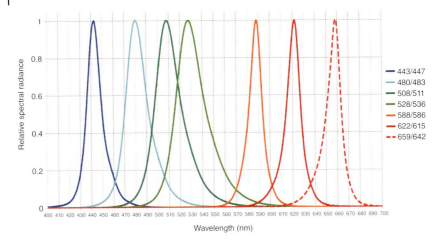

Figure 2.28 Relative spectral radiance distribution of typical colored LED light sources (non-white LEDs or colored chip LEDs). Legend: peak wavelength (λ_P in nm, see Section 2.4.1)/dominant wavelength (λ_D in nm, see Section 2.4.4.1).

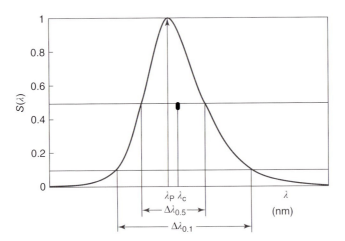

Figure 2.29 Characteristic wavelengths and wavelength intervals used to characterize the spectral radiance distributions of non-phosphor-converted LED light sources [27]. (Reproduced from Ref. [27] with permission from the CIE.)

2.4.2
Spectral Bandwidth at Half Intensity Level ($\Delta\lambda_{0.5}$)

The spectral bandwidth at half intensity level $\Delta\lambda_{0.5}$ is obtained as the difference of two wavelengths on either side of the peak wavelength λ_P where the peak intensity of the LED drops to 50%. The quantity $\Delta\lambda_{0.1}$ can be defined similarly as the difference of two wavelengths on either side of the peak wavelength λ_P where the peak intensity of the LED drops to 10% (see Figure 2.29).

2.4.3
Centroid Wavelength (λ_C)

The centroid wavelength λ_C (see Figure 2.29) is defined as the "center of gravity wavelength" [27] by Eq. (2.31).

$$\lambda_C = \frac{\int_{\lambda_1}^{\lambda_2} \lambda \cdot S(\lambda) d\lambda}{\int_{\lambda_1}^{\lambda_2} S(\lambda) d\lambda} \quad (2.31)$$

In Eq. (2.31), λ_1 and λ_2 represent the bandwidth limits of the LED emission, the LED does not emit outside the $(\lambda_1; \lambda_2)$ range. Note that $\lambda_C \neq \lambda_P$ because the relative spectral radiance distribution curve of a typical LED is asymmetric, see Figure 2.29.

2.4.4
Colorimetric Quantities Derived from the Spectral Radiance Distribution of the LED Light Source

There are two colorimetric quantities which are sometimes useful to characterize the color stimulus of colored (i.e., non-phosphor-converted) chip LEDs concerning their hue and saturation. The concept of *dominant wavelength* can be used to characterize *hue* while *colorimetric purity* helps quantify *saturation* [27]. These two quantities can be calculated from the *x, y* chromaticity coordinates of the emission of the colored chip LED. They are useless for white LEDs. For white LEDs, CCT (for hue) and the distance of their chromaticity point from the blackbody or daylight loci (for saturation) are similar useful measures, see Section 5.7.

2.4.4.1 Dominant Wavelength (λ_D)
The dominant wavelength (of a colored LED; denoted by λ_D) is equal to the "wavelength of the monochromatic stimulus that, when additively mixed in suitable proportions with the specified achromatic stimulus, matches the color stimulus considered" [13]. The achromatic stimulus is usually chosen to be the equi-energetic stimulus E (its spectral radiance is constant at every wavelength and $x = 0.333$; $y = 0.333$). An example can be seen in Figure 2.26: the filled black circle represents the chromaticity point of a colored LED. It is connected with the chromaticity point of illuminant E ($x = 0.333$; $y = 0.333$) and this line is extended toward the spectral locus. The wavelength on the spectral locus at this intersection point is the dominant wavelength ($\lambda_D = 520$ nm in this example).

2.4.4.2 Colorimetric Purity (p_C)
The colorimetric purity [13] (of a colored LED; denoted by p_C) is defined by Eq. (2.32).

$$p_C = \frac{L_d}{L_d + L_n} \quad (2.32)$$

In Eq. (2.32), L_d is the luminance of the monochromatic stimulus ($\lambda_D = 520$ nm in the example of Figure 2.26) and L_n is the luminance of the achromatic stimulus (illuminant E) that are necessary to match the color stimulus, that is, the colored LED in the example of Figure 2.26, see the filled black circle. The narrower the bandwidth of the LED, the closer its chromaticity point gets to the spectral locus and the less achromatic stimulus is needed to match its chromaticity. Consequently, narrowband LEDs have a high colorimetric purity.

2.5
Circadian Effect of Electromagnetic Radiation

Up to now, the visible effects of electromagnetic radiation have been discussed. But radiation affects not only the human visual system but also the so-called *circadian clock* which is responsible for the timing of all biological functions according to daily rhythm [28]. This effect depends on the intensity, the spatial and spectral power distribution and the timing (e.g., morning, evening, or night) and duration of visible electromagnetic radiation reaching the human eye. The aim of this section is to summarize some important experimental data on the effect of light source spectral power distributions on the circadian clock. A circadian model and its application to white LED light sources are presented in Section 5.10.

2.5.1
The Human Circadian Clock

Life has a cyclic rhythm: chronobiologists distinguish between ultradian (a couple of hours like hunger or short phases of sleep by babies), circadian (24 h, lat. circa = about, dies = day) and infradian (>24 h like seasonal changes) rhythms. For all aspects of human life, the most important rhythm is the circadian one. Figure 2.30 shows this 24 h cycle of human activity in terms of readiness to work.

As can be seen from Figure 2.30, human activity exhibits a 24 h cycle which is controlled by light. In other words, light is an important *zeitgeber* (literally from German: "time giver"). But how does this timing mechanism work in the human neural system? In 2002, the discovery of a new photoreceptor in the eye, the so-called intrinsically photoreceptive retinal ganglion cell (ipRGC), was published [30]. It turned out that 2% of all ganglion cells contain the photosensitive protein melanopsin which has a different spectral sensitivity from the sensitivity of the rods and the cones, see Figure 2.6. The ipRGCs are distributed nonuniformly across the retina: in the lower retinal area (where light sources in the upper part of the viewing field are imaged), ipRGC density is much larger than in the upper area. Light sources emitting from homogeneous *large* areas in the *upper* half of the viewing field cause an increased circadian effect (i.e., timing effect of optical radiation) compared to *punctual* light sources and those in the *lower* half of the viewing field.

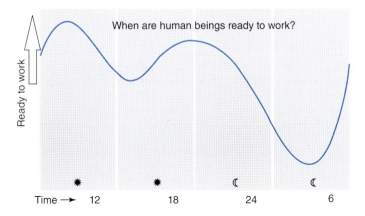

Figure 2.30 Human activity (readiness to work) in the 24 h cycle [29]. (Reproduced with permission from licht.de.)

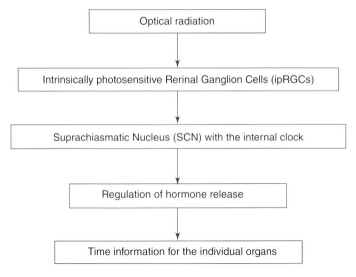

Figure 2.31 Schematic flowchart of the nonvisual photosensitive pathway.

Figure 2.31 shows the scheme of this so-called nonvisual photosensitive pathway.

As can be seen from Figure 2.31, optical radiation reaches the ipRGCs that send neural signals toward the circadian clock located at the suprachiasmatic nucleus (SCN) which, in turn, regulates hormone (melatonin) release. Melatonin is a sleep hormone that fosters sleep-in: it is an important component of the circadian clock. Exposure to light *suppresses* the production of melatonin hence alertness increases. Figure 2.32 shows the cyclic change of melatonin level together with cortisol level which is a stress hormone.

46 | *2 The Human Visual System and Its Modeling for Lighting Engineering*

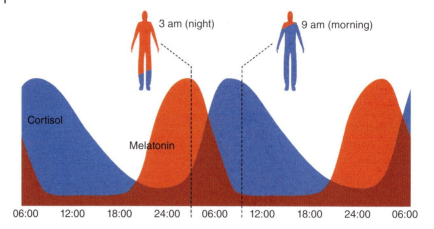

Figure 2.32 The cyclic change of melatonin (sleep hormone) level and cortisol (stress hormone) level [29]. (Reproduced with permission from licht.de.)

As can be seen from Figure 2.32, melatonin peaks at about 3o'clock in the night (in the sleeping period) while cortisol (and alertness) is maximal at about 9o'clock in the morning. In the daytime, melatonin is suppressed and this can be hardly influenced by any artificial light source. But during the dark hours, it is possible to suppress melatonin and increase nighttime work performance by using an appropriate spectral power distribution. Alternatively, the aim of spectral optimization can also be to create a relaxing atmosphere after work in order not to suppress

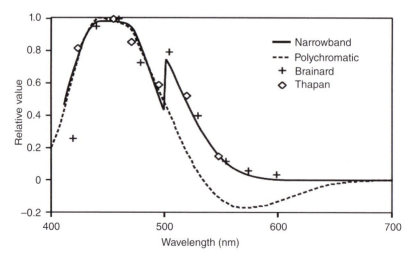

Figure 2.33 Experimental data of Brainard *et al.* [31] and Thapan *et al.* [32] on the extent of nocturnal melatonin suppression (ordinate) as a function of the peak wavelength of narrowband and broadband stimuli (abscissa). Continuous curve: fit to narrowband [31, 32] data; dash curve: fit to broadband data [33]. (Reproduced after Figure 3 in [28] with permission from the *Journal of Circadian Rhythms*.)

melatonin artificially. To do so, it is essential to know the effect of light source spectral power distribution on nocturnal melatonin suppression.

In Figure 2.33, experimental data on the extent of nocturnal melatonin suppression (ordinate) is shown as a function of the peak wavelength of narrowband and more broadband stimuli (abscissa) [28, 31–33].

As can be seen from Figure 2.33, the maximum of nocturnal melatonin suppression occurs at 460 nm and not at 480 nm which is the peak spectral sensitivity of melanopsin, the ipRGC photopigment. This indicates a possible interaction with S cones (see the ipRGC and S cone spectral sensitivity curves in Figure 2.6). Also, there is a local maximum for narrowband spectra at 510 nm while this local maximum disappears for more broadband spectra indicating that there is a spectrally opponent retinal mechanism working with cone signal differences and contributing to the nonvisual photosensitive pathway.

From the experimental data of Figure 2.33 it can be concluded [28] that multiple retinal mechanisms contribute to the light induced control of nocturnal melatonin suppression. The use of a single spectral weighting function (corresponding to a hypothetical single mechanism) to model the spectral dependence of the circadian effect of a light source is not appropriate and a more complex modeling (see Figure 5.62) is needed. This model is described in Section 5.10.

References

1. Sharpe, L.T., Stockman, A., Jägle, H., and Nathans, J. (1999) in *Color Vision: From Genes to Perception* (eds K.R. Gegenfurtner and L.T. Sharpe), Cambridge University Press, pp. 3–51.
2. Frisby, J.P. and Stone, J.V. (2010) *Seeing: The Computational Approach to Biological Vision*, 2nd edn, MIT Press.
3. Khanh, T.Q. (2014) *Student Lectures on Lighting Technology, Winter Semester 2013–2014*, Technische Universität Darmstadt.
4. Østerberg, G.A. (1935) Topography of the layer of rods and cones in the human retina. *Acta Ophthalmol.*, **6**, 1.
5. Curcio, C.A., Sloan, K.R., Kalina, R.E., and Hendrickson, A.E. (1990) Human photoreceptor topography. *J. Comp. Neurol.*, **292**, 497–523.
6. Color & Vision Research Laboratory Web Database of the Color and Vision Research Laboratory, University College London, Institute of Ophthalmology, London, www.cvrl.org (accessed 04 June 2014).
7. Stockman, A., Sharpe, L.T., and Fach, C.C. (1999) The spectral sensitivity of the human short-wavelength cones. *Vision Res.*, **39**, 2901–2927.
8. Stockman, A. and Sharpe, L.T. (2000) Spectral sensitivities of the middle- and long-wavelength sensitive cones derived from measurements in observers of known genotype. *Vision Res.*, **40**, 1711–1737.
9. Commission Internationale de l'Eclairage (CIE) (2005) *CIE 10 Degree Photopic Photometric Observer*, Publication CIE 165:2005.
10. Bodrogi, P. and Khanh, T.Q. (2012) *Illumination, Color and Imaging: Evaluation and Optimization of Visual Displays*, Wiley Series in Display Technology, Wiley-VCH Verlag GmbH.
11. Khanh, T.Q. (2004) Physiologische und psychophysische Aspekte in der Photometrie, Colorimetrie und in der Farbbildverarbeitung (Physiological and psychophysical aspects in photometry, colorimetry and in color image processing). Habilitationsschrift (Lecture Qualification thesis). Technische Universitaet Ilmenau, Ilmenau.

12. Zalewski, E.F. (2010) in *Handbook of Optics*, Vol. 2, Part 7, 3 (ed. in-chief M. Bass), McGraw-Hill.
13. Commission Internationale de l'Eclairage (CIE) (2004) *Colorimetry*, 3rd edn, Publication CIE 015:2004, CIE.
14. Stockman, A. (2004) in *The Optics Encyclopedia: Basic Foundations and Practical Applications*, vol. 1 (eds T.G. Brown, K. Creath, H. Kogelnik, M.A. Kriss, J. Schmit, and M.J. Weber), Wiley-VCH Verlag GmbH & Co. KGaA, Berlin, pp. 207–226.
15. Commission Internationale de l'Eclairage (CIE) (2011) Standard CIE S 017/E:2011 ILV: International Lighting Vocabulary.
16. Hunt, R.W.G. and Pointer, M.R. (2011) *Measuring Colour*, The Wiley-IS&T Series in Imaging Science and Technology, 4th edn, John Wiley & Sons, Ltd.
17. Hunt, R.W.G. (1977) The specification of colour appearance. I. Concepts and terms. *Color Res. Appl.*, **2** (2), 55–68.
18. Derefeldt, G., Swartling, T., Berggrund, U., and Bodrogi, P. (2004) Cognitive color. *Color Res. Appl.*, **29** (1), 7–19.
19. Comission Internationale de l'Eclairage (CIE) (2004) *A Color Appearance Model for Color Management Systems: CIECAM02*, Publication CIE 159-2004.
20. Fairchild, M.D. (1997) *Color Appearance Models*, The Wiley-IS&T Series in Imaging Science and Technology, 2nd edn, John Wiley & Sons, Ltd.
21. Fairchild, M.D. Color Appearance Models, *http://www.cis.rit.edu/fairchild/CAM.html* (accessed 04 June 2014).
22. Luo, M.R., Cui, G., and Li, C. (2006) Uniform color spaces based on CIECAM02 color appearance model. *Color Res. Appl.*, **31**, 320–330.
23. MacAdam, D.L. (1942) Visual sensitivities to color differences in daylight. *J. Opt. Soc. Am.*, **32** (5), 247–274.
24. Commission Internationale de l'Eclairage (CIE) (1993) *Parametric Effects in Color-Difference Evaluation*, CIE Publication 101-1993.
25. Li, C., Luo, M.R., Li, C., and Cui, G. (2012) The CRI-CAM02UCS colour rendering index. *Color Res. Appl.*, **37**, 160–167.
26. Bodrogi, P., Brückner, S., and Khanh, T.Q. (2011) Ordinal scale based description of colour rendering. *Color Res. Appl.*, **36**, 272–285.
27. Commission Internationale de l'Eclairage (CIE) (2007) *Measurement of LEDs*, 2nd edn, Publication CIE 127:2007, CIE.
28. Rea, M.S., Figueiro, M.G., Bierman, A., and Bullough, J.D. (2010) Circadian light. *J. Circadian Rhythms*, **8** (2), 1–10.
29. Kunz, D. (ed) (2010) *Wirkung des Lichts auf den Menschen (Effect of light on humans)*, licht.de, Frankfurt am Main, Germany licht.wissen 19, 2010.
30. Berson, D.M., Dunn, F.A., and Takao, M. (2002) Phototransduction by retinal ganglion cells that set the circadian clock. *Science*, **295**, 1070–1073.
31. Brainard, G.C., Hanifin, J.P., Greeson, J.M., Byrne, B., Glickman, G., Gerner, E., and Rollag, M.D. (2001) Action spectrum for melatonin regulation in humans: evidence for a novel circadian photoreceptor. *J. Neurosci.*, **21**, 6405–6412.
32. Thapan, K., Arendt, J., and Skene, D.J. (2001) An action spectrum for melatonin suppression: evidence for a novel non-rod, non-cone photoreceptor system in humans. *J. Physiol.*, **535**, 261–267.
33. Rea, M.S., Figueiro, M.G., Bullough, J.D., and Bierman, A. (2005) A model of phototransduction by the human circadian system. *Brain Res. Rev.*, **50**, 213–228.

3
LED Components – Principles of Radiation Generation and Packaging

Holger Winkler, Quang Trinh Vinh, Tran Quoc Khanh, Andreas Benker, Charlotte Bois, Ralf Petry, and Aleksander Zych

3.1
Introduction to LED Technology

The final, general aim of lighting technology is the development of highly efficient, intelligent, reliable luminaires, and lighting systems for a number of applications from street lighting to automotive lighting to indoor lighting with very different requirements regarding color quality, luminous intensity distribution, luminance on the area to be illuminated, and installation possibilities. In order to design these luminaires in a fundamentally correct manner and for a conception of lighting with maximal user acceptance in a defined context, it is very important to know the fundamentals of light-emitting diode (LED) light sources as double identities, that is, as semiconductor devices and as light-emitting systems with a certain degree of color quality and light quality. In this chapter, the principal technological issues of LED technology are described. The structure of the contents is based on the bottom-up principle:

1) First, the physical, mathematical, electrical, and material aspects of the generation of optical radiation in different wavelength ranges are described in Sections 3.2 and 3.3. These contents answer the question of how a semiconductor color LED emits radiation.
2) On the basis of the knowledge that the technology of white LEDs is a consequence of using blue pumping LEDs and luminescent materials, in Section 3.4, the absorption, excitation, and emission spectra of modern phosphor systems with different chemical structures are analyzed and presented.
3) Keeping the primary aim of lighting technology with maximal light-converting efficiency and best color quality for a broad range of applications with different correlated color temperatures (CCTs) in mind, Section 3.5 describes the basic ideas of how modern phosphor mixtures can be designed and manufactured in order to realize LED lighting units with predefined white point, color rendering index (CRI), and luminous efficacy.

LED Lighting: Technology and Perception, First Edition.
Edited by Tran Quoc Khanh, Peter Bodrogi, Quang Trinh Vinh and Holger Winkler.
© 2015 Wiley-VCH Verlag GmbH & Co. KGaA. Published 2015 by Wiley-VCH Verlag GmbH & Co. KGaA.

4) The possibilities of the generation of optical radiation and color engineering with different color LEDs or with a number of phosphor systems constitute only one aspect of LED technology. The most important issue (and question) in the view of the next decades is how reliable, compact, and efficient LEDs can be constructed and realized as packages with long lifetime and minimal color shift throughout their usage time. Therefore, in Section 3.6, the advantages and deficits of a variety of LED packaging designs are analyzed.

Thus, Chapter 3 establishes a basis to understand how modern LED components can be designed and also a basis for Chapter 4 in which the optical, electrical, and thermal characteristics of the LEDs have priority.

3.2
Basic Knowledge on Color Semiconductor LEDs

3.2.1
Injection Luminescence

A color semiconductor LED is a type of diode that emits light under appropriate bias conditions. Thus, color semiconductors LEDs include an abundant semiconductor material layer with positive carriers and another with negative carriers. Positive carriers (called *holes*) are particles that lack electrons because of their chemical composition. Thus, the positive material layer is called a *p-layer*. By contrast, negative carriers are electrons belonging to a negative material layer called *n-layer* because its chemical compound creates electrons in excess. Normally, there are two main states in the operation of color semiconductor LEDs, namely, an *unbiased state* and a *biased state*.

- *Unbiased state*: Under the unbiased condition, a depletion region is created and extends mainly into the p-side causing an insulating state like a lock with a tiny leakage current ($10^{-9} - 10^{-12}$ A) as shown in Figure 3.1. This state is ensured and determined by a *built-in voltage* V_0 (in V).
- *Biased state*: When an appropriate forward voltage V is applied to a junction of a color semiconductor LED, the built-in voltage reduces to $V_0 - V$ (in V). Therefore, the corresponding barrier energy reduces to $e(V_0 - V)$ in eV. Consequently, electrons can flow from the n-side toward the p-side of the color semiconductor LED and in the opposite direction for holes. Successively, if a hole and an electron have the same momentum, they can recombine and emit optical radiation with a wavelength depending on the energy bandgap E_g (in eV) of the semiconductor material as described in Figure 3.2. This process is characterized by the equilibrium photon energy described by the Planck–Einstein equation (Eq. (3.1)) and bandgap energy E_g (in eV) according to the energy conservation law.

$$E = h = \frac{hc}{\lambda} \approx E_g \text{ (eV)} \tag{3.1}$$

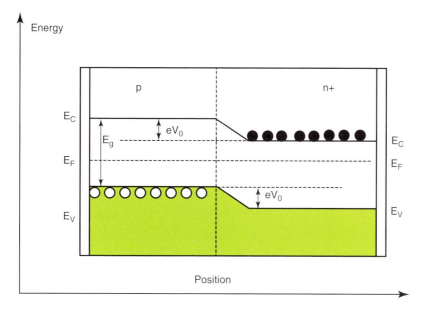

Figure 3.1 Energy diagram of an ideal color semiconductor LED in unbiased state.

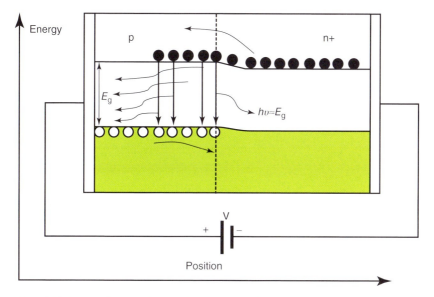

Figure 3.2 Energy diagram of an ideal color semiconductor LED in biased state.

Here h is the Planck constant ($\sim 6.626 \times 10^{-34}$ J s), c is the speed of light ($\sim 3 \times 10^{8}$ m s^{-1}), and λ (nm) is the wavelength of emitted light. Solving Eq. (3.1) yields Eq. (3.2) to calculate the emitted wavelength.

$$\lambda \approx \frac{hc}{E_g} = \frac{1.24}{E_g} \ (\mu m) \tag{3.2}$$

This process is the most important mechanism in color semiconductor LEDs. It is called *injection luminescence*.

3.2.2
Homo-Junction, Hetero-Junction, and Quantum Well

In the real operation of commercial color semiconductor LEDs in the world market, their injection luminescence is much more complicated. Intrinsic phenomena occurring inside color semiconductor LEDs depend on the chemical composition and physical structure of their junction. Generally, there are three structural levels of LED junctions, namely, homo-junction (there are no such real LEDs), hetero-junction, and quantum well.

3.2.2.1 Homo-Junction

Zukauskas et al. [1] described a mono-junction similar the one in Figures 3.1 and 3.2 as composed of a p-type semiconductor and an n-type semiconductor belonging to the same semiconductor type with a similar bandgap parameter [1]. Thus, the name homo-junction reflects this meaning. Theoretically, homo-junctions can create one conductivity region or both conductivity regions where radiative recombination can occur. In order to obtain a material for recombination, electrons and holes diffuse into the two sides of mono-junctions. Under a bias state such as the one mentioned in Section 3.2.1, majority carriers diffuse because of an external electric field creating an excess density of minority carriers on both sides of the junction layer. Successively, in that junction layer, injected carriers can recombine both radiatively and nonradiatively. Hence, injection currents can be characterized as shown in Eqs. (3.3) and (3.4).

$$I_n = I_{n,0} \cdot \left[\exp\left(\frac{V}{V_T}\right) - 1 \right] \tag{3.3}$$

$$I_p = I_{p,0} \cdot \left[\exp\left(\frac{V}{V_T}\right) - 1 \right] \tag{3.4}$$

Here, I_n and I_p (in A) are the forward currents of electrons and holes at the carrier temperature T_c in K, respectively; $I_{n,0}$ and $I_{p,0}$ (in A) are the forward currents for electrons and holes at a reference temperature $T_{c,0}$ (in K), respectively. In addition, the thermal voltage at the carrier temperature T_c equals $V_T = k_B T_c / e$ (in V) with the Boltzmann constant k_B ($\sim 1.38 \times 10^{-23}$ J K^{-1}) and the elementary electric charge e (1.602×10^{-19} C). In reality, the decrease in barrier potential when a voltage is applied is lower because voltage drops at series resistances R_s (in Ω) and a current is added because of nonradiative recombinations at the junction region.

Therefore, the real current can be seen in Eq. (3.5).

$$I = I_n + I_p + I_{nr}$$
$$= (I_{n,0} + I_{p,0}) \cdot \left[\exp\left(\frac{V - IR_s}{V_T}\right) - 1\right] + I_{nr,0} \cdot \left[\exp\left(\frac{V - IR_s}{n_{ideal} V_T}\right) - 1\right] \quad (3.5)$$

Here, $I_{nr,0}$ (in A) is the reverse nonradiative recombination current at a reference temperature $T_{c,0}$ and n_{ideal} is an ideality factor for recombination current ($1 \leq n_{ideal} \leq 2$). Unfortunately, the injection efficiency of homo-junction LEDs was very low, in the range of 0.3–0.8 [1]. Therefore, homo-junctions have never been engineered for practical LEDs.

3.2.2.2 Hetero-Junction

On the basis of practical investigations, semiconductor physicists confirmed that there were many disadvantages with homo-junction structures such as high reabsorption because of low light extraction efficiency [1]. Thus, their internal efficiency can be enhanced in only one conductive region (e.g., p-region, $I_n \gg I_p$) if doping is carried out with many more donors than acceptors. However, doping with many more donors makes the reabsorption of light increase and creates many nonradiative centers. Consequently, only the change of material composition as a function of distance can help adjust carrier injection, radiative recombination, and reabsorption in the junction layers of color semiconductor LEDs. As a result, a new junction structure established from dissimilar semiconductors with different bandgaps because of different chemical compositions was created to satisfy the earlier described requirement. The difference in the semiconductor materials used is reflected by the name *hetero*-structure. Conventionally, there are two types of hetero-structures, namely, *single* hetero-structure and *double* hetero-structure. In the single hetero-structure shown in Figure 3.3, a conductive region is composed of two semiconductor types so that the energy level of the n-type layer with bandgap $E_{g,1}$ (in eV) is lower than that of the p-type layer with bandgap $E_{g,2}$ (in eV) and the n-type conductivity region is wider than that of the p-type. This band discontinuity increases the potential barrier for the holes that diffuse toward the n-region at the energy barrier of ΔE_v (in eV). Moreover, depending

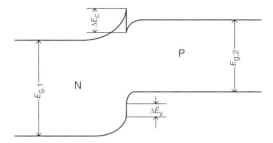

Figure 3.3 Structure of a single hetero-junction with an increase in potential barrier ΔE_v for p-carriers and a decrease in potential barrier ΔE_c for n-carriers [1]. (Reproduced from [1, Figure 4.1.6], copyright © 2002 with permission from Wiley.)

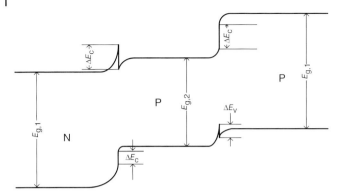

Figure 3.4 Double hetero-junction with bidirectional injection [1]. (Reproduced from [1, Figure 4.1.6], copyright © 2002 with permission from Wiley.)

on the abruptness of the interface, the potential barrier for electrons can decrease by a value in the range $0-\Delta E_c$ (in eV). Accordingly, the rate of injection currents I_n/I_p can increase (proportional to $e^{\frac{\Delta E}{k_B T}}$), where $\Delta E_v \leq \Delta E \leq \Delta E_v + \Delta E_c$. In addition, another important advantage is that the n-type layer is usually transparent for photons generated in the p-region. For this reason, the photons can escape easily and reabsorption of light emanating toward the n-side is minimal.

On the other hand, in Figure 3.4 a *double* hetero-structure is made of a narrow-gap active p-type layer kept between two wide-gap conductive regions of n-type and p-type. This special structure allows for the bidirectional injection of excess carriers into the active layer, where electrons and holes can recombine to emit light. In addition, minority carriers that diffuse through one side of the hetero-interfaces are trapped in the active layer by the remainder of the hetero-interface and cannot diffuse away. For this reason, excess carrier density increases. Therefore, the rate of radiative recombinations also increases. Moreover, both conductive layers are transparent so that the reabsorption effect is minimized in both directions although reabsorption can still occur in the active layer.

3.2.2.3 Quantum Well

The quantum well (Figure 3.5) is a special case of hetero-structures [1]. Its specific structure helps enhance the ability of carrier trapping. A double hetero-structure can be considered a single quantum well (SQW). However, the junction of high brightness LEDs is not only made of this SQW but also of double quantum wells (DQWs) or multiple quantum wells (MQWs). The optical properties of QWs are different from those of a normal semiconductor material volume. In particular, instead of free motion along the perpendicular direction (x) to a hetero-interface, discrete energy levels (E_n in eV) occur there. For an infinitely deep rectangular quantum well, these discrete energy levels separated from the bottom of their conduction band (E_c in eV) can be described by Eq. (3.6).

$$E_n - E_c = \frac{\pi^2 \hbar^2 n^2}{2 m_e a^2} \tag{3.6}$$

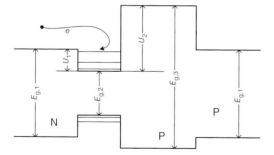

Figure 3.5 A quantum well with an electron trapped in its structure [1], reproduced from [1, Fig 4.1.10], copyright © 2002 with permission from Wiley.

Here, $n = 1, 2, 3, \ldots$ is the quantum number, the constant a (in nm) is the width of the quantum well, and m_e (in mg) is the effective electron mass. In addition, in a y–z plane, which is parallel to the hetero-interface, electron motion is not quantized so that the electron energy within a sub-band n is given by Eq. (3.7).

$$E = E_n - E_c + \frac{\hbar^2 k^2}{2m_e} = \frac{\pi^2 \hbar^2 n^2}{2m_e a^2} + \frac{\hbar^2 k^2}{2m_e} \tag{3.7}$$

Here k is the two-dimensional wave number.

For a quantum well structure which is almost always used to produce color semiconductor LEDs in today's industry, the forward current–voltage characteristics can be described more complicatedly by three current components, namely, a diffusion current, a recombination current (both radiative and nonradiative), and a temperature-independent tunnel current, see Eq. (3.8).

$$I_f(V_f) = I_D(V_f) + I_R(V_f) + I_T(V_f) \tag{3.8}$$

According to [1], the diffusion, recombination, and tunnel currents can be characterized by Eqs. (3.9)–(3.11).

$$I_D(V_f) = I_{D0} \cdot \left[\exp\left(\frac{V - IR_s}{V_T}\right) - 1 \right] \tag{3.9}$$

$$I_R(V_f) = I_{R0} \cdot \left[\exp\left(\frac{V - IR_s}{n_{ideal} V_T}\right) - 1 \right] \tag{3.10}$$

$$I_T(V_f) = I_{T0} \cdot \left[\exp\left(\frac{V - IR_s}{\frac{E_t}{e}}\right) - 1 \right] \tag{3.11}$$

Here, E_t is the characteristic energy constant with magnitude ~ 0.1 eV [1]. Current amplitudes (I_{D0}, I_{R0}, and I_{T0} in A) were not yet described and explained theoretically. But they were fitted different curves experimentally depending on the various chemical compositions and physical structures of the different color semiconductor LEDs.

3.2.3
Recombination

Section 3.2.1 described injection luminescence in a simple homo-junction while Section 3.2.2 dealt with the different junction structures of color semiconductor LEDs with different trapping abilities for carriers leading to different recombining abilities and different internal quantum efficiency. In reality, the recombination of electrons and holes in the junctions occurs more complicatedly than what was described in the earlier sections. Further different phenomena occur because of the chemical compositions and physical structures inside the semiconductor LEDs. In this section, recombinations are classified and described according to the intrinsic phenomena occurring in semiconductor structures including direct and indirect recombination, radiative and nonradiative recombination.

3.2.3.1 Direct and Indirect Recombination

According to the intrinsic structure of semiconductor materials, recombinations can be classified into *direct* and *indirect* recombinations. Indeed, direct recombination takes place in *direct* bandgap semiconductor materials where the minimum energy of their conduction band lies directly above the maximum energy of their valence band in a momentum–energy coordinate system as shown in Figure 3.6. Therefore, the electrons at the bottom of a conduction band can recombine directly with the holes at the top of the valence band and they emit light efficiently. In this case, the probability of band-to-band transition is high, leading to more radiative recombinations than in other cases because the carriers

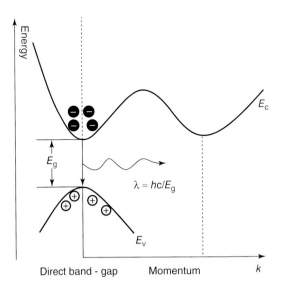

Figure 3.6 Direct recombination in a material with direct bandgap (GaN) [2]. (Reproduced from [2, Fig 1.3], copyright © 1999 with permission from Taylor & Francis Books UK.)

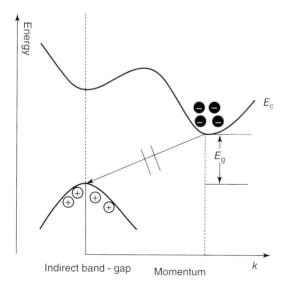

Figure 3.7 Indirect recombination in a material with an indirect bandgap (SiC) [2], reproduced from [2, Fig 1.3], copyright © 1999 with permission from Taylor & Francis Books UK.

have approximately similar momentums at the best energy position between the conduction band and the valence band where the energy level offset is the lowest.

On the other hand, *indirect* recombination occurs in indirect bandgap semiconductor materials. Here, the minimum energy of their conduction band is shifted from the maximum energy of their valance band as shown in Figure 3.7. Therefore, there is a dissimilar momentum for the electrons at the minimum energy position of the conduction band and the four holes at the maximum energy position of the valence band and this causes difficulty to achieve a radiative recombination. Thus, in order to emit light, electrons must change their momentum. Consequently, phonons are necessary to be added into the junction in order to establish the same momentum state for both the electrons and the holes. This is established by adding impurities or dopants in order to form shallow donor/acceptor states. Then, these donor/acceptor states play the role of recombination substations where electrons are captured and supplemented by an appropriate amount of momentum for their recombination. In general, self-indirect recombination is worse than direct recombination from the point of view of emitting light. However, it should be mentioned that if the indirect bandgap semiconductor material is doped with a certain type of impurity then the indirect bandgap *doped* semiconductor material can emit even more light than the direct bandgap semiconductor material *in some special cases*.

3.2.3.2 Radiative and Nonradiative Recombinations and Their Simple Theoretical Quantification

As mentioned before, in order to understand what really takes place inside LEDs, researchers assumed that there are two types of recombination, radiative

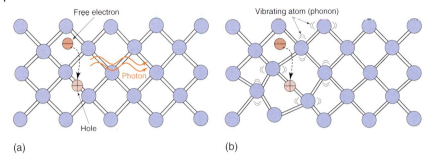

Figure 3.8 Radiative (a) and nonradiative recombination (b) in the host lattice of LEDs [3], reproduced from [3, Figure 2.5], copyright © 2003 with permission from Cambridge University Press.

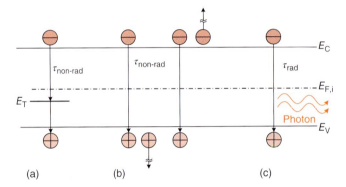

Figure 3.9 (a) Nonradiative recombination at the deep level, (b) Auger recombination, and (c) radiative recombination [3], reproduced from [3, Fig 2.6], copyright © 2003 with permission from Cambridge University Press.

recombination and nonradiative recombination. Radiative recombination enables LEDs to emit photons while nonradiative recombinations produce heat: they consume the input electrical energy of LEDs but this electrical energy is not converted into optical energy. It only creates new vibrations and increases the energy of old vibrations. This causes an increase in carrier temperature and the overall temperature of the semiconductor structure of the LEDs. These vibrations of the atoms are called *phonons*, see Figure 3.8.

There are some reasons for nonradiative recombinations such as the so-called Auger recombination, that is, the recombination at defect areas or the so-called multiphonon emission at deep impurity levels as illustrated in Figure 3.9.

Radiative recombinations and nonradiative recombinations always occur concurrently. Therefore, it is necessary to consider them simultaneously. According to [2, 3], the successive carrier injection process causes excess carrier density in the semiconductor layers. Therefore, in order to have neutral statuses after each injection cycle, the excess carrier density of n-particles (Δn) and that of p-particles (Δp) must return to the neutral statuses in a certain time cycle (τ in ns). To express

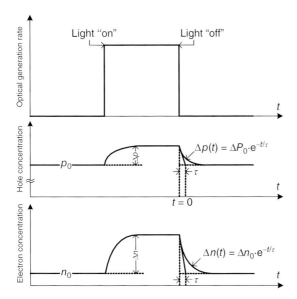

Figure 3.10 Carrier density of n- and p-particles as functions of recombination time [3], reproduced from [3, Fig 2.2], copyright © 2003 with permission from Cambridge University Press.

this in more detail, the excess carrier density decreases exponentially over time, see Eqs. (3.12) and (3.13). This is also shown in Figure 3.10.

$$\Delta n = \Delta n_0 \cdot e^{-\frac{t}{\tau}} \qquad (3.12)$$

$$\Delta p = \Delta p_0 \cdot e^{-\frac{t}{\tau}} \qquad (3.13)$$

Here, Δn_0 and Δp_0 represent initial excess carrier densities of the p- and n-particles and τ (in ns) is recombination lifetime. This recombination lifetime includes both radiative recombination lifetime τ_r and nonradiative recombination lifetime τ_{nr} (in ns) according to Eq. (3.14). These lifetimes are quite different between the low excitation case (τ = constant) and the high excitation case ($\tau(t) = t + (B\Delta n_0)^{-1}$) with B being an appropriate constant. As a result, luminescence intensity may be a stretched exponential decay function in case of high excitation ($\Delta n \gg N_A$) or it may also become a different exponential decay function in case of low excitation ($\Delta n \ll N_A$). Here N_A (in cm^{-3}) is doping concentration. Then, the total probability of a recombination is determined by the sum of the radiative and nonradiative probabilities, see Eq. (3.14).

$$\frac{1}{\tau} = \frac{1}{\tau_r} + \frac{1}{\tau_{nr}} \qquad (3.14)$$

Radiative recombination rate and nonradiative recombination rate as functions of excess carrier density and radiative recombination lifetime or nonradiative recombination lifetime can also be expressed by Eqs. (3.15) and (3.16).

$$R_r = \frac{\Delta n}{\tau_r} \tag{3.15}$$

$$R_{nr} = \frac{\Delta n}{\tau_{nr}} \tag{3.16}$$

3.2.4
Efficiency

For the different LED components going from the inside toward the outside of the LEDs, different efficiency concepts can be found [1–5] such as internal quantum efficiency (η_i), injection efficiency (η_{inj}), light extraction efficiency ($\eta_{extraction}$), external quantum efficiency (η_{ext}), radiant efficiency (η_e, see Eq. (2.13)), and luminous efficiency (η_v). These concepts are described forward.

3.2.4.1 Internal Quantum Efficiency (η_i)
Internal quantum efficiency is the ratio between radiative recombination rate and total recombination rate, see Eq. (3.17).

$$\eta_i = \frac{R_r}{R_r + R_{nr}} = \frac{1/\tau_r}{1/\tau_r + 1/\tau_{nr}} = \frac{1}{1 + 1/\tau_{nr}/\tau_r} \tag{3.17}$$

In the 1960s, internal efficiency was very low (about 1%). Gradually, lower defect density and better material impurity led to higher efficiency.

3.2.4.2 Injection Efficiency (η_{inj})
The injection of carriers into an active region is not easy. The total forward current density is designated by J_f (in A m^{-2}), the electron-current density injected into the active region by J_e (in A m^{-2}), and the hole-current density injected into an active region by J_h (in A m^{-2}). The last has a lower value compared with the original forward current density J_f. So, the injection efficiency of p-semiconductors is shown in Eq. (3.18).

$$\eta_{inj.p} = \frac{J_h}{J_e + J_h} \tag{3.18}$$

Equation (3.18) can be extended to obtain Eq. (3.19).

$$\eta_{inj.p} = \frac{D_n n L_p}{D_n n L_p + D_p p L_n} \tag{3.19}$$

In Eq. (3.19), D_n and D_p are the diffusion coefficients for electrons and holes, respectively. L_p and L_n are the minority carrier diffusion lengths. Parameters n and p represent the net electron and hole concentrations on either side of the junction layer.

3.2.4.3 Light Extraction Efficiency ($\eta_{extraction}$)
Optical radiation must escape from the radiative recombination centers located in the junction layer. However, the physical structure and geometry of material layers

3.2 Basic Knowledge on Color Semiconductor LEDs

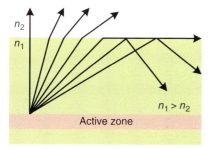

Figure 3.11 Total reflections in LED material layers.

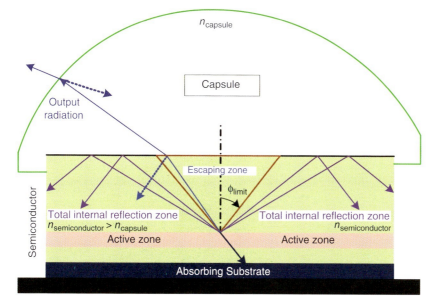

Figure 3.12 Light extraction in a typical LED structure [1], reproduced from [1, Fig 5.1.1], copyright © 2002 with permission from Wiley.

can cause total reflections and barriers against escape. In particular, some of the emitted rays have incident angles that are higher than the limiting angle Φ_{limit} of the material layer and this causes total reflection as illustrated in Figure 3.11. Then, these rays return to the inside of the junction layer and they can be reabsorbed.

Therefore, many LED manufacturers attempt to create various structural forms and diverse materials for the LED encapsulations in order to enlarge their escape zones including the enhancement of the limit angle $\Phi_{limit} = \arcsin(n_{capsule}/n_{semiconductor})$ by increasing the refractive constant of the encapsulation material or by building an appropriate encapsulation geometry for each ray, as shown in Figure 3.12. Consequently, extraction efficiency can be defined as the ratio of radiation leaving an optical surface and total radiation emitted by the junction layer, see Eq. (3.20).

$$\eta_{\text{extraction}} = \frac{\text{Output radiation on an optical surface}}{\text{Radiation emitted in the junction layer}} \qquad (3.20)$$

3.2.4.4 External Quantum Efficiency (η_{ext})

External quantum efficiency (η_{ext}) is defined as the ratio of the number of output photons and injected electrons. It can be calculated as the product of internal efficiency η_i, injection efficiency η_{inj}, and light extraction efficiency $\eta_{\text{extraction}}$, see Eq. (3.21).

$$\eta_{\text{ext}} = \eta_i \cdot \eta_{\text{inj}} \cdot \eta_{\text{extraction}} \qquad (3.21)$$

External efficiency can be enhanced if electron current dominates over hole current. In addition, in order to have a high external efficiency, semiconductor materials must be processed so as to have much more radiative recombinations than nonradiative recombinations by means of appropriate bandgap, high-quality doping in n-layers, and in p-layers with lowest resistivity and optimum encapsulation geometry with reduced total reflection and less reabsorption.

3.2.4.5 Radiant Efficiency (η_e, See Section 2.2.5, Eq. (2.13))

If the quantity Φ_e (radiant power or radiant flux) is alternatively denoted here by P_{opt}, electrical power is denoted here by $P_e = P$, and the symbols V_f and I_f denote forward voltage (in V) and forward current (in A), respectively, then radiant efficiency can be expressed by Eq. (3.22).

$$\eta_e = \frac{P_{\text{opt}}}{P_e} = \frac{P_{\text{opt}}}{V_f I_f} \qquad (3.22)$$

This efficiency is at the top level of the entire LED efficiency calculation when internal efficiency, injection efficiency, and extraction efficiency have already been accounted for in the calculation procedure.

3.2.4.6 Luminous Efficacy (η_v)

For the sake of clarity, to remember luminous flux, let us repeat Eq. (2.6) here, see Eq. (3.23).

$$\Phi_V = K_m \int_{380 \text{ nm}}^{780 \text{ nm}} \Phi_{e\lambda} \cdot V(\lambda) \cdot d\lambda \qquad (3.23)$$

According to Eq. (2.14), for an LED with V_f and I_f, luminous efficacy (of a source) can be expressed by Eq. (3.24).

$$\eta_v = \frac{\Phi_v}{P_e} = \frac{683 \text{ lm W}^{-1} \cdot \int_{380 \text{ nm}}^{780 \text{ nm}} V(\lambda) \cdot \Phi_{e\lambda} \cdot d\lambda}{V_f \cdot I_f} \qquad (3.24)$$

As of today, luminous efficacy is widely used to characterize solid-state lighting applications and plays an important role to estimate the cost and quality of LEDs regarding the energy saving aspect (see also the remarks after Eq. (2.15)).

3.2.5
Semiconductor Material Systems – Efficiency, Possibilities, and Limits

The descriptions of the internal phenomena and the efficiency of LEDs are useful for the application and evaluation of color semiconductor LEDs. The questions about what happens inside the LEDs and how they can be evaluated from the point of view of energy and efficiency aspects were described in Sections 3.2.1–3.2.4. In this section, the relationships between the intrinsic parameters of semiconductor material systems such as bandgap energy E_g (in eV) and lattice constant (in Å) and LED properties such as chromaticity, peak wavelength, and efficiency are discussed including the limits of today's semiconductor material systems and suggestions for improvements.

3.2.5.1 Possible Semiconductor Systems

Theoretically, color semiconductor LEDs can be built from various semiconductor materials such as silicon carbide (SiC), gallium arsenide (GaAs), aluminum gallium arsenide (AlGaAs), gallium arsenide phosphide (GaAsP), gallium phosphide (GaP), gallium nitride (GaN), indium gallium nitride (InGaN), and aluminum gallium indium phosphide (AlGaInP) by an appropriate growth technique such as liquid phase epitaxy (LPE, old technique and unsuitable for the Al-system), hydride vapor phase epitaxy (HVPE, an old technique and not suitable for the Al-system), molecular beam epitaxy (MBE), gas source molecular beam epitaxy (GSMBE), and metal organic vapor phase epitaxy (MOVPE) also known as *organometallic vapor phase epitaxy* (*OMVPE*) or *metal organic chemical vapor deposition* (*MOCVD*). However, up to now only a few of them are usable to fabricate color semiconductor LEDs in practice because of the limits of efficiency and of technology. In particular, conventional semiconductor materials can be grouped into three main categories: the first group with indium gallium nitride (InGaN), the second group with aluminum indium gallium phosphide (AlInGaP), and the third group with gallium phosphide (GaP), gallium arsenide phosphide (GaAsP), and gallium aluminum arsenide (GaAlAs). The first group (InGaN) is usually used to fabricate ultraviolet, royal blue, blue, and green semiconductor LEDs with high radiant efficiency and high stability; the second group (AlInGaP) is usually used to fabricate yellow, amber, orange, and red semiconductor LEDs with worse radiant efficiency and lower stability; and the last group (GaP, GaAsP, and GaAlAs) is also used to fabricate yellow, amber, orange, and red semiconductor LEDs with bad radiant efficiency and low stability. In fact, almost all current semiconductor LEDs are fabricated by thin-GaN technology for royal blue (440 nm $\leq \lambda_p \leq$ 450 nm), blue (450 nm $\leq \lambda_p \leq$ 470 nm), and green (510 nm $\leq \lambda_p \leq$ 535 nm) semiconductor LEDs and by InGaAlP-thin-film technology for red (610 nm $\leq \lambda_p \leq$ 630 nm) and deep

Table 3.1 Semiconductor material systems with typical parameters and growth techniques [1, 5, 6].

Material group	Material	Color	λ_p (nm)	η_e (%)	Growth technique	Junction layer type
I	InGaN	UV	372	7.5	MOCVD	DH
	InGaN	Blue	470	11	MOCVD	QW
	InGaN	Royal Blue	440–450	Up to about 65	Thin GaN	MQW
	InGaN	Blue	460–470	Up to about 64	Thin GaN	MQW
	InGaN	Green	510–535	About 25	Thin GaN	MQW
	InGaN	Green	520	10	MOCVD	QW
	InGaN	Amber	590	3.5	MOCVD	QW
II	AlInGaP	Red	636	About 45	InGaAlP-thin-film	MQW
	AlInGaP	Deep Red	650	About 50	InGaAlP-thin-film	MQW
	GaAlAs	Red	650	16	LPE	DH
	AlGaInP	Amber	590	10	MOCVD	DH
III	GaP:N	Yellow-green	>565	0.3	LPE	H
	GaAsP:N	Yellow	>590	0.3	VPE	H

red (630 nm $\leq \lambda_p \leq$ 680 nm) semiconductor LEDs. High quality LEDs are usually built in the form of MQWs instead of double hetero (DH), hetero (H), or single quantum well (QW) as shown in Table 3.1.

Furthermore, Müller [4] described applicable electroluminescence – semiconductors and semimetals. On the other hand, Nakamura and Chichibu [2] focused on details of nitride semiconductor blue lasers and LEDs. In the semiconductor material field, these authors mentioned the relationship among semiconductor systems, lattice constant, and bandgap energy leading to peak wavelength and efficiency calculations.

The possible semiconductor systems are shown in Figure 3.13 with a bandgap of about 0.62 eV (InSb)–6.2 eV (AlN) corresponding to about 2000 nm (IR wavelength area) down to 200 nm (UVC area). In Figure 3.13, the red and green bars denote indirect and direct junctions, respectively. Band gap energy (on the upper horizontal axis) and wavelength (on the lower horizontal axis) can be calculated on the basis of the relationship λ (nm) $= 1240/E_g$ (eV). For lighting purposes, the violet $V(\lambda)$ curve also can be considered. Thus, visible/invisible wavelengths and low/high luminance ranges can be seen. The range of 1300–1600 nm (0.7–0.9 eV) is usually applied in optical telecommunications.

3.2.5.2 Semiconductor Systems for Amber–Red Semiconductor LEDs

The semiconductor systems of amber–red LEDs ($Al_xGa_{1-x}As$, $\lambda > 570$ nm, from red to IR, composed of three components Al–Ga–As) with direct bandgap (GaAs, about 1.4 eV, IR) or indirect bandgap (AlAs, about 2.17 eV, amber), and lattice constant of about 5.65 Å (see Figure 3.14a) can achieve only low efficiency with poor stability. In contrast, the semiconductor system of the amber-red LEDs

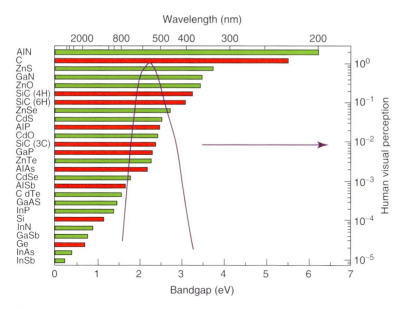

Figure 3.13 Band gaps and wavelengths of possible semiconductor systems and the $V(\lambda)$ curve as a rough measure of the human visual system's spectral sensitivity [1], reproduced from [1, Fig 4.2.1], copyright © 2002 with permission from Wiley.

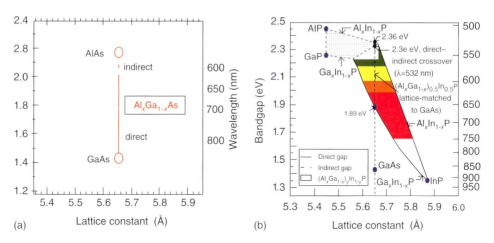

Figure 3.14 Semiconductor systems, bandgaps, and wavelengths for amber–red LEDs, reproduced from [1, Fig 4.2.2], copyright © 2002 with permission from Wiley.

$(Al_x(Ga_{1-x})_y In_{1-y} P$, $\lambda > 570$ nm, composed of four components (Al–In–Ga–P) with direct bandgap (and a lattice constant of about 5.58–5.8 Å) has a better efficiency and higher stability depending on both bandgap energy and lattice constant as shown in Figure 3.14b. Particularly, with the same bandgap and different lattice constants, the chromaticity or peak wavelength of color semiconductor

LEDs is similar but the mobility of holes and electrons is quite different, leading to different emission efficiencies. In these semiconductor systems, GaAs can play the role of a substrate, and, by available quaternary mixed crystals (those with four components) green LEDs can also be created by the use of these semiconductor material systems.

3.2.5.3 Semiconductor Systems for UV–Blue–Green Semiconductor LEDs

The semiconductor systems of UV–blue–green LEDs (AlInGaN) shown in Figure 3.15 have a lattice constant of about 3.16–3.42 Å and a bandgap of about 2.25–4.1 eV. They always have a higher efficacy and better stability compared with those of the amber–red LEDs. However, there is no available substrate for these systems and the internal reactions can affect the quality of the LEDs. By the suitable mixing of Al–In–Ga–N crystals, not only UV–blue–green semiconductor LEDs but also yellow–red-IR semiconductor LEDs can be created. But they are conventionally used to manufacture UV–blue–green LEDs.

3.2.5.4 The Green Efficiency Gap of Color Semiconductor LEDs

The relationship between semiconductor systems, wavelengths, and external quantum efficiency is shown in Figure 3.16. The efficiency of royal blue LEDs (400–450 nm) is about 0.55–0.6, that of deep-red LEDs (640–660 nm) is slightly lower (about 0.52) but much better than that of yellow–amber–orange LEDs (about 0.1–0.15 for yellow and amber LEDs and about 0.4 for orange LEDs). The worst efficiency is achieved for the case of green LEDs (at about 525–545 nm). They can only achieve an efficiency of about 0.25. Therefore, in applications that require green LEDs, green LEDs at about 520 nm should be used in order

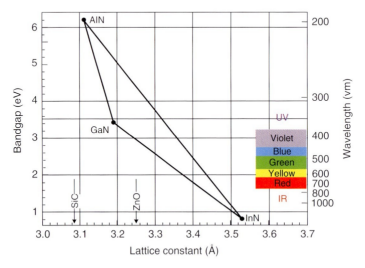

Figure 3.15 Band gap and wavelength versus lattice constant for UV – blue–green LEDs[3], reproduced from [3, Fig 8.12], copyright © 2003 with permission from Cambridge University Press.

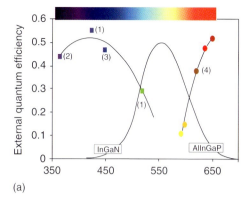 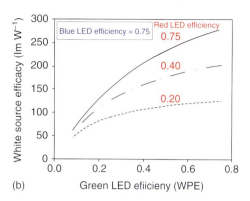

Figure 3.16 Semiconductor systems with external quantum efficiency, efficacy, and the green gap, reproduced with permission from Smithers (PGL 2007), Gerd Mueller and Regina Mueller–Mach. In diagram (a), the function in the middle is the $V(\lambda)$ function. In diagram, higher (>0.3) green efficiency values are speculative.

to reduce the green efficiency gap. Note that green–yellow–amber–orange phosphor-converted light-emitting diodes (PC-LEDs) have a chance to be developed to replace color semiconductor LEDs in hybrid LED luminaires or in special lighting applications with colored light. It can be seen from Figure 3.16b that the higher the efficiency of green and red semiconductor LEDs is, the higher the luminous efficacy of an RGB LED–based white-light source (mixed from red, blue, and green semiconductor LED) is. Unfortunately, although the efficiency of red semiconductor LEDs can be high, their temperature stability is not good. At high operating temperatures, the color of the red semiconductor LEDs is shifted strongly but their efficiency is also strongly reduced. In addition, the efficiency of green semiconductor LEDs is difficult to improve on from the current level of about 25% (Figure 3.16a). The green efficiency gap is a limit and also a good suggestion to create better semiconductor material systems, especially, new phosphors and phosphor combinations.

3.3
Color Semiconductor LEDs

The input conditions of the LEDs can be described by their forward current I_f and operating temperature. The relationship between the input condition and their output spectrum (an absolute spectral power distribution of the color semiconductor LED) is discussed in this section. An absolute spectral power distribution can be characterized by three parameters including peak wavelength λ_p, full width at half maximum (FWHM) λ_{FWHM}, and peak intensity S_p (see Section 2.4). The relationship between these parameters and the input conditions can be explained by the band structure theory where both the forward current and the operating temperature play an important role in the change of carrier density, the Boltzmann distribution, the joint density of states, and the carrier temperature as well as the probability of recombination (both radiative and nonradiative recombination) in the active region. However, the operating temperature T_s always exhibits a complicated relationship with junction temperature T_j and carrier temperature T_c. This relationship is described in detail in Chapter 4. In this section, the carrier temperature or the carrier temperature converted from junction temperature is used in the mathematical model whose basic concepts are introduced forward.

3.3.1
Concepts of Matter Waves of de Broglie

The concepte of matter waves or de Broglie waves reflect wave–particle duality of matter as proposed by de Broglie in 1924. Their wavelength is inversely proportional to their momentum and their frequency is proportional to the kinetic energy of the particle. The oscillation of a particle is described by Eq. (3.25).

$$\psi = A \cdot e^{px - \omega t} \tag{3.25}$$

In Eq. (3.25), ψ is the intensity of oscillation, p is the momentum of the particle, and $\omega = 2\pi f$ is the angular frequency. According to de Broglie,

$$f = \frac{p}{h} = \frac{\gamma m_0 \vartheta}{h} = \frac{m_0 \vartheta}{h\sqrt{1 - \frac{\vartheta^2}{c^2}}} \tag{3.26}$$

$$p = \hbar \cdot k = \frac{h}{2\pi} \cdot \frac{2\pi}{\lambda} = \frac{h}{\lambda} \tag{3.27}$$

Here, m_0 is the particle's rest mass, ϑ is the particle's velocity, γ is the Lorentz factor, c is the speed of light in vacuum, and $k = 2\pi/\lambda$ is the wave number.

3.3.2
The Physical Mechanism of Photon Emission

The physical mechanism of photon emission in LEDs is the spontaneous recombination of an electron and a hole as illustrated in Figure 3.17. Electrons in the conduction band and holes in the valence band are distributed according to two different parabolic forms in the energy diagram, see Eqs. (3.28) and (3.29).

$$E_e = E_C + \frac{\hbar^2 k^2}{2m_e^*} \tag{3.28}$$

$$E_h = E_V - \frac{\hbar^2 k^2}{2m_h^*} \tag{3.29}$$

Here, m_e^* and m_h^* are the effective masses of electrons and holes, respectively, $\hbar$ is the Planck constant divided by 2π, k is the carrier wave number and E_C and E_V represent the conduction band edge and valence band edge, respectively.

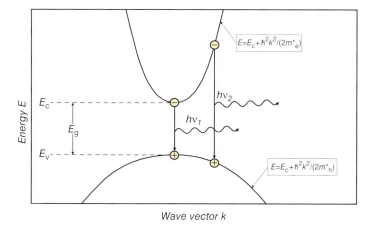

Figure 3.17 Radiative recombination mechanism of electron–hole pairs, reproduced from [3, Fig 5.1], copyright © 2003 with permission from Cambridge University Press.

Similarly, a hole with an effective mass m_h^* has the momentum as shown in Eq. (3.30).

$$p_h = m_h^* \cdot \vartheta = \sqrt{(2m_h^*) \cdot \left(\frac{1}{2}m_h^* \vartheta^2\right)} = \sqrt{(2m_h^*) \cdot (kT_j\sigma)} \tag{3.30}$$

The momentum of a photon with energy E_g can be calculated by Eq. (3.31).

$$p_{ph} = \hbar \cdot k = \frac{h}{\lambda} \approx \frac{hE_g}{hc} = \frac{E_g}{c} \tag{3.31}$$

After the recombination process, if a new particle does not oscillate any more, the momentum conservation law can be applied, see Eqs. (3.32) and (3.33).

$$p_{ph} = p_e + p_h \tag{3.32}$$

$$\frac{E_g}{c} = \sqrt{(2m_e^*) \cdot (kT_j\sigma)} + \sqrt{(2m_h^*) \cdot (kT_j\sigma)} \tag{3.33}$$

Therefore, in the transition from the conduction band to the valence band, the electron cannot change its momentum significantly. Moreover, an electron can recombine with a hole if they have the same momentum. As a result, the photon energy is shown in Eqs. (3.34) and (3.35).

$$E_{ph} = h\nu = h\frac{c}{\lambda} = E_e - E_h = E_C + \frac{\hbar^2 k^2}{2m_e^*} - E_V + \frac{\hbar^2 k^2}{2m_h^*} = E_g + \frac{\hbar^2 k^2}{2m_r^*} \tag{3.34}$$

$$\frac{1}{m_r^*} = \frac{1}{m_e^*} + \frac{1}{m_h^*} \tag{3.35}$$

3.3.3
Theoretical Absolute Spectral Power Distribution of a Color Semiconductor LED

By using the joint dispersion relation, the joint density of states can be calculated, see Eq. (3.36).

$$\rho(E) = \frac{1}{2\pi^2} \cdot \left(\frac{2m_r^*}{\hbar^2}\right)^{3/2} \cdot \sqrt{E - E_g} \tag{3.36}$$

Otherwise, the distribution of carriers in the allowable bands is given by the Boltzmann distribution, see Eq. (3.37).

$$f_B(E) = e^{-E/(kT_j \sigma)} \tag{3.37}$$

On the basis of Eqs. (3.36) and (3.37), a theoretical emission intensity distribution of a color semiconductor LED is proportional to the product of the joint density of states and the Boltzmann distribution of carriers as shown in Eq. (3.38) with the energy variable E (in eV) or the one in Eq. (3.39) with the wavelength variable λ (in nm).

$$S(E) = a \cdot \left\{ \frac{1}{2\pi^2} \cdot \left(\frac{2m_r^*}{\hbar^2}\right)^{\frac{3}{2}} \cdot \sqrt{E - E_g} \right\} \cdot \left\{ e^{-\frac{E}{\sigma k T_j}} \right\} \tag{3.38}$$

$$S(\lambda) = a \cdot \left\{ \frac{1}{2\pi^2} \cdot \left(\frac{2m_r^*}{\hbar^2}\right)^{\frac{3}{2}} \cdot \sqrt{\frac{hc}{\lambda} - E_g} \right\} \cdot \left\{ e^{-\frac{hc}{\lambda \sigma k T_j}} \right\} \tag{3.39}$$

3.3.4
Characteristic Parameters of the LEDs Absolute Spectral Power Distribution

If the function in Eq. (3.39) is drawn in a graphic form, the theoretical absolute spectral power distribution of a color semiconductor LED can be obtained, see Figure 3.18. Its intensity is maximal at a specific energy and this energy is shown in Eq. (3.40).

$$E = E_g + \frac{1}{2} k T_j \sigma \tag{3.40}$$

Likewise, based on Figure 3.18, the peak wavelength, the FWHM, and the peak intensity can also be computed, see Eqs. (3.41)–(3.43).

$$\lambda_p = \frac{hc}{E_g + \frac{1}{2} k \sigma T_j} \tag{3.41}$$

$$\lambda_{FWHM} = \frac{hc}{1.8 k \sigma T_j} \tag{3.42}$$

$$S_p = a \cdot \left\{ \frac{1}{2\pi^2} \left(\frac{2m_r^*}{\hbar^2}\right)^{\frac{3}{2}} \cdot \sqrt{\frac{hc}{\lambda_p} - E_g} \right\} \cdot \left\{ e^{-\frac{hc}{\lambda_p k \sigma T_j}} \right\} \tag{3.43}$$

3.3 Color Semiconductor LEDs

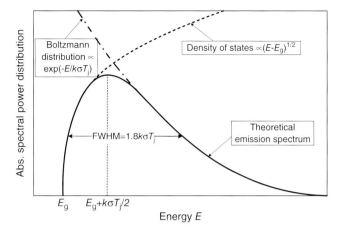

Figure 3.18 Theoretical emission spectrum of a color semiconductor LED, reproduced from [3, Fig 5.2], copyright © 2003 with permission from Cambridge University Press.

3.3.5
Role of the Input Forward Current

The simple theory about the forward current of the color semiconductor LEDs was mentioned in Section 3.2.2. It was described by Eqs. (3.3)–(3.5) and (3.8)–(3.11). According to Müller [4], the relationship between currents and effective masses can be described by Eq. (3.44).

$$\frac{I_h/A_h}{I_e/A_e} = \frac{J_h}{J_e} = \left(\frac{D_h N_h}{L_h}\right) \cdot \left(\frac{L_e}{D_e N_e}\right) \cdot \left(\frac{(m_e^* m_h^*)_n}{(m_e^* m_h^*)_p}\right)^{\frac{3}{2}} \cdot \exp\left(\frac{\Delta E_g}{k_B \sigma T_j}\right) \quad (3.44)$$

Here, J_e and J_h are the injected electron and hole current densities, D_e and D_h and L_e and L_h are the minority electron and hole diffusion coefficients and diffusion lengths, respectively, A_e and A_h are the cross-sectional areas of electrons and holes, N_e and N_h are the electron and hole doping densities, m_e^* and m_h^* are the electron and hole effective masses (on the n-side in the numerator and the p-side in the denominator), and ΔE_g is the energy gap difference between the upper confining layer and the active layer.

3.3.6
Summary

The equation system (3.28)–(3.44) constitutes the basis for the theoretical description of the relationship among three characteristic parameters of the spectral power distribution of a color semiconductor LED and the operating condition including forward current and junction temperature based on the band structure theory, the concepts of de Broglie, the energy conservation law, the momentum conservation law, the joint dispersion relation, and the Boltzmann distribution. It should be recognized that, theoretically, these relationships can be described mathematically as three-dimensional functions such as peak wavelength $\lambda_p(I_f, T_j)$, FWHM $\lambda_{FWHM}(I_f, T_j)$, and peak intensity $S_p(I_f, T_j)$. Unfortunately, many of these parameters are very difficult to measure and determine explicitly. Therefore, on the basis of the fundamental knowledge, more feasible, and practical mathematical descriptions should be investigated and offered to characterize real spectral power distributions of color semiconductor LEDs. Moreover, it is also necessary to examine whether the three characteristic parameters are good enough to describe the real absolute spectral power distributions of color semiconductor LEDs.

3.4
Phosphor Systems and White Phosphor-Converted LEDs

3.4.1
Introduction to Phosphors

In the late 1990s, the first PC-LEDs emerged after the invention of the blue LED by Nakamura and Fasol [7]. In the first years after the blue LEDs became available, PC-LEDs based on blue-emitting indium gallium nitride, (In,Ga)N, semiconductor dyes competed against RGB chip-based LEDs [8]. In the meantime, RGB chip LEDs are only a small niche in the LED market because numerous disadvantages, such as different lifetime and temperature behavior of blue or green and red chips and poor efficiency of green- and yellow-emitting semiconductors [3].

The scope of this chapter is to introduce the most important phosphor materials which are currently used in PC-LEDs. In addition, some basic concepts describing the luminescent materials' properties are discussed with reference only to LED phosphors. For in-depth theoretical details, a number of textbooks are available [9–11], where the quantum mechanical concepts of luminescence are also addressed. Recently, quite a number of scientific papers on alternative luminescent materials such as quantum dots (QDs) and phosphors activated with ions showing forbidden transitions, organic dyes have been published. According to the best knowledge of the authors, these materials are not widely used in LEDs on commercial basis. Organic dyes suffer from instability toward very high light fluxes, which are present in LEDs where near-UV or blue light excites the luminescent material. Phosphors activated with ions relying on forbidden transitions are interesting but the drawback is that the emission lifetime is usually too long for use in LEDs because of saturation effects. This aspect is addressed later in the chapter along with the luminescence mechanisms. Nonetheless, some of these concepts may find their way into LEDs in the future. Readers who are more interested in this field may refer [3, 11–13].

The number of different phosphor materials applied in the actual LEDs is pretty small. The reasons behind this are the challenging requirements and conditions for operation in LEDs. In contrast to plasma discharge lamps which excite phosphors in the UV there is only a limited number of phosphors available that can absorb efficiently in the blue (or near-UV) region of the spectrum and emit light efficiently in the visible region. In addition, the phosphor has to be thermally stable with low quenching of its emission especially for the high power LEDs and chemically stable in order to allow for high reliability under challenging environments such as high temperature and high humidity over several tens of thousands of hours of operation. In addition, high radiation loads or light fluxes generated by the LED chip of several hundred hours per square millimeter should not harm the phosphor by bleaching or by saturation of the excited state, which hinders the brightness of the whole LED.

Phosphors are usually polycrystalline inorganic materials activated with dopant ions, which are able to emit visible light after absorption of suitable light from external sources, such as LED chips. The dopants, also called *luminescent centers and activators*, are metal ions that substitute a (small) part of the host lattice cations. The process of emission takes place on such ions via electronic transitions between their energy levels located in the forbidden band (bandgap) of the host material. In addition, there are also some self-activated phosphors, where there is no dopant inside the host lattice and the emission comes from the host material itself. As materials these are not used in PC-LEDs because of the fact that the emission is usually not very intense unless the material is pumped with high energy radiation (e.g., X-rays), they are not discussed further. The phosphor's emission process is called *photoluminescence* and occurs beyond thermal equilibrium.

Common LED phosphors are compounds such as $YAG:Ce$ ($Y_3Al_5O_{12}:Ce^{3+}$) and its numerous chemical modifications (e.g., $(Y,Gd)_3Al_5O_{12}:Ce^{3+}$, $(Y,Tb)_3Al_5O_{12}:Ce^{3+}$, $Y_3(Al,Ga)_5O_{12}:Ce^{3+}$, $Y_3Al_5O_{12}:Ce^{3+},Pr^{3+}$ as well as $Lu_3Al_5O_{12}:Ce^{3+}$, $Lu_3(Al,Ga)_5O_{12}:Ce^{3+}$, $Lu_3Al_5O_{12}:Ce^{3+},Pr^{3+}$ etc.), and BOSE $(Ca,Sr,Ba)_2SiO_4:Eu^{2+}$, CASN ($CaAlSiN_3:Eu^{2+}$), all of which are discussed in detail in the following.

Luminescence occurs as a result of external stimulation of the phosphor material. This stimulation can have different forms, but we will limit ourselves to the case of photoluminescence occurring in LED phosphors. When doped with activators, the excitation light shining on the material leads to electronic transitions on the dopant ions, as mentioned already. Depending on the kind of dopant used, the resulting emission can be contained in sharp peaks or in broad bands or, alternatively, a combination of these. This is related to the nature of the energy levels involved in the electronic transitions and whether they participate in the chemical bonding or not. Schematically, the luminescence process has been depicted in Figure 3.19.

Although numerous different ions are capable of light generation once incorporated into host materials, rare-earth ions (RE) are considered the prime candidates for the use as activators in luminescent materials. This is because of their unique electronic structure, where the transitions between different energy states can lead to very narrow, atomic-like, emission lines (f–f transitions) of up to a couple of nanometers in width or broad emission bands (d–f or charge transfer (CT) transitions), covering tens of nanometers, up to even 200 nm. The distinctive sharpness of the f–f lines is related to the fact that the electrons in the 4f

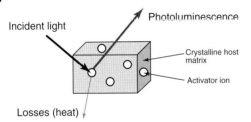

Figure 3.19 Basic operation of phosphors: incident light of suitable energy is absorbed and leads to excitation of the activator ion that is homogeneously distributed in the crystalline host matrix. The absorbed photons are down converted to lower energy and emitted by means of fluorescence, that is, photoluminescence. However, the generation of heat is a competing loss process reducing the (quantum) efficiency and thus the fluorescence intensity of the phosphor.

shell are shielded from the environment by the 5s and 5p electrons and the 4f electrons do not participate in the chemical bonding. These transitions do not vary much between different host lattices but the relative intensities of the emission lines with respect to one another as well as absence of some lines and presence of others can be modeled using, for example, the Judd–Ofelt theory [14, 15]. The nature and the appearance of emission features is totally opposite in case of the transitions involving the 5d states, which participate in the chemical bonding, and thus the energy levels related to the 5d states are strongly influenced by the surrounding ligands, that is, the nearest chemical surrounding within the host lattice. Thus changing the ligands (the chemical composition of the host material) leads to changes in the absorption and emission of the same ion (e.g., Ce^{3+} or Eu^{2+}), allowing tailoring the material's spectral properties. Just as in the case of the f–f transitions, for the f–d and d–f transitions as well there are some theoretical models available, which are helpful in predicting certain luminescence properties and can be used together with experimental data to tailor the luminescence properties of new materials [16–20]. In addition, the CT transitions in inorganic compounds received some theoretical treatment [21]. Depending on the intended application, the appropriate ion must be used, as different rare-earth cations show different luminescence properties and the luminescence relies on different electronic transitions, some of which might be allowed, partially allowed, or forbidden. This has an impact on the luminescence intensity, emission lifetime, temperature quenching of luminescence, and so on. In some cases, sensitizer ions must be used in order to observe high efficiency of luminescence from a desired ion, while overcoming its poor absorption capabilities for a desired type of the excitation energy. A very important parameter is the quantum efficiency (QE) of luminescence. Though technically simple in definition, there is often a lot of confusion in the literature as to the different values obtained by different researchers. This is partially related to the fact that there are two different kinds of QE, namely the internal QE and the external QE, QE_{int} and QE_{ext}, respectively. The internal QE is the ratio between the number of photons emitted by the material to the number of photons absorbed by the material, while the external QE is the ratio between the number of emitted photons to the number of pump photons. Thus, as the number of pump photons is always greater than the number of absorbed photons, the value for the QE_{ext} will always be smaller than the value for the QE_{int}. Of course in an ideal situation, when all the photons pumped are absorbed, these two values will be equal to each other. The two different QEs are described by the following equations:

$$QE_{int} = \frac{\text{\# photons emitted}}{\text{\# photons absorbed}} \quad (3.45)$$

$$QE_{ext} = \frac{\text{\# photons emitted}}{\text{\# pump photons}} \quad (3.46)$$

Those trying to calculate the values for QE know that the task is not trivial. Very often, a standard material needs to be used for comparison, although there is a need to take into account the reflection spectra of the standard and the material being compared as well. If no standard is used, single photon counting techniques must be applied where the absolute number of photons is collected by an integrating sphere and registered by the detector. Alternatively, one may use the emission lifetimes in order to get the value of QE. As QE can also be expressed as a ratio of the emission lifetime at the desired conditions to the value of the radiative emission lifetime, one may relatively easily calculate the quantum efficiency on the basis of the lifetime measurements. The drawback is the measurement of the radiative or, in other words, the intrinsic emission lifetime of a phosphor where theoretically all the luminescent processes happen via radiative routes and no nonradiative relaxations of any kind occur. This is very difficult, as the radiative emission lifetime can be obtained only at very low dopant concentrations, at very low temperature (preferably 4 K or below), and by directly exciting the lowest emitting state of the dopant in the host lattice in order to exclude any energy transfers and relaxations.

$$QE = \frac{\tau_{exp}}{\tau_{rad}} \quad (3.47)$$

Thus, the emission lifetime is another important factor that needs to be taken into account. The lifetime or decay time of emission is the time in which the intensity of emission decreases to a value of $1/e$ of the initial intensity. A pulsed light source (Xe flash lamp, laser, laser diode, etc.) is used as the excitation source in these measurements. The emission decay time is very important not only because it can help in the determination of the QE but also because it is a critical factor in the determination of the applicability of a particular dopant ion as an activator. In case of LEDs, there is a very high excitation light flux, as mentioned already. This means that the emission lifetime needs to be short enough to avoid the saturation of luminescence [22]. This occurs when the excitation flux is high enough to excite the available activator ions by depleting the ground state. Once this happens, there is no emission, when the lifetime of emission is too long as the electrons do not have sufficient time to recombine radiatively. Saturation can be surpassed in two ways. Either the phosphor has to be moved away from the LED chip in order to artificially reduce the flux of the excitation light that hits the phosphor particles

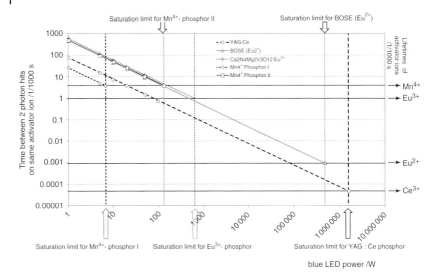

Figure 3.20 Saturation effects for activators with different emission lifetimes.

(remote phosphor concept) or the dopants that are used must be chosen among the ones showing fully allowed transitions, as these have typically short enough emission lifetimes in the range of tens of nanoseconds to a few microseconds. Decay times in the millisecond range are considered too long to be useful in LEDs, although such materials can be used in a remote configuration.

Saturation effects are shown in Figure 3.20. Depicted are the saturation effects for five different phosphors, YAG:Ce^{3+}, BOSE:Eu^{2+}, $Ca_2NaMg_2V_3O_{12}$:Eu^{3+}, and two Mn^{4+} activated phosphors, Phosphor I (high molecular weight host material) and Phosphor II (low molecular weight host material). There are a few assumptions to be made when constructing the figure. The activator concentrations were set at 2, 10, 5, and 1 at.% for Ce^{3+}, Eu^{2+}, Eu^{3+}, and Mn^{4+}, respectively. These amounts are taken from the typical LED phosphor activator concentrations. In case of Eu^{3+}, the amount given is typically on a high concentration end for any phosphor, while the concentration for Mn^{4+} originates from the fact that 1 at.% is the upper limit, where clustering of Mn^{4+} ions starts, which leads to the total quenching of luminescence. In addition, the amounts of particular phosphors in the phosphor–silicone slurry are assumed, hence a different amount of activator ions are available for excitation. This comes from experimental trials of the authors and is related to the absorption strength of the dopants, among other reasons. Thus, as trivalent cerium and divalent europium exhibit fully allowed transitions, the amount of the phosphor used is not very high as these ions absorb the incident light with ease. Trivalent europium and tetravalent manganese, on the other hand, show parity forbidden transitions and as such are not well suited to absorb the excitation light efficiently (unless sensitizers are used, which is not the case here), and significantly higher amounts (5–10 times higher) of the phosphors in the LED are required in order to improve the absorption of these materials. Also assumed are emission lifetimes of 50 ns, 1 μs, 1 ms, and 4 ms for Ce^{3+}, Eu^{2+}, Eu^{3+}, and Mn^{4+}, respectively. Last but not least is the assumption that the absorption of the excitation light takes place in the whole volume of a particle and the particle size and shape are assumed to be constant for the phosphors presented. From the figure, it can be seen that for trivalent Ce and divalent Eu dopants, the power of the LED chip that would result in saturation occurrence is well beyond anything reachable. This goes in line with the expectations that dopants with relatively short emission lifetimes are best choices for LED phosphors. Eu^{3+}, even though showing much worse behavior, seems, surprisingly, not to result in a very critical saturation, at least in the vanadate material presented. Even the tetravalent manganese activator ions do not show dramatically poor saturation behavior, although, as expected, because of the longer emission lifetime, this activator will show saturation much quicker than the other dopants. What is surprising and what has not been mentioned elsewhere in literature is the fact that there is a significant difference even when the same activator is used but in a different host lattice. This is shown for Mn^{4+}. The dopant concentration assumed is the same (1 at.%) and the emission lifetime of 4 ms is taken for both phosphors. The difference is the molecular weight of the host matrix in both cases. When a phosphor of higher molecular weight (Phosphor I) is taken for the model, the saturation effects kick in at much lower values of the LED power, that is, over an order of magnitude lower power than for the low molecular weight material (Phosphor II). This is directly related to the amount (in moles) of phosphor taken, assuming that the same mass of the phosphor is used to construct the PC-LED. In turn, the number of activator ions in the same mass is different, hence the number of dopants available for excitation as well.

In case of the PC-LEDs, the activators of choice are Ce^{3+} and Eu^{2+}, both of which rely on the allowed electric dipole transitions. Because the transitions of these ions are fully allowed, their absorption strengths are high (they absorb excitation light efficiently), the emission can be tuned in a wide range of wavelengths, and the emission lifetime is short enough to avoid the saturation effects mentioned earlier.

$$Ce^{3+} : [Xe]4f^1 \underset{\text{Emission}}{\overset{\text{Excitation}}{\rightleftarrows}} [Xe]4f^0 5d^1 \tag{3.48}$$

Trivalent cerium has a $4f^1$ electronic configuration, thus after the excitation there are no core electrons left to interact with the electron which was excited into the 5d state. This is shown in Eq. (3.48). The luminescence of trivalent cerium–activated materials is the result of a transition from the lowest excited fd state to the ^2F ground state, which is split by the spin–orbit coupling into two levels, namely, the $^2F_{5/2}$ and $^2F_{7/2}$, $^2F_{5/2}$ being the lowest in energy. These two levels are separated from each other by about 2000 cm^{-1}, which is the value of the spin–orbit coupling for Ce^{3+}. As a result, in the emission spectra there is often a double emission band observed, with the two components being separated by about 2000 cm^{-1}. Whether there is a clear separation or not an observable one depends on the value of the electron–phonon coupling of the material at hand. For larger values of electron–phonon coupling, the separation starts to disappear and the emission appears to be single band, though very broad. The mentioned separation of the emission band into a double band can also disappear when the emission occurs from multiple crystallographic sites present in the material. These sites usually differ in energy just slightly, leading to emission bands which are spectrally close to one another and can often overlap. The overlap shows in the emission spectrum as a very broad emission band. Depending on the host lattice used, the emission can span from UV all the way to the red spectral region. For more ionic host lattices the emission is more blueshifted, while more covalent ones exhibit emission in the longer wavelength region of the visible spectral range.

The typical lifetime of the Ce^{3+} emission is in the range of 40–70 ns in the visible spectral range [23]. The emission lifetime is strongly dependent on the concentration of the dopant and the defect concentration in the material as well as any energy transfer processes that might take place. In addition, the emission lifetime becomes shorter with increasing temperature, which is evidence for temperature quenching of luminescence just as the decrease in the luminescence intensity itself is. The shorter emission lifetimes are observed in case of luminescence in the bluish spectral part and become longer for emissions in the green and even longer for luminescence occurring in the red. The emission lifetime is also influenced by the presence of other optically active species in the host material, such as any codopants or sensitizers, as well as the host material itself, as there might be an interaction between the energy levels of the optically active ions and the energy states of the host lattice.

$$Eu^{2+} : [Xe]4f^7 \underset{\text{Emission}}{\overset{\text{Excitation}}{\rightleftarrows}} [Xe]4f^6 5d^1 \tag{3.49}$$

Divalent europium has a $4f^7$ electronic configuration and is isoelectronic with trivalent gadolinium. On excitation, the electronic configuration changes from $4f^7$ to $4f^6 5d^1$, which is shown in Eq. (3.49). The excited electron in the $4f^6 5d^1$ state interacts with the six electrons left in the core, leading to a more complex behavior of luminescence as well as a rich structure (depending on concentration) in the excitation spectra. Depending on the crystal field acting on the Eu^{2+} ions in a particular host lattice and how covalent or ionic the character of that lattice is, two kinds of transitions are possible, namely, the fully allowed f–d transition (for stronger crystal fields and more covalent hosts) or a parity-forbidden f–f transition (for weaker crystal fields and more ionic hosts). The latter are not discussed in more detail as there are currently no LED phosphor materials which rely on such an emission.

In case of luminescence of the divalent europium, the emission is the result of a transition from the excited $4f^6 5d^1$ (or fd state for simplicity) to its ^8S ground state. The emission occurs as a single emission band, unless it originates from Eu^{2+} located on different crystallographic sites, as has already been mentioned in case of cerium luminescence. The emission can span from UV to the red spectral region, depending on the host lattice used.

The typical lifetime of the Eu^{2+} emission is about 1 μs, although it is strongly dependent on the concentration of the dopant and the defect concentration in the material as well as any energy transfer processes that might take place [24]. Also, depending on the host lattice into which divalent europium ions are embedded, there will be influence over the lifetime of the emission, as, just as in case of Ce^{3+}, there might be interaction between the energy states of the Eu^{2+} and the host lattice. Therefore, the lifetime of emission in the red spectral region might increase to even a millisecond.

The relatively long emission lifetimes for divalent europium, even for emissions in the UV or blue spectral regions, are rather surprising. It is not totally understood why such long emission lifetimes are observed and the topic is under a constant debate in the scientific community. It seems that it might be related to the fact that Eu^{2+} is a rather large ion, leading to longer bond lengths with the ligands. This in turn will lead to a smaller radial overlap of the 5d state wave functions with the ligands, making the transition less favorable or, in other words, slower. Also, the interaction of the electron in the excited fd state with the high energy 4f states cannot be underestimated, as admixing of the different electronic states will make the normally allowed transition somewhat *"less allowed,"* thus increasing the luminescence lifetime. This is an opposite effect to the well-known occurrence in, for example, Eu^{3+}, where the emission is very intense because of the mixing of wavefunctions of the 4f states and the CT state. Thus it seems likely that the longer than expected emission lifetime of divalent europium is actually a result of at least these two contributions, with the second one becoming more significant in the case of the more ionic hosts, where there is a high lying fd state.

In the same host lattice, the divalent europium emission will be somewhat redshifted compared to the trivalent cerium luminescence [25].

3.4.2
Luminescence Mechanisms

In most LED phosphors, emission is generated on an optical center, that is, the activator ion. Usually, for LED phosphors, this kind of optical transition involves only the electronic states of the luminescent ion and as such the process is called *characteristic luminescence* or *center luminescence*. Typical LED phosphors activated with Ce^{3+} or Eu^{2+} show broad emission bands with bandwidths exceeding 50 nm. Such broad bands are present if the nature of the chemical bonds in the ground state and the excited state are different. In such a case, excitation leads to changes in the equilibrium distance between the emitting activator ion and its direct environment (the ligands) and can be explained by means of the configuration coordinate diagram (Figure 3.21). The configuration coordinate diagram comprises the potential energy of an activator ion and its nearest neighbor ions in the crystalline lattice (in the ground state and the excited state) as a function of the distance between them or, in other words, as a function of bond length between the dopant and the ligands. In this diagram, R_g and R_e are the metal–ligand distances in the ground state and the excited state, respectively. E_a and E_e are the energies at which the absorption and emission bands have their maximum intensity, respectively. The figure illustrates two boundary examples, where there is an

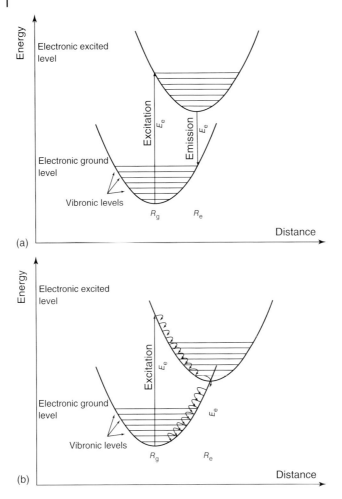

Figure 3.21 Configurational coordinate diagram. (a) No parabola crossing and emission occurs and (b) parabola crossing and no emission is observed.

emission despite changes in the bond length in the excited state (Figure 3.21a) and when the bond length for a specific energy of the excited state changes enough to prevent the emission from occurring (Figure 3.21b). From the figure, it becomes evident that for the same change of bond length in the ground and the excited state but for different energies of the excited state, there can be a parabola crossing in such a way that the electrons can cross from the excited state to the ground state without the emission of light. Alternatively, the same might occur when the energy of the excited state stays the same but the bond length changes are more pronounced (not depicted in Figure 3.21).

Thus, the diagram can be used in order to illustrate the relation between the relaxation in the excited state and the Stokes shift, which might explain poor thermal behavior of a phosphor or the lack of emission in some cases. Because

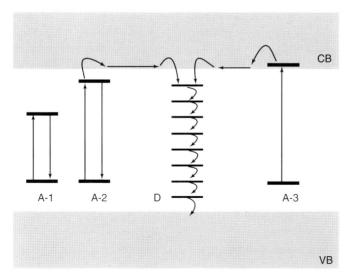

Figure 3.22 Process of photoionization. A-1, A-2, and A-3 are activator ions with different energies of the excited state (same energy of the ground state with respect to the top of the valance band assumed). D is the defect site.

of this, the diagram is very popular when describing the temperature-induced luminescence quenching (temperature quenching). By looking at Figure 3.21a, one can see that even though the parabolas do not cross at low vibronic levels, an electron might be thermally promoted to higher vibronic levels, at which levels parabola crossing might already occur. In such a case, the luminescence might still take place, although the thermal quenching can be exhibited.

Of course, the configurational coordinate diagram does not explain phenomena such as the photoionization. In such an instance, a semiconductor model of band structure is used. An example showing the thermally induced photoionization is depicted in Figure 3.22.

This is a quenching mechanism which is explained based on the proximity of the emitting level to the bottom of the conduction band. As the electron is located on the activator ion, under certain conditions it may escape that ion into the conduction band. Hence the process is called *photoionization* and it is strongly temperature driven, namely, at elevated temperatures, the probability of such an electron escape increases as the activation energy is reached. The activation energy is understood here as the energy difference between the bottom of the conduction band and the emitting level.

On the basis of Figure 3.22, one can see that when the emitting state of the activator ion is far from the bottom of the conduction band (A-1 in Figure 3.22), the emission is not hindered by thermally induced photoionization. When the energy difference between the emitting state and the bottom of the conduction band is smaller (A-2 in Figure 3.22), the emission process competes with the photoionization process and, at elevated temperature, the emission can be totally quenched. A boundary case is when the lowest fd level (in case of lanthanides) is located

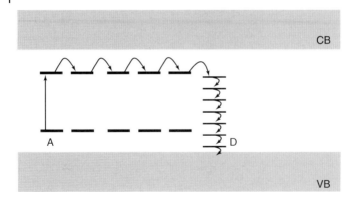

Figure 3.23 Electron hopping as a concentration and thermal mechanism for luminescence quenching.

within the conduction band of the host material. In such a case, even though the activator level can absorb the pumping light, there is no emission observed as the diffusion of the electrons through the conduction band to the quenching site(s) is much faster, preventing the emission from occurring.

Yet another quenching mechanism is the so-called electron or charge carrier hopping. Figure 3.23 shows schematically the process of electron hopping. The diagram can explain two phenomena, both of which can be coexistent. The first one is concentration quenching. It occurs when the dopant concentration in the material exceeds a certain threshold limit. In such a case, the activator ions (A in Figure 3.23) start cross-talking and because of the fact that there is a resonance between the energy levels, excited electrons can start to move from one activator ion to the other, hence the electron hopping. The quenching of luminescence happens as the chance to reach a defect (D in Figure 3.23) (quenching) site in the host material increases greatly. The second mechanism is related to the thermally induced electron hopping, as the electrons on the dopant energy levels become more mobile via gain of thermal energy. The outcome is identical, as the hopping leads to the reaching of a defect site. In this scenario, however, the concentration does not necessarily need to be very high. On the other hand, a situation where the two scenarios happen together is the most likely, namely, the concentration- and thermal-induced quenching of luminescence happens.

3.4.3
Aluminum Garnets

$Y_3Al_5O_{12}:Ce^{3+}$ (YAG:Ce) and $Lu_3Al_5O_{12}:Ce^{3+}$ (LuAG:Ce) are representatives of the aluminum garnet–type materials which became very popular not only in the field of LED phosphors but also, for example, in the field of scintillation materials for various applications where detection of high energy radiation is required. They are derived from a silicate garnet mineral grossular ($Ca_3Al_2Si_3O_{12}$) [1] by substituting calcium by yttrium or lutetium and silicon by aluminum. The garnet structure is highly organized and rigid, which leads to very temperature-stable luminescence, when the material is activated with dopants capable of light generation. The materials can be synthesized in a very broad range of temperatures, from about 800 up to 1800 °C, with the sample quality and crystallinity generally increasing with increasing temperature. The low temperature synthesis promotes the formation of nano-materials.

YAG:Ce^{3+} has been the first and by far the most successful LED phosphor to date [26]. The emission of cerium activated yttrium aluminum garnet is situated in the yellow spectral region, centered around 550 nm, which is a result of the transition from the lowest fd state of cerium to its ground state. The Stokes shift, defined as the energy difference between the maximum of the emission band and the maximum of the lowest excitation band, is about 2800 cm^{-1} [17]. The broadness of the emission band, described by the FWHM value, is 2400 cm^{-1} [17]. The exact emission wavelength as well as the Stokes shift and the FWHM value depend on the cerium concentration in the material. Thus, the emission, for example, can be redshifted with the increase in the amount of cerium used. Increasing the dopant concentration will also broaden the emission band. Ce^{3+} activated yttrium aluminum garnet materials exhibit very high quantum efficiencies in the range of 95–100%.

The phosphor's use in combination with the 450–480 nm blue (In,Ga)N LED emission yields cold white light because of the mixing of blue and yellow components [26]. In fact, this is the most common combination in the low-end commodity devices, where cold white light is sufficient, such as flashlights or gadgets.

Because of a relatively low value of Stokes shift and a relatively large optical bandgap of the material ($E_g = 180$ nm or 6.9 eV), the quenching temperature of luminescence is very high, above 700 K (based on the luminescence lifetime measurements) with a quenching onset around 600 K [27]. The small Stokes shift means that there is relatively low relaxation in the excited state, which promotes temperature stability of luminescence, while the large bandgap makes the temperature-induced photoionization less probable.

The earlier-mentioned physical, chemical, and spectroscopic properties make YAG:Ce^{3+} a material of choice for a wide range of applications as well as a standard material for the comparison of the luminescence properties, when efficiency and stability come into play. Also, because of the very good spectroscopic properties of the material and its stability, the yttrium aluminum garnet activated with trivalent cerium ions is often used for theoretical modeling of luminescence processes [19, 28], which later become grounds for investigation and analysis of other luminescent materials and a better understanding of their properties. YAG:Ce^{3+} can easily be thought of as the best known and best understood luminescent material yet (Figure 3.24).

LuAG:Ce^{3+} shows emission which is blueshifted compared to its Y analog and is centered around 500 nm. This is counter intuitive, as one expects that if the crystal structure stays the same, the emission becomes redshifted when the dopant ions accommodate on a smaller lattice site. This is different in aluminates and is most likely related to the fact that the distance between the Lu ions and the oxygen

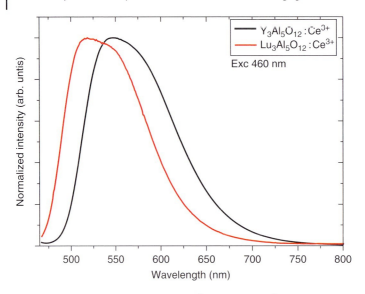

Figure 3.24 Emission spectra of YAG:Ce^{3+} and LuAG:Ce^{3+} under 460 nm excitation.

ions (the ligands) is very small. In a normal situation, it is expected that when this distance becomes smaller, the emitting state is shifted to lower energies, but there is a critical value for this parameter as well. Thus, when the critical distance is surpassed, the electrons on the dopant and the ligands start to repel strongly, leading to the rise in energy of the emitting fd state. The emission, just as in the case of YAG:Ce^{3+}, is a consequence of the transition from the excited fd state to the ground level of trivalent cerium. As the electron–phonon coupling value is smaller for LuAG:Ce^{3+} in comparison with YAG:Ce^{3+} because of the presence of Lu^{3+}, which is heavier than Y^{3+}, the emission is more clearly split into two bands, although the overlap is still rather large. The Stokes shift of about 2300 cm^{-1} [17] is considerably smaller for Ce^{3+}-activated LuAG compared to YAG:Ce^{3+}. Also, the value of 1900 cm^{-1} for the FWHM [17] of the emission band is smaller for the lutetium aluminum garnet when activated with trivalent cerium compared to the Ce^{3+} emission band in YAG. The quantum efficiency of LuAG:Ce^{3+} materials is comparable to that of the YAG:Ce^{3+} samples and is in the range of 90–100%.

As a result of a large optical bandgap (E_g=165 nm or 7.5 eV) and a significantly smaller Stokes shift compared to YAG, the quenching temperature of luminescence in LuAG:Ce^{3+} is even higher than for YAG:Ce^{3+} and is somewhere above 800 K (based on the luminescence lifetime measurements) with a quenching onset at about 700 K [27] (Figure 3.25).

Because of the exceptionally good chemical stability and relative ease of synthesis as well as the superb temperature properties of the Y and Lu aluminum garnets activated with trivalent cerium ions, it does not come as a surprise that these materials very often serve as standards for other phosphor materials. Moreover, if a yellow- or green-emitting material is to be successful in the market, its properties

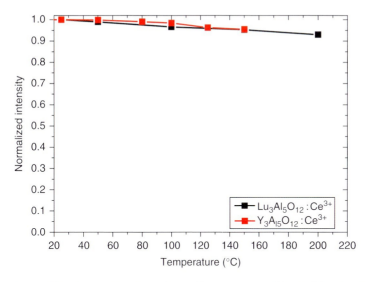

Figure 3.25 Temperature dependence of luminescence of YAG:Ce^{3+} and LuAG:Ce^{3+}.

must at least come very close to YAG:Ce^{3+}. Of course, the emission spectra of these materials can be characterized as lacking the red component (especially LuAG:Ce^{3+}, which is blueshifted compared to YAG:Ce^{3+}) [29]. Also, the so-called cyan gap (spectral region between the blue LED emission of about 450 nm and the green part of the spectrum of about 530 nm) is poorly covered by the emission of the garnets. Thus alone, YAG:Ce^{3+} and LuAG:Ce^{3+} produce light of rather poor quality, with a low value of the CRI and rather high value of the CCT. This imposes the need to use other phosphors that could act, for example, as the red emitters in the mixture together with YAG:Ce^{3+} or LuAG:Ce^{3+}. This approach, of course, brings additional issues into scope, such as reliable mixing and different sedimentation rates in the mixture with silicone. Nonetheless, phosphor blends are very common when general lighting is concerned, as the end LED device must produce light of very good properties, which emulates the sun spectrum as closely as possible.

There have been other aluminate garnets proposed for use as luminescent powders for PC-LEDs, such as $(Y,Gd)_3Al_5O_{12}:Ce^{3+}$ [30], $Y_3(Al,Ga)_5O_{12}:Ce^{3+}$ [31], or $Tb_3Al_5O_{12}:Ce^{3+}$ (TAG:Ce^{3+}) [31]. The two former materials are examples where the standard YAG:Ce^{3+} is modified such that the crystal field which is felt by the dopant ions is changed, which in turn leads to the redshift (Y,Gd) or blueshift (Al,Ga) of emission. In case of TAG:Ce^{3+}, the idea is that the material shows dual absorption of the excitation light, namely, by Ce^{3+} and Tb^{3+} which exists in large amounts as it comprises the host lattice. Even though the trivalent terbium absorption in case of LED use is forbidden (f–f transition), the large amount of Tb^{3+} ensures the improved absorption. In the end, the emission comes from Ce^{3+} only and the luminescence spectrum shows slightly improved emission in the reddish spectral part. Nonetheless, the materials show greatly worsened temperature stability of luminescence compared with the standard YAG:Ce^{3+} phosphor. In addition, terbium is a very expensive element and as there is no really satisfactory improvement of the luminescence compared to the standard material, TAG:Ce^{3+} did not become a success story in the field of LED phosphors.

3.4.4
Alkaline Earth Sulfides

Alkaline earth sulfides activated with divalent europium ions (MS:Eu^{2+}, M = Mg, Ca, Sr, Ba) have been known to the scientific community for decades [32, 33]. Their luminescent properties on activation with divalent Eu, trivalent Ce as well as other transition metal ions (e.g., Cu^+, Ag^+, Au^+, Mn^{2+}, Cd^{2+}) or the s^2 ions (e.g., Bi^{3+}, Pb^{2+}, Sb^{3+}, Sn^{2+}) [32] have been investigated for use as luminescent phosphor powders for PC-LEDs, electroluminescent powders and thin films for display applications. The emission as well as excitation bands of the activators can be tuned within these compositions by changing the host site ion and/or employing a solid solution of different host site ions, for example, (Ca,Sr)S:Eu^{2+} [34, 35]. In general, the alkaline earth sulfide materials utilizing divalent europium ions as activators lead to emission in the orange-to-red spectral region [34–37] as a result of their highly covalent character, and until the red emitting nitride materials came along, they had been suggested as potential red-emitting components in white LEDs together with, for example, YAG:Ce^{3+}. The emission observed is a single band (unless the emission originates from multiple sites) and is the result of a transition from the lowest excited fd state of divalent europium to its ground state. It is interesting to note that the emission is rather narrow (typically FWHM below 70 nm), which is of importance for enhancing the color gamut of a display device where such an LED is used as a backlighting unit (BLU). The observed FWHM values are 3200 cm^{-1} for BaS:Eu^{2+}, 1930 cm^{-1} for SrS:Eu^{2+}, 1460 cm^{-1} for CaS:Eu^{2+}, and 1120 cm^{-1} for MgS:Eu^{2+} [17]. This, combined with the high brightness and reasonably high quantum efficiencies of these materials, makes them suitable for the use as phosphor powders for PC-LEDs. The relatively small Stokes shift for the alkaline earth sulfides is also beneficial, as it leads to less heat generation on relaxation of the excited state, which has negative influence on the luminescence properties of the phosphor (Figure 3.26).

Eu^{2+}-activated alkaline earth sulfides vary in the quantum efficiencies. Ca-based material shows quantum efficiency of 53%, while SrS:Eu^{2+} has a quantum efficiency of 31% [32]. These numbers should be regarded carefully as materials and their properties and performance are greatly influenced by the preparation technique. This is because of the fact that the sulfide host lattices are prone to the loss of S^{2-} in the structure if the synthesis atmosphere is not sufficient in sulfide. As a result of the loss of sulfide ions, mixed sulfide-oxide compositions as well as materials rich in anion vacancy defects might be expected. This will hinder the performance of the phosphor materials. Thus, by making samples of superior quality, one might expect that these materials could achieve quantum efficiencies closer to 100%, as the emission is based on the allowed transition on the divalent europium

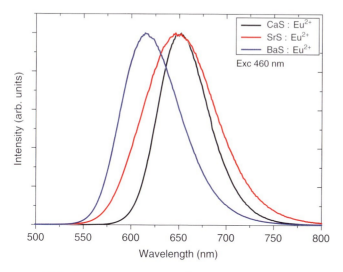

Figure 3.26 Emission spectra of CaS:Eu^{2+}, SrS:Eu^{2+}, and BaS:Eu^{2+} under 460 nm excitation.

ions. In addition, the quantum efficiencies mentioned earlier relate to the actual conditions of LED operation and it is known that the sulfides in general often have much better performance at temperature well below the room temperature. Also worth noting is the relatively low tolerance of the materials for the incorporation of the dopant ions. Compared to other sulfide materials, alkaline earth sulfides are more prone to concentration quenching and the typical amount of divalent europium ions incorporated into the materials seldom exceeds 0.5 mol%. This also has an impact on the lower light emission of these phosphors as compared to the alternative ones.

Albeit the numerous advantages of the alkaline earth sulfides, their use has been greatly limited outside the lab. This is related to a few disadvantages of these compounds compared to other LED phosphors. First of all, the synthesis of the sulfides is very problematic, as it requires the use of H$_2$S gas throughout the whole synthesis process. H$_2$S is necessary as it leads to the reduction of Eu^{3+} to its divalent state and, more importantly, it allows the formation of a pure sulfide material, which is not deficient in any S^{2-}, which in turn could lead to lowered brightness of the phosphor, as well as a none-phase pure system, because of the leftover reagents. Hydrogen sulfide requires special care, as it is highly toxic. Also, the final materials suffer from a few drawbacks. Because of the relatively low optical bandgaps of sulfides (E_g = 148 nm or 8.4 eV (calculated) for MgS, E_g = 221 nm or 5.6 eV for CaS, $E_g \approx 275$ nm or 4.5 eV for SrS, E_g = 315 nm or 3.94 eV for BaS), there is an increased risk of thermally induced photoionization. This effect leads to the decrease or, in some cases, total quenching of the luminescence on increasing the temperature of operation. As a typical high-power LED operates at around 150 °C, the temperature is sufficient to induce temperature quenching of luminescence via the process of photoionization. The poor temperature stability is apparent especially for $SrS:Eu^{2+}$ and $BaS:Eu^{2+}$, both of which have small optical bandgaps as indicated before. Finally, sulfide materials are known to be moisture and air sensitive. Alkaline earth sulfides slowly decompose on exposure to humidity and increased temperature, losing their performance. Moreover, the decomposition occurs via the formation of H_2S, which is released from the phosphor powder. H_2S in turn reacts with the Ag reflective surface or mirror of the LED, leading to the degradation of the whole device and a significant decrease of its overall performance and lifetime, as less light is out-coupled from the LED device [36]. Thus, the perceived brightness of the LED falls over time. The degradation issues might be resolved by employing a coating techniques, which might help chemically stabilize the particles of $MS:Eu^{2+}$ [32, 37–40]. Though interesting and theoretically sound, coating techniques have not been successfully utilized in the case of the alkaline earth sulfides so far.

In the scope of improvement of the sulfide materials luminescence properties, it is also worth noting that the luminescence from the alkaline earth sulfides has been investigated for Ce^{3+} doping, either on its own or as a sensitizer ion. Using Ce^{3+} on its own leads to a blueshifted emission compared to Eu^{2+}-activated materials. Doping alkaline earth sulfides with trivalent activators requires charge compensation, as there is a charge mismatch between the divalent alkaline earth metal site in the host crystal and the trivalent Ce activator ion. Such charge compensation might be introduced on purpose or the material might compensate on its own. Both scenarios have impact on the material's luminescence, which is not discussed further as it falls outside the scope of this chapter. Ce^{3+} used as a sensitizer [32, 41] for the Eu^{2+} emission in alkaline earth sulfides relies on the energy transfer process from the sensitizer to the actual activator ion intended for emission. The energy transfer from Ce^{3+} to Eu^{2+} is very efficient and does not result in changes in the emission spectra, besides increasing the luminescence intensity.

As a result of the problematic synthesis, which has influence over the luminescence properties of the final products as well as issues with chemical stability of the materials, both of which have direct influence over the luminescence properties of materials, there is no single quenching temperature value available for the alkaline earth sulfides activated with divalent Eu ions. Specimen of better quality and purity exhibit improved emission characteristics. Nonetheless, it seems that these materials emit very well at low temperatures and are subjected to significant quenching already at room temperature. As mentioned earlier, this is mostly apparent for $SrS:Eu^{2+}$ and $BaS:Eu^{2+}$, and, in both cases, the rather poor temperature stability of luminescence is related to the low bandgaps of these materials. $CaS:Eu^{2+}$, on the other hand, shows a drop of about 30% in the luminescence intensity from room temperature to about 125 °C, but at temperature higher than 125 °C the emission intensity stabilizes at a constant level. This observation can also be addressed in terms of the bandgap value, which is high for CaS. In this line, the temperature stability of $MgS:Eu^{2+}$ is expected to be very high (Figures 3.27–3.29).

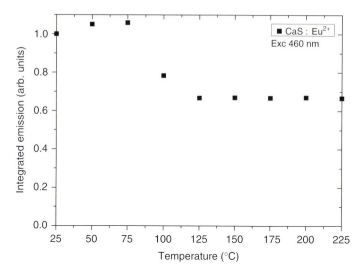

Figure 3.27 Temperature dependence of luminescence of CaS:Eu^{2+}.

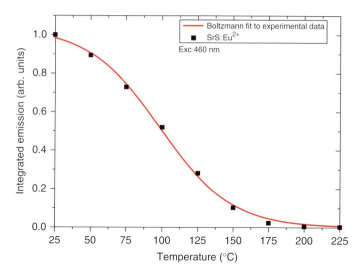

Figure 3.28 Temperature dependence of luminescence of SrS:Eu^{2+}.

3.4.5
Alkaline Earth *Ortho*-Silicates

First reported by Barry [42], M$_2$SiO$_4$:Eu^{2+} (M = Ca, Sr, Ba), alkaline earth *ortho*-silicates activated with divalent europium ions have been used as green-to-orange emitting phosphor materials for PC-LED applications. These compounds can be used as alternatives for YAG:Ce^{3+} when emitting in the yellow spectral region or as constituents of a phosphor blend together with YAG:Ce^{3+} when emitting

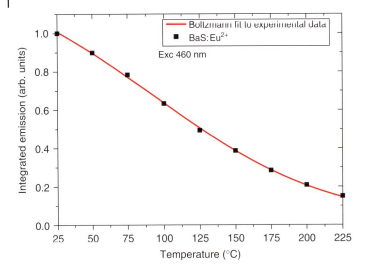

Figure 3.29 Temperature dependence of luminescence of BaS:Eu^{2+}.

in the orange-reddish spectral range, as this would increase the CRI of the mixture just as a low CCT, making it suitable for general lighting applications [43] to create warm white light. Pure Ba^{2+} composition yields emission in the bluish-green spectral region (around 500 nm), while Sr^{2+} gives orange emission (about 570 nm) and Ca^{2+} leads to deep-orange light generation (around 600 nm). Therefore, by the alignment of the chemical composition of the materials, a broad range of chromaticity can be addressed (Figure 3.30). The exact emission wavelength is strongly dependent on the divalent europium concentration in the host lattice and with the increasing amount of the dopant, one observes a redshift of luminescence. The emission is a result of the transition from the lowest excited fd state to the ground state of Eu^{2+} and appears as a single band, unless it originates from different Eu^{2+} sites in the material. The FWHM is 2360 cm^{-1} for Ba_2SiO_4:Eu^{2+}, 3200 cm^{-1} for Sr_2SiO_4:Eu^{2+}, and 3160 cm^{-1} for Ca_2SiO_4:Eu^{2+} [17].

Because of the fact that the FWHM of *ortho*-silicates is small, their emission yields high color purity. This makes them perfectly suitable for high color gamut display backlighting applications. However, different *ortho*-silicate compounds can be mixed in such a way that a broad emission band occurs which allows for high color rendition in general lighting. This is achieved on the basis of the fact that the physical properties (e.g., specific gravities and morphology) of different *ortho*-silicates are very similar, which allows for stable phosphor mixtures to be created (Figure 3.31).

Silicate chemistry is very rich. A search in a popular Internet search engine for "*silicate luminescence*" gives about four million hits. This shows the great variety of materials investigated over the years as well as their true potential. As a result of the existence of numerous variations of silicates, the synthesis of silicate materials is not trivial and often leads to the formation of other undesired silicate

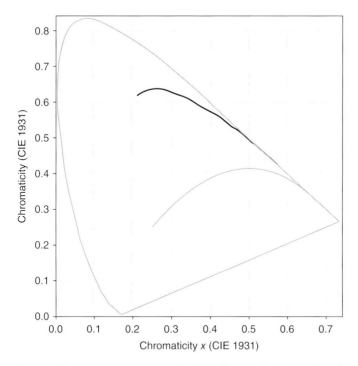

Figure 3.30 Color point variations for BOSE:Eu phosphors (thick black line) showing all the different possibilities in colors that could be obtained using the *ortho*-silicates. The thin line shows the Planckian locus.

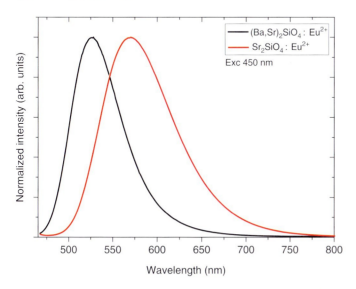

Figure 3.31 Emission spectra of $(Ba,Sr)_2SiO_4:Eu^{2+}$ and $Sr_2SiO_4:Eu^{2+}$ under 450 nm excitation.

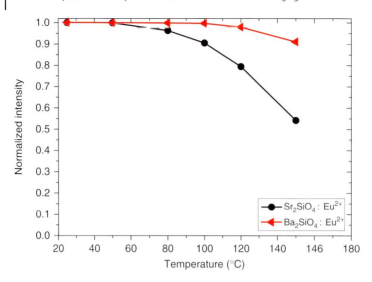

Figure 3.32 Temperature dependence of luminescence of $Sr_2SiO_4:Eu^{2+}$ and $Ba_2SiO_4:Eu^{2+}$.

by-phases in the resulting product. This, however, can be overcome by tailoring the synthesis procedure or posttreatment of the resulting particles of the material (Figure 3.32).

In spite of the difficulties in preparation, alkaline earth *ortho*-silicates have very high quantum efficiencies of 90–100%. Of course, based on the previous discussion, the sample quality will have a direct impact on these numbers. The alkaline earth *ortho*-silicates are very good phosphor materials for LED applications and they bring numerous benefits material-wise, as discussed before.

3.4.6
Alkaline Earth Oxy-*Ortho*-Silicates

$M_3SiO_5:Eu^{2+}$ (M = Ca, Sr, Ba), alkaline earth oxy-*ortho*-silicates, represent another class of materials from the rich silicate family which is of interest for PC-LED applications. Their emission is slightly redshifted compared to their *ortho*-silicate analogs [44, 45]. Emission ranges from about 570 nm for the pure Ba^{2+} material to about 600 nm when Ca^{2+} constitutes the structure. Alkaline earth oxy-*ortho*-silicates show very high quantum efficiencies of about 100%. The FWHM is 2790 cm^{-1} for $Ba_3SiO_5:Eu^{2+}$, 3850 cm^{-1} for $Sr_3SiO_5:Eu^{2+}$, and 3370 cm^{-1} for $Ca_3SiO_5:Eu^{2+}$ [17]. As usual, the exact emission wavelength and broadness of the emission depend on the dopant concentration as well as on the exact composition, as mixed alkaline earth oxy-*ortho*-silicates might be synthesized [45] in order to tune the emission color or cover a broader spectral range which is beneficial, for example, in the case of general lighting applications, where the emission spectrum must resemble the sun spectrum as well as possible for the light source to be considered of high quality. The alkaline

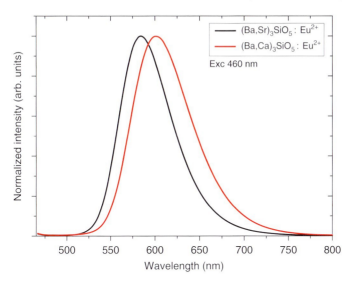

Figure 3.33 Emission spectra of $(Ba,Sr)_3SiO_5:Eu^{2+}$ and $(Ba,Ca)_3SiO_5:Eu^{2+}$ under 460 nm excitation.

earth oxy-*ortho*-silicates might also be used as starting materials for the synthesis of even more sophisticated oxy-nitrides such as $Ba_3Si_6O_{12}N_2:Eu^{2+}$, as presented by Kang *et al.* [46], where $Ba_3SiO_5:Eu^{2+}$ was used as basis for the final material synthesis (Figure 3.33).

The structure of the oxy-*ortho*-silicates is quite different from that of the normal silicates because of the fact that the "*extra*" oxygen atom present in the structure is not bound to the silicon atoms, just as in the case of the rare-earth oxy-*ortho*-silicates which are well known for their Ce^{3+} emission where there are two different oxygen atoms present in the structure, some of which are bound only to the metal ions and others which are bound to both the metal and silicon ions [47]. Thus, there are two kinds of oxygen atoms, namely, the ones bound to silicon and existing as $[SiO_4]^{4-}$ tetrahedral units and those bound only to the alkaline earth metal M^{2+}.

As these phosphors show very high quantum efficiencies they are well suited orange–red LED phosphors.

3.4.7
Nitride Phosphors

Since the discovery of nitride-based phosphors in the beginning of the current century, this phosphor class has gained a lot of importance in LED-based lighting solutions. By applying red-emitting nitride phosphors together with yellow and green phosphors, which might also consist of nitride compositions, higher portion of red in the fluorescence spectrum of a white light–emitting LED can be obtained compared to a single phosphor solution. This enables the emission of warm white light with high color rendering. Further, higher color gamut in display BLUs can be obtained.

In nitride phosphors the nitride host lattice typically consists of alkaline earth metals M, silicon Si, and nitride N to form a composition of M–Si–N or a composition with partial substitution of silicon by aluminum and/or nitrogen by oxygen in the form of M–Si–Al–O–N. The typical highly covalent bonding between nitrogen and the activator ions enables fluorescence light which is shifted to the red spectral region compared to phosphors with more ionic host lattices.

3.4.7.1 CASN

The red fluorescent phosphor $CaAlSiN_3 : Eu^{2+}$ ($CASN : Eu^{2+}$) was first reported by Uheda et al. [48].

The material crystallizes in an orthorhombic crystal structure which is related to the hexagonal wurtzite-type structure [49] (Figure 3.34). $CASN:Eu^{2+}$ shows a broad excitation band, spanning from about 250 nm all the way to the visible spectral region of almost 600 nm, with an excitation optimum at 450 nm [50] (Figure 3.35), but it can also be efficiently excited with near UV LEDs, for example, with a 405 nm emitting chip, with only a minor loss in excitability as compared to the blue wavelength excitation.

On the basis of the excitation and emission spectra of the material, Uheda et al. estimated the Stokes shift to be about 2000 cm^{-1}.

The emission band of $CASN:Eu^{2+}$ spans from 550 nm up to slightly above 800 nm with a FWHM of about 95 nm (about 2140 cm^{-1}), as can be seen in Figure 3.35. The emission peak is centered at 650 nm, which, in combination with a relatively narrow emission, makes this phosphor suitable as a red-emitting material for high color gamut BLUs and high CRI general lighting applications in combination with green phosphors when excited by a blue LED chip. The clear

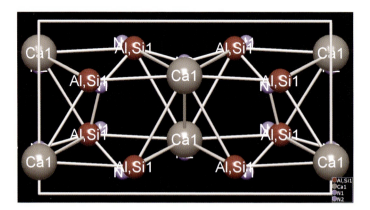

Figure 3.34 Crystal structure of $CaAlSiN_3$ viewed along the [0 0 1] direction. P. Villas, K. Cenzual, Pearsons Crystal Data – Crystal Structure Database for Inorganic Compounds (on CD-ROM), Release 2011/2012, reproduced with permission from ASM International, Materials Park, Ohio, USA.

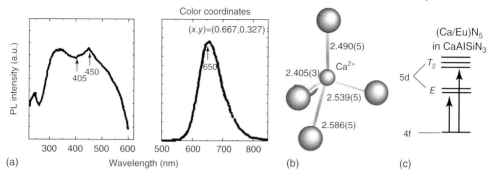

Figure 3.35 Excitation spectrum and emission spectrum (a), coordination geometry (b), and energy level diagram (c) of CaAlSiN$_3$:Eu^{2+}. (Copyright (2006) Wiley. Used with permission from [50], John Wiley and Sons.)

drawback of the phosphor, however, is that based on the emission spectrum, about 20–25% of the emission intensity falls beyond the visible spectral range, which in turn means that only about 75–80% of the emitted light can be perceived by the human eye.

The emission properties of CaAlSiN$_3$ can be modified by either Eu^{2+} concentration variation or substitution of Ca by Sr in the host lattice. By increasing the Eu^{2+} concentration, the emission wavelength can be tuned from 650 to 695 nm, although this means that majority of the emission falls into the infrared part of the electromagnetic spectrum. The optimum concentration of Eu^{2+}-activator in the host lattice in terms of quantum efficiency is around 1.6 mol%, as the introduction of greater amounts of the dopant leads to a decrease in luminescence intensity because of concentration quenching (Figure 3.36).

By replacing Ca with Sr in the host lattice, the structure remains unchanged, but an increase in cell volume by +4% [50] is experimentally observed. A significant change in excitation and emission properties occurs as a result as the crystal field strength of the material decreases. The emission maximum of Eu^{2+} is shifted by about 40–610 nm (Figure 3.37).

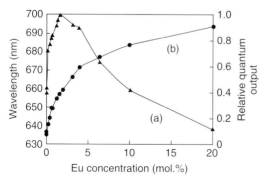

Figure 3.36 Emission properties of CASN:Eu^{2+}. (Reprinted from [48]. Reproduced by permission of ECS – The Electrochemical Society.)

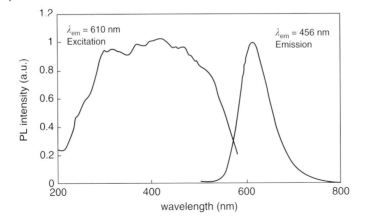

Figure 3.37 Excitation and emission spectra of SrAlSiN$_3$. (Reprinted from [51], Copyright (2008) [52], with permission from Elsevier.)

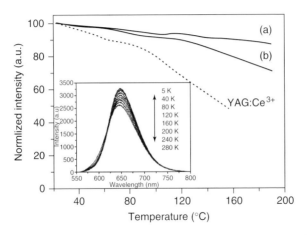

Figure 3.38 Temperature dependence of emission for Ca$_{1-x}$Eu$_x$AlSiN$_3$. (Reprinted with permission from [53], Copyright (2007), American Chemical Society.)

An important parameter for high power LED applications is the change in brightness as a function of the operating temperature, characterized by temperature quenching. Piao et al. [53] reported the temperature behavior of Ca$_{1-x}$AlEu$_x$SiN$_3$, which is given in Figure 3.38. Temperature quenching is shown for two different europium concentrations, namely, (a) $x = 0.02$ and (b) $x = 0.05$. A comparison to a YAG:Ce material is also shown in Figure 3.25. Please note that the quality of YAG:Ce shown in [53] seems rather poor (compare with Figure 3.25 in Section 3.4.3) as a much better temperature behavior for this phosphor is known. The inset of Figure 3.38 shows the emission changes in the low temperature (5–280 K) range.

The material shows a decay in brightness of about 10% for $x = 0.02$ and nearly 20% for $x = 0.05$ at an operating temperature of 160 °C. Therefore this phosphor is well suited for LED applications with a high thermal load, namely, for the use in the high power LEDs.

3.4.7.2 2-5-8-Nitrides

The alkaline earth (M) nitridosilicates, $M_2Si_5N_8:Eu^{2+}$ or 2-5-8 nitrides, show orange–red fluorescence (depending on the exact composition), which makes them important phosphors for use in white-light emitting LEDs – for example, the application of $Sr_2Si_5N_8:Eu^{2+}$ for warm white light–emitting LEDs with high CRI was reported by Mueller-Mach et al. [52]. The materials crystallize in a monoclinic structure (M = Ca) or in the orthorhombic structure (M = Sr or Ba) [54].

The photoluminescence properties of $M_2Si_5N_8:Eu^{2+}$ were reported by Hoeppe et al. [55] and Li et al. [56, 57]. The very broad excitation band of Eu^{2+} in this phosphor covers a range from about 270 to around 560 nm, with the excitation optimum in the range 400–450 nm. This makes this phosphor well applicable for near-UV and blue-emitting LED applications.

It was found that the excitation characteristics of Eu^{2+}-doped $M_2Si_5N_8$ is nearly independent of the alkaline earth metal type, whereas the maximum of the emission peak shifts to red by the exchanging of Ba to Sr and further to Ca. The Stokes shift therefore increases from M = Ba, Sr, Ca ranging from 3500, 3700, to 3800 cm^{-1}, respectively. In Figure 3.39, the effect of Sr by Ca replacement in the host lattice on the emission properties is shown. From the spectra, it is clear that even though the (Sr, Ca) composition shows the desired deep-red emission, there is a lot of emission intensity that falls outside the visible spectral range to the infrared.

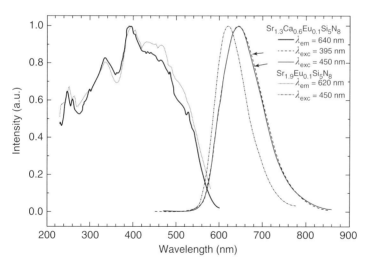

Figure 3.39 Excitation (left) and emission (right) spectra of $Sr_{1.9}Eu_{0.1}Si_5N_8$ and $Sr_{1.3}Ca_{0.6}Eu_{0.1}Si_5N_8$. (Reprinted from [56] with permission of Elsevier.)

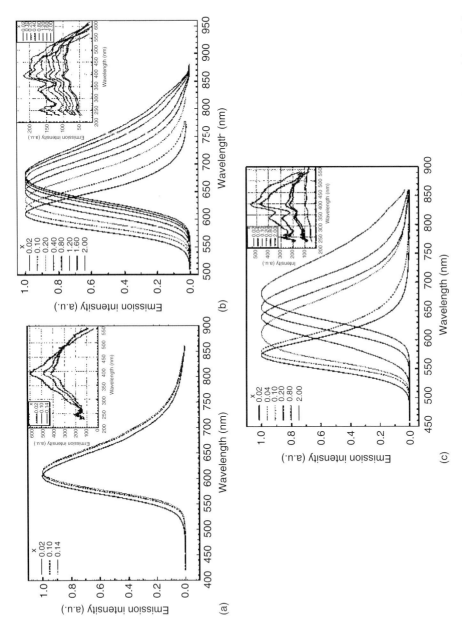

Figure 3.40 Excitation (inset) and emission spectra of (a) $Ca_{2-x}Eu_xSi_5N_8$, (b) $Sr_{2-x}Eu_xSi_5N_8$, (c) $Ba_{2-x}Eu_xSi_5N_8$. (Reprinted from [57] with permission of Elsevier.)

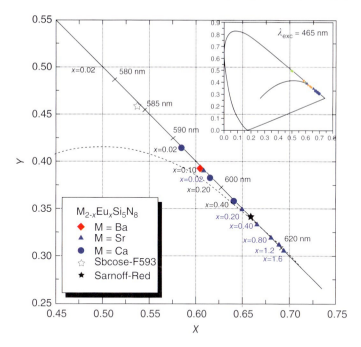

Figure 3.41 Chromaticity coordinates of different $M_{2-x}Eu_xSi_5N_8$ phosphors. (Reprinted from [57] with permission of Elsevier.)

Pure barium containing $M_{2-x}Eu_xSi_5N_8$ shows emission maximum ranging from 590 nm for low Eu^{2+} concentration up to 680 nm for high europium content (Figure 3.40).

As the crystal field splitting and average energy of the split 5d levels remain constant, a redshift of emission with higher Eu concentration is observed [57].

In summary, by modifying either the host lattice or change of Eu^{2+} activator concentration a wide range of different colors of fluorescence can be observed, which is shown in Figure 3.41 using a CIE 1931 chromaticity diagram.

The reported luminescence intensities of the different compositions differ significantly. It was reported by Li et al. [57] that $Sr_2Si_5N_8:Eu^{2+}$ showed the highest conversion efficiency compared to the Ca and Ba analogs by excitation with the 465 nm light.

Furthermore, this material shows very low thermal quenching. Xie et al. [58] reported that depending on the synthesis procedure, the luminescence intensity at 160 °C is still at 85–90% relative to the room temperature intensity, as can be seen in Figure 3.42. This makes the 2-5-8 nitrides activated with divalent europium very attractive phosphors for the use in the high power LEDs.

3.4.7.3 1-2-2-2 Oxynitrides

The oxy-nitride phosphors $MSi_2N_2O_2:Eu^{2+}$ or simply the 1-2-2-2 oxy-nitrides show, as a function of the host lattice composition, emission bands from

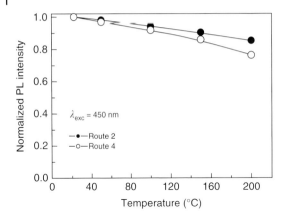

Figure 3.42 Temperature-dependent fluorescence of $Sr_{1.98}Eu_{0.02}Si_5N_8$ which is differently processed under blue light excitation. (Reprinted with permission from [58], Copyright (2006), American Chemical Society.)

bluish to yellowish-green. This is related partially to the fact that the crystal structure changes from monoclinic (M = Ca) to triclinic (M = Sr) and most likely orthorhombic (M = Ba), as well as to the fact that the crystal field experienced by the dopant ions on sites of different sizes is different. These effects result in the differences in the energy of the excitation and emission bands. The dependence of spectroscopic properties on composition was studied by Li et al. [59].

The excitation spectrum of $MSi_2N_2O_2$: Eu^{2+} (Figure 3.43) shows a number of bands which correspond to the different crystal field components of the 5d energy level of the Eu^{2+} ion. The emission spectrum of $CaSi_2N_2O_2$: Eu^{2+} shows a band with a peak maximum at 560 nm and a FWHM of about 100 nm. In comparison, the emission of $SrSi_2N_2O_2$: Eu^{2+} is shifted to 545 nm with a significantly lower FWHM of about 85 nm. In contrast to this, $BaSi_2N_2O_2$: Eu^{2+} has an emission band with a very small FWHM of about 40 nm with a maximum which is located at 495 nm.

The difference in peak emission position for different alkaline earth metal 1-2-2-2 oxy-nitrides is caused by different crystal structures of these materials, which results in different crystal field strengths, as has been pointed out earlier. The Stokes shift decreases from 4200 for M = Ca [60] to 3450 cm^{-1} for M = Sr and to 1130 cm^{-1} for M = Ba [61].

Bachmann et al. [61] showed in detail the effect of Eu^{2+} concentration on emission thermal quenching properties as well as quantum efficiencies for different $MSi_2O_2N_2$ phosphors. By increasing the Eu^{2+} concentration in the host lattice from 0.5 at.% upto 16 at.%, a shift of the emission peak maximum from 535 to 554 nm is observed, while the quantum efficiency decreases from 81% to 58%. Bachmann et al. argued that Eu^{2+} occupies four different sites in the structure in case of the Ca- and Sr-based materials, and, based on this fact, at higher doping concentrations an energy transfer to europium ions which show longer wavelength emission occurs. This then results in redshift of emission.

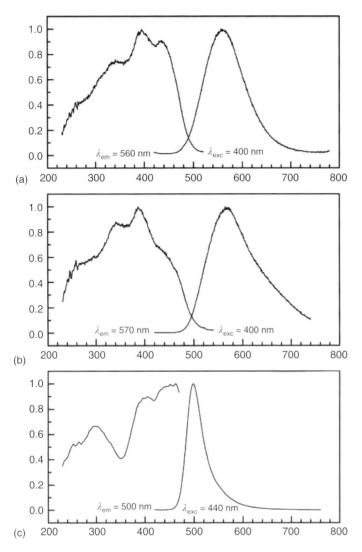

Figure 3.43 Excitation and emission spectra of $MSi_2N_2O_2:Eu^{2+}$ for different alkaline earth metal. (a) M = Ca, (b) M = Sr, and (c) M = Ba. (Reprinted with permission from [59], Copyright (2005), American Chemical Society.)

Furthermore, with higher temperature, an increasing overlap of excitation and emission band is observed, which leads to faster energy migration and enhanced concentration quenching.

In general, $Sr_{0.98}Eu_{0.02}Si_2N_2O_2$ shows very good thermal-quenching properties. Even at 200 °C, an integrated emission intensity of 90% relative to the room temperature value was observed (Figure 3.44).

By gradually replacing Sr with Ba or Ca, the emission wavelength can be tuned to the desired chromaticities. In the case of $Sr_{1-x-y}Ca_xEu_ySi_2O_2N_2$, the emission

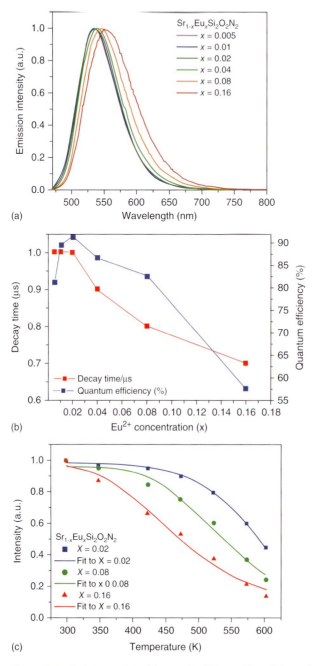

Figure 3.44 Emission spectra (a), emission lifetime (b), and thermal quenching of $Sr_{1-x}Eu_xSi_2N_2O_2$ (c). (Reprinted with permission from [61], Copyright (2009), American Chemical Society.)

wavelength can be continuously tuned between the extremes of $x = 0$ and 1. Up to $x = 0.5$, solid solutions are formed, whereas for the larger x values, two distinct structures were observed in the XRD patterns. In contrast to this, for $Sr_{1-x-y}Ba_xEu_ySi_2O_2N_2$, the emission wavelength can be continuously redshifted from 538 to 564 nm by change of x to 0.75. A change in the crystal structure is observed above $x = 0.75$, which explains the significant blueshift to 495 nm for $x = 1$, when the structure corresponds to a pure Ba 1-2-2-2 oxy-nitride.

3.4.7.4 β-SiAlON

In recent years, a new class of so-called β-SiAlON phosphors came into focus for high color gamut backlighting applications, as this material class offers green fluorescence emission with a small FWHM of about 1690 cm^{-1}. In principle, β-SiAlON has a hexagonal crystal structure and is based on a solid solution of β-Si$_3$N$_4$ in which Si and N atoms are partially replaced by Al and O. Because of restricted solubility, the compositions are limited to $Si_{6-z}Al_zO_zN_{8-z}$ ($0 < z < 4.2$) [62].

In 2005, Hirosaki et al. [63] first reported on the luminescence properties of β-SiAlON. β-SiAlON shows a broad excitation band, which spans from 300 to 500 nm and shows maxima at 303 and 405 nm. It can also be efficiently excited at 450 nm with only a minor loss in excitability compared to the 405 nm excitation. The emission band is located at 535 nm with an FWHM of about 55 nm (Figure 3.45).

The Stokes shift can be estimated to be around 2700 cm^{-1}. Xie et al. [64] reported about the dependency of Si–N-substitution z on the emission wavelength. A redshift in emission ranging from 528 to about 550 nm up to $z = 2.0$ was observed. The redshift is based on lattice expansion following the substitution of shorter Si–N with longer Al–O bonds. This results in a less rigid lattice and

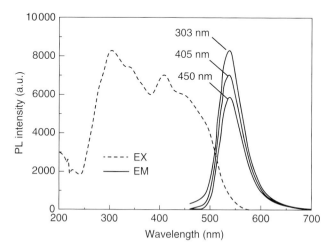

Figure 3.45 Excitation and emission properties of β-SiAlON. (Reprinted with permission from [63], Copyright (2005), AIP Publishing LLC.)

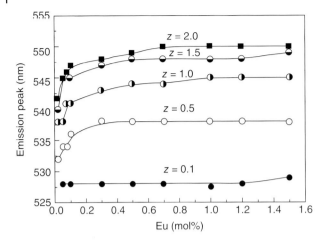

Figure 3.46 Effect of the Eu^{2+} concentration and z-value on the emission wavelength of β-sialon. (Reprinted from [64]. Reproduced by permission of ECS – The Electrochemical Society.)

larger Stokes shift. The critical Eu^{2+} concentration with respect to concentration quenching depends on the value of z. By trend, z-values above 1 show a lower optimal Eu concentration of about 0.3 mol% [64]. In principle, the emission band can be shifted to deeper red by the enhancement of Eu^{2+} concentration. This, in particular, is true for higher z-values, where the effect of activator concentration on emission is much stronger compared to low z-values, especially for $z = 0.1$ – as can be seen in Figure 3.46.

In general, β-SiAlON shows very good thermal-quenching properties. Even at 200 °C, an integrated emission intensity of 75–85% relative to the room temperature value was observed for β-SiAlON with $z = 1.0$ and Eu^{2+} concentrations varying from 1.5 at.% to 0.02 at.%. Therefore, β-SiAlONs are well suited for application in high power LEDs (Figure 3.47).

3.4.8
Phosphor-Coating Methods

The next three sections deal with the question how PC-LEDs could be manufactured and which aspects need to be considered for this.

The most important component (besides the blue-emitting InGaN semiconductor LED chip itself) of a PC-LED is the phosphor layer. The material selection of the phosphor layer defines the color point and has a high influence on the overall efficiency and the reliability of the LED package.

In general, there exist three approaches of bringing a phosphor layer onto an LED package [65]:

- Freely dispensed coating
- Conformal coating
- Remote coating

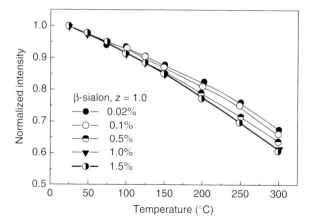

Figure 3.47 Effect of Eu^{2+} concentration on the temperature dependent emission of β-sialon. (Reprinted from [64]. Reproduced by permission of ECS – The Electrochemical Society.)

In the first and the third approaches the phosphor material is usually brought onto the LED by volumetric dispensing methods. The phosphor, which exists in a powder form, is homogenously dispersed in an optical transparent binder material. The binder material consists of a transparent silicone in most cases. The resulting dispersion is then brought onto the LED package by means of suitable dispensing techniques, for example, jet dispensing. After the dispensing step, the optical transparent binder material must be cured under elevated temperature conditions for a certain time. The actual curing conditions depend on the nature of the binder phase material. After curing, the LED package may undergo further production steps such as combination with optics and soldering on circuit boards.

In the case of the second approach (conformal coating), a very thin phosphor layer is uniformly coated directly onto the surface of the LED chip [65]. The key feature of this kind of coating process is the very high precision of the phosphor layer thickness that can be controlled in the micrometer range. One example for a conformal coating process is the electrophoretic method [66]. This method uses charged phosphor particles which are deposited on the LED chip surface. Film thickness is controlled by voltage and deposition time. According to [65], the reduced thickness of the phosphor layer causes a shortened propagation length of the light that in turn decreases the number of redundant scattering. The latter results in an increased lumen output of the LED.

Instead of conformal coating of the LED chip surface with a phosphor powder, there exists also the possibility to apply the phosphor layer in terms of thin preformed sheets. Such a phosphor sheet may consist of a silicone–phosphor foil which is stuck onto the LED chip surface [67]. Besides the flexible phosphor foil, the phosphor layer can also consist of a monolithic ceramic plate that is brought onto the chip [68]. Both methods also allow for very uniform light output because of the high level of homogeneity (thickness, phosphor concentration) of the mentioned phosphor layers.

As aforementioned, phosphor-binder dispersion is used for the remote coating configuration, too. The difference between the freely dispensed coating and conformal coating methods is that the phosphor layer is placed at a certain distance from the chip surface. This configuration leads to increased conversion efficiency [69] and therefore to higher lumen output.

3.4.9
Challenges of Volumetric Dispensing Methods

A disadvantage of the phosphor-coating methods, which are related to volumetric dispensing of a binder–phosphor mixture, is that exact volumetric control of the binder–phosphor-mixture during dispensing is challenging [70]. The volume variations, and therefore the mass variations of the phosphor during the application of the binder–phosphor mixture onto the LED dye lead to a certain spread in the color coordinates of the resulting PC-LEDs. In Figure 3.48, it can be seen that even for high performance dispensing technique (jet dispensing with a piezo valve), the color points of LED packages vary although the dispensed mass of phosphor–silicone mixture is set to identical values in the machine setup.

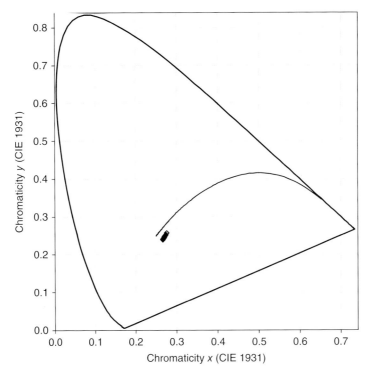

Figure 3.48 Color point variation for jetting dispense of silicone and YAG:Ce phosphor mixture. Color point variation $\Delta x = 0.010$; $\Delta y = 0.018$.

Another behavior, which may lead to a spread of the color points of PC-LEDs, is the tendency of the phosphor particles to settle inside the binder matrix because of gravity, as long as the matrix material is not polymerized. The settling behavior will affect the local concentration of the phosphor that is excited by the blue-emitting LED chip and thus will influence the optical behavior of the PC-LED [71, 72]. However, according to [71], the effect is negligible if the LED dye is conformally coated with a silicone–phosphor matrix, where the concentration variations because of the settling effect are <1%. This effect may need to be taken into account more in case of the syringe with the binder–phosphor-mixture that is used in the dispensing equipment. So phosphor-dispensing velocity increase or enlargement of the syringe volume may be the possible solutions for improving the optical homogeneity of the resulting PC-LEDs.

3.4.10
Influence of Phosphor Concentration and Thickness on LED Spectra

As stated in the previous section, phosphor mass variations affect the color of the emitted light of an LED package. This effect is utilized for the color adjustment during LED manufacturing.

Figure 3.49 [65] shows the variation of a PC-LED spectrum with different phosphor concentrations or thicknesses.

In general, the spectrum of a PC-LED consists of two parts: the blue pumping peak of the LED chip and the phosphor emission part, that is, the photoluminescence emission spectrum of the phosphor under blue excitation.

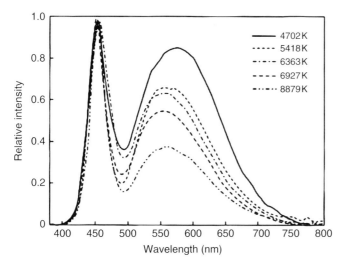

Figure 3.49 Spectra of PC-LEDs with one phosphor. Increasing phosphor concentration or thickness leads to a color point shift toward the color point of the pure phosphor. (Reprinted from [65], © 2011 Chemical Industry Press; Reprinted with permission of John Wiley & Sons, Singapore.)

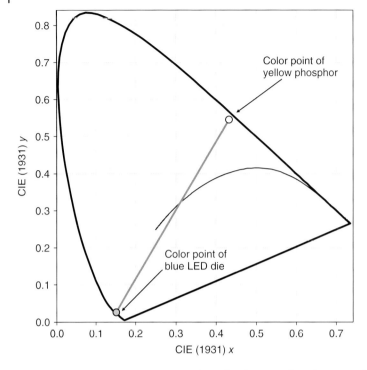

Figure 3.50 Schematic representation of the color addition concept of a PC-LED. The only possible color points of the resulting PC-LED are located on the interconnection line between the color point of the blue LED dye and the color point of the pure yellow phosphor.

The color addition of the blue exciting light and the light emitted from the phosphor (in Figure 3.49 it comes from a yellow YAG : Ce phosphor) turns out to be white light to the human observer, if the resulting color point is on or close to the Planckian locus in the CIE 1931 chromaticity diagram. The phosphor fraction in the PC-LED spectrum depends directly on the amount of phosphor over the LED chip. The amount of phosphor in turn depends on the phosphor concentration in the transparent binder material or the thickness of the phosphor layer, respectively. Figure 3.50 illustrates this fact: with increasing phosphor amount, the color point of the resulting PC-LED spectrum moves from the color point of the the pure blue LED dye toward the color point of the pure phosphor along the interconnection line between both color points.

Not only color points but also the luminous efficiency, depends highly on phosphor concentration and phosphor thickness [73]. The highest luminous efficiencies can be obtained with the combination of higher phosphor thickness and lower phosphor concentration. A similar result is given in [74], where the authors came to the conclusion that a low-concentration phosphor mixture has advantages in output luminous flux of the tested LED package. The behavior is explained by the lower trapping efficiency and less backscattering of light in case of the combination of lower phosphor concentration and higher phosphor thickness [73]. As self-absorption of the low-concentration phosphor mixture is reduced, the luminous flux is increased [74].

Considering the development of PC-LED packages, it is therefore important to find the right compromise between phosphor layer thickness and concentration for the particular color point in order to obtain the highest possible luminous efficiency. However, this compromise is also affected by the constraints of the LED package geometry, optics, radiation power of the pumping light source, and thermal management aspects of the LED package type used.

3.5
Green and Red Phosphor-Converted LEDs

The aim of general lighting is to develop a technologically, and visually effective, environmentally friendly illumination system. The combination of phosphors with blue-emitting LED is a promising solution. Phosphors exhibit broad band emission resulting in a filled spectrum in the visible wavelength range. Associated with blue-pumping LEDs, this technology enables an efficient lighting system and provides light of high quality. The use of several phosphors improves the color rendering of the LEDs. According to these considerations, a selection of four phosphors was chosen to be applied in PC-LEDs. The characteristics of these LEDs concerning the phosphor requirements and the influence of the phosphor on the PC-LEDs are described in this section.

3.5.1
The Phosphor-Converted System

All PC-LEDs include a blue LED which is necessary to produce the emission source coated by the phosphor layer. This chapter considers the configuration of a mixture of phosphors which enhances the chromaticity and color quality of the lighting device. The blue LED excites the phosphors by its emission wavelengths. Emission and absorption properties of the phosphors interact intrinsically with each other and also have an impact on the PC-LED's emission spectrum. The influencing parameters of the PC-LEDs are shown in Figure 3.51. Several interactions occur between the LED and the phosphor and this results in several possible ways of optimization of the PC-LED.

First, the photoluminescence properties of the phosphor are dealt with for a better understanding of the phosphor interactions within the PC-LED. With one phosphor, it is possible to reproduce every color situated on the line joining the blue LED and the phosphor in the CIE x, y chromaticity diagram. Mixing several phosphors offers the advantage of a wider emission spectrum of colors, expanding the color gamut of the light source. Increasing the complexity of the system by introducing several phosphors represents consequently a challenge for the comprehension of the induced interactions. The sample phosphor

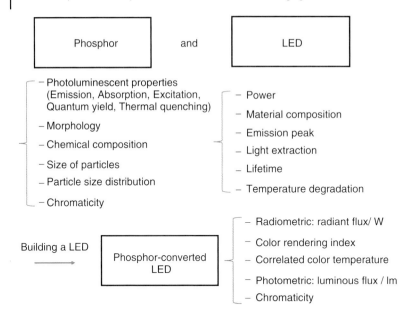

Figure 3.51 Influencing factors of a phosphor-converting system. Optimization target parameters for best visual performance are shown in the right-hand bottom corner.

combinations described in this chapter are composed of two green (G1 and G2) and two red–orange (R1, R2) emitters:

- Lutetium in garnet structure LuAG : Ce, green 1 (G1);
- Orthosilicate, green 2 (G2);
- Oxyorthosilicate, red 1 (R1);
- Oxynitride, red 2 (R2).

The phosphor powder's spectroscopic characteristics determine the PC-LED's radiation. The emission, excitation, and absorption properties of the phosphors are measured separately in the powder state. Emission characteristics are shown in Figure 3.52 for G1, G2, R1, and R2. Phosphors were excited at 440 nm, which corresponds to the wavelength peak of LED emission used during the experiments. G1 exhibits the typical broad emission of LuAG : Ce. The emission maximum peaks at 514 nm. G2 shows a narrower green emission known for *ortho*-silicates with a maximum peak situated at 524 nm. The emission shapes of both red–orange phosphors are similar. They differ by their emission peak located at 615 nm for R1 and 600 nm for R2. The combination of these phosphors provides the possibility to fill the visible emission spectrum continuously.

Excitation spectra were recorded every nanometer and are shown in Figure 3.53. The garnet phosphor G1 differs in the UV by its lower excitation values in comparison to the green *ortho*-silicate G2, oxy-*ortho*-silicate R1, and oxynitride R2. The LuAG (G1) has only one excitation peak in the UV at 345 nm, whereas the other phosphors exhibit high excitation potentials in the UV range. G1 shows an

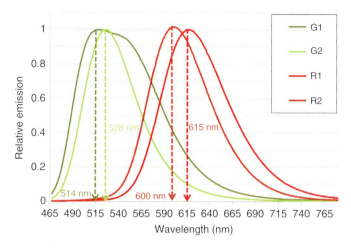

Figure 3.52 Emission spectra of G1, G2, R1, and R2, excitation at 440 nm [75].

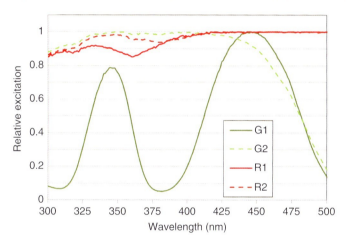

Figure 3.53 Excitation spectra of G1, G2, R1, and R2 [75].

excitation peak in the blue domain at 445 nm. Excitation intensity decreases for G1 after its maximum at 445 nm and for G2 after its maximum at 415 nm in the visible range. In contrast, oxy-*ortho*-silicate and oxynitride show high excitation potential in the visible range and reach their maximum excitation at approximately 420 nm. The emission of the green phosphors is situated in the wavelength range of the excitation of the red–orange phosphors.

3.5.2
Chromaticity Considerations

Four different phosphor mixtures were obtained from the four phosphors described before. The color points of each phosphor are represented in

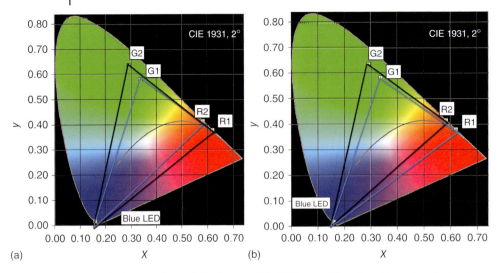

Figure 3.54 x, y chromaticity diagram (CIE, 1931) containing the color gamut obtained from the combinations G1/R2/Blue and G2/R1/Blue (a), G1/R1/Blue, and G2/R2/Blue (b).

Figure 3.54 in the chromaticity diagram. The triangles connecting the green and red phosphors in addition to the blue LED constitute the color gamut illustrating the combination of phosphors. The larger the triangle, the more different chromaticities become possible to be created from these phosphor combinations.

The size of the triangle gives a first feel of the possible quality of the light produced by these phosphor compositions. At first sight, the G2 and R1 combination suggests the best color quality.

3.5.3
Phosphor Mixtures for the White Phosphor-Converted LEDs

PC-LEDs based on these four phosphor combinations were built in the laboratory as real LED light sources. The following procedure was used to produce these PC-LEDs. The phosphors were first poured and stirred homogeneously together with the help of a planetary centrifuged mixer. The appropriate silicone binder quantity was then introduced. The slurry was stirred and spread in the ring on the top of the chip on board (COB) LED emitting at 440 nm. The method used was conventional coating. After one hour curing time, photometric and radiometric LED properties were measured.

White phosphor-converted light-emitting diodes (WLEDs) are defined in this section by their chromaticity points and their CCT. LEDs made from different phosphor combinations are compared: warm white LEDs of CCT from 2700 K are compared with cool white LEDs with CCTs of about 8000 K. Table 3.2 summarizes the matrix of phosphors used to build the PC-LEDs and the corresponding CCT values. The following constraints are required concerning the chromaticity points

Table 3.2 CCT and number of the PC-LEDs (in parentheses) made of green phosphors G1, G2, and red phosphors R1, R2.

	G1	G2
R2	2700 K (3)	2700 K (3)
	3000 K (5)	3000 K (2)
	3500 K (1)	3500 K (1)
	4000 K (1)	4000 K (4)
	5000 K (1)	5000 K (2)
	5500 K (1)	5800 K (1)
	6500 K (1)	6500 K (2)
	7900 K (1)	7500 K (1)
R1	2700 K (2)	2700 K (2)
	3000 K (3)	3000 K (3)
	4000 K (2)	3500 K (1)
	4500 K (2)	4000 K (4)
	5000 K (1)	4500 K (3)
	5500 K (1)	5000 K (1)
	6000 K (2)	6000 K (1)
	6500 K (1)	8000 K (2)
	9000 K (3)	

of these LEDs. A deviation of 100 K to the target CCT is accepted and a distance $\Delta u'v' \leq 0.001$ to the Planckian locus is tolerated. For a CCT exceeding 5000 K, $\Delta u'v'$ is calculated related to a phase of daylight. These tight restrictions considerably reduced the number of those LEDs that complied with these requirements. The number of LEDs built is also indicated in parentheses in Table 3.2.

The influence of the quantity of phosphor plays an important role in the PC-LED's lighting properties. The amount of phosphor in the silicone binder layer, that is, its concentration, determines the chromaticity of the PC-LEDs. Phosphor concentration combines three components described by two parameters: a certain ratio of green/red phosphor and the amount of silicone binder. These significant factors depend on the target CCT. The ratio of green and red phosphors and their dilution in the silicone binder have to be taken into consideration and be determined experimentally. The first step consists in finding the appropriate green and red proportion to match the CCT. The line joining the blue LED and the PC-LED's chromaticity point in the *x*, *y* chromaticity diagram is plotted in Figure 3.55. Depending on the position where this line crosses the Planckian locus, the phosphor ratio had to be calculated. When the line intersects the target *CCT* on the Planckian locus, the correct ratio is reached. In a second step, the phosphor mixture concentration in the silicone has to be determined. On the same principle as for PC-LED with one phosphor, the PC-LED color point moves along the line between the blue LED and the phosphor mixture (see the line [Blue LED-A] in Figure 3.55). Now, the phosphor mixture is considered as a "new" phosphor

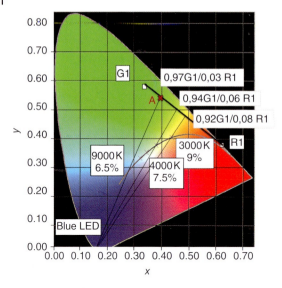

Figure 3.55 CIE diagram of the required G1/R1 phosphor weight contributions in the PC-LEDs for 3000, 4000, and 9000 K and the weight percentage of green and red phosphor mixture in the silicone layer.

having a color point situated between the green and the red phosphors, see the line [G1R1] in Figure 3.55. Decreasing the phosphor mixture concentration in the slurry (with silicone) shifts the PC-LED's color point along this line toward the blue LED. All parameters are finalized when the chromaticity difference $\Delta u'v'$ is less than or equal to 0.002.

Figure 3.55 shows the green and red ratios and their concentrations in the silicone slurry required for the PC-LEDs based on G1 and R1 in the chromaticity diagram. The phosphor concentration depends on the target CCT: for higher CCTs, the green–red phosphor mixture has an increasing green ratio. The ratio increases from 3000 to 9000 K from a weight fraction of 0.92 G1 and 0.08 R1 to 0.97 G1 and 0.03 R1. The proportion in which the green and red phosphors have to be mixed to silicone binder determines the position of the PC-LED chromaticity point on the line linking the color points of the specific phosphor mixture (point A in Figure 3.55) and the blue LED. The closer the white point is to the blue LED, the less phosphor mixture is required. 6.5 wt% is necessary to achieve 9000 K in contrast to 9 wt% for 3000 K.

The dependence of green phosphor concentration as a function of CCT for the different phosphor combinations is plotted in Figure 3.56. Green phosphor proportion increases with the CCT for every phosphor system. Indeed, the warmer the light, the more red component is required and therefore the lower the green concentration in the ratio. A clear predominance of the green proportion in the ratio green/red can be noted, down to 50% for the lowest CCT (or warmest white PC-LEDs). This is because of the photoluminescence properties as green emission is absorbed by the red phosphor. Figure 3.56 shows that G1 concentrations

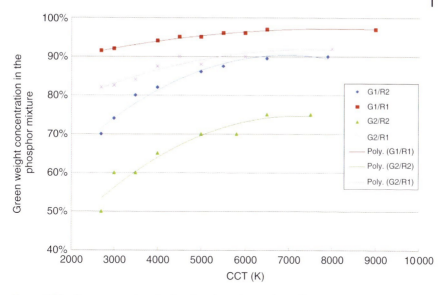

Figure 3.56 Green proportions in the phosphor mixture depending on CCT.

are higher than G2 concentrations required to reach every target CCT. Comparing red phosphor concentrations, we find that the contribution of R2 in PC-LEDs is higher than the contribution of R1.

3.5.4
Colorimetric Characteristics of the Phosphor-Converted LEDs

The efficiency and suitability of the phosphor powders and their mixtures was tested after the LEDs had been built. Efforts were taken that PC-LEDs differ only because of the different phosphors: PC-LEDs were built with the same blue LEDs; they had the same driving currents and similar CCTs in order to better compare the matrix of phosphor systems. Spectral distributions of some PC-LEDs are illustrated as examples in Figures 3.57 and 3.58 for G2/R1 and G2/R2.

Warm white LEDs were obtained by increasing the amount of red phosphor. This leads to a predominant red component in the spectral power distribution. As red phosphors were excited by and absorbed green phosphor radiation, the amount of green phosphor still had to be high in order to achieve the right color temperature. Both phosphors required the excitation from the blue LED and absorbed its emission. This resulted in a very low blue emission. When the white tone of the LEDs becomes colder, less red phosphor is necessary. The green emission increases in importance and the blue emission even more so. Cool white LEDs are characterized by a reduction of the red part; thus blue, green, and red emissions come closer to equilibrium.

Light and color quality characteristics are listed in Tables 3.3 and 3.4. As can be seen, the white points of the PC-LEDs were located between $\Delta u'v' = 0.4 \times 10^{-3}$

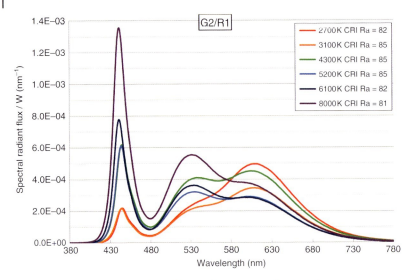

Figure 3.57 Spectral radiant flux of the PC-LEDs made of the G2/R1 system for several CCTs [75].

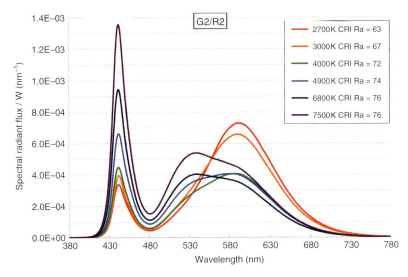

Figure 3.58 Spectral radiant flux of the PC-LEDs made of the G2/R2 system for several CCTs.

and 7×10^{-3} from the Planckian locus. The CRI R_a depends on the type of the mixture, see Table 3.5.

As can be seen from Tables 3.4 and 3.5, the red phosphor R1 exhibits higher values for CRI R_a and R_9 than the red phosphor R2. The CRI R_9 is a critical parameter indicating how accurate the red tones can be rendered by illumination with the

Table 3.3 Luminous efficacy, distance to Planckian locus, and color quality for G2/R1 [75].

CCT (K)	2700	3000	3500	4000	4500	5000	6000	8000
$\Delta u'v' \cdot 10^{-3}$	1.37	1.02	1.22	0.47	0.66	4.98	3.41	2.71
CRI R_a	82	84	84	85	85	85	82	81
R_9	14	23	32	38	43	48	48	49

Table 3.4 Luminous efficacy, distance to Planckian locus, and color quality for G2/R2 [75].

CCT (K)	2700	3000	3500	4000	5000	5800	6500	7500
$\Delta u'v' \cdot 10^{-3}$	1.72	1.00	1.56	1.07	6.31	2.02	2.02	3.72
CRI R_a	63	67	70	72	74	75	75	76
R_9	−58	−28	−39	−31	−20	−16	−11	−4

light source under test. Shifting the red emission from 600 to 620 nm improves substantially the quality of the light source. The combination of G2 and R1, an *ortho*-silicate that peaks at 524 nm and an oxynitride that peaks at 620 nm, gives better color-rendering results.

Only the phosphor combinations G2/R1 at all CCT except 2700, 6000, and 8000 K exhibit moderate to good general color rendering (≥84). This is because of the fact that the two red–orange phosphors (R1, R2) do not emit enough in the red–deep red spectral range (610–660 nm) hence they do not provide enough red spectral content to illuminate important reddish objects. The influence of the red phosphor has a significant impact on the quality of the white light. The combination of G1 and R2 with a peak wavelength for the red phosphor which is shifted from 600 to 630 nm (toward longer wavelength), increases the CRI R_a values at 2700 K from 61 (low color rendering, see Table 5.5) to 86 (at the lower limit for good color rendering, see Table 5.5) and the CRI for red surface objects R_9 can be enhanced from −69 to 72 (moderate).

To summarize Section 3.5, the use of a selected combination of phosphors enabled to fulfill white point requirements enhances the CRI (up to a moderate–good level: for high quality color rendering, advanced red phosphors emitting at longer wavelengths can be used) and enforce the required color temperature. The choice of broadband emitting phosphors distributed over the whole visible range of wavelengths assures the possibility to fine-tune the CCT of the PC-LEDs by varying phosphor concentration. The spectral emission characteristics of the phosphors used in the mixture represent a relevant factor to enhance the color quality of the white PC-LEDs.

Table 3.5 Color rendering index (CRI R_a) and CCT of the G1, G2, R1, and R2 phosphor systems [75].

	G1		G2	
	CCT (K)	CRI R_a	CCT (K)	CRI R_a
R2	2700 K	61	2700 K	63
	3000 K	65	3000 K	67
	3500 K	70	3500 K	70
	4000 K	71	4000 K	72
	5000 K	73	5000 K	74
	5500 K	75	5800 K	75
	6500 K	76	6500 K	75
	7900 K	76	7500 K	76
R1	2700 K	78	2700 K	82
	3000 K	79	3000 K	84
	4000 K	83	3500 K	84
	4500 K	82	4000 K	85
	5000 K	83	4500 K	85
	5500 K	83	5000 K	85
	6000 K	83	6000 K	82
	6500 K	81	8000 K	81
	9000 K	81		

3.6
Optimization of LED Chip-Packaging Technology

From 1997, the luminous efficacy, optical efficiency, and color quality have been rapidly increasing allowing broader applications for the LED devices and not only in display and signaling technology. The main application field of LED technology has become general lighting, made possible by the development of LED chip technology, phosphor technology, and packaging technology because of the optimization of both the materials and the manufacturing processes. In order to develop reliable, long-term stable, economic, and environment-friendly LEDs for high-quality applications with acceptable lighting and color quality, both in industrial nations and developing countries, the following aspects must be taken into account:

- LED systems must allow a lifetime of longer than 12 years, also under harsh weather and ambient conditions. For street lighting, it means a lifetime of about 50 000 h. In order to fulfill these requirements, the LED die, the package, and all other components of the LED light source (electronics, mechanical accessories, optics, and luminaire housing) have to be optimized for the highest reliability.
- A LED luminaire or a general LED lighting system can be regarded only as a replacement for a conventional lighting system if the luminous efficiency, the lighting distribution on the illuminated area, and the color quality are better

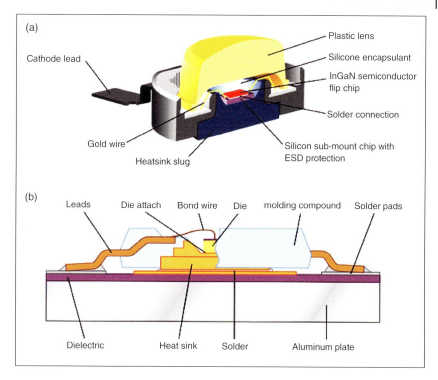

Figure 3.59 Structure of an LED device. (reproduced with permission from OSRAM Opto Semiconductors)

than those of the conventional system. Therefore, the optimization of quantum efficiency, conversion efficiency, and luminous intensity distribution of the light source is compulsory.
- The price and investment for a lighting installation should fulfill the expectation of the users so that, parallel to the improvement of reliability and luminous efficacy, cost optimization of the materials, fabrication equipment, and process organization in LED technology can have the highest priority.

To achieve these goals, one main focus of LED optimization in semiconductor industry is *packaging technology*. Generally, the structure of an LED package consists of the following components (see Figure 3.59):

- LED die which is soldered or mounted on a substrate (a submount with ESD protection) with a silver adhesive;
- substrate as ceramic or metal (silicon) functioning as heat spreader conveying the generated thermal energy away from the chip location;
- substrate mounted on the heat sink (ceramic, copper, aluminum);
- bond wires as LED interconnection components for electrical connections (silver, gold);

- silicone encapsulant which has the task of protecting the chip against external mechanical and environmental influences. In case of a white LED, the phosphor systems (powder) and the viscose silicone encapsulant are homogenously mixed and mounted in front of the bare chip surface;
- primary optics as plastic lens.

This general structure provides the reader an overview about how an LED package can be built in a conventional manner. In the process of the improvement of the technical performance and cost of LEDs, the following questions arise:

- How can the optical radiation of the LED die be extracted and which methods and tools can be used to achieve maximal extraction?
- How can the phosphor system – together with silicone encapsulant – be placed in front of the LED chip so that the absorption, thermal energy, and temperature inside the phosphor layer and the interactions between the phosphor system and the dark surface of the chips be minimized as well as the radiant power (or its corresponding photometric quantity, luminous flux) and the angular distribution homogeneity of CCT be maximized?
- How can the reliability of the LED system and its lifetime be improved by the reduction of thermal load inside the system and the thermal resistance from the junction through the heat sink and the surroundings?

In this context, the next section deals with the characteristics of the LED chip, the phosphor systems (form, geometry), and the substrate with different configurations of the chip/phosphor arrangements.

3.6.1
Efficiency Improvement for the LED Chip

The optical radiation is generated by combinations between electrons and holes and emitted mainly in two directions to the upper interface between the semiconductor material and the air and to the lower surfaces between the semiconductor material and the substrate which can be very dark and substantially absorb the oncoming radiation, for example, in the case of the GaAs substrate. On the upper side, the rays fall on the surface at different incidence angles. Because the refractive index of most semiconductor materials is between 2.4 and 3.5 and higher than the refractive index of the air, total reflection is expected, see Figures 3.11 and 3.12. The refraction law of Snellius at the interface between two optical media is shown in Eq. (4.50).

$$n_1 \cdot \sin \alpha_1 = n_2 \cdot \sin \alpha_2 \tag{4.50}$$

With $n_1 = 3.5$, $n_2 = 1.0$ and with $\alpha_2 = 1.0$ in case of the first total reflection, the critical angle (or limit angle) for total reflection is shown in Eq. (4.51).

$$\alpha_1 = \alpha_{\lim} = 17° \tag{4.51}$$

Table 3.6 Refractive index of materials used in white LEDs [76, 77].

Material	Refractive index
Encapsulant silicone	1.5–1.6
p-GaN	about 2.45
n-GaN	about 2.42
Phosphor	1.8

From a theoretical emission angle of 90°, only 17° or only 5% of the maximally possible radiant power can be extracted from the surface between the material and air while the remainder undergoes a light guide path and is absorbed on the way to the side of the material. Coming to the side surface, the effect of total reflection can also be seen so that only a small fraction of the radiation can pass through the side surface into the surroundings. In order to improve extraction efficiency, four methods have been applied in the past decade:

- Molding of a silicone encapsulant with a refractive index being between the one of the semiconductor material and of the air. The angle range with no total reflection is increased. The refractive indices of some LED materials are listed in Table 3.6.
- The upper surface is engineered with optical or chemical methods to achieve a roughness with a certain profile so that the number of angles smaller than 17° is increased to permit much more radiation to be emitted into the surroundings (see Figure 3.60).
- The dark and absorbing substrate GaAs can be replaced by a transparent substrate (e.g., GaP) with less absorption or a mirror layer between the semiconductor layer, and the carrier substrate can be coated to reflect the radiation coming from the pn-junction back to the interface between the epitaxy layer and the air (see Figure 3.60). Modern LED devices have

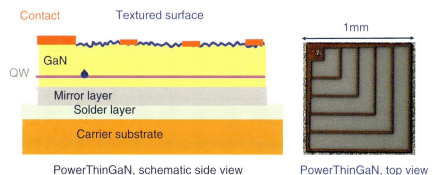

Figure 3.60 Profiling the chip surface to reduce the total reflection and metallization layer between the epitaxy layer and substrate. (reproduced with permission from OSRAM Opto Semiconductors)

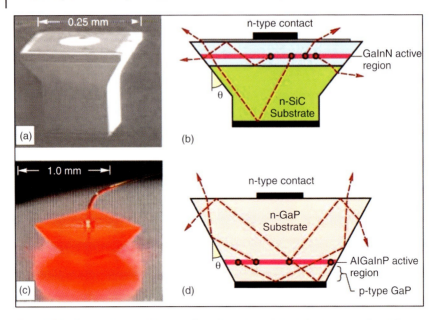

Figure 3.61 Inverted pyramid forming the side surface of the LED chip, reproduced from [78], copyright © 2006 with permission from OSRAM Opto Semiconductors

silver as mirror material because of its high reflectance in the visible range.
- In order to decrease the incidence angle for the optical rays arriving at the side surface, the form of the chip can be configured like an inverted pyramid, according to Figure 3.61.

At the lower side of a conventional LED, the whole substrate is electrically conductive and on the upper surface, radiation is emitted from only a part of the area, because the electrode occupies a fraction of the area on the surface. In order to make the whole upper surface free for the emission of light, the electrode must be mounted to the substrate side below. This so-called flip-chip technology is illustrated in Figure 3.62 (right). By using numerous pin electrodes, the total current of the LED (350 or 700 mA) is more uniformly distributed over the whole active area of the chip so that the current density can be reduced, improving the probability of a radiative recombination between electrons and holes. The consequence of this positive effect is an approximate proportionality between current and luminous flux. This means that a current increase raises the luminous flux almost proportionally.

3.6.2
Molding and Positioning of the Phosphor System

For a predefined set of chromaticity coordinates (x, y) and therefore, for a fixed CCT, the necessary masses of the phosphor system can be calculated (this is a

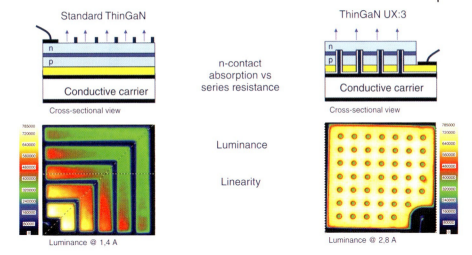

Figure 3.62 Schemes of INGaN/GaN thin-film flip-chip LED, reproduced from [79], copyright © 2012 with permission from OSRAM Opto Semiconductors.

one-dimensional correspondence between the blue LED and one phosphor or a two-dimensional correspondence between a blue LED and a red/green phosphor mixture); this phosphor mixture can be mixed with the silicone encapsulant and then dispensed onto the LED chip. In principle, four configurations of phosphor molding and positioning can be taken into consideration:

1) *Gob-top configuration*: The phosphor mixture builds a hemispherical volume on the LED chip (Figure 3.63). The disadvantage of this "*gob-top*" configuration is the different pathway length of the optical radiation for different angular directions so that the pumping blue LED radiation is absorbed in a different manner and degree causing an angle-dependent inhomogeneity of color characteristics.
2) The *in-cup configuration* of the phosphor mixture can be seen in Figure 3.64a. As can be seen from Figure 3.64a, the space for filling the phosphor mixture has the form of a trapeze (i.e., a cup) in which the chip can be mounted and bonded. The side surface of the trapeze can be coated with highly reflecting materials (e.g., metal, PTFE). In the cup volume, the absorption of the blue LED radiation is also different, depending on the emitting direction which causes an inhomogeneous color and CCT distribution. An alternative version (so-called remote phosphor) can be seen in Figure 3.64b. The latter is discussed later.
3) *Conformal coating configuration* (Figure 3.65a): In order to improve the uniformity of the angular color distribution and to keep the chromaticity coordinates of the manufactured white LEDs within certain limits despite the different peak wavelengths of the blue-pumping LEDs, a thin plate of phosphor mixture with a predefined phosphor concentration can be positioned directly

124 | *3 LED Components – Principles of Radiation Generation and Packaging*

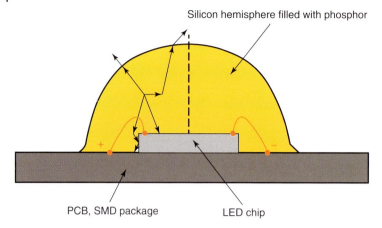

Figure 3.63 Gob-top configuration, redrawn after [80].

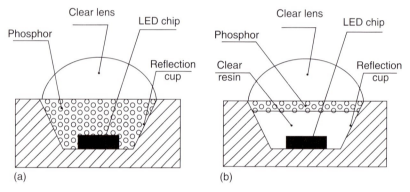

Figure 3.64 In-cup configuration (a) and remote phosphor (b), reproduced from [81], copyright © 2009 with permission from IEEE.

in front of the LED chip surface. The thickness of this plate is selected automatically depending on the peak wavelength of the blue LED so that the deviations of the chromaticity coordinates of all white LEDs within a batch can be kept lower than a defined limit. This configuration represents a large number of current LED types available on the market (see Figure 3.65b).

Before moving on to the next (fourth) phosphor configuration (the remote phosphor), how the absorption inside the phosphor/silicone mixture and the interaction between the phosphor layer and the LED chip takes place is explained. The blue radiation of the LED chip is emitted in the 2π sr hemispherical room and enters into the phosphor/silicone volume, see Figures 3.64 and 3.66. A small fraction of the radiation is absorbed in the silicone layer. A greater amount is absorbed by the phosphor particles and a part is scattered by reflections at the interfaces between the phosphor particles and silicone and undergoes a multiple reflection/absorption at the other phosphor particles or returns back onto the

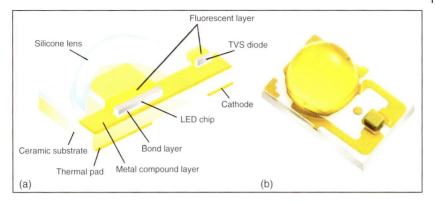

Figure 3.65 Conformal coating method (a) and the Rebel© LED of Philips/Lumileds (b).

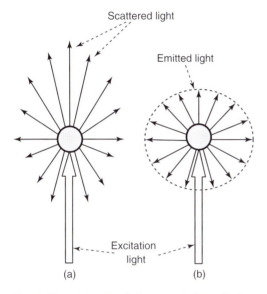

Figure 3.66 Interaction between excitation radiation and phosphor particles, reproduced from [81], copyright © 2009 with permission from IEEE.

dark surface of the LED chip contributing to the additional thermal load of the chip caused by absorption. The absorbed energy in the phosphor particles is converted partly to optical radiation with longer wavelengths (e.g., red, green, or yellow radiation) and to thermal energy. The energy of the blue radiation is distributed in the LED light source as follows: one fraction of it remains blue; another part is converted into radiation of other (visible) wavelengths; and the last part of it is transformed into thermal energy.

Generally, the thermal energy exchange between the phosphor/silicone volume and the surroundings is performed only by a little part in terms of convection and radiation but by a greater part in terms of conduction. Because silicone has

Table 3.7 Thermal conductivity of different materials used in LED packaging technology. [S. Kobilke, Excelitas Technologies Elcos GmbH, private communication, 2014.]

No.	Materials	Thermal conductivity k (W (m K)$^{-1}$)
1	Conductive glue A	1.7
2	Conductive glue B	10
3	Solder	50
4	Al_2O_3 ceramic	25
5	AlN ceramic	160
6	Cu	400
7	Al	240
8	Dielectric layer A	2.2
9	Dielectric layer B	22

Source: Reproduced with permission from Excelitas Technologies Elcos GmbH.

a rather low thermal conductivity (see Table 3.7), the phosphor/silicone volume quickly acquires a high temperature and can dispose of its thermal energy only at the conductive part, that is, through the chip. If the chip has a higher temperature because of a higher current, the phosphor layer temperature is correspondingly high and overloads the chip causing the accelerated yellowing effect of the silicone encapsulant, on the one hand, and the aging of the chip, on the other hand. Solutions to this problem include the following:

1) The phosphor layers from the chip are decoupled so that the thermal energy of the phosphor layer escapes through a different path and not through the chip.
2) The phosphor system can be optionally mixed with the silicone encapsulant with low thermal conductivity, with plastic materials such as polycarbonate (PC) with a high yellowness index, with glass of rather low thermal conductivity or, possibly in the future, with ceramic as a carrier material.

Turning back to the remote phosphor issue, its principal arrangement (based on the above ideas) is illustrated in Figure 3.64b.

In Figure 3.67, the energy budget and temperature of an example of conformal coating, in-cup configuration, and remote phosphor are shown.

The following can be concluded from Figure 3.67:

1) The temperature of the junction is nearly the same in the three configurations.
2) The temperature of the phosphor mixture is generally much higher than that of the junction.
3) In case of *conformal coating*, the phosphor plate is relatively thin so that the thermal energy has only a short pathway to be transported to the chip material with higher thermal conductivity. In the *in-cup* configuration, the phosphor volume is thick so that the absorbed energy in the center of the cup causes a very high temperature.

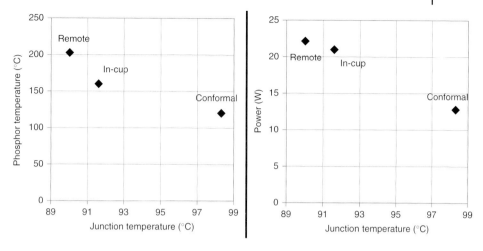

Figure 3.67 Phosphor temperature, junction temperature, and power output with different phosphor configurations after [76].

4) The temperature of the remote phosphor can be very high if thermal conduction is not optimized. It can be improved by a proper construction of the remote phosphor holder ring and by intelligent selection of the phosphor-carrying materials.

3.6.3
Substrate Technology – Integration Degree

Two important tendencies in the focus of the development of LED technology can be formulated as follows:

- The current state of the art is *single LED* in *one* housing, see Figure 3.65b. According to this principle, the LED's luminous flux is limited (e.g., a single LED with 700 mA, 3.1 V, and with 130 lm W^{-1} can deliver only a luminous flux of 280 lm). In order to achieve more luminous flux for the development of LED retrofit lamps for household, a street light luminaire, or especially for a vehicle headlamp unit, the trend goes to *multichip LEDs* with four or six chips with several thousand lumen (e.g., 1000 or 2000 lm).
- If the illumination tasks require considerably more luminous flux (e.g., for a down light with 5000 lm, for an industrial hall or an airfield lighting system with 20 000 lm, for TV-studio lighting systems replacing a HMI discharge lamp with 575 W and with 57 500 lm), LED users need a luminous flux package which can be realized only if using an array of more than 50–100 LED chips mounted in a densely integrated way. For efficient optical imaging, the active emitting area of the light source has to be as small as possible even with the high number of the LED devices. Therefore, an arrangement of the LED housings as a group is not meaningful. This consideration resulted in the idea of the so-called

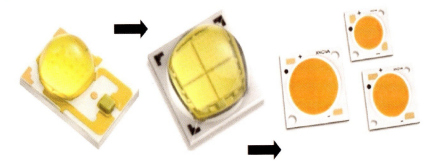

Figure 3.68 Single chip, multichip, and chip-on-board or COB. (Image source: datasheets of Philips/Lumileds, Cree, Citizen.)

COB technology. The principle of higher integration degree is illustrated in Figure 3.68.

In a single chip, only one chip is mounted. In a multiple-chip configuration, several chips are mounted in a housing, while in the COB configuration, an array of chips is mounted on a board. Besides the higher integration of the electronics and interconnects between the chips, thermal energy is much higher for the whole board and *thermal energy density* increases rapidly; consequently, the thermal load for the adhesive, solder material, board materials, and connectors increases. If the thermal resistance of the whole chain from the chips to the board surroundings is high then junction temperature can be higher than the maximally permissible temperature for the chips.

From these reasons, optimizations of the layer structure inside the LED package and of the materials are absolutely necessary. For the optimization of the LED package (single chip, multichip, or COB) the following materials are available (see Table 3.7 and Figure 3.69).

Analyzing the data in Table 3.7 and in Figure 3.69, some conclusions can be made:

1) Silicone and epoxy have a very low thermal conductivity. This fact has already been mentioned in the discussions on the phosphor/silicone encapsulant.
2) Sapphire and silicone as the submounts for the die have different thermal conductivities.
3) The LED die can be mounted on the board with a conductive glue of lower thermal conductivity or better with soldering having a higher conductivity.
4) As heat sink and heat spreader materials, copper is better than aluminum with the higher specific weight density.
5) FR4 has a very low thermal conductivity and is not suitable as a material for printed circuit boards (PCBs) of high-power LEDs. The standard configurations of many LED- PCBs are metal-core boards with different dielectric materials and structures.

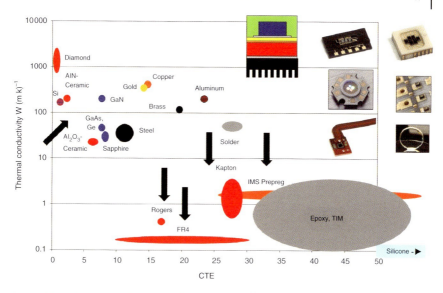

Figure 3.69 Thermal conductivity and coefficient of thermal expansion (CTE) of different LED materials. Reproduced with permission from Excelitas Technologies Elcos GmbH

In a thermal chain network from the inner side with the junction layer of the die to external components such as the Al heat sink, there is a series of thermal resistances, and the bottleneck of thermal transport through this chain is the relatively low thermal conductivity of the adhesive, the isolation, and the conductive glue although the heat sink made of copper or aluminum can also be optimally used in this chain. The principle of reducing thermal resistance is using materials that have both a good electrical isolating feature and high thermal conductivity. One of the materials used today is ceramic in different chemical compositions, see Table 3.7. The AlN ceramic is relatively expensive but it has a much higher thermal conductivity than the Al_2O_3 ceramic.

To compare different packaging materials for the realization of a COB module (e.g., with four LED chips, 1 A per chip and 14 W total power, see Figure 3.70), it can be seen that

1) there is a significant temperature difference ($dT = 15–20\,°C$, see the ordinate of Figure 3.70) for the case of Al_2O_3-ceramic and AlN ceramic with standard chip contact (i.e., with the conductive glue).
2) with chips soldered on the AlN ceramic, however, the lowest temperature difference can be achieved ($dT < 5\,°C$).

In modern COB realizations, the phosphor–silicone encapsulant mixture is molded on the chip array surface. The next steps for reducing the thermal load will be the use of a remote phosphor system optimized for an efficient separate thermal conduction and arranged at a certain vertical distance from the chip array plane.

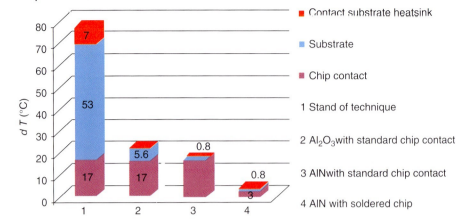

Figure 3.70 Comparison of the different packaging materials. (Reproduced with permission from Excelitas Technologies Elcos GmbH.)

References

1. Zukauskas, A., Shur, M.S., and Gaska, R. (2002) *Introduction to Solid State Lighting*, John Wiley & Son, Inc., pp. 37–55.
2. Nakamura, S. and Chichibu, S.F. (1999) *Introduction to Nitride Semiconductor Blue Lasers and Light Emitting Diodes*, CRC Press, pp. 1–25.
3. Schubert, E.F. (2003) *Light Emitting Diodes*, Cambridge University Press, pp. 1–112.
4. Müller, G. (2000) *Electroluminescence I – Semiconductors and Semimetals*, Vol. 64, Academic Press, pp. 61–66.
5. Müller, G. and Müller, R. (2007) Color conversion of LED light. Phosphor Global Summit, 2007, pp. 4–5.
6. Windisch, R., Buntendeich, R., Illek, S., Kugler, S., Wirth, R., Zull, H., and Strubel, K. (2007) *IEEE Photonics Technol. Lett.*, **19** (10), 774–776.
7. Nakamura, S. and Fasol, G. (1997) *The Blue Laser Diode: GaN Based Light Emitters and Lasers*, Springer, Berlin.
8. Nakamura, S. *et al.* (1995) *Jpn. J. Appl. Phys.*, **34**, L797.
9. Shionoya, S. and Yen, W.M. (1998) *Phosphor Handbook*, CRC Press, New York.
10. Blasse, G. and Grabmaier, B.C. (1994) *Luminescent Materials*, Springer, Berlin.
11. Ronda, C. (2007) *Luminescence: From Theory to Application*, Wiley-VCH Verlag GmbH, Weinheim.
12. Katai, A. (2008) *Luminescent Materials and Applications*, John Wiley & Sons, Ltd.
13. Qasim, K., Lei, W., and Li, Q. (2013) *J. Nanosci. Nanotechnol.*, **13**, 3173.
14. Judd, B.R. (1962) *Phys. Rev.*, **127**, 750.
15. Ofelt, G.S. (1962) *J. Chem. Phys.*, **37**, 511.
16. Dorenbos, P. (2000) *Lumin. J.*, **91**, 155.
17. Dorenbos, P. (2003) *Lumin. J.*, **104**, 239.
18. Dorenbos, P. (2008) *Lumin. J.*, **128**, 578.
19. Zych, A., Ogieglo, J., Ronda, C., de Mello Donegá, C., and Meijerink, A. (2013) *Lumin. J.*, **134**, 174.
20. Thiel, C.W., Cruguel, H., Wu, H., Sun, Y., Lapeyre, G.J., Cone, R.L., Equall, R.W., and Macfarlane, R.M. (2001) *Phys. Rev. B*, **64**, 085107.
21. Dorenbos, P. (2005) *Lumin. J.*, **111**, 89.
22. Smet, P.F., Parmentier, A.B., and Poelman, D. (2011) *J. Electrochem. Soc.*, **158**, R37–R54.
23. Lempicki, A. and Wojtowicz, A.J. (1994) *Lumin. J.*, **60–61**, 942.
24. Tavernier, S., Gektin, A., Grinyov, B., and Moses, W.W. (eds) (2006) *Radiation Detectors for Medical Applications*, Springer, Berlin.

25. Dorenbos, P. (2003) *J. Phys. Condens. Matter*, **15**, 4797.
26. Schlotter, P., Baur, J., Hielscher, C., Kunzer, M., Obloh, H., Schmidt, R., and Schneider, J. (1999) *Mater. Sci. Eng., B*, **59**, 390.
27. Ivanovskikh, K.V., Ogieglo, J.M., Zych, A., Ronda, C.R., and Meijerink, A. (2013) *ECS J. Solid State Sci. Technol.*, **2**, R3148.
28. Zych, A., de Lange, M., de Mello Donegá, C., and Meijerink, A. (2012) *J. Appl. Phys.*, **112**, 013536.
29. Müller-Mach, R. and Müller, G.O. (2000) *Proc. SPIE*, **3938**, 30.
30. Moriga, T., Sakanaka, Y., Miki, Y., Murai, K.-I., and Nakabayashi, I. (2006) *Int. J. Mod. Phys. B*, **20**, 4159.
31. Blasse, G. and Bril, A. (1967) *J. Chem. Phys.*, **47**, 5139.
32. Smet, P.F., Moreels, I., Hens, Z., and Poelman, D. (2010) *Materials*, **3**, 2834.
33. TOM Phosphor Information Leaflet, https://www.fh-muenster.de/fb1/downloads/personal/juestel/juestel/Phosphor_Information_Leaflet_L-S3-05-Eu_MgS-Eu_.pdf (accessed 06 June 2014).
34. Xia, Q., Batentschu, M., Osvet, A., Winnacker, A., and Schneider, J. (2010) *Radiat. Meas.*, **45**, 350.
35. Hu, Y., Zhauang, W., Ye, H., Zhang, S., Fang, Y., and Huang, X. (2005) *J. Lumin.*, **111**, 139.
36. Shin, H.H., Kim, J.H., Han, B.Y., and Yoo, J.S. (2008) *Jpn. J. Appl. Phys.*, **47**, 3524.
37. Avci, N., Musschoot, J., Smet, P.F., Korthout, K., Avci, A., Detavernier, C., and Poelman, D. (2009) *J. Electrochem. Soc.*, **156**, J333.
38. Yoo, S.H. and Kim, C.K. (2009) *J. Electrochem. Soc.*, **156**, J170.
39. Park, I.W., Kim, J.H., Yoo, J.S., Shin, H.H., Kim, C.K., and Choi, C.K. (2008) *J. Electrochem. Soc.*, **155**, J132.
40. Lin, J., Huang, Y., Bando, Y., Tang, C.C., and Golberg, D. (2009) *Chem. Commun.*, **43**, 6631.
41. Jia, D.D. and Wang, X.J. (2007) *Opt. Mater.*, **30**, 375.
42. Barry, T.L. (1968) *J. Electrochem. Soc.*, **115**, 1181.
43. Yamada, M., Naitou, T., Izuno, K., Tamaki, H., Murazaki, Y., Kameshima, M., and Mukai, T. (2003) *Jpn. J. Appl. Phys.*, **42**, L20.
44. Park, J.K., Lim, M.A., Choi, K.J., and Kim, C.H. (2005) *J. Mater. Sci.*, **40**, 2069.
45. Jang, H.S., Won, Y.-H., Vaidyanathan, S.D., Kim, H., and Jeon, D.Y. (2009) *J. Electrochem. Soc.*, **156**, J138.
46. Kang, E.-H., Choi, S.-W., and Hong, S.-H. (2012) *ECS J. Solid State Sci. Technol.*, **1**, R11.
47. Aitasalo, T.M., Hölsä, J., Lastusaari, M., Niittykoski, J., and Pellé, F. (2005) *Opt. Mater.*, **27**, 1511.
48. Uheda, K., Hirosaki, N., Yamamoto, Y., Naito, A., Nakajima, T., and Yamamoto, H. (2006) *Electrochem. Solid-State Lett.*, **9**, H22–H25.
49. Xie, R.J. and Hirosaki, N. (2007) *Sci. Technol. Adv. Mater.*, **8**, 588–600.
50. Uheda, K., Hirosaki, N., and Yamamoto, H. (2006) *Phys. Status Solidi*, **203**, 2712–2717.
51. Watanabe, H., Yamane, H., and Kijima, N. (2008) *J. Solid. State Chem.*, **181**, 1848–1852.
52. Mueller-Mach, R., Mueller, G., Krames, M.R., Höppe, H.A., Stadler, F., Schnick, W., Juestel, T., and Schmidt, P. (2005) *Phys. Status Solidi A*, **202**, 1727–1732.
53. Piao, X., Machida, K., Horikawa, T., Hanzawa, H., Shimomura, Y., and Kijima, N. (2007) *Chem. Mater.*, **191**, 4592–4599.
54. Schlieper, T. and Schnick, W. (1995) *Z. Anorg. Allg. Chem.*, **621**, 1037–1041.
55. Hoeppe, H.A., Lutz, H., Morys, P., Schnick, W., and Seilmeier, A. (2000) *J. Phys. Chem. Solids*, **61**, 2001–2006.
56. Li, Y.Q., de With, G., and Hintzen, H.T. (2008) *J. Solid State Chem.*, **181**, 515–524.
57. Li, Y.Q., van Steen, J.E.J., van Krevel, J.W.H., Botty, G., Delsing, A.C.A., DiSalvo, F.J., de With, G., and Hintzen, H.T. (2006) *J. Alloys Compd.*, **417**, 273–279.
58. Xie, R.-J., Hirosaki, N., Suehiro, T., Xu, F.-F., and Mitomo, M. (2006) *Chem. Mater.*, **18**, 5578–5583.
59. Li, Y.Q., Delsing, A.C.A., de With, G., and Hintzen, H.T. (2005) *Chem. Mater.*, **17**, 3242–3248.

60. Gu, Y., Zhang, Q., Li, Y., and Wang, H. (2009) *J. Phys. Conf. Ser.*, **152**, 012083.
61. Bachmann, V., Ronda, C., Oeckler, O., Schnick, W., and Meijerink, A. (2009) *Chem. Mater.*, **21**, 316–325.
62. Wakihara, T., Saito, Y., Tatami, J., Komeya, K., Meguro, T., Fukuda, Y., Matsuda, N., and Asai, H. (2009) *Key Eng. Mater.*, **403**, 141–144.
63. Hirosaki, N., Xie, R.-J., Kimoto, K., Sekiguchi, T., Yamamoto, Y., Suehiro, T., and Mitomo, M. (2005) *Appl. Phys. Lett.*, **86**, 211905.
64. Xie, R.-J., Hirosaki, N., Li, Q.Y., and Mitomo, M. (2007) *J. Electrochem. Soc.*, **154**, J314–J319.
65. Liu, S. and Luo, X. (2011) *LED Packaging for Lighting Applications*, John Wiley & Sons, Ltd, Singapore, ISBN 978-0-470-82783-3.
66. Colins, W.D., Krames, M.R., Verhoeckx, G.J., and van Leth, N.J.M. (2001) Using electrophoresis to produce a conformally coated phosphor-converted light emitting semiconductor. US Patent 6,576,488, Lumileds Lighting.
67. Holzer, P., Ewald, K.M., Eberhard, F., and Richter, M. (2004) Lumineszenzdiodenchip mit einer Konverterschicht und Verfahren zur Herstellung eines Lumineszenzdiodenchips mit einer Konverterschicht. German Patent DE 10 2004 047 727 A1, OSRAM Opto Semiconductor.
68. Mueller, G.O., Mueller-Mach, R.B., Krames, M.R., Schmidt, P.J., Bechtel, H.-H., Meyer, J., de Graaf, J., and Kop, T.A. (2004) Luminescent Ceramic for a light emitting device. US Patent 7, 361, 938, Philips Lumileds Lighting Company.
69. Kim, J.K., Luo, H., Schubert, E.F., Cho, J., Sone, C., and Park, Y. *Jpn. J. Appl. Phys.*, **44** (21), L 649–L 651.
70. Quinones, H., Sawatzky, B., and Babiarz, A. (2013) Silicone-Phosphor Encapsulation for High Power White LEDS. ASYMTEK, *http://www.nordson.com/en-us/divisions/asymtek/Documents/Papers/2008_01_SiPhosphor%20Encap%20for%20HiPower%20White%20LEDs_Nordson ASYMTEK.pdf* (accessed 19 September 2013).
71. Hu, R., Luo, X., Feng, H., and Liu, S. (2012) *J. Lumin.*, **132**, 1252–1256.
72. Sommer, C., Reil, F., Krenn, J.R., Hartmann, P., Pachler, P., Tasch, S., and Wenzl, F.P. *J. Lightwave Technol.*, **26** (21), 3226–3232.
73. Tran, N.T. and Shi, F.G. *J. Lightwave Technol.*, **28** (22), 3556–3559.
74. Masui, H., Nakamura, S., and DenBaars, S.P. *Jpn. J. Appl. Phys.*, **45** (34), L 910–L 912.
75. Bois, Ch., Bodrogi, P., Khanh, T.Q., and Winkler, H. (2014) Measuring, simulating and optimizing current LED phosphor systems to enhance the visual quality of lighting. *J. Solid State Lighting*, **1**, 5. doi: 10.1186/2196-1107-1-5
76. Yan, B., You, J.P., Tran, N.T., and Shi, F.G. (2013) Influence of phosphor configuration on thermal performance of high power white LED array. IEEE International Symposium on Advanced Packaging Materials, 2013, pp. 274–289.
77. Chen, K.J., Chen, H.C., Lin, C.C., Yeh, C.C., Tsai, H.H., Chien, S.H., and Shih, M.H. (2013) *J. Disp. Technol.*, **9**, 915–920.
78. Linder, N. (2006) *Advanced Industrial Design Methods for LEDs NUSAD 2006*, OSRAM Opto Semiconductors, Regensburg.
79. Osram Opto Semiconductors (2012) Osram Ostar for Projection, September 2012, pp. 1–27.
80. Pachler, P. (2014) LED module with improved light output. USPTO Application #20120061709, 2012.
81. Tran, N.T., You, J.P., and Shi, F.G. (2009) Effect of Phosphor Particle Size on Luminous Efficacy of Phosphor-Converted White LED. *J. Lightwave Technol.*, **27** (22), 5145–5150.

4
Measurement and Modeling of the LED Light Source

Quang Trinh Vinh, Tran Quoc Khanh, Hristo Ganev, and Max Wagner

4.1
LED Radiometry, Photometry, and Colorimetry

The spectral, radiant, colorimetric, and photometric quantities of LED products are necessary in many stages of the product value chain from development departments over production quality test laboratories of LED manufacturers and LED luminaire companies through national laboratories for LED product certification and market supervision. The huge variety of LED products can be divided into four categories representing different levels of system integration and requiring (partially) different types of physical measurements:

1) *LED components and optical devices (lenses, reflector optics)*: In this stage, the electrical, optical, and thermal quantities of the LEDs have to be measured. Spectral radiant flux (in W nm^{-1} units) in the visible wavelength range is needed in order to calculate the color rendering index (CRI, Section 5.2), the correlated color temperature (CCT in K), the chromaticity coordinates (x, y; or u', v'), the luminous flux (Φ_v in lm), and the radiant power or, in other words, radiant flux (Φ_e, see Section 2.2.1, its alternative denotation is P_{opt}; its unit is W). From absolute spectral radiant flux, thermal power (P_{th} in W) also can be determined when the electrical power (P_{el} in W) is known. This knowledge is important for the selection of suitable LED types for the development of LED luminaires. To develop the optical system, luminance (in cd m^{-2}) and its spatial distribution should be known.

2) *LED modules (as so-called Zhaga modules)*: The LED modules from different companies should have compatible optical, photometric, electrical, mechanical, and thermal characteristics so that the LED luminaire manufacturers can develop consistent, reliable, and reproducible new LED luminaires or they can replace parts of the luminaires in the course of using these products (often several years or a decade) for indoor and outdoor lighting. An institute founded to ensure compatibility is the so-called Zhaga Consortium which "is developing specifications that enable the interchangeability of LED light sources made by multiple different manufacturers."

3) *LED luminaires*. At this system integration level, the measurement of photometric and colorimetric values are important to be able to define and design luminaire groups, to develop prototypes, to ensure product quality and to carry out product marketing involving datasheets and catalogues. Luminous flux, luminous intensity distribution, CCT, and luminous efficacy are the most important parameters for a successful communication with the users.
4) *LED installations for indoor and outdoor lighting*: If the LED product is installed in exterior or interior lighting systems then the measurement of photometric and colorimetric values at the illuminated user areas is needed in order to specify lighting quality – also for periodic maintenance measurements. Luminance on the road and wall surfaces and illuminance at workplaces in a room are essential.

In this section, the radiometry, colorimetry, and photometry of LED *components* is described. In Chapters 7 and 8, the methods and systems for the determination of the photometric and colorimetric parameters of LED *modules* and *luminaires* are given priority.

Because during their production, LED components have to be quantified and selected according to their forward voltage, chromaticity coordinates, and luminous flux before delivering them to luminaire companies, spectral radiant flux ($W\,nm^{-1}$) measurements are essential for LED manufacturers. Having purchased these LED light sources, engineers of the development departments of luminaire industry must design a suitable optics and, for this, they need to know the *spatially resolved* luminance and chromaticity distributions of the LED light sources.

4.1.1
Spatially Resolved Luminance and Color Measurement of LED Components

Since more than 20 years, spatially resolved luminance and color measurements of the light sources (recently including LED devices) have become widely used methods in lighting technology. Image resolution has many advantages – and it is used in the following applications:

1) *Lighting installations*: In a room in indoor lighting or on the road in outdoor lighting, luminance distributions are not uniform: luminance changes from location to location so that an average luminance of, for example, a wall or on a working table is not very useful. For research purposes, to qualify lighting quality and also to design and plan the arrangement of luminaires in a room or above a road, spatially resolved luminance and color measurement is necessary, see Figure 4.1.
2) *Evaluation and visualization of LED products*: Many products like car rear lamps, displays, light-emitting surfaces of LED luminaires with prism optics or with diffuse glasses or plastic plates are often evaluated visually: the question is whether their color uniformity is acceptable. For technical documentation, spatially resolved luminance and/or color measurements are necessary, see Figure 4.2.

4.1 LED Radiometry, Photometry, and Colorimetry

Figure 4.1 Luminance distribution on a road illuminated with LED luminaires. (Image source: Technische Universität Darmstadt.)

Figure 4.2 Luminance distribution of a symbol for the nighttime illumination design in a car. (Data source: Technoteam, Ilmenau/Germany.)

3) *Optical system design*: As the size of most LED components is small and their optics can be placed very close to the LED light source, near-field photometry is essential and the light source cannot be regarded as a point: the LED has a light-emitting area with many emitting elements whose emission distribution also depends on the viewing angle. Such luminance or color distributions constitute a 3D ray data set for each measured LED light source, see Figure 4.3.

The device for spatially resolved luminance and color measurements is called luminance camera or color camera working on the principles of CMOS or CCD semiconductor sensor technology [1]. The CCD sensor has the advantage of low dark current. It converts photons into charges which are transformed in an analog voltage before they are digitized. The camera involves an optimized

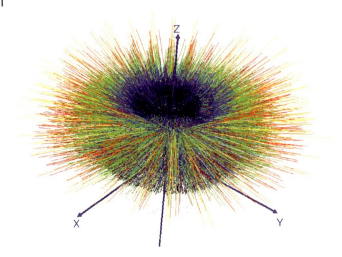

Figure 4.3 Ray data of an LED light source. (Data source: Technoteam, Ilmenau/Germany.)

lens system with a filter wheel containing separate filter sets to transform the spectral sensitivity of the CCD sensor (silicon semiconductor) and the spectral transmittance of the lenses into the $V(\lambda)$ – function or the color matching functions $\bar{x}(\lambda), \bar{y}(\lambda), \bar{z}(\lambda)$, see Figure 4.4. There may be, however, systematic errors and stochastic errors that cause deviations from the optimal operating conditions of the camera. Systematic errors include:

1) *Calibration uncertainty*: The camera is calibrated with a luminance standard with homogenous luminance distribution and with an illuminant having a CCT of about 2700–2800 K. The calibration standard itself has an uncertainty of about ±1%.

2) *Spectral matching uncertainty*: As the number of color filters is limited and as the thickness of these filters may deviate from specifications, a spectral deviation from $V(\lambda) = \bar{y}(\lambda)$ and from $\bar{x}(\lambda), \bar{z}(\lambda)$ can be expected, see Figure 4.5. In Table 4.1, a measure of these deviations, the so-called spectral mismatch factor (f_1') is listed for the spectral matching of the channels $x(\lambda), V(\lambda), z(\lambda)$ and also for $V'(\lambda)$ (a fourth channel for nighttime vision). The spectral mismatch factor f_1' is defined according to the German DIN 5032 standard and CIE Publ. No. 53.

3) *Sensor nonlinearity*: The sensor and its electronics (amplifier, A/D-converter) are not linear. Their output electronic signal is not proportional to their input radiant flux.

4) *Nonuniformity of the sensor elements (pixels)*: Pixels exhibit different relative and absolute sensitivities for different wavelengths and different locations on the sensor array.

5) *Shadings and aberration distortions*: they are caused by the lens.

6) *Stray light:* Stray light emanates from the glass/air and glass/optical silicone interfaces, the aperture edges, the inside walls of the housing, and the optical

4.1 LED Radiometry, Photometry, and Colorimetry

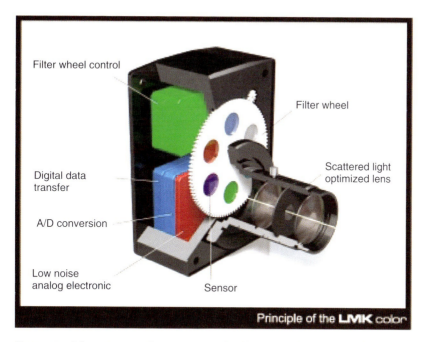

Figure 4.4 Schematic view of a luminance and color camera. (Image source: Technoteam, Ilmenau/Germany.)

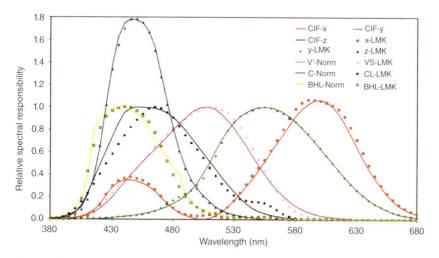

Figure 4.5 Relative spectral sensitivities of a spatially resolved (imaging) color measurement system in comparison with the color matching functions. (Image source: Technoteam, Ilmenau/Germany.)

Table 4.1 Spectral mismatch factors (f'_1) for a typical high-end luminance and color camera [1].

Function	f'_1
$\bar{x}(\lambda)$	3.9
$\bar{y}(\lambda)$	2.1
$\bar{z}(\lambda)$	5.4
$V'(\lambda)$	6.2

mountings as well as the sensor surface (its reflectance being in the range of 30–60% for silicone materials).

Stochastic uncertainties arise because of noise. This can be reduced by cooling the sensor or by shooting a series of multiple images and averaging pixel code values.

To investigate measurement uncertainty, two commercial luminance cameras were compared to a radiance type spectroradiometer (Konica-Minolta® CS 2000 A) in the Laboratory of Lighting Technology of the Technische Universität Darmstadt (Germany) [2]. The measured object was the inner wall of an integrating sphere (or, in other words, Ulbricht sphere) homogenously illuminated by the optical radiation of a number of different color LEDs with peak wavelengths of 458 nm (blue LEDs), 531 nm (green LEDs), 598 nm (orange LEDs), and with a warm white LEDs with a CCT of 2957 K. Relative luminance deviations compared to the spectroradiometer as a reference device can be seen in Table 4.2.

In order to produce ray data for an LED component or an LED group (e.g., a chip-on-board or COB LED), the luminance or color camera can be mounted on a goniophotometer that rotates the camera around different axes while the light source is positioned at the center. At each angular position, one luminance image (or color image) is then recorded so that a ray data set $L(x, y, z, \delta, \phi)$ can be established from the different images at the different angular positions, see Figure 4.6, [3]. This ray dataset is not only advantageous for optical system design (lenses and reflector optics). It can also be used for luminous intensity calculations for far-field photometric applications and to compute luminous flux as well as far-field/near-field distance limits.

Table 4.2 Relative luminance deviations of two commercial luminance cameras from the spectroradiometer Konica-Minolta® CS 2000 A as a reference device [2].

Camera|LED→	LED 458 nm	LED 531 nm	LED 598 nm	Warm white LED, 2957 K
Luminance camera 1	−10.8%	1.7%	1.7%	2.2%
Luminance camera 2	−10.9%	0.0%	1.0%	2.3%

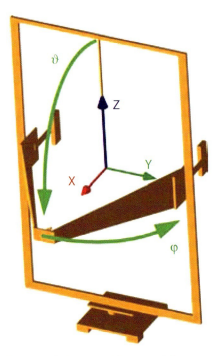

Figure 4.6 Principle of the near-field goniophotometer. (Image source: Technoteam, Ilmenau/Germany.)

4.1.2
Integrating Sphere Based Spectral Radiant Flux and Luminous Flux Measurement

For certain applications, spectral radiant flux (in W nm^{-1}), luminous flux (in lm) and luminous efficacy (in lm W^{-1}) measurements can be done with an integrating sphere because of the short measuring time [4]. This is important in LED production for LED binning (Section 5.8) and for LED measurements for quality insurance in the laboratories of LED luminaire manufacturers.

Luminous flux is measured with an integrating sphere (also known as Ulbricht sphere) with a typical diameter of 300 mm equipped with an auxiliary LED, see Figure 4.7.

The LED device under test (DUT) is placed on the inner side of the wall of the sphere. It emits optical radiation into the sphere in the 2π sr geometry [5]. Separated from the direct beam of the LED by a so-called "baffle" (see Figure 4.7), a detector is placed inside the wall of the sphere. If only luminous flux measurement is planned then the detector is a $V(\lambda)$ photometer head which is cosine corrected by a diffusor. To measure spectral radiant flux, a cosine diffusor input and an optical fiber bundle transmitting the LED's radiation to the entrance slit of a spectroradiometer can be mounted at the detector port of the integrating sphere, see Figure 4.48.

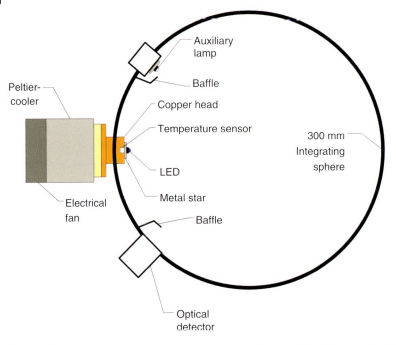

Figure 4.7 Measuring geometry of an integrating (or Ulbricht) sphere for spectral radiant flux and luminous flux measurements. LED: LED device under test.

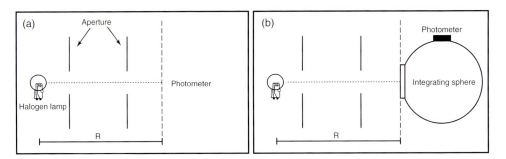

Figure 4.8 Calibration of an integrating sphere as a luminous flux measuring unit using an external light source.

When measuring the luminous flux of the LED by an integrating sphere, the *calibration reference* of this luminous flux measurement is important. Usually, the integrating sphere can be calibrated by using a tungsten filament standard, a tungsten halogen lamp or a standard LED calibrated by a national institute of standards. There are, however, only a few standard LEDs for luminous flux measurement available on the market so an alternative integrating sphere calibration method is described as follows, see Figure 4.8.

The method depicted in Figure 4.8 (actually a more sophisticated version than depicted here) is also used at the NIST (National Institute for Standard and Technology/USA). A photometer is placed at a known distance in front of a halogen lamp and the illuminance E is measured. Then the sphere is placed at the same distance in front of the lamp and then the illuminance E is reproduced at the sphere entrance (sphere port). The luminous flux Φ_N entering the integrating sphere can be calculated by Eq. (4.1).

$$\Phi_N = E \cdot A \qquad (4.1)$$

In Eq. (4.1), A is the opening area of the sphere at which the beam of the lamp enters the sphere. When measuring LEDs with an integrating sphere in this manner, the *combination* of the spectral sensitivity of the photometer *and* the spectral transmittance property of the sphere can be adapted to $V(\lambda)$: the photometer and the sphere form *one single* measuring unit. The spectral transmittance factor of the sphere (τ) multiplied by the $V(\lambda)$ adapted spectral sensitivity of the photometer represents the overall relative spectral responsibility $s_{rel}(\lambda)$ of this measuring unit which never matches the $V(\lambda)$ function perfectly. Figure 4.9 shows the relative spectral responsibility $s_{rel}(\lambda)$ of a real measuring unit (as an example).

The spectral emission curves of incandescent lamps and color or white LEDs are very different. Whereas the spectrum of an incandescent lamp is a broadband spectrum, the spectrum of a color LED is a narrowband one. Measuring a broadband source with a slightly inaccurate $V(\lambda)$ adapted measuring unit is not a big problem. However, for narrowband color LEDs, the LED peak (see e.g., Figure 2.28.) can be located in a badly $V(\lambda)$-approximated spectral range (see the right ordinate of Figure 4.9) and this results in high measurement error. This spectral mismatch error is particularly serious for red and blue LEDs, up to 25% when measuring luminous flux. Table 4.3 shows the expected measurement error for

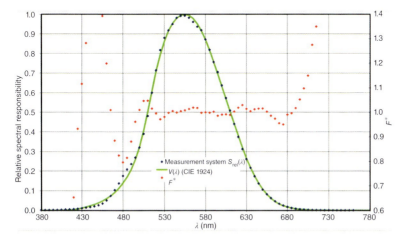

Figure 4.9 Relative spectral responsibility $s_{rel}(\lambda)$ of an integrating sphere used as a luminous flux measuring unit (left ordinate) and its deviation (in%) from $V(\lambda)$ (right ordinate).

Table 4.3 Measurement errors of luminous flux for colored LEDs without correction.

Color	Error
Red (peak 640 nm)	20%
Amber (peak 605 nm)	14%
Warm white (3000 K)	2% (!)
Green (peak 505 nm)	10%
Blue (peak 440 nm)	25%

different LED colors if there is no suitable correction. These errors apply not only to the sample measuring unit used in this section but also to all the units available on the market.

The spectral mismatch error of the measuring unit can be corrected by the so-called spectral mismatch correction factor F. To calculate this factor, the spectral distribution of the LED, the so-called DUT, and the relative spectral responsibility of the measuring unit have to be known. The spectral mismatch correction factor can then be calculated by Eq. (4.2).

$$F = \frac{\int S_{\text{LED}}(\lambda)V(\lambda)d\lambda \cdot \int S_{\text{cal}}(\lambda)S_{\text{rel}}(\lambda)d\lambda}{\int S_{\text{LED}}(\lambda)S_{\text{rel}}(\lambda)d\lambda \cdot \int S_{\text{cal}}(\lambda)V(\lambda)d\lambda} \quad (4.2)$$

where

$S_{\text{LED}}(\lambda)$ is the spectral power distribution (SPD) of the DUT.
$S_{\text{cal}}(\lambda)$ is the SPD of the used standard source.
$s_{\text{rel}}(\lambda)$ is the relative spectral responsibility of the measuring unit.

The corrected luminous flux Φ_{corr} is calculated by Eq. (4.3)

$$\Phi_{\text{corr}} = \Phi \cdot F \quad (4.3)$$

To measure *spectral* radiant flux in the visible wavelength range (between 380 and 780 nm), the photometer arrangement of Figure 4.8 can be changed to the spectroradiometer arrangement shown in Figure 4.48. In this case, the entrance optics at the detector port guides the LED's radiation into a spectroradiometer. The spectral calibration of the detector unit consists of the calibration of the sphere with all of its components and the spectroradiometer with the entrance optics, glass fiber bundle, monochromator optics. The following steps are necessary for spectral calibration:

First step: Position the entrance opening of the sphere at a distance of 70 cm along the optical axis of a *standard lamp* for spectral irradiance (unit: $W (nm\, m^2)^{-1}$; lamp type: FEL; 120 V/8 A). The spectral irradiance $E_N(\lambda)$ at this distance is known from the calibration report of the lamp. The most widely used standard lamp for the visible wavelength range is the FEL lamp with a calibration uncertainty in the range of $\pm 1.8-2.3\%$ if calibrated by a national standard laboratory [6].

Second step: Position the sphere to be calibrated at this distance with the optical axis perpendicular to and going through the center of the entrance port of the sphere. The spectral radiant flux $\Phi_N(\lambda)$ going into the sphere can be calculated by applying Eq. (4.1): $\Phi_N(\lambda) = E_N(\lambda)A$. With the *standard lamp*, the spectroradiometer's sensor generates the signal $E_N^*(\lambda)$.

Third step: Place the LED to be measured to the sphere, see Figure 4.7. After multiple reflections, the LED's radiation reaches the entrance optics of the detector and it is measured by the spectroradiometer that supplies the spectral signal $E_x(\lambda)$ so that the spectral radiant flux $\Phi_x(\lambda)$ of the measured LED can be calculated according to Eq. (4.4).

$$\Phi_x(\lambda) = \Phi_N(\lambda) \cdot E_x(\lambda)/E_N^*(\lambda) \tag{4.4}$$

Note that, if the size of the detector port, the opening port and the baffles is large or the materials of the LED holding exhibit significant absorption then a more sophisticated correction procedure using an *auxiliary* LED shown in Figure 4.7 can be applied with a more complex procedure.

4.2
Thermal and Electric Behavior of Color Semiconductor LEDs

In Section 3.2.5, semiconductor material systems were discussed according to their efficiency, technological possibilities, and limits. In this section, the radiant and colorimetric properties of four typical color semiconductor LEDs, the temperature and current dependence of their radiant efficiency η_e (in W/W or in%), radiant flux Φ_e (in W), and chromaticity difference ($\Delta u'v'$) caused by temperature and current fluctuations are investigated. A temperature range between 40 and 85 °C and a current range between 100 and 700 mA are considered.

4.2.1
Temperature and Current Dependence of Color Semiconductor LED Spectra

4.2.1.1 Temperature Dependence of Color Semiconductor LED Spectra
In Figure 4.10, the temperature dependence of four sample color semiconductor LEDs is shown in the range 40–80 °C, at 350 mA. As can be seen from Figure 4.10, when the operating temperature increases from 40 to 80 °C, the emission spectra of the orange and red semiconductor LEDs (InGaAlP LEDs) shift toward longer wavelengths and their luminous flux decreases (because of less overlap with $V(\lambda)$). Results indicate that the spectral shift of the red and orange semiconductor LEDs (InGaAlP LEDs) is more explicit than that of the blue and green semiconductor LEDs (InGaN LEDs).

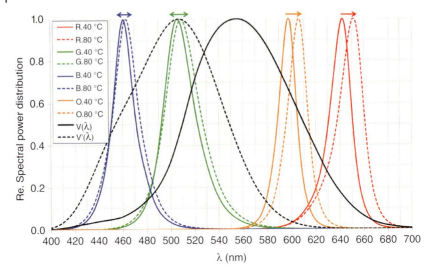

Figure 4.10 Temperature dependence of the relative spectral power distribution of four color semiconductor LEDs operated at 350 mA.

4.2.1.2 Current Dependence of Color Semiconductor LED Spectra

Figure 4.11 shows the current dependence of four typical color semiconductor LEDs from 350 to 700 mA, at 80 °C. As can be seen from Figure 4.11, when the forward current increases from 350 to 700 mA, the spectra of the red and orange semiconductor LEDs (InGaAlP LEDs) shift strongly toward longer wavelengths. There is nearly no spectral shift for the peak wavelengths of the blue and green

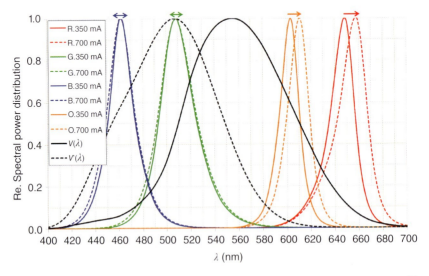

Figure 4.11 Current dependence of the spectra of four color semiconductor LEDs at 80 °C.

semiconductor LEDs (InGaN LEDs) but the bandwidths of the spectra extend more toward shorter wavelengths for 700 mA.

These findings imply that the chromaticity difference caused by temperature and current change of InGaAlP LEDs is significantly higher than that of InGaN LEDs. But these findings are not conclusive about the radiant flux, radiant efficiency, luminous flux, or luminous efficacy of these semiconductor LEDs. Particularly, when forward current increases, carrier density becomes higher in the active region of the junction layer but carrier distribution is also changed and this causes an asymmetric change in the shape of the color semiconductor LEDs' spectra according to the theory described in Section 3.3.

According to theory, more radiative recombinations because of higher carrier density help color semiconductor LEDs emit more radiant flux. In contrast, more nonradiative recombinations and an adverse change of carrier distribution make color semiconductor LEDs produce less radiant flux. These conflicting processes take place concurrently and in a different manner between InGaN and InGaAlP LEDs. Consequently, when the forward current increases, InGaN LEDs emit more but their radiant flux is not proportional to the increase of forward current and some InGaAlP LEDs sometimes become even darker, see Section 4.2.2.

4.2.2 Temperature and Current Dependence of Radiant Flux and Radiant Efficiency of Color Semiconductor LEDs

In this section, radiant flux and radiant efficiency are converted into relative units so that their temperature and current change induced differences and changes become more obvious.

4.2.2.1 Temperature Dependence of Radiant Flux and Radiant Efficiency of Color Semiconductor LEDs

The temperature dependence of radiant flux and radiant efficiency of the four typical color semiconductor LEDs described is shown in Figure 4.12. Results imply that when the operating temperature increases from 40 to 80 °C at 350 mA then radiant flux and radiant efficiency decrease in a similar way. Particularly, the loss of both radiant flux and radiant efficiency of the orange semiconductor LED is about 95%, that of the red semiconductor LED is about 28%, and that of both the blue and the green semiconductor LEDs is about 10%. Thus, the loss of the orange semiconductor LED is 3.39 times higher than the one of the red semiconductor LED and 9.5 times higher than the one of the blue and green semiconductor LEDs. Therefore, because of the big losses of both radiant flux and radiant efficiency of the orange semiconductor LEDs compared with other semiconductor LEDs, special attention must be paid to these LEDs during LED selection and LED combination in the design of hybrid LED systems.

146 | *4 Measurement and Modeling of the LED Light Source*

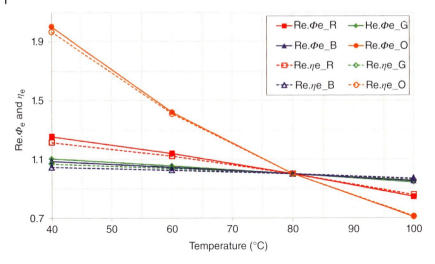

Figure 4.12 Temperature dependence of radiant flux and radiant efficiency at 350 mA for four color semiconductor LEDs, R: red; G: green; B: blue; O: orange.

4.2.2.2 Current Dependence of Radiant Flux and Radiant Efficiency of Color Semiconductor LEDs

The current dependence of radiant flux and radiant efficiency at 80 °C is shown in Figure 4.13. It can be seen that when the forward current increases from 350 to 700 mA at 80 °C, the radiant flux of the blue and green semiconductor LED increases about 65% and 50%, respectively. In contrast, radiant efficiency decreases about 25% for the blue semiconductor LED and about 30% for the

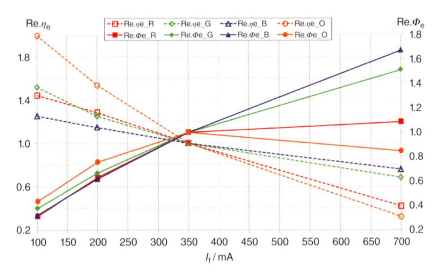

Figure 4.13 Current dependence of radiant flux and radiant efficiency at 80 °C for four color semiconductor LEDs, R: red; G: green; B: blue; O: orange.

green semiconductor LED. Especially, with similar increase of forward current (350 mA vs 700 mA) at 80 °C, the radiant flux of the red semiconductor LED increases only insignificantly and the one of the orange semiconductor LED does not increase. But there is a big reduction of radiant efficiency of both the red semiconductor LED (about 60%) and the orange semiconductor LED (about 70%). These results supplement the relative spectral evaluations of Section 4.2.1 so that the instability of the orange and red color semiconductor LEDs can be corroborated.

4.2.2.3 Conclusion

On the basis of the earlier analysis, it can be concluded that both the temperature dependence and the current dependence of the red and orange semiconductor LEDs are very significant while the characteristics of the blue and green LEDs can be accepted under these operating conditions. Therefore, if these color LEDs are used in a solid state lighting application for color mixing, a compensation method can be elaborated to counterbalance their temperature and current dependence. Although operating temperature cannot be set arbitrarily, forward current can be chosen to be appropriately low. Current shall be adjusted to a level that is still able to yield the required radiant flux but without losing too much radiant efficiency. For this problem, the orange semiconductor LED in Figures 4.12 and 4.13 is a very demonstrative example. When the forward current of the orange semiconductor LED is increased from 350 to 700 mA then its radiant flux does not increase but it decreases to about 85% at 700 mA compared with that at 350 mA and its efficiency decreases intensively to about 30% at 700 mA compared with its efficiency at 350 mA. In other words, 70% of energy is lost without any benefit (15% loss of radiant flux).

4.2.3 Temperature and Current Dependence of the Chromaticity Difference of Color Semiconductor LEDs

If the chromaticity binning current and temperature (350 mA and 80 °C) at which the chromaticity of the color LED is specified is fixed as a reference chromaticity point then the $\Delta u'v'$ chromaticity difference metric (defined after Eq. (2.23)) can be used to evaluate the chromaticity shift of the color LED from this reference chromaticity point when temperature and/or current are changing. These chromaticity differences supplement and help further quantify spectral evaluations (Section 4.2.1).

4.2.3.1 Temperature Dependence of the Chromaticity Difference of the Color Semiconductor LEDs

In Figure 4.14, the temperature dependence of the chromaticity difference $\Delta u'v'$ of the four typical color semiconductor LEDs is shown when the operating temperature changes from 40 to 100 °C at 350 mA. Indeed, the color difference of the

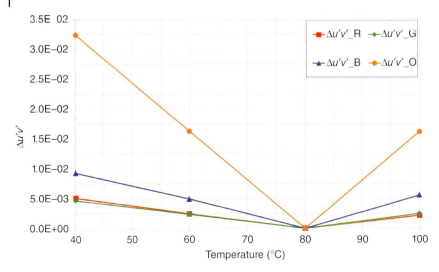

Figure 4.14 Color difference of the four color semiconductor LEDs as a function of temperature, R: red; G: green; B: blue; O: orange.

orange semiconductor LED exhibits the worst behavior. If the three-MacAdam-ellipse limit (about $\Delta u'v' < 0.003$) is assumed here as the limit for visually acceptable chromaticity differences (see e.g., [7]) then the tolerable operating temperature range of the orange semiconductor LED is about $80 \pm 2.5\,°C$, the one of the blue semiconductor LED extends from about 70 to about 90 °C and the one of the green semiconductor LED and the red semiconductor LED ranges between about 60 and 100 °C. Thus, it can be recognized that the stabilization of the chromaticity difference of the orange semiconductor LED within an acceptable limit in its operating process is nearly impossible, unless a high-cost temperature control system is applied to adjust the operating temperature for this color semiconductor LED type. Otherwise, the blue semiconductor LED is allowed to operate only in a quite narrow region, whereas, the color difference of the green and the red semiconductor LEDs can be ensured in a wider operating temperature range. Therefore, the orange semiconductor LED should be avoided in the LED combination for hybrid LED systems.

4.2.3.2 Current Dependence of Chromaticity Difference of the Color Semiconductor LEDs

In Figure 4.15, the current dependence of the chromaticity differences of the four typical color semiconductor LEDs is shown. With the similar three-Mac Adam-ellipse criterion in mind, chromaticity difference behavior ameliorates among the color LEDs in the following order: orange (worst), blue, green, and red (best). According to this rank order, the widest operating current range around the binning point (350 mA and 80 °C) extends between 100 and 600 mA for the red semiconductor LED. The worst operating current range is only about 350 ± 20 mA for the orange semiconductor LED. It is about 220 to 600 mA for the

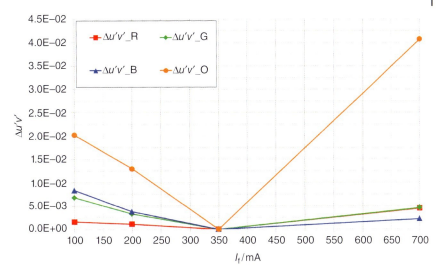

Figure 4.15 Chromaticity difference of the four color semiconductor LEDs as a function of forward current, R: red; G: green; B: blue; O: orange.

green semiconductor LED and about 220 to 700 mA for the blue semiconductor LED. On the basis of these results, it can be confirmed that the blue semiconductor LED is the best selection for operating in a high current range while the orange semiconductor LED represents the worst case because of its large chromaticity differences arising in its entire operating current range. Therefore, blue semiconductor LEDs are the most appropriate color semiconductor LEDs for the fabrication of high-brightness PC-LEDs following their excellent color stability and high excitation efficiency for the most important luminescent materials (LED phosphors). Orange semiconductor LEDs, however, should be substituted by yellow or orange spectral components of appropriate luminescent materials in the color mixing design of hybrid LED systems for high quality solid state lighting applications in order to avoid the instability of their chromaticity. But the substitution of orange semiconductor LEDs in vehicle signaling lamps is a subject of future research.

4.3
Thermal and Electric Behavior of White Phosphor-Converted LEDs

4.3.1
Temperature and Current Dependence of Warm White PC-LED Spectra

In this section, six typical warm white PC-LEDs with typical relative SPDs illustrate the spectral changes arising following the change of operating temperature (Figure 4.16) and the change of forward current (Figure 4.17).

150 | *4 Measurement and Modeling of the LED Light Source*

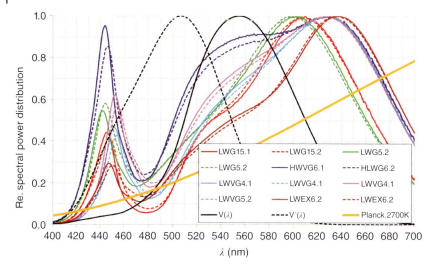

Figure 4.16 Temperature dependence of six typical warm white LEDs (2700–3500 K) at 350 mA. Solid lines correspond to 40 °C and dashed lines correspond to 80 °C.

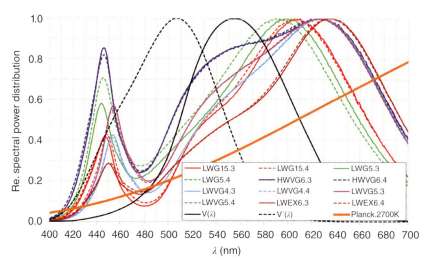

Figure 4.17 Current dependence of six typical warm white LEDs (2700–3500 K) at 80 °C. Solid lines correspond to 350 mA and dashed lines correspond to 700 mA.

4.3.1.1 Temperature Dependence of Warm White PC-LED Spectra

In Figure 4.16, the temperature dependence of warm white PC-LED spectra is shown in the operating range from 40 to 80 °C at 350 mA. Here, the indexes "1" and "2" (attached to the name of each LED in Figure 4.16) indicate the spectra at 40 °C and at 80 °C, respectively. In these warm white PC-LEDs, a luminescent material system with two different phosphor types (such as the mixture of a green phosphor and a red phosphor or a mixture of a yellow/orange phosphor and a red phosphor) is built on a blue LED chip. Therefore, in the investigation of the warm white PC-LED's spectra, three spectral components (the spectral component of the blue chip, the green/orange/yellow phosphor and the red phosphor) must be considered. Hence, based on the measured spectra of the warm white PC-LEDs in Figure 4.16, it can be recognized that the peak wavelength of the red phosphor component of the LEDs with good CRI (such as LWVG4 and HWVG6) and the one of the LEDs with excellent good CRI (such as LWEX6) always occur at a longer wavelength, about 630–640 nm. For the case of the LEDs with a lower CRI, the red maximum is located at about 600 nm (for LWG5) or about 610 nm (for LWG15). In addition, a comparison of the solid curves (for the spectra at 40 °C) and the dashed curves (for the spectra at 80 °C) in Figure 4.16 shows that the temperature stability of the spectral components belonging to the red phosphors is much better than that of the spectral components belonging to green, yellow, or orange phosphors.

The four LEDs HWVG6, LWG5, LWVG5, and LWEX6 represent examples of bad temperature stability. In case of HWVG6, LWVG5, and LWEX6, the short wavelength spectral components and the middle wavelength spectral components decrease rapidly with increasing temperature compared with the long wavelength spectral components. Therefore, their white point is shifted much into the red direction (this is called a red shift) and this causes a strong decrease of their CCT.

But in case of LWG5, there is nearly no change for the middle wavelength spectral component, while the short wavelength spectral component increases strongly and the long wavelength spectral component shifts toward the left in a very wide spectral range. As a result, its white point shifts toward blue direction (it is called blue shift) and this causes a strong increase of its CCT. Furthermore, these spectral changes also make HWVG6, LWG5, and LWEX6 exhibit big chromaticity differences compared with their binning point except for LWVG5 as shown in Figure 4.22. Especially, the less spectral change of LWVG5 in the blue gap (450–490 nm) makes its chromaticity difference more acceptable than for of the cases HWVG6, LWG5, and LWEX6.

4.3.1.2 Current Dependence of Warm White PC-LED Spectra

The current dependence of the warm white PC-LED spectra is shown in Figure 4.17 in the operating range from 350 to 700 mA at 80 °C. Here, the indexes "3" and "4" denote 350 and 700 mA, respectively. The most important issue is the simultaneous increase of the three spectral components when forward current increases. Particularly, if the increase is proportional to forward current, then there is no chromaticity difference and no CCT change. Indeed, in almost all cases, the increase of the spectral components is proportional to forward current except for LWG5 and LWG15. Therefore, relative spectral changes are quite small when forward current increases from 350 (denoted by solid curves) to 700 mA (denoted by dashed curves) in Figure 4.17. In the worst case of LWG5, a significant spectral change takes place concerning all three spectral components (the short, middle, and long wavelength spectral components). Thus, this is an extreme demonstration example for a low quality warm white LED in which both the blue chip and the phosphor system have neither good temperature stability nor good current stability. In addition, although the spectral change of LWG15 is less than that of LWG5, this spectral change takes place in some "sensitive" spectral ranges such as the blue gap (450–490 nm) and the amber – red region (550–600 nm) resulting in an adverse chromaticity difference behavior for this PC-LED.

4.3.2
Current Limits for the Color Rendering Index, Luminous Efficacy, and White Point for Warm White PC-LEDs

4.3.2.1 General Considerations

In Section 4.3.1, the temperature and current dependence of six LED types with typical spectra were discussed. In this section, the CRI, luminous efficacy and the white point of a bigger sample of warm white PC-LEDs are investigated in order to determine the current limits as the most important design targets for their optimization and stabilization. Concerning the CRI, although in many cases the general CRI (R_a) of white PC-LEDs is high, some special CRIs (such as CRI R_9) are low (see Section 5.2). Therefore, a more representative solution is to use the average CRI from the special color rendering indices of all 14 CIE standard test color samples (denoted by AVR_{1-14}) to evaluate the color quality of these warm white PC-LEDs. In addition, the CRIs of the warm white PC-LEDs were sorted into three categories including "good CRI" (80–85), "very good CRI" (86–90), and "excellent CRI" (90–100). In Figure 4.18, CRIs (R_a, R_9, and AVR_{1-14}) on the

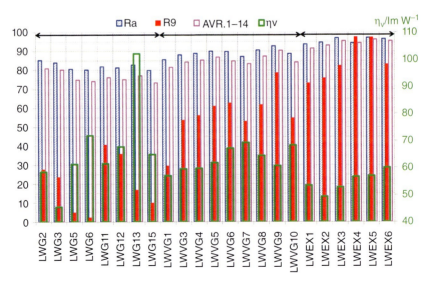

Figure 4.18 Color rendering indices (R_a, R_9, and AVR_{1-14}) and luminous efficacy for a set of current warm white PC-LEDs (2700–3000 K) at 350 mA and 80 °C.

left ordinate and luminous efficacy η_v (in lm W^{-1}) on the right ordinate are shown together. All results were measured at the hot binning point (350 mA and 80 °C).

4.3.2.2 Comparison of Color Rendering Index and Luminous Efficacy

Figure 4.18 shows an opposed tendency between color rendering indices and luminous efficacy for these warm white PC-LEDs. Particularly, when CRI is high, luminous efficacy is low and vice versa. Indeed, the luminous efficacy of excellent CRI warm white LEDs (2700–3000 K) is always low. Moreover, the ones with very good CRI also have lower efficiencies (lower than 70 lm W^{-1}). Only a few good CRI warm white LEDs have a high luminous efficacy level of about 105 lm W^{-1} such as LWG13. But their AVR$_{1-14}$ reaches about 77 only. These values are really important to establish an LED combination solution with color semiconductor LEDs and white PC-LEDs for a hybrid LED system in order to improve both lighting quality (CRIs) and luminous efficacy. The aim of hybrid LED system development is that the achieved lighting quality of the hybrid LED system will be higher or comparable with that of the warm white PC-LEDs.

4.3.2.3 White Point of the Warm White PC-LEDs

In Figure 4.19, the white point quality of the warm white LEDs (2700–3000 K) is shown in terms of $\Delta u'v'_{CCT}$, the distance between the chromaticity of the LED and its reference white point on the blackbody locus (the shorter this distance, the better the white point, see Figure 2.26 and Section 5.7.1). It can be seen from Figure 4.19 that working with, for example, the "three Mac Adam ellipses" criterion (about $\Delta u'v'_{CCT} < 0.003$) [7], there are many unacceptable LEDs such as LWG2, LWG3, LWVG1, LWEX2, LWEX4, and LWEX6. These LEDs get (visually disturbing) yellow, orange, amber, or purple tones (instead of being a visually acceptable shade of warm white) depending on their location relative to the blackbody locus. Such tones are not suitable for high quality solid state lighting applications. Therefore, it is very important to optimize and stabilize the white point so that the hybrid LED systems become *true* warm white, neutral white, or cold white light sources without visually disturbing color shadings in their light. This issue is discussed in more detail in Sections 5.7 and 5.8.

4.3.3
Temperature and Current Dependence of the Luminous Flux and Luminous Efficacy of Warm White PC-LEDs

In this section, six typical warm white PC-LEDs are chosen as examples to illustrate the temperature dependence and current dependence of the luminous flux and luminous efficacy of warm white PC-LEDs. These results are important to foster communication between LED light source manufacturers and LED luminaire manufacturers, to select a suitable operating forward current for hybrid LED systems and to improve their luminous flux and luminous efficacy.

154 | *4 Measurement and Modeling of the LED Light Source*

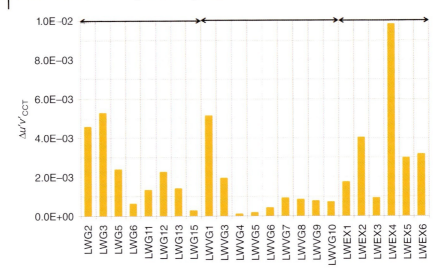

Figure 4.19 White points (ordinate: chromaticity distance from the blackbody locus, this will be minimal) of a set of warm white PC-LEDs (CCT = 2700–3000 K) at 350 mA and 80 °C.

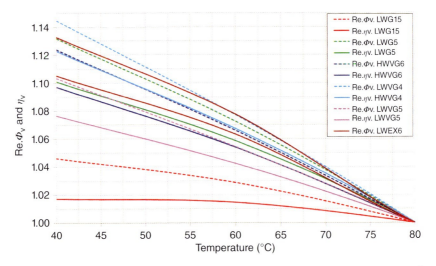

Figure 4.20 Temperature dependence of luminous flux and luminous efficacy of six typical PC-LEDs at 350 mA.

4.3.3.1 Temperature Dependence of the Luminous Flux and Luminous Efficacy of Warm White PC-LEDs

In Figure 4.20, the temperature dependence of relative luminous flux and relative luminous efficacy of these six warm white PC-LEDs is shown in the operating range from 40 to 80 °C at 350 mA compared to the binning point (350 mA and 80 °C) which corresponds to 1.00. Figure 4.20 shows that luminous flux decreases from 1.046 (min relative value, for LWG15) to 1.000 (binning reference) or from 1.146 (max relative value, for LWVG4) to 1.000 (binning reference) and luminous efficacy decreases from 1.018 (min for LWG15) to 1.000 and from 1.123 (max for LWVG4) to 1.0000 when the operating temperature increases from 40 to 80 °C. The consequence is that, with 14.6% luminous flux decrease and 12.3% luminous efficacy decrease in the worst case, LED luminaire manufacturers who buy LWVG4 from the LED manufacturer cannot specify the correct operating parameters in their user's guide because the changes are too high when an inevitable variation of the operating temperature takes place. In addition, if the PC-LEDs of the LED manufacturers are selected using the cold binning condition (350 mA and 25 °C in a 25 µs measurement) then the design parameters will deviate very significantly from the operating parameters of the LED luminaires in a real solid lighting application because the operating temperature will never be 25 °C. This is a serious problem for the production and design of LED luminaires.

4.3.3.2 Current Dependence of Luminous Flux and Luminous Efficacy of Warm White PC-LEDs

In Figure 4.21, the current dependence of the luminous flux and luminous efficacy of six typical warm white PC-LEDs is shown in the operating range from 100 to 700 mA at 80 °C. As can be seen, there are opposite (nonlinear) tendencies between luminous flux and luminous efficacy when forward current increases: luminous flux increases with the increase of forward current but luminous efficacy decreases. The different extent of luminous efficacy decrease for the different PC-LEDs is because of blue chip and luminescent material system differences and the different interactions between them. Sometimes, although there is a significant

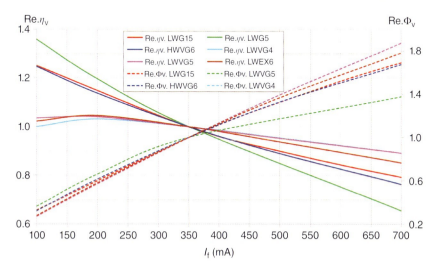

Figure 4.21 Current dependence of luminous flux (right ordinate, relative units) and luminous efficacy (left ordinate, relative units) for six typical PC-LEDs at 80 °C.

increase of forward current, the blue chip cannot radiate much more. Or, the blue chip can radiate more but the luminescent material cannot absorb the increase of radiation. Even if the phosphor system can absorb more radiation, it may not be able to reradiate the increment but convert it into more thermal energy hence the phosphor layers and the complete PC-LED becomes warmer. Therefore, it is necessary to achieve a compromise between the increase of luminous flux and the decrease of luminous efficacy for an appropriate selection of the white PC-LEDs and their forward current.

4.3.4
Temperature and Current Dependence of the Chromaticity Difference of Warm White PC-LEDs

In this section, the temperature and current dependence of the chromaticity difference $\Delta u'v'$ of warm white PC-LEDs between the actual chromaticity and the hot binning (350 mA/80 °C) chromaticity as a reference is discussed.

4.3.4.1 Temperature Dependence of the Chromaticity Difference of Warm White PC-LEDs

In Figure 4.22, the temperature dependence of the chromaticity difference of the earlier-mentioned six sample warm white LEDs is shown in the operating range from 40 to 80 °C at 350 mA. As can be seen from Figure 4.22, to comply with the earlier-mentioned $\Delta u'v' < 0.003$ visual white point criterion, the acceptable operating temperature will be higher than 45 °C for LWEX6, LWG5, and HWVG6.

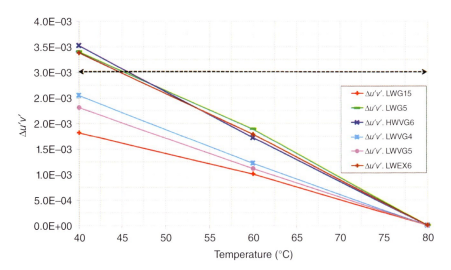

Figure 4.22 Temperature dependence of the chromaticity difference $\Delta u'v'$ between the actual chromaticity and the chromaticity at the binning reference of 80 °C for six typical warm white PC-LEDs at 350 mA.

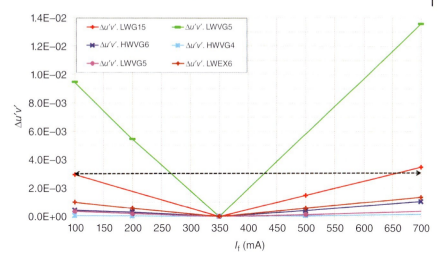

Figure 4.23 Current dependence of color differences of six typical warm white PC-LEDs at 80 °C.

It can also be seen that the other three LEDs meet this criterion in the whole range between 40 and 80 °C. This result corresponds to the temperature induced *spectral* changes, as seen in Figure 4.16. It is very important to optimize and stabilize LED lighting systems.

4.3.4.2 Current Dependence of the Chromaticity Difference of Warm White PC-LEDs

In Figure 4.23, the current dependence of the chromaticity difference of warm white LEDs is shown in the operating range 100–700 mA at 80 °C. It can be seen that the chromaticity difference of four LEDs never reaches the earlier-mentioned criterion value ($\Delta u'v' < 0.003$) so that the entire current range can be used for them. But this is not true for the case of LWG5 and LWG15. The available forward current range is only about 100–650 mA for LWG15 and about 275–425 mA for LWG5. These results correspond to the current dependence of the warm white PC-LED's spectra described in Section 4.3.2.

4.4 Consequences for LED Selection Under Real Operation Conditions

4.4.1 Chromaticity Differences Between the Operating Point and the Cold Binning Point

Figure 4.24 shows the chromaticity difference between the operating point and the so-called cold binning point (350 mA/25 °C) as a function of temperature and current for a warm white LED as an example. To comply with the $\Delta u'v' < 0.003$

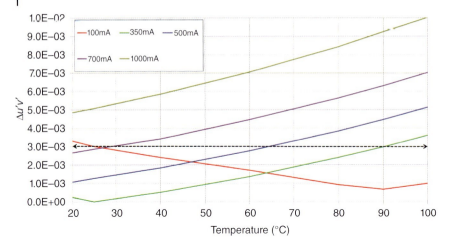

Figure 4.24 Chromaticity difference $\Delta u'v'$ between the actual temperature/current and the cold binning condition (350 mA/25 °C) for a sample warm white LED.

white point criterion, operating temperature can be limited to be above 25 °C in case of 100 mA, below 90 °C in case of 350 mA and below about 64 °C in case of 500 mA. This warm white PC-LED is not allowed to operate at 700 mA or higher to comply with the above white point criterion.

These chromaticity differences are listed in Table 4.4. Yellow cell backgrounds indicate visually low chromaticity differences (compare with Figure 5.51) for which the human visual system can hardly notice the chromaticity difference between the operating point and the binning point. Blue cells indicate in general tolerable chromaticity differences (for a more refined discussion, see Section 5.8.2). The chromaticity differences in the other cells (light orange, orange, and brown) are no more tolerable. In real applications, the results of Table 4.4 are important for LED developers to recognize the operating ranges with low and tolerable chromaticity differences, in this example for all temperatures at 100 mA, with operating temperatures below 100 °C at 350 mA, with operating temperatures below 60 °C at 500 mA, with operating temperatures below 40 °C at 700 mA, and no operation for 1000 mA. Consequently, LED developers can select the appropriate materials, electrical and mechanical elements, forward currents, and operating temperatures so that the chromaticity difference between the operating point and the binning point remains visually acceptable.

4.4.2
Chromaticity Difference Between the Operating Point and the Hot Binning Point

The binning point (i.e., the reference condition for the LED's specifications) in the previous example was the so-called "cold" binning condition with 350 mA

4.4 Consequences for LED Selection Under Real Operation Conditions | 159

Table 4.4 Chromaticity difference $\Delta u'v'$ between the actual temperature/current and the cold binning condition (350 mA/25 °C) for a sample warm white LED.

$\Delta u'v'$ (10^{-3})	100 mA	350 mA	500 mA	700 mA	1000 mA
25 °C	2	0	1	3	5
40 °C	2	1	2	3	6
60 °C	2	1	3	4	7
80 °C	1	2	4	6	8
90 °C	1	3	4	6	9
100 °C	1	4	5	7	10

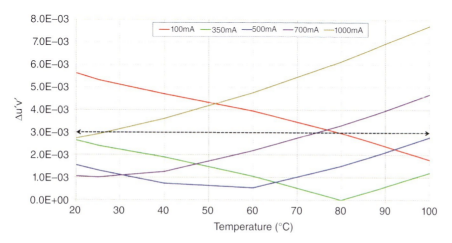

Figure 4.25 Chromaticity difference $\Delta u'v'$ between the actual temperature/current and the hot binning condition (350 mA/80 °C) for a sample warm white LED.

and 25 °C while the LED was measured for 25 μs. Nowadays, certain LED manufacturers also apply a so-called "hot" binning condition such as 350 mA and 80 °C or 85 °C or 700 mA and 80 °C or 85 °C (with a 25 μs measurement). Similar to Figure 4.24, Figure 4.25 shows the chromaticity difference of the same sample warm white PC-LED for the hot binning condition 350 mA/80 °C as a reference chromaticity. As can be seen, this hot binning operating range is broader than the previous one obtained for cold binning. At 350 and 500 mA, feasible operating temperatures range between 20 and 100 °C. In addition, at 700 mA, the PC-LED is permitted to operate until about 75 °C. Note that the forward currents 100 and 1000 mA (too low and too high currents) will not be considered for the design of LED luminaires because the operation at a low current requests a high number of LEDs to be built in and causes a high cost for the LED luminaire while the use of a high current makes a fast degradation of the LED luminaires happen.

Table 4.5 Chromaticity difference $\Delta u'v'$ between the actual temperature/current and the hot binning condition (350 mA/80 °C) for a sample warm white LED.

$\Delta u'v'/10^{-3}$	100 mA	350 mA	500 mA	700 mA	1000 mA
25 °C	5	2	1	1	3
40 °C	5	2	1	1	4
60 °C	4	1	1	2	5
80 °C	3	0	2	3	6
90 °C	2	1	2	4	7
100 °C	2	1	3	5	8

In more detail, the chromaticity differences for the hot binning condition are listed in Table 4.5 (similar to Table 4.4). As can be seen from Table 4.5, more operating points can be used to obtain an acceptable white point than for the case of cold binning and this represents a big advantage of the hot binning condition.

4.5
LED Electrical Model

In this section, the electrical model of the LED is discussed with several approaches. The LED's electrical model is assumed to be a multiple-input-single-output system (MISO system). The input of the MISO system is the junction temperature (T_j/°C) and the forward current (I_f/mA). The output is the forward voltage (V_f/V), see Figure 4.26.

4.5.1
Theoretical Approach for an Ideal Diode

The theoretical approach corresponds to the ideal abrupt junction described in Section 3.2. It is assumed that there is no compensation for dopants caused by unintentional impurities and defects. Thus, the ideality factor n_{ideal} equals 1 for an ideal diode. Mathematical modeling starts with the well-known Shockley equation, see Eqs. (4.5) and (4.6).

$$I_f = A \cdot J_f = I_s \cdot \left[\exp\left(\frac{eV}{n_{ideal} k_B T}\right) - 1 \right] = A \cdot J_s \cdot \left[\exp\left(\frac{eV}{n_{ideal} k_B T}\right) - 1 \right] \quad (4.5)$$

$$J_s = e \cdot \left[\sqrt{\frac{D_n}{\tau_n} \cdot \frac{n_i^2}{N_D}} + \sqrt{\frac{D_p}{\tau_p} \cdot \frac{n_i^2}{N_A}} \right] \quad (4.6)$$

Here, I_f is the forward current (in mA), I_s is the saturated current (in mA), A is the cross-sectional area (in m²), e is the electron charge ($\sim 1.6 \times 10^{-19}$ C),

Figure 4.26 A multi-input single-output (MISO) system of the LED electrical model.

n_{ideal} is the ideal constant, k_B is the Boltzmann constant ($\sim 1.38 \times 10^{-23}$ J K^{-1}), T is the absolute temperature (in K), D_n is the diffusion constant of free electrons (in m^2 s^{-1}), D_p is the diffusion constant of free holes (in m^2 s^{-1}), τ_n is the lifetime of electron minority (in s), τ_p is the lifetime of hole minority (in s), N_D is the density of donors (in cm^{-3}), N_A is the density of acceptors (in cm^{-3}), $n_i = \sqrt{p \cdot n}$ is the intrinsic concentration (in cm^{-3}), p is the concentration of free holes (in cm^{-3}), n is the concentration of free electrons (in cm^{-3}), and V is the voltage at the junction (in V). The space charge region has a potential called built-in voltage or diffusion voltage (in V) which is denoted by V_D, see Eq. (4.7).

$$V_D = \frac{kT}{e} \cdot \ln\left(\frac{N_A \cdot N_D}{n_i^2}\right) \tag{4.7}$$

The width (W_D/m) of the depletion region is determined by the Poisson equation for reverse current bias state and low current forward bias state, see Eq. (4.8).

$$W_D = \sqrt{\frac{2\epsilon}{e} \cdot (V_D - V_A) \cdot \left(\frac{1}{N_A} + \frac{1}{N_D}\right)} \tag{4.8}$$

Here, $\epsilon = \epsilon_r \epsilon_0 = (1 + \chi) \cdot \epsilon_0$ is the dielectric permittivity (in F m^{-1}), ϵ_r is the relative permittivity, ϵ_0 is the vacuum permittivity ($\epsilon_0 \sim 8.85 \times 10^{-12}$ F m^{-1}), χ is the electric susceptibility of the material, and V_A is the voltage (in V) applied to the junction. Furthermore, the free electron concentration (n/cm^{-3}) and the free hole concentration (p/cm^{-3}) can be determined by Eq. (4.9).

$$n = N_C \cdot Exp\left(\frac{E_C - E_F}{k_B T}\right) \text{ and } p = N_V \cdot Exp\left(-\frac{E_V - E_F}{k_B T}\right) \tag{4.9}$$

Here N_C is the effective state density of the conduction band edge (in cm^{-3}), N_V is the effective state density of the valence band edge (in cm^{-3}), E_C is the energy level of the conduction band (in eV), and E_V is the energy level of the valence band (in eV). Effective state densities can be calculated by Eqs. (4.10) and (4.11).

$$N_C = 2 \cdot \left(\frac{2\pi m_{en} k_B T}{h^2}\right)^{\frac{3}{2}} \cdot M_C \tag{4.10}$$

$$N_V = 2 \cdot \left(\frac{2\pi m_{ep} k_B T}{h^2}\right)^{\frac{3}{2}} \tag{4.11}$$

Here, m_{en} and m_{ep} are the effective mass of electrons and of holes for each state, respectively, h is the planck constant, M_c is the number of equivalent minima in the conduction band. So, the intrinsic concentration n_i can be computed by Eq. (4.12).

$$n_i = \sqrt{N_C \cdot N_V \cdot \left\{\exp\left(\frac{E_C - E_V}{k_B T}\right)\right\}} = \sqrt{N_C \cdot N_V \cdot \left\{\exp\left(\frac{E_g - eV_D}{k_B T}\right)\right\}} \quad (4.12)$$

According to the energy conservation law, the energy equivalent can be considered as described Eq. (4.13).

$$eV_D - E_g + (E_F - E_V) + (E_C - E_F) = eV_D - E_g + E_C - E_V = 0 \text{ or } E_C - E_V = E_g - eV_D \quad (4.13)$$

Under the reverse bias condition, the LED's current saturates and this saturation current is given by the factor preceding the exponential function in the Shockley equation (Eq. (4.5)) while n_{ideal} is assumed to be equal 1. Therefore, Eq. (4.14) arises.

$$I_r = I_s \cdot \left[\exp\left(\frac{eV}{k_B T}\right) - 1\right] \text{ with } I_s = eA \cdot \left[\sqrt{\frac{D_n}{\tau_n}} \cdot \frac{n_i^2}{N_D} + \sqrt{\frac{D_p}{\tau_p}} \cdot \frac{n_i^2}{N_A}\right] \quad (4.14)$$

Under a typical forward bias condition, the LED's voltage is $V_f \gg k_B T/e$ and n_{ideal} is assumed to be equal 1. Hence I_f can be approximated by Eqs. (4.15) and (4.16).

$$\left[\exp\left(\frac{V}{\frac{k_B T}{e}}\right) - 1\right] \approx \exp\left(\frac{eV}{k_B T}\right) = \exp\left[\frac{e(V_A - V_D)}{k_B T}\right] \quad (4.15)$$

$$I_f = eA \cdot \left(\sqrt{\frac{D_p}{\tau_p}} \cdot N_A + \sqrt{\frac{D_n}{\tau_n}} \cdot N_D\right) \cdot \exp\left[\frac{e(V_A - V_D)}{k_B T}\right] \quad (4.16)$$

The current starts with a very strong increment if the applied voltage is approximately equal to the built-in voltage (or diffusion voltage) V_D. Therefore, the diffusion voltage is also called threshold voltage (V_{th}). If the energy distance between the conduction band and the valence band ($E_C - E_V$) ~ 0, the relationship between the band gap energy of the semiconductor material and the diffusion voltage can be determined by Eq. (4.17).

$$V_{th} \approx V_D \approx \frac{E_g}{e} \quad (4.17)$$

The equation system (4.5)–(4.17) describes and quantifies the electrical phenomena that take place inside an ideal diode. However, the above approximations ignored many important factors that influence the operation of real LEDs. In addition, many parameters such as the diffusion constant, lifetime of hole minority and electron minority, effective mass, acceptor concentration, donor concentration, free electron concentration, free hole concentration, number of the equivalent minima in the conduction band, dielectric permittivity, and others are very difficult to be determined explicitly under the real operating conditions of LEDs. Although the classical theoretical approach helps understand the fundamental electrical nature of the LEDs, it is unable to describe the electric behavior of real LEDs in a usable way.

4.5.2
A LED Experimental Electrical Model Based on the Circuit Technology

Although there are many investigations reported in [8–43] about the electrical processes in the semiconductor layers inside LEDs, an accurate qualification is difficult (as mentioned before) because there are several nonmeasureable parameters related to minority carriers in tiny charge volumes. Applicable measures include the measured V–I characteristics and AFM (Atomic Force Microscopy)/TEM (cross-sectional Transmission Electron Microscopy) images [8–43]. Any attempt to establish a link between such measured quantities and the theoretical predictions (i.e., the theoretical hetero-junction models) turned out to be nearly unfeasible. Consequently, such theoretical models can be used only to show the way for more feasible modeling methods. According to this idea, several different models were developed including a so-called tunneling current model for low bias [16, 42], a model using defects, dislocations, and impurities with ideality constants for low currents and low temperatures [29, 30] and a diffusion-recombination current model for high biases and high temperatures with the ideality constant of 1 for the diffusion component and the ideality constant of 2 for limited space-charge recombination.

In the approach to be presented in this section (Baureis model [10]), the so-called small signal scattering (S) parameters with the high frequency approximation (>50 MHz) are used to determine the model parameters. The electrical model is established based on this idea shown in Figure 4.27 where the parameters C_1 and C_2 are used to quantify the capacitive coupling of the diode related to the ground plane of the test fixture. The parameter L_s is used to describe bond wire and package inductance, R_s stands for series resistance, C_j for the combination of the diffusion and junction capacitance, R_j for differential junction resistance and the parameters L_{skin}, R_{skin}, and k are used to take the skin effect into account. On the basis of the measured and simulated results, the author estimated the model parameters for a sample LED, see Table 4.6.

4.5.3
An Example for a Limited Electrical Model for LEDs

4.5.3.1 Limited Operating Range
First, the operating range of high power LEDs is defined. For today's usual LEDs in solid state lighting applications, the operating current ranges between 150 and 1000 mA and the operating temperature is about 40–100 °C. In order to achieve a

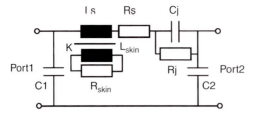

Figure 4.27 Small signal equivalent circuit for an InGaAlP LED at 0.9 mA. Reproduced from [10], copyright © 2005 with permission from IEEE.

Table 4.6 Parameters of the Baureis LED model estimated by means of the high frequency technique for a sample LED [10].

R_s/Ω	1.59	C_1/fF	150	L_{skin}/nH	1.00
L_s/nH	1.72	C_2/fF	330	R_{skin}/Ω	15.0
C_j/pF	76.2	R_j/Ω	72.3	K	0.25

higher usability of LED modeling, this operating range will be slightly extended to 100–1000 mA and 25–100 °C. Reverse and low voltage ranges and low temperatures (below 25 °C) are not necessary to be considered for the modeling in this section because no solid state lighting system operates under such conditions.

4.5.3.2 Mathematical Description of the LEDs Forward Current in the Limited Operating Range

In the earlier-defined operating range, thermal diffusion – recombination is the predominant carrier transport mechanism. Other mechanisms can be ignored. Thus, the corresponding mathematical descriptions for the LED's forward current are available in Eqs. (4.5) and (4.16). They can be rewritten as

$$I_f = eA \cdot \left(\sqrt{\frac{D_p}{\tau_p}} \cdot N_A + \sqrt{\frac{D_n}{\tau_n}} \cdot N_D \right) \cdot \left\{ \exp\left[\frac{e(V_A - V_D)}{n_{ideal} k_B T} \right] \right\} \quad (4.18)$$

$$I_f = I_s \cdot \left\{ \exp\left[\frac{e(V_A - V_D)}{n_{ideal} k_B T} \right] \right\} \quad (4.19)$$

From Eq. (4.19), the following equations can be obtained:

$$V_A = \left[\frac{n_{ideal} k_B}{e} \right] T_j \ln(I_f) + \left[\frac{273.15 \ln(I_s) n_{ideal}}{e} \right] \ln(I_f) - \left\{ \left[\frac{n_{ideal} k_B}{e} \right] \ln(I_s) \right\} T_j$$
$$+ \left\{ V_D - \left[\frac{273.15 \ln(I_s) n_{ideal}}{e} \right] \ln(I_s) \right\} \quad (4.20)$$

$$V_f = V_A = aT_j \ln(I_f) + b\ln(I_f) - cT_j + d \quad (4.21)$$

$$a = \left[\frac{n_{\text{ideal}} k_{\text{B}}}{e}\right] \tag{4.22}$$

$$b = \left[\frac{273.15 \ln\left(I_{\text{s}}\right) n_{\text{ideal}}}{e}\right] \tag{4.23}$$

$$c = \left\{\left[\frac{n_{\text{ideal}} k_{\text{B}}}{e}\right] \ln(I_{\text{s}})\right\} \tag{4.24}$$

$$d = \left\{V_{\text{D}} - \left[\frac{273.15 \ln\left(I_{\text{s}}\right) n_{\text{ideal}}}{e}\right] \ln(I_{\text{s}})\right\} \tag{4.25}$$

The meaning of the symbols in the above equations was explained in Section 4.5.1. In addition, for the specific semiconductor structure in the defined operating range, the parameters a, b, c, and d can be assumed as constants. Moreover, based on Eq. (4.20), it can be recognized that the forward voltage of the LEDs has a linear dependence on junction temperature T_{j} and a logarithmic dependence on forward current I_{f}.

4.5.3.3 An Example for the Application of the Limited Electrical Model

In order to prove the usability of the above equations, 20 typical cold white LEDs (5000 K) were selected. Their junction temperature, forward voltage, and forward current were measured. From this measurement, the relationship between forward voltage and junction temperature (minimal, average, maximal values, and 95% confidence intervals for these 20 cold white LEDs) is shown in Figure 4.28. Similarly, the relationship between their forward voltage and their forward current is shown in Figure 4.29.

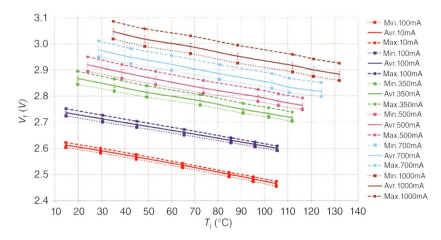

Figure 4.28 Forward voltage versus junction temperature with the minimal, average, and maximal values of 20 similar cold white LEDs (5000 K) for different currents.

166 | *4 Measurement and Modeling of the LED Light Source*

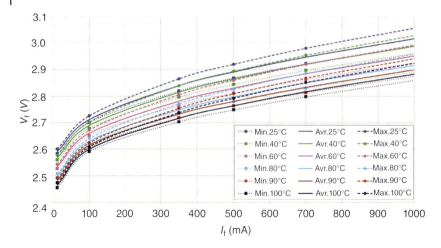

Figure 4.29 Forward voltage versus forward current with the minimal, average, and maximal values of 20 similar cold white LEDs (5000 K).

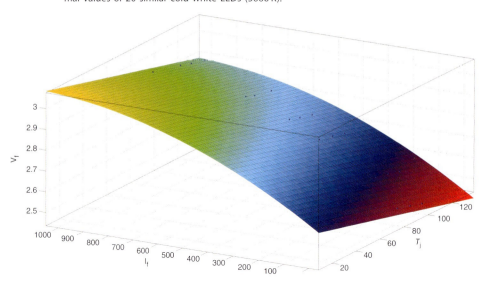

Figure 4.30 Three-dimensional representation of the electrical model for cold white LEDs: forward voltage (ordinate, V) versus junction temperature (abscissa, °C) and forward current (abscissa, mA).

The results shown in Figures 4.28 and 4.29 agree well with Eq. (4.21). Thus, it can be confirmed that the usage of a simple mathematical formula with a feasible LED measurement in the predefined operating range is a good solution to overcome the difficulties of concurrent approaches. Finally, based on the results of junction temperature, forward current and forward voltage measurements and the mathematical formula of Eq. (4.21), this electrical model for cold white LEDs can be considered valid within this limited operating range. To make the

Table 4.7 Parameter values of the model of Eq. (4.21) and the approximate model of Eq. (4.26) for cold white LEDs. R^2, goodness-of-fit to experimental data.

Parameters in Eq. (4.21)	Value for Eq. (4.21)	Parameters in Eq. (4.26)	Value for Eq. (4.26)
a	1.134×10^{-4}	a_1	-1.637×10^{-3}
b	8.329×10^{-2}	b_1	-3.287×10^{-7}
c	2.077×10^{-3}	c_1	7.465×10^{-4}
d	2.421	d_1	2.657
R^2	0.9238	R^2	0.9807

model intuitively more accessible, its three-dimensional representation is shown in Figure 4.30 (its parameters are listed in Table 4.7).

4.5.3.4 Evaluation and Improvement of the Electrical Model

Equation (4.21) can be approximated by the following equation which is more straightforward hence more convenient for the control algorithm in the microcontroller of an LED lighting system:

$$V_f = a_1 T_j + b_1 I_f^2 + c_1 I_f + d_1 \qquad (4.26)$$

Parameter values of the electrical model for cold white LEDs according to Eq. (4.21) and the approximate Eq. (4.26) are listed in Table 4.7.

As can be seen from Table 4.7, the approximate electrical model of Eq. (4.26) has a R^2 value of 0.9807 which is even better than that of the original model of Eq. (4.21). Note that the approximate model of Eq. (4.26) can also be applied to other LED types (not only cold white) with appropriate parameters.

4.6
LED Spectral Model

4.6.1
Spectral Models of Color Semiconductor LEDs and White PC-LEDs

The SPD of the LEDs is an important characteristic to determine luminous flux, chromaticity, color quality, and many other parameters. Therefore, since about 2005, several publications have appeared about its modeling. Until now, there have been two main approaches including a purely mathematical approach and a combined approach between mathematical descriptions and LED measurements. In the spectral model to be described in this section, the LED is assumed to be a multiple-input-multiple-output system (MIMO system, see Figure 4.31). The input of the MIMO system is junction temperature ($T_j/°C$) and forward current (I_f/mA). The output is the LED's emission spectrum, that is, spectral intensity S_p

Figure 4.31 A multi-input multi-output (MIMO) system for the LED's spectral model.

(W nm^{-1}) which can be roughly characterized by its peak wavelength (λ_p/nm) and full width at half maximum (λ_{FWHM}/nm).

4.6.1.1 Mathematical Approach

Gaussian Function In a purely mathematical approach, the authors in [44, 45] assumed that a first order Gaussian function can describe LED's emission spectrum. This function has peak wavelength (λ_p/nm) and full width at half maximum (λ_{FWHM}/nm) as parameters, see Eq. (4.27).

$$S(\lambda_p, \lambda_{FWHM}, \lambda) = \exp\left\{\frac{-2.7725(\lambda - \lambda_p)^2}{\lambda_{FWHM}^2}\right\} \quad (4.27)$$

This Gaussian function has been used continuously until now in many documents about LED spectra although its simulating quality is not good enough. In addition, the current dependence and the temperature dependence of the LED's spectrum are not reflected by this approach.

Further Approaches Developed from the Gaussian Function The theoretical background about LED SPDs was described in Section 3.3. As can be seen from Eqs. (3.38)–(3.39), LED SPDs are asymmetric. Indeed, there is a difference between the left side and the right side of the LED spectrum: the left distribution has a square root form ($S_{left}(E) \sim \sqrt{E - E_g}$) while the right distribution has an exponential form ($S_{right}(E) \sim e^{-E/(kT_j\alpha)}$). Thus, the rising edge of the left distribution is faster while the falling edge of the right distribution is slower. Gaussian functions, however, are symmetric. Therefore, the authors F. Reifegerste (Private communication, 2014[1]), Ohno [45] and Ohno [46] attempt to improve the Gaussian function form. Authors in [47] support a double Gaussian function with two different sets of parameters including power, peak wavelength, and full width at half maximum. All parameters are functions of both junction temperature and forward current. The estimated $M \times N$ spectral matrix S for color semiconductor LEDs is described in Eq. (4.28).

$$S = G + G\prime \quad (4.28)$$

Here, $G = (gg_1, gg_2, \ldots, gg_M)^T$ and $G' = (gg'_1, gg'_2, \ldots, gg'_M)^T$ are Gaussian spectral matrices. The matrix G has M spectral vectors gg with N sampling wavelengths. From the nth point of the mth row vector gg_m (denoted by g_{mn}), its value

---

1) Technische Universität Dresden, Institut für Feinwerktechnik und Elektronik-Design, Dresden, Germany.

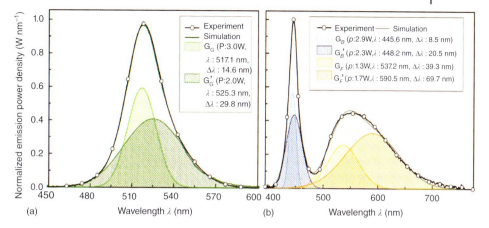

Figure 4.32 Double Gaussian model for a green semiconductor LED (a) and a white PC-LED (b) at $T_j = 25\,°C$ and $I_f = 350\,mA$. For the PC-LED, the blue and phosphor-converted spectral components are considered individually by two double functions G_B and G'_B and G_F and G'_F, respectively. (Reproduced from [47] copyright © 2012 with permission from the Optical Society of America.)

can be calculated by Eq. (4.29).

$$gg_{mn} = p_m \exp\left\{\frac{[\lambda_n - (\lambda_p)_m]^2}{(\lambda_{FWHM})_m^2}\right\} \quad (4.29)$$

The parameters p_m, $(\lambda_p)_m$, and $(\lambda_{FWHM})_m$ correspond to the mth power, the mth peak wavelength, and the mth full width at half maximum and these values can be obtained by carrying out the minimization in Eq. (4.30).

$$\arg\min\,[|S_m - S_m|^2, \{p_m, (\lambda_p)_m, (\lambda_{FWHM})_m, p'_m, (\lambda'_p)_m, (\lambda'_{FWHM})'_m\}] \quad (4.30)$$

Here, s_m and $s_m = gg_m + gg'_m$ are the mth row vectors of S and S, respectively. To validate the accuracy of this approach, simulated and measured spectra of (a) a green semiconductor LED and (b) a PC-LED are compared in Figure 4.32.

Concerning the alternative mathematical forms shown in Table 4.8, F. Reifegerste (Private communication, 2014[1])), assumes that the functions Nos. 1–4 should be ignored because they are symmetric. Function No. 10 approximates the spectra very well, but its parameters cannot be interpreted directly. In addition, this function is defined piecewise and cannot be extended for varying temperatures and currents. Thus, it is not suitable for modeling. Other forms that consider the asymmetry of the LED spectra are the equations with two ranges – one for the left side and one for the right side of the LED spectra like the functions Nos. 2 and 8. Unfortunately, these mathematical forms are too cumbersome for optimization. In function No. 9, two skew parameters are used but these parameters cannot be interpreted.

Table 4.8 Model functions for LED spectra.

Ord.	Name	Number of Parameters	Function $f(\lambda)$
1	Gaussian	3	$f(\lambda) = Ae^{-\left(\frac{\lambda-C}{W}\right)^2}$
2	Split Gaussian	4	$f(\lambda) = Ae^{-\left(\frac{\lambda-C}{W}\right)^2}$ with $W = W_1$, for $\lambda < C$ and $W = W_2$ otherwise
3	Sum of Gaussian	6	$f(\lambda) = A_1 e^{-\left(\frac{\lambda-C_1}{W_1}\right)^2} + A_2 e^{-\left(\frac{\lambda-C_2}{W_2}\right)^2}$
4	Second Order Lorentz	3	$f(\lambda) = \dfrac{A}{\left(1+\left(\frac{\lambda-C}{W}\right)^2\right)^2}$
5	Logistic Power Peak	4	$f(\lambda) = \dfrac{A}{S}\left(1+e^{-\frac{\lambda-C+W\ln(S)}{W}}\right)^{\frac{-S-1}{S}} \left\{e^{-\frac{\lambda-C+W\ln(S)}{W}}\right\} (S+1)^{\frac{S+1}{S}}$
6	Asymmetric Logistic Peak	4	$f(\lambda) = A\left(1+e^{-\frac{\lambda-C+W\ln(S)}{W}}\right)^{-S-1} \left\{e^{-\frac{\lambda-C+W\ln(S)}{W}}\right\} S^{-S}(S+1)^{S+1}$
7	Pearson VII	4	$f(\lambda) = \dfrac{A}{\left(1+\left(\frac{\lambda-C}{W}\right)^2\left(2^{\frac{1}{S}}-1\right)\right)^S}$
8	Split Pearson VII	6	$f(\lambda) = \dfrac{A}{\left(1+\left(\frac{\lambda-C}{W}\right)^2\left(2^{\frac{1}{S}}-1\right)\right)^S}$ with $W = W_1$, $S = S_1$ for $\lambda < C$, otherwise $W = W_2$, $S = S_2$
9	Asymmetric double sigmoidal	5	$f(\lambda) = \dfrac{A}{1+e^{-\frac{\lambda-C+\frac{W}{2}}{S_1}}}\left(1-\dfrac{1}{1+e^{-\frac{\lambda-C+\frac{W}{2}}{S_2}}}\right)$
10	Piecewise third order polynomial (Spline)	4n	Piecewise: $f(\lambda) = a_3 x^3 + a_2 x^2 + a_1 x^1 + a_0$, piecewise defined for n ranges $x_{k-1} \leq x < x_k$, $k = 1, \ldots, n$

Source: After [48], reproduced from F. Reifegerste (Private communication, 2014[1]), copyright © 2014 with permission from Dr. Frank Reifegerste.

Present authors consider only functions Nos. 5–7 as adequate. Especially, the parameters A and C can be interpreted directly as intensity and peak wavelength, respectively. In the next step, various LEDs with different semiconductor material systems were investigated in practice to validate these mathematical formulae and the present authors concluded that function No. 5 was the most suitable one for the approximation of the color semiconductor LED's spectra. Particularly, function No. 5 with its extension can be rewritten as follows:

$$f(\lambda) = \frac{A}{S}\left(1+e^{-\frac{\lambda-C+W\ln(S)}{W}}\right)^{\frac{-S-1}{S}} \left\{e^{-\frac{\lambda-C+W\ln(S)}{W}}\right\} (S+1)^{\frac{S+1}{S}} \tag{4.31}$$

$$A(T_j, I_f) = a_0 T_j^{a_T} I_f^{a_I} \tag{4.32}$$

$$C(T_j, I_f) = c_0 + c_T T_j + c_I \log(I_f) \tag{4.33}$$

$$S(T_j, I_f) = s_0 + s_T T_j + s_I \log(I_f) \tag{4.34}$$

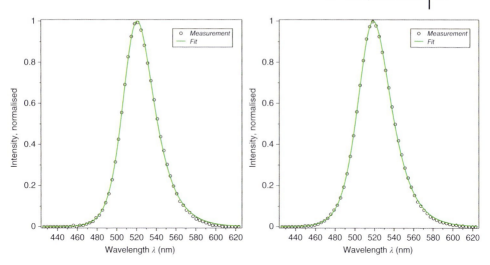

Figure 4.33 Comparison between simulated (Eq. (4.31), continuous curve) and measured (empty diamond symbols) emission spectra for a sample color semiconductor LED at 10 mA and 76.8 °C (a) and at 20 mA and 104.7 °C (b). After [48], reproduced from F. Reifegerste (Private communication, 2014[1])), copyright © 2014 with permission from Dr. Frank Reifegerste.

$$W(T_j, I_f) = w_0 + w_T T_j + w_I I_f \tag{4.35}$$

The accuracy of Eq. (4.31) was proved by a comparison between simulated and measured spectra at 10 mA and 76.8 °C and 20 mA and 104.7 °C, see Figure 4.33. As can be seen from Figure 4.33, the mathematical form of Equation 4.31 represents a very good approximation for this color semiconductor LED. However, this is a pure mathematical approach. Therefore, the aim of Section 4.6.1.b is to describe a combined approach that takes also the physical nature of the color semiconductor LED into consideration.

The authors of [49] propose a combined approach between the above mathematical description and a consideration of the physical nature of color semiconductor LEDs. Latter issue is not only to be considered by the distribution of the joint density of states and the Boltzmann distribution but also by the relationship between junction temperature (T_j) and carrier temperature (T_c) and the band gap energy shift and the increment in the nonradiative recombination rate with junction temperature. These authors [49] start with the mathematical form containing the combination of two exponentials as follows:

$$\Phi_{e,\lambda} = \frac{2S_{p0}}{\exp\left[\frac{\lambda - \lambda_p}{B(\lambda)}\right] + \exp\left[-\frac{\lambda - \lambda_p}{B(\lambda)}\right]} \tag{4.36}$$

Here, S_{p0} is spectral intensity at the peak wavelength λ_p and $B(\lambda)$ is defined piecewise by asymmetrical line width. Then, the authors [49] insert the Boltzmann behavior into Eq. (4.36) to obtain

$$\Phi_{e,v} = \frac{1}{S_1 \exp[-a(v-v_p)] + S_2 \exp\left[\frac{h}{kT_C}(v-v_p)\right]} \quad (4.37)$$

Here, v_p is the experimental peak frequency and S_1, S_2, a, and T_c are four positive fitting parameters. Eq. (4.37) was not able to model the complete shape of the spectrum. Therefore, the authors [49] improved it by the addition of a Gaussian function as follows:

$$\Phi_{e,v} = \frac{1}{S_1 \exp[-a(v-v_p)] + S_2 \exp\left[\frac{h}{kT_C}(v-v_p)\right] + S_3 \exp\left[-\left(\frac{v-v_G}{b}\right)^2\right]} \quad (4.38)$$

Here, S_3, b and the peak frequency of the Gaussian function (v_G) act as three additional parameters. Successively, the authors [49] assumed that Eq. (4.38) can be accepted for a single condition (25 °C and 350 mA) as this was verified for typical red, green, and blue semiconductor LEDs with R^2 values exceeding 99%. Carrier temperature was higher than junction temperature, as expected. Then, it was attempted to add three temperature dependent terms to Eq. (4.38) including carrier temperature, peak wavelength, and absolute power. The carrier temperature term is shown in Eq. (4.39).

$$T_C(T_j) \approx c(T_j - T_{ref}) + T_{C,ref} = c\Delta T + T_{C,ref} \quad (4.39)$$

Here, T_{ref} is 300 K, $T_{c,ref}$ is the reference temperature of the carrier, and c is a positive fitting parameter. As the second term, based on the Varshni equation, the temperature dependence of the energy band gap can be rewritten as follows:

$$E_g(T_j) = E_g(0) - \frac{aT_j^2}{T_j + b} \quad (4.40)$$

$$E_g(T_j) \approx E_{g,ref} - \alpha'(T_j - T_{ref}) \quad (4.41)$$

Here, $E_{g,ref}$ is the band gap energy at 300 K and α' is a positive fitting constant. The temperature dependence of the peak frequency and the Gaussian peak frequency is calculated by Eqs. (4.42) and (4.43), respectively.

$$v_p(T_j) \approx v_{p,ref} - \gamma_p(T_j - T_{ref}) = v_{p,ref} - \gamma_p \Delta T \quad (4.42)$$

$$v_G(T_j) \approx v_{G,ref} - \gamma_G(T_j - T_{ref}) = v_{G,ref} - \gamma_G \Delta T \quad (4.43)$$

Here, $\gamma_p = \alpha'/h$ and γ_G plays a similar role as γ_p for the Gaussian element; $v_{p,ref}$ and $v_{G,ref}$ are the peak wavelength and the Gaussian peak wavelength at 300 K, respectively. Eq. (4.44) shows the temperature dependence of spectral radiant power at a constant forward current based on the energy relationship.

$$\Phi_e(T) \exp\left(-\frac{T_j - T_{ref}}{T_0}\right) = \exp\left(-\frac{\Delta T}{T_0}\right) \quad (4.44)$$

Here, T_{ref} is the reference temperature and the denominator T_0 is called characteristic temperature. When the three above temperature dependence terms are added, the SPD of the color semiconductor LED described in Eq. (4.38) can be rewritten in the new form shown in Eq. (4.45).

$$\Phi_{e,v}(T) = \left\{ \begin{array}{c} \dfrac{1}{S_{1,ref} \exp\left[-a\left(v - v_{p,ref} + \gamma_p \Delta T\right)\right]} \\ + S_{2,ref} \exp\left[\dfrac{h(v - v_{p,ref} - \gamma_p \Delta T)}{k(c\Delta T + T_{c,ref})}\right] \\ + S_{3,ref} \exp\left[-\left(\dfrac{v - v_{G,ref} + \gamma_G \Delta T}{b}\right)^2\right] \end{array} \right\} \exp\left(-\dfrac{\Delta T}{T_0}\right) \quad (4.45)$$

Here, $v_{p,ref}$, $S_{1,ref}$, a, $S_{2,ref}$, $T_{c,ref}$, $v_{G,ref}$, $S_{3,ref}$, and b are determined from one reference spectrum at the reference temperature T_{ref} and $\Delta T = T_j - T_{ref}$. In order to determine the parameters γ_p, γ_G, c, and T_0, a spectrum obtained at an additional temperature must be available. Eq. (4.45) is considered as the most appropriate equation which combines a plausible mathematical hypothesis with the knowledge on the physical nature of semiconductor LEDs. Note that current dependence is not fully accounted for by this model.

The accuracy of the model in terms of chromaticity difference ($\Delta u'v'$) and luminous flux difference between the modeled and measured spectra at 350 mA and 340 K for six sample LEDs is shown in Table 4.9. As can be seen, the high chromaticity difference $\Delta u'v' = 20.7 \times 10^{-3}$ for the LED R_2 is not acceptable.

Table 4.9 Comparison between modeled (Eq. (4.45)) and measured results at 350 mA and 340 K for six sample color semiconductor LEDs.

LED	Measured			Simulated			Deviation	
	Φ_v/lm	CIE x	CIE y	Φ_v/lm	CIE x	CIE y	$\Delta\Phi_v$/%	$\Delta u'v'$ /10^{-3}
R_1	20.1	0.6939	0.3055	19.6	0.6935	0.3064	2.4	1.5
R_2	26.9	0.6941	0.3053	24.5	0.7037	0.2962	8.9	20.7
G_1	29.2	0.2045	0.6714	28.7	0.2049	0.6844	1.7	2.8
G_2	55.4	0.1671	0.6679	54.7	0.1709	0.6649	1.3	1.7
B_1	4.5	0.1513	0.0284	4.5	0.1511	0.0270	0.0	3.8
B_2	18.0	0.1367	0.0537	17.9	0.1363	0.0534	0.5	0.7

Source: Reproduced from [49], copyright © 2010 with permission from the Journal of Applied Physics.

4.6.2
An Example for a Color Semiconductor LED Spectral Model

Two approaches for the modeling of color semiconductor LEDs and white PC-LEDs were described in Section 4.6.1. Also, some other LED spectral models from literature were compared and their advantages, disadvantages, accuracy, and applicable ranges were reviewed. In this section, the multi-Gaussian LED spectral model (i.e., the sum of two or more Gaussian functions, similar to Eq. (4.28)) is applied to a typical blue semiconductor LED, as an example.

4.6.2.1 Experiments on Spectral Models for Color Semiconductor LEDs

Description of the Subject   A blue semiconductor LED with a dominant wavelength of 450 nm was used in this investigation. This LED was measured at the forward currents 350, 700, and 1000 mA and the layout temperatures 25, 55, 85, and 100 °C. Measured spectra are shown in Figure 4.34. Model parameters of the multi-Gaussian spectral model were calculated for this blue semiconductor LED. Applying the procedure described in Section 4.8, the junction temperature of the blue semiconductor LED could be determined.

Multi Gaussian Spectral Model   The multi-Gaussian model was investigated from first order (one Gaussian only) up to sixth order (sum of six Gaussians). The result of curve fitting for the sample blue LED's spectrum with the forward current of

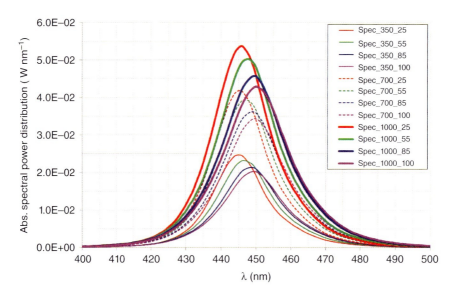

Figure 4.34 Absolute spectral power distributions of the investigated blue semiconductor LED at the forward currents of 350, 700, and 1000 mA and the layout temperatures of 25, 55, 85, and 100 °C.

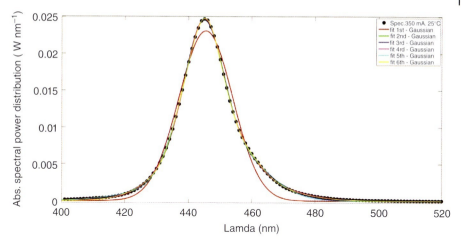

Figure 4.35 Curve fitting to the spectrum of the blue semiconductor LED at 350 mA and 25 °C using different Gaussian functions (first order–sixth order).

350 mA and the layout temperature of 25°C is illustrated in Figure 4.35. Curve fitting results show that the first order Gaussian function is not usable as the difference between the measured and the fitted spectrum is too high. But there is no significant improvement of fitting quality from the second order on.

The parameters of the Gaussian function were obtained by the Matlab® curve fitting toolbox, see Table 4.10. Although the value of R^2 was always greater than 0.99, only the parameters of the second order Gaussian function could be predicted from forward current and junction temperature reasonably well. Therefore, the second order Gaussian function was selected to establish the LED's spectral model for this blue semiconductor LED.

In a real LED lighting system, microcontrollers perform these modeling computations. For the microcontroller it is important to specify the dependence of the six parameters of the double (second order) Gaussian function (Eq. (4.28)) on forward current and junction temperature. To this aim, the polynomials of Eqs. (4.46)–(4.51) can be used.

$$a_1 = p_{00,a1} + p_{10,a1} \cdot I_f + p_{01,a1} \cdot T_j + p_{20,a1} \cdot I_f^2 + p_{11,a1} \cdot I_f \cdot T_j + p_{02,a1} \cdot T_j^2 \quad (4.46)$$

$$a_2 = p_{00,a2} + p_{10,a2} \cdot I_f + p_{01,a2} \cdot T_j + p_{20,a2} \cdot I_f^2 + p_{11,a2} \cdot I_f \cdot T_j + p_{02,a2} \cdot T_j^2 \quad (4.47)$$

$$b_1 = p_{00,b1} + p_{10,b1} \cdot I_f + p_{01,b1} \cdot T_j + p_{20,b1} \cdot I_f^2 + p_{11,b1} \cdot I_f \cdot T_j + p_{02,b1} \cdot T_j^2 \quad (4.48)$$

$$b_2 = p_{00,b2} + p_{10,b2} \cdot I_f + p_{01,b2} \cdot T_j + p_{20,b2} \cdot I_f^2 + p_{11,b2} \cdot I_f \cdot T_j + p_{02,b2} \cdot T_j^2 \quad (4.49)$$

$$c_1 = p_{00,c1} + p_{10,c1} \cdot I_f + p_{01,c1} \cdot T_j + p_{20,c1} \cdot I_f^2 + p_{11,c1} \cdot I_f \cdot T_j + p_{02,c1} \cdot T_j^2 \quad (4.50)$$

$$c_2 = p_{00,c2} + p_{10,c2} \cdot I_f + p_{01,c2} \cdot T_j + p_{20,c2} \cdot I_f^2 + p_{11,c2} \cdot I_f \cdot T_j + p_{02,c2} \cdot T_j^2 \quad (4.51)$$

Table 4.10 Fitting parameter values of the second order Gaussian function to the sample blue LED (see Figure 4.35).

I_f/mA	350	350	350	350	700	700	700	700	1000	1000	1000	1000
T_j/°C	35	64	95	110	41	70	103	120	54	80	112	128
T_s/°C	25	55	85	100	25	55	85	100	25	55	85	100
V_f/V	2.905	2.856	2.806	2.783	3.034	2.980	2.927	2.901	3.114	3.057	3.002	2.977
$a_1/10^{-3}$	16	14	13	12	27	24	21	20	34	30	27	25
$a_2/10^{-3}$	8.56	9.04	8.77	8.65	14.9	15.4	15.4	14.2	19.7	19.9	18.8	17.8
b_1	444.7	446.3	448.1	448.9	444.9	446.6	448.4	449.3	445.4	447.1	448.9	449.8
b_2	447.8	448.8	450.1	450.7	447.7	448.9	450.2	451.0	447.9	449.3	450.7	451.4
c_1	8.3	8.5	8.9	9.1	8.8	9.0	9.3	9.8	9.1	9.4	9.9	10.3
c_2	19.0	19.4	20.2	20.5	19.7	20.2	20.7	21.5	20.2	20.8	21.7	22.2

Table 4.11 Fit parameter values for the constants of the second order Gaussian function to describe the dependence on forward current and junction temperature via Eqs. (4.46)–(4.51).

Parameters	$P_{00}/10^0$	$P_{10}/10^{-4}$	$P_{01}/10^{-4}$	$P_{20}/10^{-7}$	$P_{11}/10^{-7}$	$P_{02}/10^{-6}$
a_1	0.005	0.421	−0.561	−0.055	−1.051	0.225
a_2	−0.002	0.251	0.844	−0.060	−0.087	−0.546
b_1	442.9	−9.034	594.3	0.8182	60.14	−41.09
b_2	447.1	−19.44	333.1	4.379	112.3	8.056
c_1	7.98	16.39	−94.65	−3.704	6.286	131.6
c_2	18.22	18.88	−6.104	−2.415	3.282	144.8

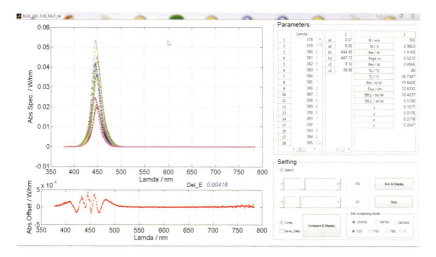

Figure 4.36 Interface of the evaluation program for the blue semiconductor LED model.

Fit parameter values for the sample blue LED in Figure 4.35 are listed in Table 4.11.

Evaluation of the Accuracy of the Model In order to evaluate the accuracy of the model, the equation system (Eqs. (4.46)–(4.51)) and the parameter values of Table 4.11 were programmed, see Figure 4.36. The input was forward current and layout temperature and the output were the absolute SPD, the constants of the LED spectral model, forward voltage, electrical power, optical power, junction temperature, thermal resistance, luminous flux, luminous efficiency, radiant efficiency, and chromaticity. In addition, the program calculated the difference between modeled and measured spectra, see the lower diagram of the user interface titled "Abs. Offset" in Figure 4.36. The $\Delta u'v'$ chromaticity difference between the two emission spectra was also computed. Latter values are listed in

Table 4.12 Chromaticity differences ($\Delta u'v'$) between modeled and measured spectra for the blue semiconductor LED.

I_f/mA	350	350	350	350	700	700	700	700	1000	1000	1000	1000
T_s/°C	25	55	85	100	25	55	85	100	25	55	85	100
$\Delta u'v'/10^{-3}$	4.18	4.2	4.7	5.01	4.2	4.6	5.06	5.14	4.4	4.9	5.15	4.9

Table 4.12. As can be seen from this table, the minimal chromaticity difference is 4.18×10^{-3} for the forward current of 350 mA and layout temperature of 25 °C and the maximal chromaticity difference is 5.15×10^{-3} for the forward current of 1000 mA and the layout temperature of 85 °C. As mentioned above, these chromaticity differences are well noticeable (see Section 5.8) so the accuracy of this model can be considered moderate.

4.6.3
An Example for a PC-LED Spectral Model

In this section, the multi-Gaussian spectral model is applied and verified for white PC-LEDs.

4.6.3.1 Experiments for the Spectral Models of White PC-LEDs

Description of the Subject A typical cold white PC-LED (5000 K) was chosen for this investigation. The blue chip in this cold white PC-LED had a peak wavelength of about 445 nm. When the phosphor is built on the blue chip, its peak wavelength is shifted a little toward longer wavelengths because of internal chemical, physical, and geometrical interactions between the phosphor and the chip structure. Moreover, the phosphor of this cold white PC-LED is a type of YAG phosphor with a peak wavelength of about 560 nm. However, the peak wavelength of the YAG phosphor is also shifted toward longer wavelengths in the cold white PC-LED because of the interaction between the phosphor and the chip structure. In addition, with changing operating conditions (forward current and temperature), the blue spectral component of the blue chip and the yellow spectral component of the YAG phosphor change their shape and intensity. Indeed, Figure 4.37 illustrates these spectral changes of the cold white PC-LED under different operating conditions. Operating conditions in this example include 350, 700, and 1000 mA (forward current) as well as 25, 55, 85, and 100 °C (layout temperature). Applying the procedure described in Section 4.8, the junction temperature of the cold white PC-LED could be determined.

Selection of the Appropriate Spectral Model In order to establish the spectral model for the cold white PC-LED shown in Figure 4.37, two approaches were considered. In the first approach, the spectrum of the cold white PC-LED was separated into

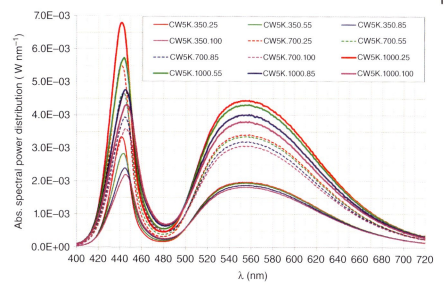

Figure 4.37 Spectral power distribution of the sample white PC-I FD (5000 K) for different values of forward current and operating temperature.

smaller spectral pieces such as a blue spectral piece (400–450 nm) and a YAG phosphor spectral piece (450–740 nm), or a blue spectral piece (400–450 nm), a cross spectral piece (450–500 nm), a green-amber spectral piece (500–620 nm), and a red spectral piece (620–740 nm). From the mathematical viewpoint, the blue, cross, and red spectral pieces could be modeled by double Gaussian, polynomial, and power functions, respectively. The green spectral pieces and the YAG phosphor spectral pieces could also be modeled by polynomial, Gaussian, or Fourier functions.

The authors in [47] described an LED model for a cold white PC-LED with a four-Gaussian-component function (or, in other words, a fourth-order Gaussian function) including a first two-Gaussian-component function (also called double Gaussian function) for the blue spectral piece and a second two-Gaussian-component function for the YAG phosphor spectral piece, see Figure 4.32b. The fitting for the YAG phosphor spectral piece with the two-Gaussian-component function or any other higher order Gaussian function turned out to be not accurate enough. Furthermore, other spectral separations with other mathematical forms (polynomial, Gaussian, or Fourier) could not be accepted, because the constants had no correlation with junction temperature or forward current.

On the basis of numerous trials with diverse combinations of spectral separations and mathematical functions, only the spectral separation with the blue spectral piece and the YAG phosphor spectral piece yielded an acceptable result. A second order Gaussian function and an eighth-order Fourier function were used as fit functions, see the blue and brown curves Figure 4.38. In this piecewise approach, the LED's spectral model has a problem at the limiting point (450 nm)

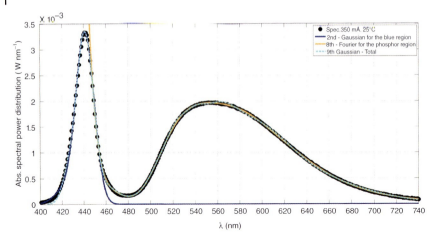

Figure 4.38 Curve fitting to the emission spectrum of the sample white PC-LED at 350 mA and 25 °C.

between the two spectral pieces: the left-to-450 nm range belongs to the blue spectral model and the right-to-450 nm range belongs to the YAG phosphor spectral model, and these two parts change dissimilarly under different operating conditions. Unfortunately, the model parameters of these spectral functions do not have a usable functional relationship with forward current and junction temperature. Therefore, this approach should not be used for spectral modeling.

In order to avoid these problems, the total measured spectrum was fitted by a ninth-order Gaussian function in a second approach, see the cyan curve in Figure 4.38. Fortunately, a good functional relationship was found between the parameters of this ninth-order Gaussian function and forward current and junction temperature.

Evaluation of the Accuracy of the White PC-LEDs Ninth-Order Gaussian Spectral Model
For the parameters $a_1 - a_9$, $b_1 - b_9$, and $c_1 - c_9$ of the ninth-order Gaussian function, the functions $a_1 = f_{a1}(I_f, T_j)$, ..., $a_9 = f_{a9}(I_f, T_j)$, $b_1 = f_{b1}(I_f, T_j)$, ..., $b_9 = f_{b9}(I_f, T_j)$, and $c_1 = f_{c1}(I_f, T_j)$, ..., $c_9 = f_{c9}(I_f, T_j)$ can be established. A computer program was written to compare the modeled and measured spectra, see Figure 4.39. Particularly, its input is the forward current and layout temperature and the output is the emission spectrum of the PC-LED, constants of the LED's spectral model, forward voltage, electrical power, optical power, thermal power, junction temperature, thermal resistance, luminous flux, luminous efficacy, radiant efficiency, CCT, white point chromaticity (x, y, u', v'), color rendering indices ($R_1 - R_{14}$, R_a, AVR_{1-14}), peak wavelength, and full width at half maximum.

Chromaticity differences between the modeled and measured spectra for different operating conditions are listed in Table 4.13. It can be seen that the biggest chromaticity difference is $\Delta u'v' = 3.70 \times 10^{-3}$ and the smallest chromaticity difference is $\Delta u'v' = 0.62 \times 10^{-3}$. These white chromaticity differences (except for $\Delta u'v' = 3.70 \times 10^{-3}$) belong to the generally tolerable range (for a more detailed

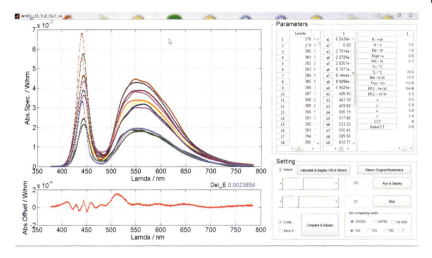

Figure 4.39 Interface of the program for the cold white PC-LED model.

Table 4.13 Chromaticity differences ($\Delta u'v'$) between the simulated and measured spectra in different operating conditions.

I_f/mA	350	350	350	350	700	700	700	700	1000	1000	1000	1000
T_s/°C	25	55	85	100	25	55	85	100	25	55	85	100
$\Delta u'v'/10^{-3}$	2.39	2.67	2.99	3.70	1.05	2.66	2.09	2.29	1.68	1.48	0.62	0.72

analysis, the reader is advised to consult Section 5.8). There were some spectral shape differences at some operating points such as 350 mA and 55 °C or 350 mA and 85 °C (despite tolerable white chromaticity differences) that may cause CRI errors. Therefore, an advanced modeling method is needed for those applications that require higher accuracy.

4.7 Thermal Relationships and Thermal LED Models

4.7.1 Thermal Relationships in LEDs

In order to identify and establish a thermal LED model, the material layers built in the finished LED have to be investigated in more detail. There is abundant literature (e.g., [50–52]) about LED structure and LED thermal properties which is summarized in this section.

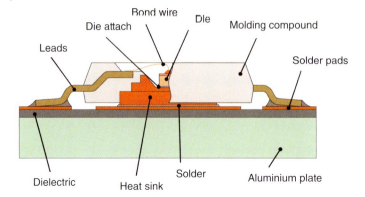

Figure 4.40 Thermal structure of an Osram Golden Dragon® LED [52]. Reproduced with permission from OSRAM Opto Semiconductors.

4.7.1.1 Thermal Structure of a Typical LED

LEDs of different LED manufacturers with various materials have different thermal properties. However, the data of the Golden Dragon® LEDs in [50] and [52] can be used as an example for the thermal structure of a typical LED, see Figure 4.40. As can be seen from Figure 4.40, a typical LED is composed of dielectric, heat sink, solder, solder pads, molding compound, die, bond wire, die attach, aluminum plate, and leads. Particularly, it is explained in [50] and [52] that the thermal power $P_{th,chip}$ from a junction layer of a chip is first transported via the package and the substrate by a heat conduction mechanism and then it is transported successively from the free surfaces to the outside environment by radiation and convection mechanisms. The Golden Dragon® LED consists of a chip mounted on a chip carrier (called *heat spreader*) by soldering or a bonding adhesive. The heat spreader is made of a highly conductive material such as copper. In this LED structure, a phosphor layer does not appear because it is only a color semiconductor LED.

4.7.1.2 A Typical Equivalent Thermal Circuit

On the basis of the above structure of the Golden Dragon® LED, an associated static equivalent thermal circuit diagram can be drawn, see Figure 4.41. The power dissipation $P_D = P_{th,chip}$ that occurs close to the chip surface is represented by a current source. The resistance network of the LED is equivalent to a serial connection to the ambient temperature T_a. The parallel-connected thermal resistance of the plastic housing can be neglected. The thermal circuit can be analyzed as follows: First, there is thermal power dissipation P_D playing the role of a "current source." This is indicated by the junction temperature T_j or the temperature offset $\Delta T = T_j - T_a$ between the junction temperature T_j and the ambient temperature T_a. Second, the thermal resistances of the die, die attach, heat sink, and the solder point with small values play the role of *internal* thermal resistances

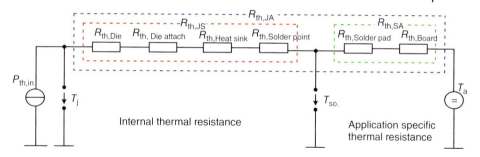

Figure 4.41 Equivalent thermal circuit of a Golden Dragon® LED [52]. Reproduced with permission from OSRAM Opto Semiconductors.

$R_{th,JS}$ (in K W^{-1}). Third, the thermal resistances of the solder pad and the board with higher values play the role as partial thermal resistances $R_{th,SA}$. In practice, $R_{th,SA}$ should be specified as two thermal resistances, $R_{th,SB}$ from the solder point to the board and $R_{th,BA}$ from the board to the ambience. The value of $R_{th,SB}$ is usually strongly influenced by various factors such as solder pad design, component placement, printed board material, and board construction. Finally, the thermal resistances can be grouped into three thermal resistance types including the internal thermal resistance $R_{th,JS}$ from the junction to the solder point, the middle thermal resistance $R_{th,SB}$ from the solder point to the board, and the external thermal resistance $R_{th,BA}$ from the board to the ambience. Or, they can also be thought of as being only one external thermal resistance, the sum of $R_{th,SB}$ and $R_{th,BA}$.

4.7.1.3 External Thermal Resistance

On the basis of the earlier analysis, two thermal resistance components ($R_{th,SB}$ and $R_{th,BA}$) are described more in detail.

First, the thermal resistance $R_{th,SB}$ depends on substrate technology. In order to ensure the best efficiency and a good thermal management, R_{thSB} has to be optimized. Particularly, the printed circuit board (PCB) technology of the LED has to be considered. Indeed, besides standard substrates such as the FR4 PCB material, new thermally enhanced substrate technologies became available. Some of these technologies such as MCPCB (Metal Core Printed Circuit Board), flexible PCB laminated on aluminum, and enhanced FR4 PCB glued to aluminum are established in the Golden Dragon® LEDs, see Table 4.14.

The structure of a single layer MCPCB is shown in Figure 4.42a: the typical thickness of the copper layer is about 35–200 µm, the thickness of the dielectric layer is about 75–100 µm, and the thickness of the metal layer is about 1–3 mm.

For the thickness of 100 µm for the dielectric layer, the thickness of 1.5 mm for the metal core of an aluminum alloy and the thickness of 35 µm for the copper layer, the lowest thermal resistance can be reached by the use of an enhanced thermal dielectric layer. In a simulation [52], the thermal conductivity of the substrate

Table 4.14 Parameters of the PCB substrate in a Golden Dragon® LED [52].

Outer optical circuit board dimensions (L × W)	Board material	Material for solder pads	Power distribution	Reference temperature
25 × 25 mm² FPC on Aluminum	FR4 on Aluminum MCPCB	35 µm Cu	1 W	25 °C

Source: Reproduced with permission from OSRAM Opto Semiconductors.

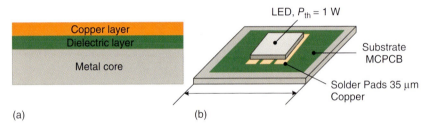

(a) (b)

Figure 4.42 (a) Structure of a single layer MCPCB and (b) structure of a heat sink component of an LED [52]. Reproduced with permission from OSRAM Opto Semiconductors.

Table 4.15 $R_{th,SB}$ of a Golden Dragon® LED with different substrate technologies [52].

Substrate technology	MPCB with enhanced dielectric	MPCB with FR4 dielectric	FPC on aluminum with standard PSA	FPC on aluminum with thermal enhanced PSA	FR4 – PCB on aluminum with thermal vias
$R_{th,SB}$	3.4 K W⁻¹	7.3 K W⁻¹	9.5 K W⁻¹	7.6 K W⁻¹	9.7 K W⁻¹

Source: Reproduced with permission from OSRAM Opto Semiconductors.

was about 1.3 W m⁻¹K⁻¹. Moreover, an analysis of other structures which are valid only for the earlier-mentioned PCB constructions and boundary conditions can be seen in Table 4.15.

Second, the thermal resistance $R_{th,BA}$ from the board to the ambience depends on heat sink properties. The heat transfer from a solid body in a fluid can be enhanced by extending the surface of the solid body. In fact, for a given heat dissipation, the temperature difference is controlled by the heat transfer coefficient h_{tr} (in W mm⁻²K⁻¹) and the heat surface area A (in mm²). Unfortunately, increasing the heat transfer coefficient h_{tr} is not easy and this depends on the nature of the material used. Therefore, it is more feasible to achieve a desired thermal improvement by increasing the heat transfer surface area A.

4.7.2
One-Dimensional Thermal Models

In this section, the mathematical background of thermal circuits is described in order to establish the equations of LED thermal models. The feasibility of the classical modeling method is evaluated in more detail. The thermal model was assumed to be a single-input-single-output system (SISO system, see Figure 4.43). The input is the thermal power ($P_{th,in}$/W) and the output is the temperature difference (ΔT/K) between the junction temperature (T_j/°C) and the directly measureable ("sensor") temperature (T_s/°C).

4.7.2.1 The First Order Thermal Circuit

In Figure 4.44, the so-called first order equivalent thermal circuit is shown. Here, the thermal property of a general material layer can be quantified by two parameters including thermal resistance R_{th} (in K W^{-1}) and thermal capacitance C_{th} (in W s K^{-1}).

Thermal resistance R_{th} is a parameter to characterize the ability of a material layer that resists to heat flow P_{th} through it when a temperature difference $\Delta T = T_j - T_s$ occurs between the two sides of this material layer. Here, T_j is the temperature of the hot side of the general material layer or the junction temperature in the case of LEDs while T_s is the temperature of the cold side of a general material layer or the encapsulant temperature in case of LEDs, and P_{th} is the thermal power that has to be transported from the hot side to the cold side, see Eq. (4.52).

$$R_{th} = \frac{\Delta T}{P_{th}} = \frac{T_j - T_s}{P_{th}} \tag{4.52}$$

On the other hand, in order to heat up a mass m of a specific material layer from T_s to T_j, it is necessary to feed a thermal energy Q_{th} into the material layer,

Figure 4.43 A single-input single-output system (SISO system) for LED thermal models.

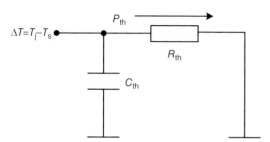

Figure 4.44 First order equivalent thermal circuit for a general material layer.

see Eq. (4.53).

$$Q_{th} = mC(T_j - T_s) = \rho V_{vol.} C(T_j - T_s) \tag{4.53}$$

Hence, thermal capacitance C_{th} can be defined as a parameter that quantifies the heat accumulation ability of a material volume, see Eq. (4.54).

$$C_{th} = \rho V_{vol.} C = \frac{Q_{th}}{\Delta T} = \frac{\int P_{th} dt}{\Delta T} \tag{4.54}$$

The material can also be characterized by three parameters including volume $V_{vol.}$ (in m³), mass m (in kg), specific heat capacity C (in J kg^{-1} K^{-1}), and bulk density ρ (in kg m^{-3}). On the basis of the definitions of thermal resistance and thermal capacitance, for a specific material layer, the thermal processes including heating and cooling can be described mathematically. Particularly, if an initial thermal energy $Q_{th,int}$ (in J) has to be dissipated when going from the hot side to the cold side, the reduction of thermal power in a unit of time will be P_{th} (in J s^{-1}). Then, the heat sink process over time can be described by Eqs. (4.55)–(4.57).

$$Q_{th}(t) = Q_{th,int} - \int_0^t P_{th}(t) dt \tag{4.55}$$

$$P_{th} = -\frac{dQ_{th}(t)}{dt} = -C_{th}\frac{d\Delta T(t)}{dt} = \frac{\Delta T(t)}{R_{th}} \tag{4.56}$$

$$\Delta T(t) + R_{th} C_{th} \frac{d\Delta T(t)}{dt} = 0 \tag{4.57}$$

Solving Eq. (4.57), the cooling process can be determined by using Eq. (4.58) (called *cooling equation*) with the thermal system time τ_{sys} or T_{sys} shown in Eq. (4.59).

$$\Delta T = \Delta T_0 \left[\exp\left(\frac{-t}{\tau_{sys}}\right)\right] \tag{4.58}$$

$$\tau_{sys} = T_{sys} = R_{th} C_{th} \tag{4.59}$$

The reverse process, *heating*, is shown in Eq. (4.60).

$$\Delta T = \Delta T_\infty \left[1 - \exp\left(\frac{-t}{\tau_{sys}}\right)\right] \tag{4.60}$$

The maximal amplitudes of these two processes are shown in Eqs. (4.58) and (4.60) ($\Delta T_\infty = \Delta T_0 = \Delta T_{max}$ in K). These values represent the saturation value for the heating process and the initial value for the cooling process.

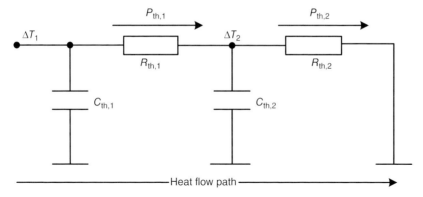

Figure 4.45 Second order equivalent thermal circuit for two material layers in a serial connection.

4.7.2.2 Second Order Thermal Circuit

If the two material layers are in a series connection then this so-called second order equivalent thermal circuit with two stage RCs is illustrated in Figure 4.45. Here, ΔT_i, C_{thi}, R_{thi}, and P_{thi} represent the ith temperature difference, thermal capacitance, thermal resistance, and thermal power at the stage i. The index i runs from 1 to 2.

The analysis of the thermal circuit in Figure 4.45 with the temperature difference variables results in Eqs. (4.61)–(4.63).

$$\begin{cases} \Delta T_1 = \Delta T_2 + R_{th,1} P_{th,1} \\ \Delta T_2 = R_{th,2} P_{th,2} \end{cases} \quad (4.61)$$

$$\begin{bmatrix} P_{th,1} \\ P_{th,2} \end{bmatrix} = \begin{bmatrix} 1/R_{th,1} & -1/R_{th,1} \\ 0 & 1/R_{th,2} \end{bmatrix} \begin{bmatrix} \Delta T_1 \\ \Delta T_2 \end{bmatrix} \quad (4.62)$$

$$\begin{bmatrix} P_{th,1} \\ P_{th,2} \end{bmatrix} = \begin{bmatrix} g_1 & -g_1 \\ 0 & g_2 \end{bmatrix} \begin{bmatrix} \Delta T_1 \\ \Delta T_2 \end{bmatrix} \quad (4.63)$$

and:

$$\begin{cases} \Delta P_{th,1} = -C_{th,1} s \Delta T_1 + P_{th,in} \\ \Delta P_{th,2} = -C_{th,1} s \Delta T_1 - C_{th,2} s \Delta T_2 + P_{th,in} \end{cases} \quad (4.64)$$

$$\begin{bmatrix} P_{th,1} \\ P_{th,2} \end{bmatrix} = \begin{bmatrix} -C_{th,1} s & 0 \\ -C_{th,1} s & -C_{th,2} s \end{bmatrix} \begin{bmatrix} \Delta T_1 \\ \Delta T_2 \end{bmatrix} + \begin{bmatrix} P_{th,in} \\ P_{th,in} \end{bmatrix} \quad (4.65)$$

$$\begin{bmatrix} P_{th,1} \\ P_{th,2} \end{bmatrix} = \begin{bmatrix} -1/f_1 & 0 \\ -1/f_1 & -1/f_2 \end{bmatrix} \begin{bmatrix} \Delta T_1 \\ \Delta T_2 \end{bmatrix} + \begin{bmatrix} P_{th,in} \\ P_{th,in} \end{bmatrix} \quad (4.66)$$

Combining Eqs. (4.63) and (4.66), Eqs. (4.67) and (4.68) arise.

$$\begin{bmatrix} g_1 + 1/f_1 & -g_1 \\ 1/f_1 & g_2 + 1/f_2 \end{bmatrix} \begin{bmatrix} \Delta T_1 \\ \Delta T_2 \end{bmatrix} = \begin{bmatrix} P_{th,in} \\ P_{th,in} \end{bmatrix} \quad (4.67)$$

188 | 4 Measurement and Modeling of the LED Light Source

$$\begin{bmatrix} Tf_1 \\ Tf_2 \end{bmatrix} = \begin{bmatrix} \frac{\Delta T_1}{P_{th,in}} \\ \frac{\Delta T_2}{P_{th,in}} \end{bmatrix} = \begin{bmatrix} g_1 + 1/f_1 & -g_1 \\ 1/f_1 & g_2 + 1/f_2 \end{bmatrix}^{-1} \begin{bmatrix} 1 \\ 1 \end{bmatrix} \tag{4.68}$$

The solving of Eq. (4.68) yields the transfer functions between the ith output temperature difference ΔT_i and the input thermal power $P_{th,in}$, see Eqs. (4.69) and (4.70).

$$Tf_1 = \frac{\Delta T_1}{P_{th,in}} = \frac{(R_{th,1} + R_{th,2}) + R_{th,1}R_{th,2}C_{th2}s}{1 + (R_{th,1}C_{th,1} + R_{th,2}C_{th,2} + R_{th,2}C_{th,1})s + R_{th,1}R_{th,2}C_{th,1}C_{th,2}s^2} \tag{4.69}$$

$$Tf_2 = \frac{\Delta T_2}{P_{th,in}} = \frac{R_{th,2}}{1 + (R_{th,1}C_{th,1} + R_{th,2}C_{th,2} + R_{th,2}C_{th,1})s + R_{th,1}R_{th,2}C_{th,1}C_{th,2}s^2} \tag{4.70}$$

Similarly, the transfer functions between the ith output local thermal power $P_{th,i}$ and input thermal power $P_{th,in}$ can be written as follows:

$$Tg_1 = \frac{P_{th,1}}{P_{th,in}} = \frac{1 + R_{th,2}C_{th,2}s}{1 + (R_{th,1}C_{th,1} + R_{th,2}C_{th,2} + R_{th,2}C_{th,1})s + R_{th,1}R_{th,2}C_{th,1}C_{th,2}s^2} \tag{4.71}$$

$$Tg_2 = \frac{P_{th,2}}{P_{th,in}} = \frac{1}{1 + (R_{th,1}C_{th,1} + R_{th,2}C_{th,2} + R_{th,2}C_{th,1})s + R_{th,1}R_{th,2}C_{th,1}C_{th,2}s^2} \tag{4.72}$$

4.7.2.3 The nth Order Thermal Circuit

Figure 4.46 illustrates the so-called nth order equivalent thermal circuit for n RC stages that match n material layers in a serial connection of LEDs. Here, the symbols ΔT_i, $C_{th,i}$, $R_{th,i}$, and $P_{th,i}$ represent the ith temperature difference, thermal capacitance, thermal resistance, and thermal power at the stage i. In this case the index i runs from 1 to n.

Applying a similar analysis as for the second order equivalent thermal circuit, a general matrix of the transfer functions between the output temperature difference ΔT_i and the input thermal power $P_{th,in}$ and between the output local thermal power $P_{th,i}$ and the input thermal power $P_{th,in}$ can be seen in Eqs. (4.73) and (4.74).

$$\begin{bmatrix} Tf_1 \\ Tf_2 \\ Tf_3 \\ \cdots \\ Tf_{n-1} \\ Tf_n \end{bmatrix} = \begin{bmatrix} g_1 + 1/f_1 & -g_1 & 0 & 0 & \cdots & 0 \\ 1/f_1 & g_2 + 1/f_2 & -g_2 & 0 & \cdots & 0 \\ 1/f_1 & 1/f_2 & g_3 + 1/f_3 & -g_3 & \cdots & 0 \\ \cdots & \cdots & \cdots & \cdots & \cdots & \cdots \\ 1/f_1 & 1/f_2 & 1/f_3 & 1/f_4 & \cdots & -g_{n-1} \\ 1/f_1 & 1/f_2 & 1/f_3 & 1/f_4 & \cdots & g_n + 1/f_n \end{bmatrix}^{-1} \begin{bmatrix} 1 \\ 1 \\ 1 \\ \cdots \\ 1 \\ 1 \end{bmatrix} \tag{4.73}$$

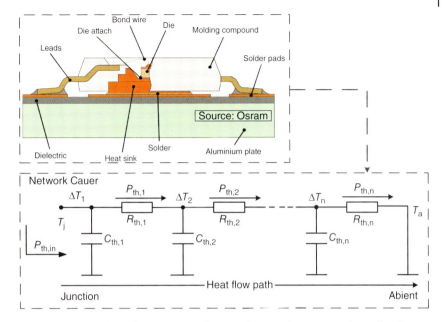

Figure 4.46 Top: thermal components of an LED; bottom: the corresponding nth order equivalent thermal circuit for n material layers in a serial connection. Reproduced with permission from OSRAM Opto Semiconductors.

$$\begin{bmatrix} Tg_1 \\ Tg_2 \\ Tg_3 \\ \cdots \\ Tg_{n-1} \\ Tg_n \end{bmatrix} = \begin{bmatrix} R_{th,1}+f_1 & R_{th,2} & R_{th,3} & R_{th,4} & \cdots & R_{th,n} \\ -f_2 & R_{th,2}+f_2 & R_{th,3} & R_{th,4} & \cdots & R_{th,n} \\ 0 & -f_3 & R_{th,3}+f_3 & R_{th,4} & \cdots & R_{th,n} \\ \cdots & \cdots & \cdots & \cdots & \cdots & \cdots \\ 0 & 0 & 0 & 0 & \cdots & R_{th,n} \\ 0 & 0 & 0 & 0 & \cdots & R_{th,n}+f_n \end{bmatrix}^{-1} \begin{bmatrix} f_1 \\ 0 \\ 0 \\ \cdots \\ 0 \\ 0 \end{bmatrix}$$

(4.74)

4.7.2.4 The Transient Function and Its Weighting Function

Concerning the temporal domain, for the n-stage thermal circuit illustrated in Figure 4.46, combining the solution of Eq. (4.60) and the linear system property, the transient function $Z_{th}(t) = \frac{\Delta T_1(t)}{P_{in}(t)}$ of the temperature difference $\Delta T_1(t)$ as a function of time can be written as

$$z_{th}(t) = A_1\left[1 - e^{-\frac{t}{T_1}}\right] + A_2\left[1 - e^{-\frac{t}{T_2}}\right] + \cdots + A_n\left[1 - e^{-\frac{t}{T_n}}\right] \quad (4.75)$$

Then, the weighting function $g(t) = \frac{dz_{th}(t)}{dt}$ can be obtained:

$$g(t) = \frac{A_1}{T_1}e^{-\frac{t}{T_1}} + \frac{A_2}{T_2}e^{-\frac{t}{T_2}} + \cdots + \frac{A_n}{T_n}e^{-\frac{t}{T_n}} \quad (4.76)$$

As a result, the transfer function $G(s)$ or $Z_{th}(s)$ between the output temperature difference ΔT_1 and the input thermal power P_{in} can be determined:

$$G(s) = Z_{th}(s) = \frac{\Delta T_1(s)}{P_{in}(s)} = \frac{A_1}{T_1 s + 1} + \frac{A_2}{T_2 s + 1} + \cdots + \frac{A_n}{T_n s + 1} \qquad (4.77)$$

4.7.2.5 Conclusions

Based on the above analysis, it can be confirmed that the thermal description of an LED system needs a higher order linear system with no overshoot, no derivation component (D), and no integration component (I) but one proportional component (P). Therefore, the thermal system approaches to a saturation state when the time variable converges to infinity. Otherwise, it can also be recognized that, with Eqs. (4.75)–(4.77), it is very difficult to determine the explicit values of the partial thermal resistances and partial thermal capacitances of an n-stage thermal circuit. Also, the order of the total system is not easy to determine accurately. Therefore, a good theory of the identification and decoding of this higher order thermal transient LED model has to be proposed. This theory should be as simple as possible to get a feasible thermal LED model.

4.8 Measurement Methods to Determine the Thermal Characteristics of LED Devices

4.8.1 Measurement Methods and Procedures

4.8.1.1 Selection of an Available Measurement Method

There exist different LED junction temperature measurement methods. As the simplest method, a thermal camera can be used to measure junction temperature. However, this measurement is only qualitative and not accurate. Other options include indirect optical test methods (OTMs) such as the light power method (POP) with the relationship $P_{opt} = f(T_j)$ between the optical power P_{opt} and the junction temperature T_j, the wavelength shift method (WSM) with the relationship $\lambda_p = g(T_j)$ between the peak wavelength λ_p and the junction temperature T_j or the relationship $\lambda_{FWHM} = h(T_j)$ between the full width at half maximum wavelength λ_{FWHM} and the junction temperature T_j. The LED junction temperature can also be determined based on electrical test methods (ETMs) such as the single current method (SCM) and the dual current method (DCM) discussed by Siegal in [53]. Especially, Gu and Nadrean in [54] described a noncontact method for the junction temperature measurement of white PC-LEDs based on the ratio W/B of the spectral white component W and the blue spectral component B. These authors criticize the accuracy of the WSM method. Similarly, Schubert et al. in [55, 56] explain and compare WSM and ETMs.

Among all these methods, the method based on forward voltage (belonging to ETMs) is very well known and preferred in practical applications because of its accuracy, simplicity, and feasibility. Particularly, the linear relationship

$V_f = aT_j + b$ between forward voltage V_f and junction temperature T_j was proved in [56–59]. Thus, based on the measured voltage V_f, the junction temperature T_j can be determined easily by means of the relationship $T_j = (V_f - b)/a$. Nevertheless, the constants a and b cannot be determined so easily. Therefore, in order to obtain these constants, two measurement procedures based on thermal transient measurement were invented including the procedure based on the heating process and another procedure based on the cooling process. In more detail, the heating procedure is described in [56, 57, 60, 61] while the idea of the cooling procedure appeared in [53, 62]. The cooling procedure is preferred in real laboratory conditions because of its accuracy.

4.8.1.2 Description of the Cooling Measurement Procedure

The cooling procedure was developed to determine the junction temperature T_j indirectly based on the voltage difference $\Delta V_f = (V_{f\text{-cooling-begin}} - V_{f\text{-cooling-end}})$ captured during the cooling process. The cooling procedure is shown in Figure 4.47. It includes a transient measurement and a DC measurement. In the transient measurement, first a specific forward current I_f is supplied for the LED during a long enough time in order to reach its full thermal charge. Successively, a step-down current pulse is created by shutting from a higher current down to a lower current (e.g., 1 mA). At this lower current, there is nearly no more thermal power and a cooling process takes place gradually. Then, the previously charged thermal power P_{th} is transported from the LED junction layer into the external environment. This cooling process is characterized by means of the change of forward voltage $V_{f\text{-cooling}}$ or the forward voltage difference ΔV_f. The thermal power P_{th}, case temperature T_s and the constant a_0 of the voltage relationship $V_f = a_0 T_j + b_0$ at low current can be determined easily during the DC measurement. Therefore, the temperature difference $\Delta T = T_j - T_s$ can be calculated from the relationship $\Delta T = \Delta V_f / a_0$. This difference is an important characteristic parameter of the LED's thermal property. Finally, the junction temperature T_j and the thermal resistance $R_{th,\Sigma}$ can be determined by means of the relationships $T_j = T_s + \Delta T$ and $R_{th,\Sigma} = \Delta T / P_{th}$, respectively. In addition, the theory of system identification can be applied to identify this thermal transient process. On the basis of the identified functions, partial thermal parameters such system time $T_{sys,i}$, thermal resistance

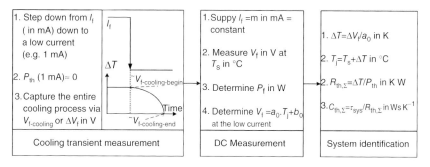

Figure 4.47 Cooling measurement procedure.

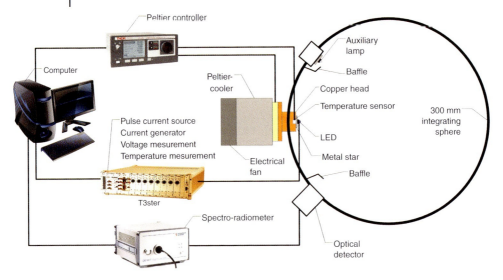

Figure 4.48 Measurement system of the LED's thermal, electrical, and optical properties.

$R_{th,i}$, and thermal capacitance $C_{th,i}$ can be determined (the index i runs between 1 and k while the value of k depends on the LED being measured). Or, more simply, total system time τ_{sys} can also be determined to help calculate the total equivalent thermal capacitance $C_{th,\Sigma}$ based on the relationship $C_{th,\Sigma} = \tau_{sys}/R_{th,\Sigma}$.

4.8.2
Description of a Typical Measurement System and Its Calibration

In Section 4.8.1, several measurement methods and procedures of the LED junction temperature determination were briefly overviewed. In real operations, the noncontact measurement method via the LED forward voltage based on the cooling procedure shown in Figure 4.47 is preferred because it is simple and exact. The entire measurement system of an application laboratory contains electrical, optical, and thermal components, see Figure 4.48.

4.8.2.1 Components and Structure of the Measurement System

Figure 4.48 shows a 300 mm optical integrating sphere with an auxiliary lamp, a baffle, and an optical detector. This arrangement is used to measure the absolute SPDs of the LEDs. LEDs are soldered to a PCB of a star form. Under this star, a temperature sensor is attached to it directly. Then, the entire metal star and the temperature sensor are positioned on a copper head that is attached on the cold side of a Peltier module. The hot side of the Peltier module is connected to a Peltier cooler system including a heat sink and an electrical fan so that it can transfer the thermal energy P_{th} quickly into the surrounding environment. A controller connected to a computer is used to control the Peltier module in order to keep the temperature under the metal star constant at any desired level. In addition, a spectroradiometer

4.8 Measurement Methods to Determine the Thermal Characteristics of LED Devices

is used to measure and convert the electrical signals from the optical detector (behind the baffle) and then send them to the computer. The so-called T3ster® device supplies the desired current pulses and measures the forward voltage of the LEDs. More details about T3ster® can be found in literature [63–70].

4.8.2.2 Determination of Thermal Power and Calibration Factor for Several LEDs

According to the conservation law of energy, the original electrical power P_{el} is converted first by an internal semiconductor structure into the optical power $P_{opt,chip}$ of the blue chip represented by the original blue spectrum and the first chip thermal power $P_{th,chip}$. Then, $P_{opt,chip}$ is converted by the luminescent material (phosphor) into the optical power $P_{opt,phosphor}$ corresponding to the color phosphor spectral component and the second phosphor thermal power $P_{th,phosphor}$. Finally, the output optical power $P_{opt,out}$ is the sum of the optical power $P_{opt,rest.blue}$ of the remaining blue spectral component and the optical power $P_{opt,phosphor}$ of the luminescent material. These relationships can be described mathematically by Eqs. (4.78)–(4.81).

$$P_{el} = P_{opt,chip} + P_{th,chip} \tag{4.78}$$

$$P_{opt,chip} = P_{th,phosphor} + P_{opt,phosphor} + P_{opt,rest.blue} \tag{4.79}$$

$$P_{opt,out} = P_{opt,rest.blue} + P_{opt,phosphor} \tag{4.80}$$

$$P_{el} = (P_{th,chip} + P_{th,phosphor}) + (P_{opt,rest.blue} + P_{opt,phosphor}) \tag{4.81}$$

There are complicated chemical and physical interactions between the numerous components inside the LED. In order to analyze, identify, and decode a thermal map of the LED, the total effective thermal power $P_{th,in}$ that influences the junction temperature T_j can be expressed as follows:

$$P_{th,in} = \alpha \cdot P_{th,chip} + \beta \cdot P_{th,phosphor} \tag{4.82}$$

Here, α and β are the effective percentages of blue chip thermal power and phosphor thermal power at the junction temperature T_j, respectively. The constants α and β are difficult to be determined. However, the optical instrument in Figure 4.48 can measure the emission spectra of the LEDs and the optical power $P_{opt,out} = P_{opt,rest.blue} + P_{opt,phosphor}$ can be determined by the integration of the emission spectra. The electrical instrument in Figure 4.48 can measure the forward voltage and the current of the LEDs and then the electrical power P_{el} can be calculated. Consequently, the thermal power $P_{th,in} = P_{el} - P_{opt,out}$ can be computed. On the other hand, the thermal instrument can change the temperature of the LEDs and the electrical instrument can measure the corresponding voltages. Therefore, the calibration factor a_0 (mentioned in the cooling procedure in Figure 4.47) can be determined for each LED from the measurement.

Figure 4.49 shows the thermal power and the calibration factor for several LEDs based on this procedure. Results indicate that the calibration factors are quite different among the LEDs although they belong to the same binning group. Thermal powers are also entirely dissimilar depending on the luminescent

194 | *4 Measurement and Modeling of the LED Light Source*

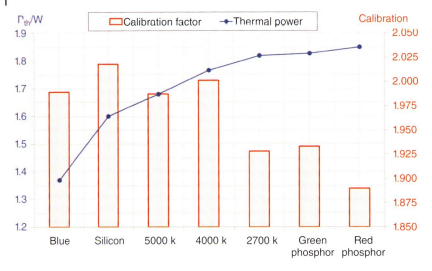

Figure 4.49 Thermal power and calibration factor for a set of sample LEDs.

material mixed and built onto the blue semiconductor LED. The red PC-LED has the highest thermal power and the green PC-LED obtains the second ranking. The thermal power reduces from the warm white LED (2700 K) to the blue semiconductor LED. Moreover, the difference between the thermal power of the blue semiconductor LED and the corresponding PC-LEDs (that use the same type of blue semiconductor LEDs) is quite significant. Consequently, thermal power and calibration factors can be determined for each measurement individually.

4.8.3
Methods of Thermal Map Decoding

4.8.3.1 Decoding of the Thermal Map by the Method of the Structure Function

In order to decode an LED's thermal map with an array of partial thermal resistances and partial thermal capacitances, the transformation from a thermal transient process into a structure function is a powerful but a complicated solution described in [71, 72]. The following steps can be carried out:

$$\xi = \ln(T_i) \text{ or } T_i = e^{\xi} \tag{4.83}$$

$$z = \ln(t) \text{ or } t = e^{z} \tag{4.84}$$

Then, Eq. (4.75) can be transformed into

$$Z_{th}(z) = \sum_{i=1}^{n} R_{th}(\xi) \left[1 - e^{-\frac{e^z}{e^\xi}} \right] \int_{-\infty}^{+\infty} R_{th}(\xi)[1 - e^{-e^{(z-\xi)}}] d\xi \tag{4.85}$$

$$\frac{dZ_{th}(z)}{dz} = \int_{-\infty}^{+\infty} R(\xi)[-e^{-e^{(z-\xi)}} \cdot -e^{(z-\xi)}] d\xi \tag{4.86}$$

4.8 Measurement Methods to Determine the Thermal Characteristics of LED Devices

$$\frac{dZ_{th}(z)}{dz} = \int_{-\infty}^{+\infty} R(\xi)[e^{(z-\xi)-e^{(z-\xi)}}]d\xi \qquad (4.87)$$

On the basis of the definition of convolution, the following can be written:

$$(f * g)(t) \quad \int_{-\infty}^{+\infty} f(\tau)g(t-\tau)d\tau \quad f(j\omega)g(j\omega) = f(s)g(s) \qquad (4.88)$$

$$\frac{dZ_{th}(z)}{dz} = R(z) * W(z) = R(z) * \exp(z - e^z) \qquad (4.89)$$

From Eq. (4.89), the thermal resistance equation $R(z)$ can be converted as follows:

$$R(z) = \text{deconv}\left\{\frac{dZ_{th}(z)}{dz}, \exp[z - \exp(z)]\right\} = \text{deconv}\left\{\frac{dZ_{th}(z)}{dz}, W(z)\right\} \qquad (4.90)$$

The deconvolution denoted by the function "deconv" in Eq. (4.90) yields the mathematical form of the $R(z)$ function and this is a Foster network. It causes an inaccuracy for the transformation. Then, a successive transformation of the Foster network to the Cauer network causes a wide range of time constants. Concerning the real performance of the method, these inaccuracies can be improved only by good numerical algorithms of deconvolution and numerical transformations. The determination of a structure function needs the theory of one-dimensional heat-flow model [73]. Thus, the authors in [72] proposed that a reasonable approximation for a given structure based on this theory would indicate the correspondence between the different sections of the Cauer network and the different parts of the physical structure of the LED.

4.8.3.2 Thermal Map Decoding by the Euclidean Algorithm

Besides the earlier-mentioned method of structure function, some other approximation methods of the Foster to Cauer transformation are discussed in [74, 75]. Here, the method based on the Euclidean algorithm is preferred because of its simplicity and effectiveness. The implementation of the cooling procedure for an LED is illustrated in Figure 4.50. At the first step, a voltage transient progress corresponding to the cooling process is captured from the transient measurement. At the second step, the temperature difference is transformed from the voltage progress based on the determined calibration factor. At the third step, the time axis t is converted into the new variable $z = \ln(t)$ so that the material layers can be identified via the visually obtained boundaries. Finally, at the fourth step, the fitting with the function of Eq. (4.75) yields the partial thermal resistances and capacitances matching directly the Foster network in Figure 4.51.

Unfortunately, the Foster network does not reflect the physical nature of the material layers in series connection according to the description in Section 4.7.2 (which is purely a mathematical concept). Consequently, the Euclidean algorithm can be applied to perform the Foster to Cauer transformation: Eq. (4.77) is transformed in the Laplace impedance function (Eq. (4.87)) and then the Euclidean Eqs. (4.91) and (4.92) directly match the partial impedances $Z_1 - Z_{n-1}$ of the

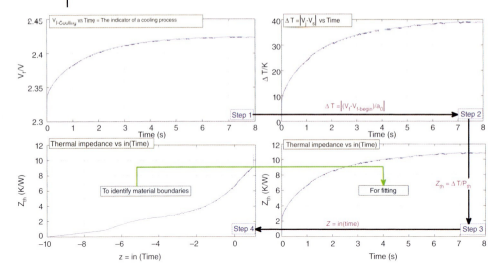

Figure 4.50 Steps of the cooling procedure (a real example).

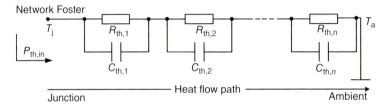

Figure 4.51 The structure of the Foster network.

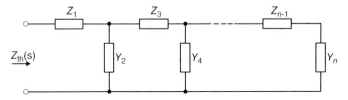

Figure 4.52 Correspondence between the Euclidean algorithm and the Cauer network.

thermal resistors and the partial conductivities $Y_1 - Y_n$ of the thermal capacitors in the Cauer network in Figure 4.52.

$$Z_{th}(s) = \frac{a_0 s^n + a_1 s^{n-1} + \cdots + a_n}{b s^m + b_1 s^{m-1} + \cdots + b_m} \tag{4.91}$$

$$Z_{th}(s) = Z_1 + \cfrac{1}{Y_2 + \cfrac{1}{Z_3 + \cfrac{1}{Y_4 + \cdots \cfrac{1}{Z_{n-1} + \cfrac{1}{Y_n}}}}} \tag{4.92}$$

4.9
Thermal and Optical Behavior of Blue LEDs, Silicon Systems, and Phosphor Systems

4.9.1
Selection of LEDs and Their Optical Behavior

In order to investigate the thermal properties of LEDs, a blue semiconductor LED (441 nm), a cold white (5000 K), a warm white (2700 K), a phosphor-converted green LED, and a phosphor-converted red LED were selected and measured. A LuGaAG phosphor and a red phosphor (650 nm) were used, see Figure 4.53. Best temperature stability was obtained for the phosphor-converted green and red PC-LEDs. The temperature stability of the blue semiconductor LED was also good whereas the temperature stability of the warm white (2700 K) and cold white (5000 K) PC-LEDs in which the same phosphors were built in and the same blue chip type was used was worse because of internal chemical and physical interactions inside the LED structure. These interactions represent a big challenge for phosphor coating and LED packaging technology.

4.9.2
Efficiency of the LEDs

The radiant efficiency of the LEDs is shown as a function of junction temperature (at 600 mA) in Figure 4.54. It can be seen that radiant efficiency diminishes going from the blue semiconductor LED to the phosphor-converted red PC-LED. At 600 mA, the radiant efficiency of the blue semiconductor LED is about 35%. But it is only about 10% in case of the phosphor-converted red PC-LED. This result

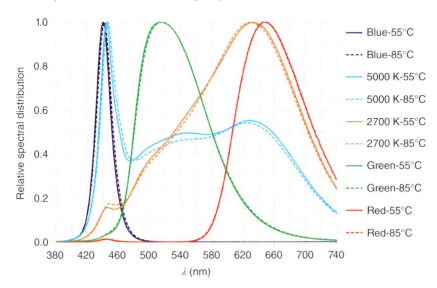

Figure 4.53 Emission spectra of the sample LEDs at 600 mA; 55 and 85 °C.

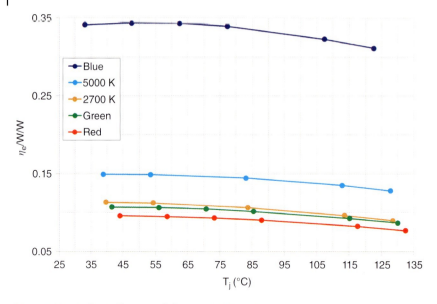

Figure 4.54 Radiant efficiency of the sample LEDs at 600 mA.

indicated the extreme optical absorption and thermal loss of the red phosphor. In addition, the cold white PC-LED exhibited a better efficiency than the warm white or phosphor-converted colored PC-LEDs. The reason is that the junction layer has a higher temperature in case of the phosphor-converted green and red LEDs, see Figure 4.55.

4.9.3
Results of Thermal Decoding by the Structure Function Method

Applying the method of structure functions described in Section 4.8.3, the structure functions of the PC-LEDs with different phosphor systems were decoded, see the thermal resistance–thermal capacitance diagram of Figure 4.56. It can be seen that the thermal resistance of the different LEDs with phosphor systems is nearly similar but there is a difference in their thermal capacitance. This causes the problem of the speed of the transport of thermal energy from the internal to the external components of these PC-LEDs. The PC-LEDs with higher amounts of the red phosphor such as the warm white LED or the phosphor-converted red phosphor LED have a higher temperature than the others. The influence of silicon on the thermal properties of PC-LEDs must also be investigated with different amounts of silicon because the silicon mass in the phosphor mixtures of the cold white, neutral white, and warm white PC-LEDs is different. This, in turn, causes dissimilar influences on the thermal properties of the PC-LEDs. Consequently, the relationship between thermal power or other thermal properties and silicon mass is investigated in more detail. In addition, blue chips belonging to the same binning group have very dissimilar thermal

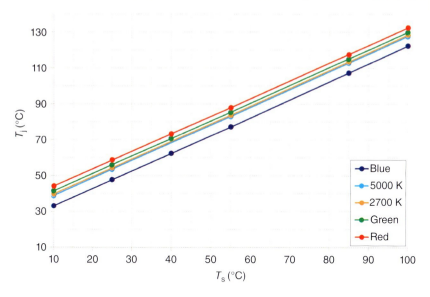

Figure 4.55 Junction temperature T_j of the sample LEDs at 600 mA and at different sensor temperatures T_s.

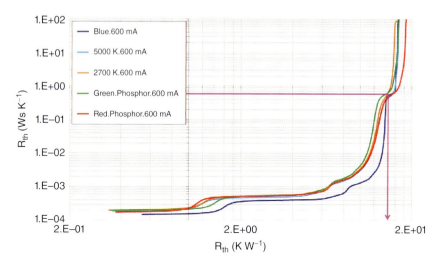

Figure 4.56 Thermal structure functions of the sample LEDs at 600 mA and 27.7 °C sensor temperature.

properties. Therefore, the thermal measurement for the original (standalone) blue chip is essential to shed light on the thermal properties. Finally, note that the use of remote phosphor LEDs helps reduce the thermal influence of silicon and the phosphor system on PC-LEDs leading to higher efficacy and longer lifetime.

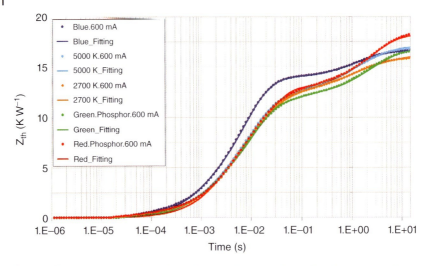

Figure 4.57 Identification of the thermal process by using the sixth order exponential function according to Eq. (4.75).

4.9.1
Results of Thermal Decoding by the Method of the Euclidean Algorithm

Applying the method of the Euclidean algorithm described in Section 4.8.3, the thermal impedance function of the PC-LEDs can be defined by Eq. (4.75) for the case of six material layers in series connection. The measurement points and the fitting curves are illustrated in Figure 4.57 for the five sample LEDs. Results indicate very good fitting accuracy so that the Eqs. (4.91) and (4.92) can be applied for these PC-LEDs. On the basis of these equations, the corresponding thermal maps (see Figure 4.52) can be decoded easily. As a result, the partial thermal resistances and capacitances of the Foster and Cauer networks were obtained, see Tables 4.16 and 4.17. The Foster thermal resistances and capacitances are purely mathematical values. In contrast, the Cauer thermal resistances and capacitances describe real thermal properties of real material layers inside the LEDs. However, the issue of which value corresponds to which material layer can be figured out by combining simulation results and the LED material structure. Results in Tables 4.16 and 4.17 only confirm that the conclusions from the above structure function are correct and, at this aging state, the specific thermal map of the LEDs can be outlined by using the resulting values. Note that, during the investigation of LED aging (see Section 4.10), the thermal maps of the LEDs at different aging time periods can be compared in order to identify the aging level or the aging limit of the material layers. This information is important for both LED manufacturers and LED designers in the various solid state lighting applications.

Table 4.16 Thermal capacities and resistances for the five sample LEDs based on the Foster network.

Name	Blue	5000 K	2700 K	Green	Red
$R_{th,1}/\text{K W}^{-1}$	0.04502	7.685	1.354	2.832	1.332
$R_{th,2}/\text{K W}^{-1}$	2.21800	2.194	3.231	8.005	7.224
$R_{th,3}/\text{K W}^{-1}$	1.08500	0.8833	4.317	0.7308	1.026
$R_{th,4}/\text{K W}^{-1}$	6.62700	2.196	2.061	1.511	1.054
$R_{th,5}/\text{K W}^{-1}$	4.98000	2.654	1.720	2.531	2.037
$R_{th,6}/\text{K W}^{-1}$	1.89100	1.398	3.464	1.306	5.497
$C_{th,1}/\text{W s K}^{-1}$	0.21863	0.00116	4.89108	0.00042	1.69892
$C_{th,2}/\text{W s K}^{-1}$	0.00035	0.35360	0.00498	0.00134	0.00287
$C_{th,3}/\text{W s K}^{-1}$	5.81855	0.06038	0.00052	1.89000	0.80087
$C_{th,4}/\text{W s K}^{-1}$	0.00083	1.68657	0.47061	0.94370	1.95783
$C_{th,5}/\text{W s K}^{-1}$	0.00286	0.00046	0.04362	2.38013	2.41000
$C_{th,6}/\text{W s K}^{-1}$	0.49515	0.02968	0.00329	0.05407	0.00049
$R_{th,\Sigma}/\text{K W}^{-1}$	16.846	17.010	16.147	16.916	18.170

Table 4.17 Thermal capacities and resistances for the five sample LEDs based on the Cauer network.

Name	Blue	5000 K	2700 K	Green	Red
$R_{th,1}/\text{K W}^{-1}$	4.978	5.076	6.700	4.779	7.380
$R_{th,2}/\text{K W}^{-1}$	7.455	6.240	4.880	6.509	5.469
$R_{th,3}/\text{K W}^{-1}$	1.521	1.662	0.814	1.197	3.532
$R_{th,4}/\text{K W}^{-1}$	0.026	0.172	0.700	1.232	0.606
$R_{th,5}/\text{K W}^{-1}$	2.098	2.657	2.120	3.168	1.529
$R_{th,6}/\text{K W}^{-1}$	2.870	3.860	3.055	3.202	1.182
$C_{th,1}/\text{W s K}^{-1}$	2.274E−04	3.227E−04	4.056E−04	3.164E−04	4.162E−04
$C_{th,2}/\text{W s K}^{-1}$	6.112E−04	2.809E−02	2.321E−03	1.324E−03	3.337E−03
$C_{th,3}/\text{W s K}^{-1}$	7.854E−03	2.809E−02	3.534E−02	6.568E−02	3.714E−01
$C_{th,4}/\text{W s K}^{-1}$	3.791E−01	2.636E−01	5.137E−02	4.838E−01	1.213E+00
$C_{th,5}/\text{W s K}^{-1}$	4.776E−01	3.452E−01	4.834E−01	1.573E+00	4.380E+00
$C_{th,6}/\text{W s K}^{-1}$	2.043E+00	7.960E−01	1.974E+00	1.088E+00	2.441E+00

4.10
Aging Behavior of High-Power LED Components

In the LED products, LED components are driven under certain operating conditions as current, pulse width modulation (PWM)-duty cycle and at ambient temperatures which change depending on weather and locations. The LED components are soldered or mounted on the circuit board together or they are separated from other electronic units (driving and controlling electronics). All these components and structure units undergo gradual degradations over time causing a change of the lighting, thermal, mechanical, and electrical parameters. Modern LED products are often praised as being user and environment friendly at a high user acceptance level. Such a statement can become true only if the LED products are long-lasting, highly efficient, and reliable. In addition, LED lighting systems must be designed in a proper manner taking the specifications of all used LED components into account.

In a research project for the German luminaire industry, the present authors studied the aging behavior of high power LEDs, medium power LEDs, and COB LEDs from different well-known international LED manufacturers. The issues of LED degradation behavior can be grouped into the following categories:

1) *Depreciation of the photometric quantities*: From all photometric quantities, luminous flux depreciation is the most often investigated effect. In some literature articles, the terminology *lumen maintenance* is often used in order to describe this effect. From the point of view of optical designers, however, the change of absolute luminance and its distribution over the emitting area and emission direction is much more important.
2) *Change of spectral and color characteristics of the emitted LED radiation*: For most users and luminaire designers of indoor lighting, the color shift (or chromaticity shift) is a key criterion for evaluating the acceptance and usability of the LED products. For applications like museum and shop lighting, stage, and film productions the change of spectral distributions is of essential relevance.
3) *Change of the electrical and thermal characteristics of the LED components*: At a constant LED current, the change of forward voltage is an indication for the change of the electrical resistance of LED chip and/or packaging (change of the contact location between bond wires and chip and the structural change of the pn-depletion layer and of the substrates). From the thermal chain analysis, as described in Section 4.9, the locations where thermal resistance changes take place can be located and studied.

In this section, different degradation mechanisms are presented briefly before moving on to the detailed description of the aging behavior for high-power LEDs. These LEDs were aged 7000 h in a test room of the Technische Universität Darmstadt (Germany).

4.10.1
Degradation and Failure Mechanisms of LED Components

Figure 4.58 shows the cross-section of a modern high-power LED with the transparent primary optics and the phosphor plate on the LED die. The die is connected with a ceramic substrate for mechanical stability and for better heat spreading. The underlying side of the ceramic substrate is soldered on a PCB. Generally, the LED component as a system can be divided into two subsystems: the LED chip itself and

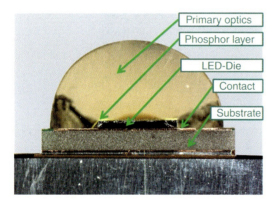

Figure 4.58 Cross-section of a modern high-power LED. (Image source: University of Applied Science, Hannover/Germany.)

its packaging. The chip emits the radiation and the packaging has the tasks to convert and distribute the optical radiation (primary optics and phosphor layer), to bring the contact nearer (wire bonds inside the optics), and to electrically connect to the chip (die attach).

Regarding the statistics on the total number of failures reported in literature [76], it can be concluded that the majority of the problems have their roots in the packaging part and only a small fraction of them should be searched for inside the chip structures. In comparison to other semiconductor products, LEDs seldom exhibit sudden failures and the degradation is expressed in a slower depreciation of their optical performance (e.g., luminous flux).

For the LED chip, the total failure is possible only if the chip breaks in a mechanically overloaded application. The most prominent failure reason in the chip is the relatively high defect density in the semiconductor material (e.g., InGaN) the LED die is made of. This high defect density can be discovered by measurements in the I–V-diagram in reverse mode. At the beginning of use, these defect locations are relatively small but they become enlarged by negative pulses or by ESD (electrostatic discharge) with the consequence that they establish internal shunts in the chip. These internal shunts are similar in function to parallel resistances and cause dark lines or dark spots in and on the chips, see Figure 4.59. Because these dark lines have a relatively high current density, they grow larger with time and optical radiation is reduced gradually.

It can also be seen from Figure 4.58 that damages can arise between the packaging and the chip. There are dark locations at the left and right sides indicating the places of ionic contaminations. This effect is an indication of the leakage in the LED housing or during the packaging process. This also damages the phosphor layer at the center of the chip, see Figure 4.58.

The causes of defect because of packaging can be described as follows:

1) Contact problems of the bond contact (ball form), delamination of the die attach, delamination of the phosphorus layer.

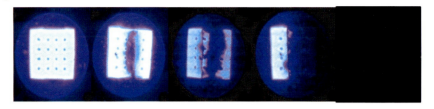

Figure 4.59 Phenomena of the dark lines caused by ESD. (Image Source: Technische Universität Darmstadt.)

2) Contact problem between the transparent lens and the bond wire which have different thermal expansion coefficients. In applications with high thermal load or with fast changing thermal cycles these different thermal characteristics will have an impact and lead to bond break or damage of the bond location.
3) In most SMD-LEDs, the primary optics is a silicone lens and can therefore transport humidity which can be forced if the LED's temperature or the ambient temperature is high.
4) In packaging technology, silver-containing adhesives are often used which leads to the generation and growth of dendrite between the conductive electrical layers inside the LED causing a short circuit.
5) In a white LED, a blue light source pumps short wavelength radiation which can cause the yellowing of the primary optics and of the silicone encapsulant as the binding material for the phosphor mixture, see also Section 3.5.
6) The silicone encapsulant material and the silicone lens consist of silicone polymers and are reactive to a number of chemical substances and gases [77] which can be found in the surrounding air (e.g., in a tunnel in LED based tunnel lighting, in the rooms of a chemical company, or as street lighting in an industrial area) or in the materials used (O-ring, glues, conformal coating) for the luminaire housing and for electrical cables. The consequence of these reactions is the yellowness and darkening of the LED structure and a substantial reduction of optical power, luminous flux, and absolute spectral distribution leading to a color shift. In Figure 4.60, the darkening area at the center of the LED is the result of a reaction with PVC cables inside the luminaire unit.

4.10.2
Research on the Aging Behavior of High-Power-LEDs

In the time period between 2011 and 2014, the present authors performed a comprehensive aging test series for high-power LEDs, COB-LEDs, and medium power LEDs from a number of well-known LED manufacturers with the aims to deliver company-independent aging data to LED luminaire industry and to analyze the

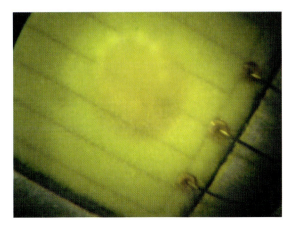

Figure 4.60 Chemical incompatibility of LED components to a PVC cable. (Image source: University of Applied Science, Hannover/Germany.)

causes and the consequences of LED aging for the subsequent luminaire development. The basis of the test methodology was the American standard IES LM-80-08 with the test and measuring conditions described in Table 4.18.

In the research reported here, two types of high-power white LEDs with a CCT of 4000 K from two different LED manufacturers were aged and analyzed. Table 4.19 represents the specifications of this study. The measurement was performed by the spectroradiometer system described in Section 4.8.2 (see Figure 4.48). For each case temperature and current, 21 LEDs were built electrically onto each LED aging unit, with each LED on a separate PCB (Figure 4.61). On this PCB, the thermocouple, pin connector for voltage, and current supply as well as the LEDs were arranged.

4.10.2.1 Change of Spectral Distribution and Chromaticity Coordinates

In Figure 4.62, the change of the spectral distribution of two examples of LED type 1 aged at 700 mA and $T_s = 85\,°C$ is shown. These emission spectra consist of blue chip radiation together with a phosphor system emitting a maximum at 600 nm. In order to analyze the change at different parts of the wavelength range, the ratio of the absolute spectra after 7000 h and at the beginning of the aging period is illustrated in Figure 4.63. It can be seen that the spectral change is very small until 570 nm and is about − 5% in the wavelength range 570–700 nm. Obviously, the yellowing process of the silicone encapsulate has taken place – but at a slow pace.

For the same aging conditions, two examples of LED type 2 were also analyzed. This is illustrated in Figure 4.64. Comparing the spectra of LED type 1 and LED type 2, it can be assumed that, for both cases, a similar phosphor system with a maximum at 600 nm is in use. Dividing the absolute spectra of these two samples after 7000 h and at the beginning of aging, the result can be seen in Figure 4.65.

Table 4.18 Test and measurement conditions for the LED lifetime analysis according to LM-80, after [78].

Requirements	LM80
Case temperature with 2 °C uncertainty, monitored during the lifetime testing	55 °C, 85 °C, and at a higher selectable temperature
Temperature of the surrounding air	Within −5 °C of the case temperature during testing
Humidity	Less than 65 RH throughout the life test
Airflow	To be minimized
Operating orientation	Convection is made possible
Input voltage	Ripple voltage shall not exceed 2% of the DC output voltage
Line voltage wave shape	The total harmonic distortion not to exceed 3% of the fundamental
Input current regulation	Input current monitored and regulated within ±3% of the rated rms value during life testing and to ±0.5% during photometric measurements
Case temperature	Thermocouple measurement system monitoring case temperature at the manufacturer-designated case temperature measurement point
Elapsed time determination	To be measured via video monitoring, current monitoring with an uncertainty within ±0.5%
Recording failure	Checking for LED light source failures either by visual observation or automatic monitoring
Forward voltage monitoring or measurement	Not required
Measurement	Total spectral radiant flux measurement is recommended for the determination of luminous flux and chromaticity

Table 4.19 Test and measurement specifications at the aging laboratory of the Technische Universität Darmstadt.

Issues	Data
Number of the LEDs for each working point	21 samples
Case temperature	55, 85, and 95 °C
Current	350, 700, and 1000 mA
Measurement conditions	$T_c = 25\,°C$, $t = 20\,ms$
Time interval	Every 1000 h
Measured quantities	$I-V$-diagrams, spectral radiant flux, luminous flux, chromaticity coordinates, CCT, thermal resistance

Figure 4.61 Arrangement of the LEDs to be aged.

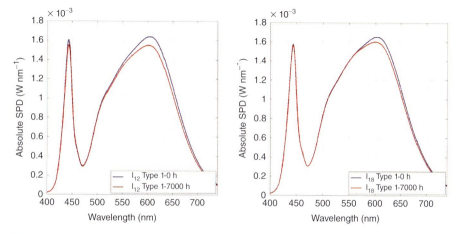

Figure 4.62 Spectral distribution change of two examples of LED type 1.

The following can be concluded from Figures 4.62–4.65:

1) Two examples of the same type of LED from the same batch under the same aging conditions can exhibit very different spectral degradation behaviors.
2) The LED No. J_3 (Figure 4.65) shows a remarkable reduction of its absolute spectrum down to about 70% after only 7000 h at 700 mA and $T_s = 85\,°C$.
3) The most significant reduction can be seen in the blue wavelength range until 480 nm indicating that the blue chip itself undergoes a very strong degradation.

Comparing LED type 1 and LED type 2 with nearly the same phosphor systems but with different aging behaviors, it can be concluded that their blue chip technology and the process control of molding the phosphor mixture on the chip surface are different. This difference can also be seen from the chromaticity shifts ($\Delta u'v'$) in Figure 4.66.

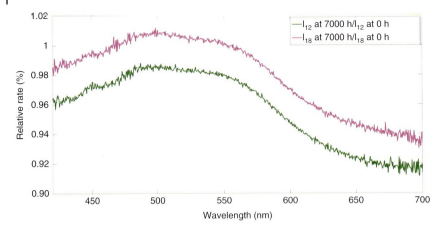

Figure 4.63 Ratio of the absolute emission spectra of two LEDs (both type 1) after 7000 h and at the beginning of the aging period.

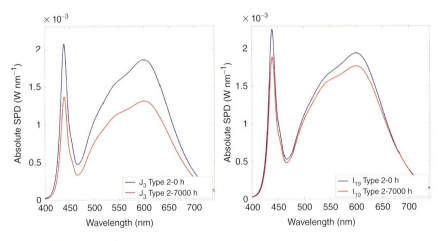

Figure 4.64 Spectral distribution change of two samples of LED type 2.

The following conclusions can be drawn:
- After 7000 h and with the hardest age conditions of 95 °C and 1000 mA, LED type 1 shows a chromaticity difference (shift) of only $\Delta u'v' < 0.0033$.
- For the LED type 2, the chromaticity difference is gradually remarkable and very high for the condition $I_f = 1000$ mA. For increasing sensor temperatures (T_s) from 55 °C over 85 °C up to 95 °C, this difference increases from 0.0061 over 0.0083 to 0.0097. From the description in Section 5.8, it can be learned that a color shift of greater than $\Delta u'v' = 0.005$ is no more acceptable. For this type of LED, an LED current of 1000 mA for the temperature range between 55 and 95 °C or higher has to be avoided.

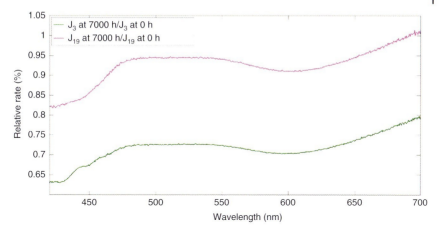

Figure 4.65 Ratio of the absolute emission spectra of two LEDs (type 2) after 7000 h and at the beginning of the aging period.

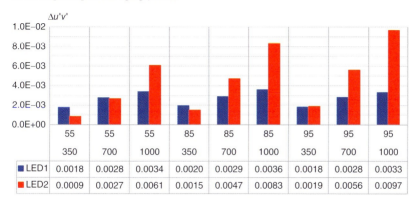

Figure 4.66 Chromaticity differences of the two aged LED types (averaged values from 20 LED samples) after 7000 h compared to the chromaticity coordinates at the beginning of aging.

4.10.2.2 Change of Electrical and Thermal Behavior

At constant current conditions, the aged LEDs were periodically measured concerning their forward voltage. These results are illustrated for LED type 1 in Figure 4.67 and for LED type 2 in Figure 4.68.

Following can be seen from Figures 4.67 and 4.68:

- LED type 2 shows – independent of the aging condition – a remarkable reduction of voltage at constant currents from about 3.15 V down to 2.9 V. The standard deviation among the 20 samples is relatively small. At constant current and reduced voltage, electrical power is less and, if a constant luminous efficacy (in lm W^{-1}) is assumed, then the luminous flux must be reduced and if luminous efficacy itself becomes less during aging because of the change in the chip structure and darkening of the silicone encapsulant then luminous flux will decrease

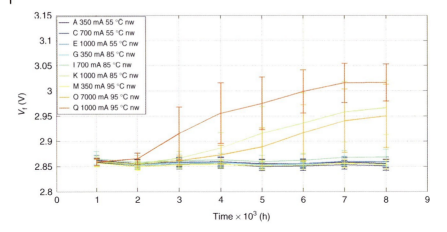

Figure 4.67 Change of forward voltage of LED type 1 with the averaged value and standard deviation among 20 examples for each aging condition.

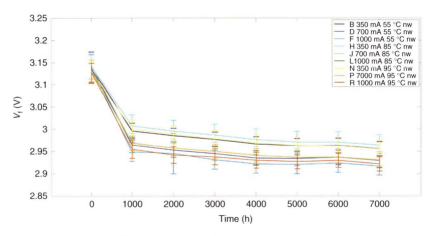

Figure 4.68 Change of forward voltage of LED type 2 with the averaged value and standard deviation among 20 examples for each aging condition.

at a higher pace. This aspect can be considered when using luminous flux at the beginning as a reference value for the design of an LED luminaire.

- LED type 1 shows surprisingly an increase of forward voltage for the samples aged under the conditions of 1000 mA and T_s of 85 and 95 °C. This phenomenon leads to the assumption that the electrical resistance of LED packaging increases because of the higher defect density in the chip and/or to the worse electrical contact between the LED die and the packaging. Because of the differences of voltage, electrical power, and aging behavior, the luminous efficacy of the two LED types will also be subject to change as illustrated in Figure 4.69.

It can be seen from Figure 4.69 that the luminous efficacy of LED type 1 decreased about 10% during the 7000 h of aging. This means that, for the same

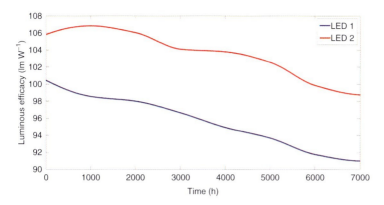

Figure 4.69 Gradual reduction of luminous efficacy of the two LED types (700 mA and $T_S = 85\,°C$).

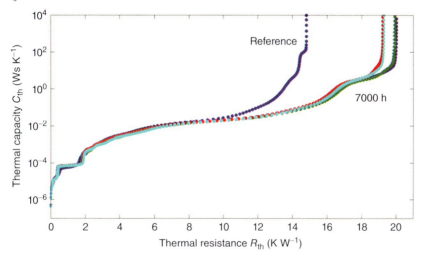

Figure 4.70 Thermal network analysis of a new LED (reference) and of four aged LEDs of the same type [79].

electrical power, the users and the design engineers of the LED luminaire with LED type 1 have to take 10% reduction of light output after 7000 h into account.

In order to qualify and answer the question whether and where the thermal characteristics as the thermal resistance of the whole chain of an LED packaging can be changed, four LEDs of the same type in a new state and after 7000 h have also been thermally measured and analyzed according to the process and mathematical analysis of Section 4.9. These results are shown in Figure 4.70.

It can be concluded from Figure 4.70 that the two types of curves (new and aged) are similar in the first part of the diagram with similar thermal resistance and capacity values. This means that the thermal structure of the LED dies remained almost unchanged after 7000 h of aging time. The difference between the two types of curves can be seen clearly from $10\,\text{K}\,\text{W}^{-1}$ on. This implies a

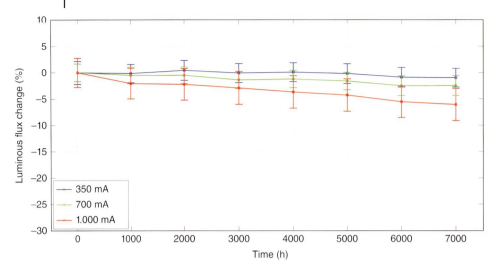

Figure 4.71 Luminous flux degradation of the LED type 1 at 85 °C (mean and 95% confidence intervals for 20 samples).

possible change of the packaging materials (e.g., in the adhesive layers and PCB material) in the course of aging and confirms the general experience that the materials and processing of the packaging, in comparison to the chip, have a higher contribution to LED aging behavior.

4.10.2.3 Change of Luminous Flux – Lumen Maintenance

The aging data for LED type 1 for $T_s = 85\,°C$ and three currents, 350, 700, and 1000 mA are shown in Figure 4.71. For this LED type under these aging conditions, the degradation rate is moderate. The curves for 350 and 700 mA do not substantially point toward any remarkable change. At the 1000 mA condition, the degradation takes place slowly and there are greater standard deviations among the 20 LED samples.

In Figure 4.72, the aging data on luminous flux for LED type 2 under the same aging conditions are illustrated.

Only the aging condition with 350 mA showed a moderate aging rate. At 700 and 1000 mA, the depreciation of luminous flux is rapid with −17% at 7000 h and a continuously increasing standard deviation among the 20 LEDs. For the case of 85 °C, the current of 1000 mA and perhaps also the current of 700 mA are not recommended for use in long-term applications (e.g., street lighting installations).

Using the aging data of 20 LEDs of LED type 2 at $I_f = 700\,mA$ and $T_s = 85\,°C$, the lumen maintenance time (L_{70}) can be calculated according to the algorithm of the IESNA TM-21-11 standard [80] which is described in more detail in Section 4.11. It can be seen from Figures 4.73 and 4.74 that the aging behavior of the 20 LEDs of the two types is different: their lifetime ($L_{70}B_{50}$) equals 46 900 h (type 1) and 11 600 h (type 2), respectively. This means that the well-known lifetime of 50 000 h or more can be fulfilled only with type 1 and a current lower than 700 mA or at a lower case temperature.

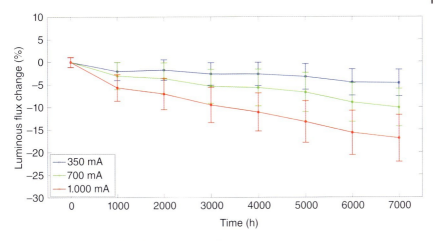

Figure 4.72 Luminous flux degradation of LED type 2 at 85 °C.

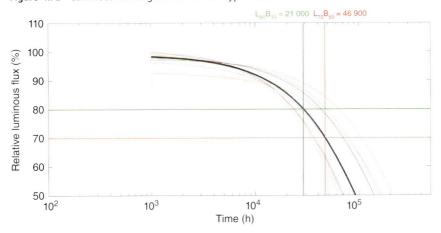

Figure 4.73 Extrapolation for the determination of the L70B50 lifetime (Section 4.11) for LED type 1 at 700 mA and 85 °C.

The lifetime of LED type 2 (only 11 600 h) is far from the general expectation of 50 000 h. For the requirement of 700 mA to have enough luminous flux for a defined application at a condition of 85 °C or higher, this LED type cannot be used.

In the course of the LED degradation study, some samples exhibited strong deformations because of the high temperature or/and current density. In Figure 4.75, an example of LED type 1 aged at 1000 mA and 95 °C can be seen in which a curved crack of the silicone lens can be observed. Roughly, it can be calculated that, at 1000 mA and at a forward voltage of about 3.5 V, an electrical power of 3.5 W is applied from which an optical power of 0.87 W (at 25% optical efficiency) and a thermal power of 2.62 W is generated. If the thermal resistance of the LED from the pn-depletion layer to the thermal pad is about 7 K W^{-1} and an additional 7 K W^{-1} can be assumed for the chain from the thermal pad to the case

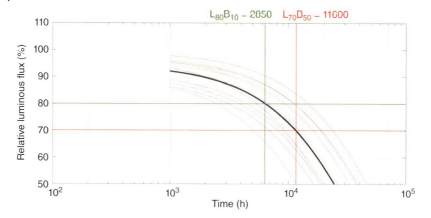

Figure 4.74 Extrapolation for the determination of the L70B50 lifetime (Section 4.11) for LED type 2 at 700 mA and 85 °C.

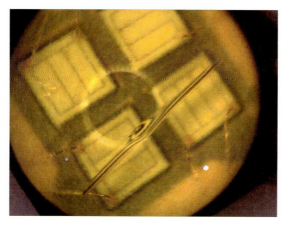

Figure 4.75 Crack in the silicon primary optic because of thermal overstress.

temperature measurement location, a total thermal resistance of 14 K W^{-1} can be estimated. Consequently, the temperature difference between the pn-depletion layer and the temperature measuring point is around 37 K (=2.62 W 14 K W^{-1}). As a result, the temperature of the pn-depletion layer is estimated to be about 132 °C (= 95 °C + 37 K). This is a value approximate to the absolute maximum value of the allowable temperature range.

4.11
Lifetime Extrapolation

The degradation curve of each LED type with a certain packaging quality is caused by temperature and current load as well as ambient influences like dirt and humidity. The form and slope of the curve are also influenced by the absolute uncertainty and reproducibility of measurement in the aging and quality testing laboratories of the LED manufacturers.

As the time to market for each new LED product is limited, only a limited time is available in the manufacturer's laboratory for the aging investigation of the new LED product before it is introduced to the market. In order to receive information on the lifetime of the LED product with a certain probability, extrapolations based on the measured degradation curves (measured within a limited time interval only) are necessary. This extrapolation can be purely mathematical or partly based on chemical and physical relationships. In this chapter, current interpolation concepts are presented.

4.11.1
TM 21-Method

IES TM-21-11 suggests nine different engineering-based models for the analysis of model fitting for LED luminous flux decay life projection [80, 81]. These models are based on the decay rate and can also be described by a function, as a closed-form solution. Linear, exponential, logarithmic, and rational functions are proposed to describe and extrapolate LED aging behavior. The most popular and often used equation is an exponential fit function (mathematical fit function (MFF-EXP)) with two parameters (Φ_0 and α):

$$\Phi(t) = \Phi_0 \cdot e^{-\alpha t} \tag{4.93}$$

This function describes a process with a decay rate that is proportional to the amount present at every time. The luminous flux Φ_0 at the beginning decreases with the decay rate α and is running toward zero. So, it is assumed that luminous decay has only one single rate and will last until the LED does not emit any longer. Many physical phenomena exhibit this behavior, for example, the decay of radioactive materials. As the LED is a complex construction, it would be surprising if its degradation followed only one exponential function. Even if there are some parts of the LED that show this typical behavior, it will be most likely a *sum* of decay functions that have different parameters.

The lifetime of an LED is defined as the point of time when 70% or 80% of its initial luminous flux is emitted. That is why lifetime curves are often extrapolated until this time. As an exponential function going to zero is nearly equal to a linear function at the beginning, in some cases it would be nearly the same to predict the lifetime with a linear curve, see Figure 4.76 (the data for the TM-21 graph were taken from the official TM-21 calculator program).

4.11.2
Border Function (BF)

Another way to get a value for the LED's lifetime is a method with a so-called border function (BF) [81]. This method is shown in Figure 4.77.

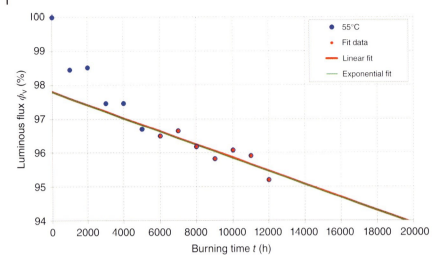

Figure 4.76 Comparison of exponential and linear fit and extrapolation functions for an LED luminous flux decay measurement data until 12 000 h.

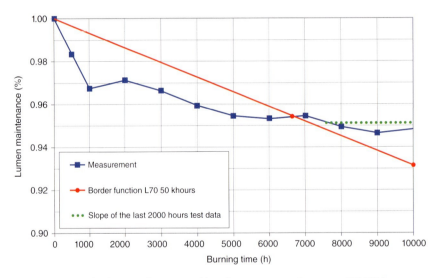

Figure 4.77 Border function (L70; 50 000 h) and measurement data up to 100 000 h, after [81].

The border function is also a TM-21 exponential function with fixed values for lifetime and the percentage of luminous flux. Real measurement data are plotted in the same graph as the border function. The course of the measurement curve at the right-hand side of the cross point (blue dot in Figure 4.77) can be analyzed. If the slope of the last 2000 h' data is higher than the border function's one, it can be assumed that the LED will last at least as long as the border function predicts.

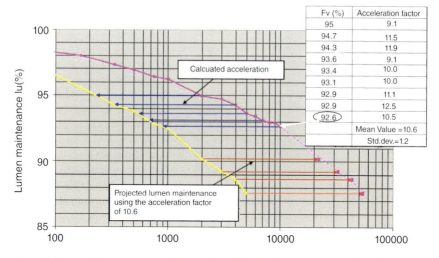

Figure 4.78 Acceleration vectors between the yellow (high temperature) and pink (low temperature) curve, after [81].

4.11.3
Vector Acceleration (Temperature Acceleration – Vector Method – Denoted by TA–V)

To use the method of vector acceleration, LEDs have to be aged at least at two different temperatures [81]. A vector is spanned between the time values of the two measurement curves at one fixed percentage. The division of these two time values gives the acceleration factor; in the example of Figure 4.78, this is the mean of nine such factors. Values of the graph corresponding to a higher temperature are multiplied by the acceleration factor so that the graph of the lower temperature can be extended. By this method, the shape of the base function is projected to the other graph.

The comparison of the two methods, the border function method and the vector acceleration method, leads to different lifetimes (up to several thousand hours).

4.11.4
Arrhenius Behavior

The Arrhenius behavior is an essential law that plays an important role in chemistry mentioned in both [80, 81]: a reaction runs faster when temperature increases [80, 81].

For any arbitrary dimension y the Arrhenius equation is:

$$y(T) = A \cdot \exp\left(-\frac{E_A}{k_B \cdot T}\right) \tag{4.94}$$

Logarithmic values over inverse temperature yield a linear function with a slope of $m = E_A/k_B$:

$$\ln(y(T)) = \ln(A) - \frac{E_A}{k_B} \cdot \frac{1}{T} \tag{4.95}$$

So, the activation energy E_A can be determined by this method. This activation energy represents the dominating reason for degradation. It can also be used for the temperature acceleration by the Arrhenius method which is denoted by TA–A in [81].

4.11.5
Groups of LEDs

A decreasing trend can follow a linear, concave, or convex function or a combination of those. Looking at the measurement data, LED aging behavior can be classified into two main groups. In one group, the course of the decrease is a convex curve so it runs toward a fix value or zero. The other group follows a concave function and its negative slope increases with time. Especially, this behavior is hard to describe with a simple exponential function (Figure 4.79).

4.11.6
Exponential Function (Belonging to the Definition "Other Mathematical Fit Functions, Flexible (MFF-FLEX)")

As many phenomena in physics show an exponential behavior, the first guess should be an exponential function [81]. The difference to the TM-21 function is that a third parameter is used that indicates the value the function converges to. This is the case if one part of the LED light source is aging with its own exponential trend until this aging step stops. Because the other part of the LED still ensures that light is emitted, its luminous flux does not decrease anymore but remains at a fixed value (Figure 4.80):

$$\Phi(t) = (\Phi_0 - \Phi_E) \cdot e^{-at} + \Phi_E \tag{4.96}$$

4.11.7
Root Function (Belonging to the Definition "Other Mathematical Fit Functions, Flexible (MFF-FLEX)")

The root function can also be chosen because it is related to diffusion. Its coefficient depends on temperature hence it is comparable to the Arrhenius equation [81]:

$$D(T) = D_0 \cdot \exp\left(-\frac{E_A}{k_B \cdot T}\right) \tag{4.97}$$

4.11 *Lifetime Extrapolation* | 219

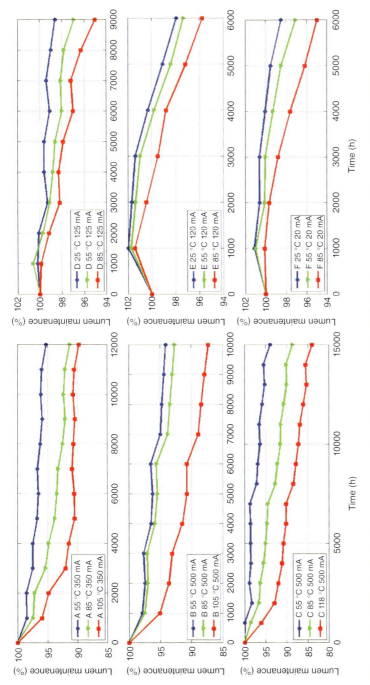

Figure 4.79 LED aging behavior groups: convex degradation curves (left – A, B, C) and concave degradation curves (right – D, E, F).

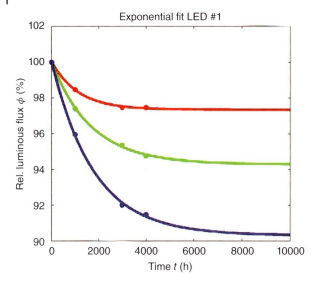

Figure 4.80 Exponential fit for three temperatures (55, 85, and 105 °C).

The diffusion length can be described by a square root function:

$$L = \sqrt{D \cdot t} \tag{4.98}$$

As the diffusion of defects runs until a certain value and can be responsible for less light emission, a possible function for the decrease of luminous flux is:

$$\Phi(t) = \Phi_0 \cdot (1 - \sqrt{D \cdot t}) \tag{4.99}$$

Further investigations should be made in order to find out more about possible light traps in an LED chip.

4.11.8
Quadratic Function (Belonging to the Definition "Other Mathematical Fit Functions, Flexible (MFF-FLEX)")

For concave curves (see Figure 4.81), it is not reasonable to use a linear or even an exponential function that is convex but a quadratic term in the function could try to reproduce the accelerating negative inclination [81]:

$$\Phi(t) = \Phi_0 \cdot (1 - (\alpha \cdot t)^2) \tag{4.100}$$

The Arrhenius fit (in the right diagram of Figure 4.81) for the three temperatures (red, green, and blue curves in the left diagram of Figure 4.81) yields the activation energy value of 0.069 eV in this example.

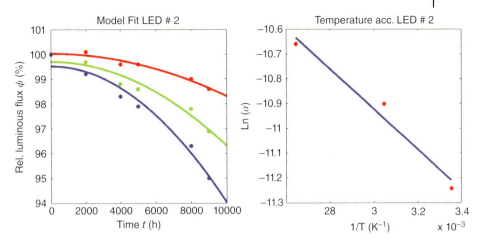

Figure 4.81 Quadratic fit functions for concave degradation tendencies and the Arrhenius plot for parameter α.

4.11.9
Limits of the Extrapolation Procedure

Because of the complex structure of an LED package, there are several components that can age in different ways. So, it is difficult to use only one function to describe the LED's aging at every time. Figure 4.82 shows a root fit for three temperatures (55, 85, and 105 °C). The red, green, and blue filled dots in Figure 4.82 (corresponding to the different temperatures) were used to get the fit functions. But the stable region afterward cannot be described by the blue line which represents the aging curve until 4000 h. So, the extrapolated lifetimes are not representative of the measurement data. Even if the trend had been represented by the root function until L_{80}, the extrapolated lifetime would have been extremely high, see Table 4.20.

4.11.10
Conclusions

The extrapolation of the LEDs' luminous flux degradation curves is a complex issue. Degradation curves exhibit different shapes and they can be grouped. If the right extrapolation function is taken then a good result can be obtained. Arrhenius plots are useful to determine the extrapolation curve parameters. As not all of

Table 4.20 Lifetimes by extrapolation with a root function.

Temperature	55 °C	85 °C	105 °C
L_{70}/h	412 740	124 940	42 970
L_{80}/h	223 600	56 550	20 400

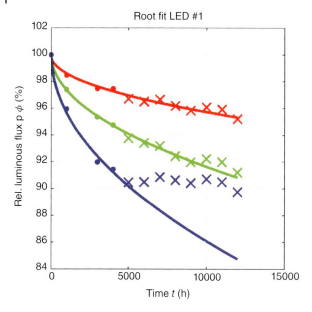

Figure 4.82 Root function up to 4000 h.

the data are measured with a high accuracy it is recommended to filter out some points if necessary. As in the aging process, more than one aging mechanism is possible; aging curves may show a more complex form which cannot be described with a single exponent as every part of the complex LED system has its own aging behavior. For such aging curves, it is difficult to calculate the time of the LED's stabilization period.

4.12
LED Dimming Behavior

The history of the technological development of LEDs can be divided into four steps:

1) *In the first step*: the material's characteristics, chip structure, production reliability, and test methodology were analyzed and optimized.
2) *In the second step*: Different phosphor systems and packaging technology involving the improvement of the silicone encapsulant and bonding technology were developed and characterized.
3) *In the third step*: Luminaire technology with LED components, PCB-technology, electronics (driving and regulating electronics), cooling systems, and optical design were established and improved.
4) *And in the last step*: The topic of smart lighting and adaptive lighting has been regarded as the fourth step in the development. Smart and adaptive lighting

technology includes the following aims and aspects from the point of view of lighting technology:

a. Adaptive light and color regulation depending on weather conditions (rain, fog, ambient brightness), on geometry, usage density, and the topology of the areas to be illuminated.
b. Adaptive light and color adjustment depending on clock time, season, and needs/personal characteristics of the individual users of the lighting systems.

The concepts of light and color regulation can be designed according to the following possibilities:

- Adaptive variation of absolute luminous intensity for the whole angular distribution or the adjustment of luminous intensity for predefined angular directions and parts of the used areas.
- Adaptive variation of CCT and luminance depending on clock time. This so-called *dynamic lighting* is based on the knowledge from the research on *light & health of human beings* and can be used in offices, schools, hospitals, and houses for elderly persons.
- Adaptive adjustments of spectral distributions of the optical radiation and light without changing its chromaticity coordinates in order to maximize the CRI and other color quality issues for a predefined colored object group (skin tones, foods, shoes, textiles, books and magazines, furniture). This idea can be realized in shop lighting, museum lighting, or TV and film lighting.

Smart lighting concepts have been worked out in the time of conventional light source technologies. They can be entirely realized by LED systems because the LED components can be switched very fast within the microsecond range and dimmed continuously without problems with the lifetime of the light source, in contrast to conventional discharge lamps. In this section, the methods of LED dimming and their influence on the change of luminous flux and color characteristics are described.

4.12.1
Overview on the Dimming Methods

There are two dimming methods used in the current LED technology:

1) *The method of PWM method*: The LED components are operated with a series of rectangular pulses which enables switching on and off in a very fast manner. With a switching frequency higher than 100 Hz, the human visual system cannot resolve the intensity variation temporally. Brightness is perceived as an average value. In the PWM method, the amplitude of the pulse remains constant and perceived brightness can be varied by changing the ratio of pulse width to the period of the whole pulse (see Figure 4.83).
2) *Constant current reduction (CCR -method):* The luminous flux and the luminance of the LED component can be dimmed or increased by the variation

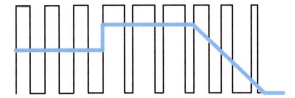

Figure 4.83 Principle of pulse width modulation (PWM).

of LED current. The variation of LED current defines the number of recombinations between electrons and holes (positive charges) and, consequently, photon energy is generated. The electronics for the CCR method is more complicated than for the PWM method and it is therefore more expensive. The advantages of the CCR method include flicker-free operation and the absence of stroboscopic effects.

4.12.2
Experiments: Setup and Results

In the Laboratory of Lighting Technology of the Technische Universität Darmstadt (Germany), a series of measurements comparing two dimming methods were carried out [82]. The aims of these experiments were:

- Study the dimming methods for white LEDs (for automotive lighting, indoor, and street lighting) and for red LEDs (for car rear lamps and for the development of hybrid luminaires containing white and color LEDs for changing CCT);
- The parameters to be analyzed included luminous flux, luminous efficacy, CCT, and dominant wavelength.

4.12.2.1 **Experimental Setup**

For the tests, the following conditions were established in order to receive comparable test results:

- All LED components were temperature-stabilized at a certain constant temperature for a long time in order to have a thermodynamic equilibrium.
- The LED voltage was measured directly at the LED pins by means of the four-pole technique in order to avoid the transition resistance between the cables and connectors.

The spectral radiant flux of the LEDs was measured by a calibrated and spectrally characterized Ulbricht sphere (in other words: integrating sphere) in order to calculate the CCT for the white LEDs and the peak wavelength and dominant wavelength for the color LEDs.

The LED to be measured was soldered on an MCPCB which was mounted by screws on a heat spreader with copper for maximal thermal conduction, see Figure 4.84. A temperature sensor was positioned inside the copper block directly under the MCPCB. By means of a powerful Peltier element and regulation

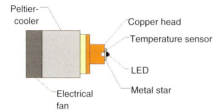

Figure 4.84 Temperature stabilization and regulation for the LEDs with Peltier cooling (TEC), thermoelectric cooling.

electronics, the temperature of the LEDs could be kept constant at a regulated value and independent of the thermal load of the measured LED. The system for the absolute spectral radiant flux measurement has already been shown in Figure 4.48.

4.12.2.2 Test Results for White LEDs

Typical white LEDs used in automotive lighting and indoor and outdoor lighting were selected for the tests:

- For the applications in automotive daytime running lights, the white LEDs labeled by LED1 and LED2 were used.
- LED2 were also used for outdoor and interior lighting. For each one of the two LED types, 10 LED examples were measured and evaluated so that the value reported here can be regarded as the average value of this LED batch. The LED1 type was operated with a current of 120 mA and the LED2 type with a current of 350 mA.
- For automotive front lighting systems (low beam and high beam head lamps), LED3 came into consideration. For this measurement, two examples were operated at a nominal current of 700 mA.
- All LEDs under measurement were temperature stabilized at 50 °C.

The results for the linearity test of their luminous flux are illustrated in Figure 4.85 which shows a very good linearity of luminous flux for the case of the PWM method and a slight super-proportionality for the CCR method.

At constant current amplitude, the voltage and therefore the electrical power are constant in case of the PWM method. This causes a constant value of luminous efficacy, see Figure 4.86. In case of the CCR method, current is reduced leading to the correspondingly reduced current density (current per chip volume unit) so that the efficiency of recombination between the electrons and the holes is increased which delivers more photons hence more luminous flux. The luminous efficacy of LED2 increases from 90 lm W^{-1} at 100% current to more than 120 lm W^{-1} at 13% of the maximal current, see Figure 4.86.

The CCT change when operating with the CCR and PWM methods depends on the LED type under consideration, see Figure 4.87. For LED1, no clear difference between the two dimming methods could be registered for the dimming range between 100% and 3%. For LED2 for indoor and street lighting, the CCT value

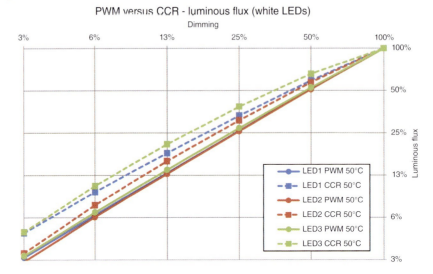

Figure 4.85 Linearity test results of white LEDs for the CCR and PWM methods.

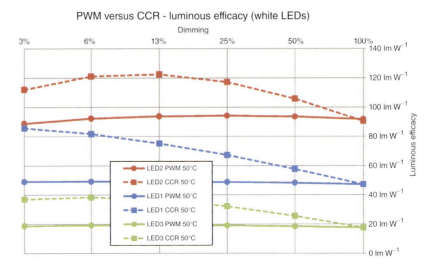

Figure 4.86 Luminous efficacy of white LEDs for the CCR and PWM methods.

changed 60 K between 100% and 3% with the PWM method and about 340 K (from 5820 to 5480 K) with the CCR method. For the same dimming range, LED3 lost about 470 K (from 6050 to 5580 K) with the PWM method and about 800 K with the CCR method. The opinion that color characteristics do not change if the PWM method is used cannot be confirmed here. With the PWM method, the current is constant but, with the variation of the pulse width, the generated heat or thermal energy dissipated in the semiconductor structure changes and therefore the pn-depletion layer's temperature also changes.

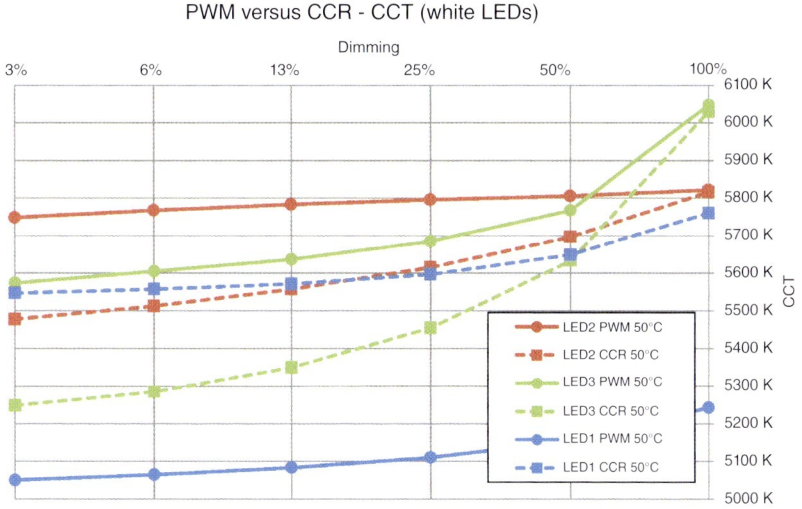

Figure 4.87 Correlated color temperature of white LEDs for the CCR and PWM methods.

From these results, some consequences and recommendations for dimming white LED light sources can be derived:

- For many applications as street lighting, object lighting, industrial lighting (industrial hall, store), and standard office lighting, the CCR method can be used resulting in a color shift of some hundred Kelvin and in a higher luminous efficacy in the case of dimming.
- In most cases, the dimming factor will not be reduced from 100% to 3%. Dimming to 13% or 20% should be the most widely used values so that the color shifts are lower.

4.12.2.3 Test Results for Red LEDs

Red LEDs built on the base of the AlInGaP material system are currently used as light sources for signaling purposes (rear lamps of the car, traffic lamps, computer, and TV display) and in the future also as hybrid lighting systems together with blue, green, and white LEDs to be able to vary CCT and luminous flux. In the research reported here, the red LED1 was used with 10 examples operated with 50 mA at the temperature of the cooling system (50 °C, see Figure 4.84). From the measured values of the 10 sample LEDs, an average value can be calculated.

In Figure 4.88, the linearity of luminous flux is illustrated when the two dimming methods are applied. The PWM method shows a slight super-proportionality and the result with the CCR method indicates a visible under-proportionality which explains the luminous efficacy results in Figure 4.89. In any case, luminous efficacy is not constant over the analyzed dimming range above a dimming factor of 6%.

To evaluate the color change characteristics of color LEDs, it is important to know the shift of dominant wavelength, see Figure 4.90.

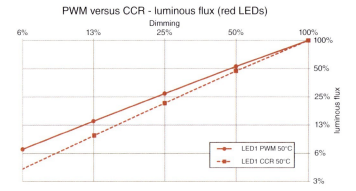

Figure 4.88 Linearity test results of red LEDs for the CCR and PWM methods.

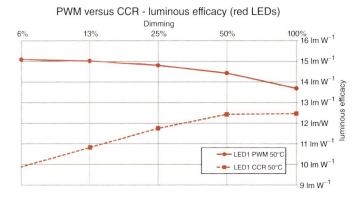

Figure 4.89 Luminous efficacy of red LEDs for the CCR and PWM methods.

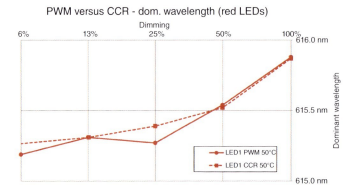

Figure 4.90 Dominant wavelength shift of red LEDs for the CCR and PWM methods.

The shift of dominant wavelength is 0.7 nm for the PWM method and 0.4 nm for the CCR method so that it can be concluded that the color shifts are small.

References

1. Krüger, U. and Schmidt, F. (2003) Applied image resolved light- and color measurement – introduction and application examples. Progress in Automobile Lighting, Technische Universität Darmstadt, PAL 2003, September 23–24, 2003, pp. 567–577.
2. Groh, A. (2011) Photometrie von LEDs, Vortrag auf der 43. Jahrestagung des Deutsch-Schweizerischen Fachverbandes für Strahlenschutz e. V. NIR2011, Dortmund 19m, September 19–21, 2011.
3. Bredemeier, K., Schmidt, F., and Jordanov, W. (2005) Ray data of LEDs and arc lamps. ISAL 2005 Symposium, Darmstadt University of Technology, pp. 1030–1037.
4. Singer, C., Brückner, S., and Khanh, T.Q. (2007) Methods and techniques for the absolute and accurate determination of the luminous efficiency of today high power LEDs Proceedings of the International Symposium for Automotive Lighting, Darmstadt, September 25–26, 2007 (ed. Khanh T.Q.), pp. 642-648.
5. Commission Internationale de l'Eclairage (CIE) (2007) *Measurement of LEDs*, 2nd edn, Publication CIE 127:2007, CIE.
6. National Physical Laboratory (0000) Certificate of Calibration, Reference no E09010185/B1-09-1, National Physical Laboratory, Teddington, Middlesex, *http://fsf.nerc.ac.uk/calibration/pdf_files/E09010185_B1-09.pdf* (accessed 05 June 2014).
7. Bieske, K., Wolf, S., and Nolte, R. (2006) Wahrnehmung von Farbunterschieden von Licht-und Körperfarben. Vortrag auf der Lichttagung der Deutschsprachigen Länder, Bern, Schweiz, September 10–13, 2006, pp. 63-64.
8. Ihringer, J. (2010) Modellierung moderner Hochleistungs – Leuchtdioden im Pulsebetrieb. Master thesis. Technische Universität Darmstadt, pp. 21–35.
9. Nottrodt, O. (2009) Entwicklung eines Messsystems zur Bestimmung des thermischen Widerstandes. Master thesis. Technische Universität Darmstadt, pp. 37–40.
10. Baureis, P. (2005) Compact modeling of electrical, thermal and optical LED behavior. Proceedings of ESSDERC, Grenoble, France, University of Applied Sciences Würzburg, 2005, pp. 145–148.
11. Shockley, W. and Read, W.T. Jr., (1952) Statistics of the recombinations of holes and electrons. *Phys. Rev.*, **87** (5), 835–842.
12. Shockley, W. and Queisser, H.J. (1961) Detailed balance limit of efficiency of pn junction solar cells. *J. Appl. Phys.*, **32** (3), 510–519.
13. Zhe, B. (1982) W. Shockley's equation and its limitation. *Appl. Math. Mech.*, **3** (6), 827–832.
14. Kasap, S.O. (2001) pn Junction – The Shockley Equation, e-Booklet, *http://Materials.Usask.ca* (accessed 05 June 2014).
15. Nelson, D.F., Gershenzon, M., Ashkin, A., D'Asoro, L.A., and Sarace, J.C. (1963) Bandfilling model for gaas injection luminescence. *Appl. Phys. Lett.*, **2** (9), 182–184.
16. Dumin, D.J. and Pearson, G.L. (1965) Properties of gallium arsenide diodes between 4.2 and 300 K. *J. Appl. Phys.*, **36** (11), 3419–3426.
17. Donnelly, J.P. and Milnes, A.G. (1966) Current/voltage characteristics of p-n Ge-Si and Ge-GaAs heterojunctions. *Proc. IEEE*, **113** (9), 1468–1476.
18. Riben, A.R. and Feucht, D.L. (1966) *Solid – State Electronics*, Pergamon Press, pp. 1055–1065.
19. Nelson, H. and Dousmanis, G.C. (1964) Effect of impurity distribution on simultaneous laser action in gaas at 0.84 and 0.88 μm. *Appl. Phys. Lett.*, **4** (11), 192–194.
20. Casey, H.C., Muth, J., Krishnamkutty, S., and Zavada, J.M. (1996) Dominance

21. Forrest, S.R., Didomenico, M., Smith, R.G., and Stocker, H.J. (1980) Evidence for tunneling in reverse-biased III – V photodetector diodes. *Appl. Phys. Lett.*, **36** (7), 580–582.
22. Lester, S.D., Ponce, F.A., Craford, M.G., and Steigerwald, D.A. (1995) High dislocation densities in high efficiency GaN – based light – emitting diodes. *Appl. Phys. Lett.*, **66** (10), 1249–1251.
23. Mártil, I., Redondo, E., and Ojeda, A. (1997) Influence of defects on electrical and optical characteristics of blue LEDs based on III – V nitrides. *J. Appl. Phys.*, **81** (5), 2442–2444.
24. Chen, Y., Taheuchi, T., Amano, H., Akasaki, I., and Yamada, N. (1998) Pit formation in GaInN quantum wells. *Appl. Phys. Lett.*, **72**, 710–712.
25. Hino, T., Tomiya, S., Miyajima, T., Yanashima, K., and Hashimoto, S. (2000) Characterization of threading dislocations in GaN epitaxial layers. *Appl. Phys. Lett.*, **76** (23), 3421–3423.
26. Cherms, D., Heney, S.J., and Ponce, F.A. (2001) Edge and screw dislocations as nonradiative centers in InGaN – GaN quantum well luminescence. *Appl. Phys. Lett.*, **78** (18), 2691–2693.
27. Manasreh, O. (2000) *III – Nitride Semiconductors – Electrical, Structural and Defects Properties*, Elsevier Science B.V., pp. 17–45.
28. Hsu, J.W.P., Manfra, M.J., Lang, D.V., Richter, S., and Chu, S.N.G. (2001) Inhomogeneous spatial distribution of reverse bias leakage in GaN Schottky diodes. *Appl. Phys. Lett.*, **78** (12), 1685–1687.
29. Cao, X.A., Topol, K., Shahedipour-Sandvik, F., Teetsov, J., LeBoeuf, S.F., Ebong, A., Kretchmer, J., Stokes, E.B., Arthur, S., Koloyeros, A.E., and Walker, D. (2002) Influence of defects on electrical and optical characteristics of GaNInGaN-based LEDs. *Proc. SPIE*, **4776**, 105–113.
30. Cao, X.A., Stokes, E.B., Sandvik, P.M., LeBoeuf, S.F., and Walker, D. (2002) Diffusion and tunneling currents in GaN – InGaN multiple quantum well LEDs. *IEEE Electron Device Lett.*, **23** (9), 535–537.
31. Park, J. and Lee, C.C. (2005) An electrical model with junction temperature for light – emitting diodes and impact on conversion efficiency. *IEEE Electron Device Lett.*, **26** (5), 308–310.
32. Fuchs, D. and Sigmud, H. (1986) Analysis of the current – voltage characteristic of solar cells. *Solid-State Electron.*, **29** (8), 791–795.
33. Graeme, J.G. (1995) *Photodiode Amplifiers – OP Amp Solutions*, McGraw-Hill, pp. 1–19.
34. Ray, L.L. and Chen, Y.F. (2009) Equivalent circuit model of light – emitting – diode for system analyses of lighting drivers. IEEE Industry Applications Society Annual Meeting, 2009. IAS 2009, Houston, TX, IEEE.
35. Stringfellow, G.B. and Craford, M.G. (1966) *High Brightness Light Emitting Diodes*, Semiconductors and Semimetals, vol. **48**, Academic Press, pp. 83–95.
36. Ren, F. and Zolper, J.C. (2003) *Wide Energy Band – Gap Electronic Devices*, World Scientific Publishing, pp. 34–108.
37. Piprek, J. (2005) *Optoelectronic Devices – Advanced Simulation and Analysis*, Springer, pp. 40–41.
38. Lumileds Lighting (2002) Advanced Electrical Design Models, Application brief AB20-3A.
39. Calow, J.T., Deasley, P.J., Owen, S.J.T., and Webb, P.W. (1967) A review of semiconductor heterojunction. *J. Mater. Sci.*, **2**, 86–89.
40. Anderson, R.L. (1962) Experiments on Ge-GaAs heterojunctions. *Solid-State Electron.*, **5**, 341.
41. McAfee, D.B., Ryder, E.J., Schockley, W., and Sparks, M. (1964) *Phys. Rev.*, **83**, 534.
42. Moll, J.L. (1964) *Physics of Semiconductors*, MCGraw-Hill Book Compay, Inc, New York.
43. (a) Keldysh, L.V. (1958) Behavior of non-metallic crystals in strong electric fields. *Sov. Phys. JETP*, **6**, 763; (b) Keldysh, L.V. (1958) *Sov. Phys. JETP*, **7**, 665; (c) Kane, E.O. (1961) Theory of tunneling. *J. Appl. Phys.*, **32**, 83.

44. Khanh, T.Q. (2004) Ist die LED – Photometrie mit V(λ) – Funktion wahnehmungsgerecht. Licht 06/2004, www.lichtnet.de (accessed 05 June 2014).
45. Ohno, Y. (1999) LED Phometric Standards, Lecture 1999.
46. Ohno, Y. (2005) Spectral design considerations for white LED color rendering. *Opt. Eng.*, **44** (11), 111302.
47. Chien, M.C. and Tien, C.H. (2012) Multispectral mixing scheme for LED clusters. *J. Opt. Soc. Am*, **20** (S2, Optics Expresss), 245–254.
48. Reifegerste, F. (2009) *Modellierung und Entwicklung neuartiger halbleiterbasierter Beleuchtungssysteme*, Fortschritt-Berichte VDI, Reihe 21, Nummer 386, VDI-Verlag, Düsseldorf.
49. Keppens, A., Ryckaert, W.R., Deconinck, G., and Hanselaer, P. (2010) Modeling high power light-emitting diode spectra and their variation with junction temperature. *J. Appl. Phys.*, **108**, 043104.
50. Dunn, T. and Stich, A. (2007) *Driving the Golden Dragon LED*, Application Note, OSRAM Opto Semiconductors GmbH.
51. Karlicek, R. (2012) *The Evolution of LED Packaging*, Smart Lighting Engineering Research Center.
52. Huber, R. (2008) *Thermal Management of Golden Dragon LED*, Application Note, OSRAM Opto Semiconductors GmbH.
53. Siegal, B. (2006) *Practical Considerations in High Power LED Junction Temperature Measurements*, IEMT.
54. Gu, Y. and Nadrean, N. (2004) A non – contact method for determining junction temperature of phosphor-converted white LEDs. Third SPIE, Vol. 5187, pp. 107–114.
55. Schubert, E.F. (2003) *Light Emitting Diodes*, Cambridge University Press, pp. 1–112.
56. Chhajed, S., Xi, Y., Gessmann, Th., Xi, J.-Q., Sha, J.M., Kim, J.K., and Schubert, E.F. (2005) Junction Temperature in Light-Emitting Diodes Assessed by Different Methods.
57. Xi, Y. and Schubert, E.F. (2004) Junction-temperature measurement in GaN ultraviolet light – emitting diodes using diode forward voltage method. *Appl. Phys. Lett.*, **85** (12), 2163–2165.
58. Hulet, J. and Kelly, C. (2008) Measuring LED junction temperature. *Photonics Spectra*, **42**, 73–75.
59. Keppen, A., Reyckaet, W.R., Deconinck, G., and Hanselaer, P. (2008) High power light – emitting diode junction. *J. Appl. Phys.*, **104**, 093104.
60. Zong, Y., Ohno, Y. (2008) New practical method of measuring high – power LEDs. CIE Expert Symposium on Advances in Photometry and Colorimetry, No. CIE x033:2008, 2008, pp. 102–106.
61. Isuzu Optics Copr (2010) ITR-180 User Manual, Version 1.0.
62. Ma, Z., Zheng, X., Liu, W., Lin, X., and Deng, W. (2005) Fast thermal resistance measurement of high brightness LED. IEEE, 6th International Conference on Electric Packaging Technology, 2005
63. Mentor Graphics (2012) Measurement Control – Reference Guide.
64. Poppe, A., Zhang, Y., Wilson, J., Farkas, G., Szabó, P., and Parry, J. (2006) Thermal Measurement and Modeling of Multi-die Packages, pp. 1–6.
65. Farkas, G., Poppe, A., Kollar, E., and Stehouwer, P. (2003) Dynamic Compact Models of Cooling Mounts for Fast Board Level Design, pp. 1–8.
66. Poppe, A. (2009) Thermal Management – When Designing with Power LEDs, Consider their Real Thermal Resistance, LED Professional, 2009, pp. 2–5.
67. Poppe, A. (2010) Ten Good Reasons Why Thermal Measurements are Important to Your Design. MentorGraphics White Paper 2010, pp. 1–8.
68. Zahner, T. (2010) *T3ster – Thermal Transient Tester Helps Design and Manufacture Better High – Performance Light-Emitting Diodes (LEDs)*, Osram.
69. Mentor Graphics (2013) T3ster – Thermal Transient Tester – General Overview.
70. Mentor Graphics (2013) T3ster – Thermal Transient Tester – Technical Information.
71. Farkas, G., Haque, S., Wall, F., Paul, S., Poppe, A., Vader, Q.V.V, and Bognár, G (2004) Electric and thermal transient

effects in high power optical devices. 20th IEEE SEMI-THERM Symposium, San Jose, CA, March 9-11, 2004.
72. Székeley, V. and Tran, V.B. (1988) Fine structure of heat flow path in semi-conductor device – A measurement and identification method. *Solid-State Electron.*, **31** (9), 1363–1368.
73. Carlaw, H.S. and Jaeger, J.C. (1959) *Conduction of Heat in Solids*, 2nd edn, Clarendon, Oxford.
74. Weinberg, L. (1962) *Network Analysis and Synthesis*, McGraw-Hill, New York.
75. Leonowicz, Z. (2008) Selected problems of circuit theory 1, Lecture 2008, pp. 11–22.
76. Jacob, P. (2014) Ausfalluntersuchungen bei LEDs- Übersicht über Gehäuse- und Chip-basierte Fehlerquellen, Electrosuisse, Bulletin 1/2014, pp. 41–44.
77. Cree® (0000) XLamp® LEDs Chemical Compatibility, support document, *www.cree.com/xlamp* (accessed 05 June 2014).
78. Illuminating Engineering Society (IES) (2008) LM-80-08: IES Approved Method for Measuring Lumen Maintenance of LED Light Sources (September 2008).
79. Wagner, M. and Khanh, T.Q. (2013) Thermal Analysis of LED, Project-Meeting, Peking, China, 2013.
80. IES (2011) TM-21-11: Projecting Long Term Lumen Maintenance of LED Light Sources, July 2011.
81. International Electrotechnical Commission (2013) Technical Committee No. 34: Lamps and related equipment, Sub-Committee 34A: Lamps – Document for comment: LED Packages – Long-Term Luminous Flux Maintenance Projection, 34A/1723/DC, 11-01-2013, Mr. A Patel – Secretary IEC/SC 34A, amit.pate@bsigroup.com.
82. Brückner, S. and Khanh, T.Q. (2011) *Dimmung von Hochleistungs-LEDs*, Zeitschrift Licht, Heft 3, Pflaum Verlag GmbH, München, pp. 44–49.

5
Photopic Perceptual Aspects of LED Lighting

Peter Bodrogi, Tran Quoc Khanh, and Dmitrij Polin

5.1
Introduction to the Different Aspects of Light and Color Quality

To ensure a high level of user acceptance and user satisfaction for modern LED light sources in interior lighting, it is essential to consider several visual and cognitive lighting quality aspects influenced by the spatial and spectral radiance distribution of the LED light source and the LED luminaire. A keyword list of such aspects was prepared by Kronqvist [1] in a detailed review of office lighting literature. This list included among others the following items: comfort, performance, fatigue, glare, well-being, alertness, task lighting, biological clock, flicker, ambient lighting, illuminance, spectral sensitivity, visual acuity, luminance, surroundings, brightness, and color.

In the so-called ELI (Ergonomic Lighting Indicator) system [2, 3], lighting quality assessment criteria are grouped in the following way: visual performance, vista (compliance with the user's expectations), visual comfort, vitality (well-being), empowerment (use of automatic, dynamic light scenes). Present authors propose an alternative grouping of lighting quality aspects (including color quality aspects, see Figure 5.2). This grouping scheme can be seen in Figure 5.1.

As can be seen from Figure 5.1, *perceptual aspects* arise from visual perception directly while a subsequent *aesthetic-emotional* assessment of the color perceptions of the colored objects in the illuminated scene takes place to arrive at the color preference, color harmony, and long-term color memory aspects. Photobiological safety is a basic requirement of lighting quality to avoid the hazard of photochemical damage to the retina (so-called blue light hazard). This issue is not be dealt with here in detail. It should be mentioned, however, that white LED (WLED) light sources are no more hazardous than other lighting technologies that have the same correlated color temperature (CCT) [4].

At the cognitive level of human visual information processing, semantic interpretations of perceptions, or aesthetic/emotional assessments occur: the subject develops a categorical judgment about whether and to what extent (e.g., very good, good, mediocre, low, or bad) a certain aspect is acceptable for a certain lighting application. Understanding the nature of semantic interpretation categories

---

LED Lighting: Technology and Perception, First Edition.
Edited by Tran Quoc Khanh, Peter Bodrogi, Quang Trinh Vinh and Holger Winkler.
© 2015 Wiley-VCH Verlag GmbH & Co. KGaA. Published 2015 by Wiley-VCH Verlag GmbH & Co. KGaA.

Perceptual aspects
   -Visual performance (character recognition, visual search, speed, and error rate in office tasks)
   -Visual comfort (glare-free field of view, shadowiness of illuminated objects)
   -Whiteness, white preference (light source, white objects)
   -Spatial brightness of the illuminated scene
   -Visual clarity I. (large global brightness contrasts)
   -Visual clarity II. (perception of fine spatial detail, colored texture, continuous color shading)
   -Color fidelity (color appearance compared to a reference light source)
   -Color gamut (presence of saturated object colors, large color contrasts among the colored objects in the illuminated scene)
   -Color differences (perceptibility of small color differences among colored objects of similar spectral reflectance)
Aesthetic-emotional aspects
   -Color preference (whether an object exhibits a preferred color appearance under the current light source – preference depends on age, personality, culture and profession of the individual)
   -Color harmony (preference of object color combinations)
   -Similarity to long-term memory colors (whether the color appearance of an object under the current light source is similar to what the subject remembers)
Cognitive aspect
   -Semantic interpretation (a cognitive judgment of the subject about whether a certain aspect is acceptable for a certain lighting application, see Section 5.3)
Light and health aspects
   -Circadian effect (5.10)
   -Sleeping quality
   -Concentration, alertness, arousal, emotional well-being
   -Winter depression

Figure 5.1 Aspects of lighting quality (including color quality) for interior lighting. These aspects depend on the spectral and spatial radiance distribution of the light source. Color quality aspects are illustrated by Figure 5.2.

(see Section 5.3) and predict them mathematically from instrumentally measured quantities is very important for LED lighting design in order to ensure a broad user acceptance of LED lighting technology as the purchase decision of the user occurs at the cognitive (semantic) level.

Light and health aspects include the regulation of the Circadian rhythm by light because of suppressed melatonin levels (Section 5.10) and the disorders of the Circadian rhythm that have severe consequences on sleeping quality and, in turn, on concentration and alertness during work hours and on emotional well-being. It was shown that, in addition to the Circadian brain pathway, the light stimulus can also use the so-called amygdala (a part of the brain) to send signals to the cortex and regulate light-induced alertness as well as provoke and modulate emotions [5].

There is an important implication from Figure 5.1 for lighting engineers: specifications on the LED lighting product datasheet should be expressed in terms of appropriate numeric indices (that are able to describe the behavior of the human visual system according to Figure 5.1) for every aspect. These indices (also called correlates or numeric descriptor quantities of perceptions or aesthetic/emotional assessments) should be *aspect specific* and *mathematically predictable*. Desirable

Figure 5.2 Illustration of color quality aspects: a still life or tabletop arrangement of real colored objects in a viewing booth whose light source can be varied. (Reproduced from Ref. [6], with permission from Wiley-VCH.)

ranges of the indices (e.g., for the category "very good" or "good") depend on the application, for example, for museum lighting we really need "very good" color fidelity, see Section 5.3.

The different aspects in Figure 5.1 are further analyzed below. As can be seen from Figure 5.1, perceptual aspects of lighting quality include the quality of perceived whiteness (see the white standard plate in the middle of Figure 5.2 above the water colors). This aspect is associated with the fact whether the light emitted by the LED luminaire has a shade of white that is perceived without disturbing greenish or purplish tones, see Section 5.7 while yellowish tones are preferred in some applications like home lighting in Western culture.

The general (spatial) brightness impression of the illuminated room is dealt with in Section 5.6 in detail. Spatial brightness means that the observer assesses his/her brightness impression across large areas while he/she is immersed in the illuminated scene. Color quality aspects [7–10] incorporate those lighting quality aspects in Figure 5.1 that are related to the *color appearance* of the objects illuminated by the light source: visual clarity in the sense of clear spatial color perception (visual clarity II), color fidelity, color gamut, color differences, color preference, color harmony, and the long-term color memory aspect.

It is interesting to compare the scheme of Figure 5.1 with some aspects that were found to be important by Viénot *et al.* [11]: (i) fidelity of the object colors as rendered by the actual light source; (ii) fine discrimination of nuances of color (e.g., on a masterpiece in a museum; this is visual clarity II, i.e., clear perception of spatial changes, shadings of color); and (iii) natural color appearance (which is a color preference related aspect). Visual comfort, pleasantness, and visual performance for interior lighting were also mentioned in this study as further aspects of interior lighting quality [11].

Aspects such as the absence of discomfort glare from windows and from WLED light sources [12] in an office depend not only on the *spectral* radiant intensity distribution but also, more strongly on the *spatial* radiant intensity distribution of the LED luminaire. It should be mentioned here that current interior lighting design practice uses the so-called UGR (Unified Glare Rating) metric to avoid glare with UGR values ranging between 7, which means imperceptible glare perception, and 31, which means intolerable glare perception. But UGR was not found to correlate well with subjective glare ratings [13].

Concerning the workflow of the *spectral* engineering of LEDs, first, an appropriate shade of white should be achieved by balancing the spectral power distribution (SPD) of the LED light source across the visible range of the spectrum and, second, the spectral *reflectance* of the illuminated colored objects should be taken into consideration in order to enhance one desired aspect of color quality or combinations of two or more aspects [8, 14].

The color appearance of colored objects (e.g., skin, flowers, leaf, textiles, and watercolors, see Figure 5.2) can change drastically if the SPD of the light source changes because of the interaction of light source emission spectra and object reflectance spectra, see Section 5.4. Such a change is illustrated in Figure 5.3.

For *general* lighting purposes, the most important aspect of color quality is color *fidelity*. Color fidelity is the similarity of the color appearance of the illuminated colored objects under a test light source (e.g., an LED light source) to their reference color appearance under a reference light source which is a phase of daylight or a Planckian (blackbody) radiator of the same CCT as the test light source. Color fidelity is illustrated by Figure 5.3: the color appearance of the colored objects of the still life of Figure 5.2 is similar to the appearance under a reference light source (a Planckian radiator) on the left side of Figure 5.3 hence color fidelity is high. Just the opposite is true for the color appearance on the right side.

Figure 5.3 Change of color appearance of colored objects if the spectral power distribution of the light source changes following the interaction of light source emission spectra and object reflectance spectra (Section 5.4). Left: high color fidelity, right: low color fidelity (illustration). (Left: Reproduced from Ref. [6], with permission from Wiley-VCH.)

In other words, the color fidelity aspect means that the perceived color *difference* between two color appearances of the same colored object (under the current LED light source and its reference light source) should be minimal. According to the standard definition of color fidelity (also called *color rendering* but fidelity is a more recent, unambiguous term), the color fidelity judgment of the observer results from the "conscious or subconscious comparison of the color appearance of the objects with their appearance under the reference illuminant" [15].

Some visual tasks require that the light source alleviates the perception of *small color differences* among different colored objects of slightly different colors, for example, among the members of sets of wires, yarns, paints, petals, and so on (see the slightly different reddish–purplish shades of dahlia petals in the top left corner of Figure 5.2). This aspect corresponds to the color discrimination property, a further aspect of light source color quality.

Visual clarity I (i.e., visual clarity in the first sense) is related to the general spatial brightness sensation of the light source user when he or she enters the room or looks at a still life arrangement like the one in Figure 5.2. This sensation depends on the presence of large enough global lightness contrasts among the illuminated objects [16]. Visual clarity is also associated with another meaning related to *spatially resolved* color or clear spatial color perception. This is visual clarity II (visual clarity in the second sense): the clear perception of fine color shadings with continuous color transitions or clearly visible textures on the surface of the object [8], see for example, the pullover hanging in the right-hand side of Figure 5.2.

LED light sources of more spectral content in the greenish-blue range (i.e., those with higher CCT, e.g., cool white at 6500 K) increase visual clarity II. The reason is that under such a light source, the pupil area is smaller and, although a small pupil area reduces retinal illumination, "it increases the depth of field and it reduces distortion of the retinal image by spherical and chromatic aberrations" [17].

The *color gamut* (the extent of the set of all possible perceived colors, even saturated colors of any hue under a given light source) aspect is related to the perception of large color differences and the presence of all possible (saturated) colors among the objects illuminated by the light source, see for example, the color gamut of the watercolors or the ColorChecker® chart in Figure 5.2.

Color preference includes an *aesthetic* judgment of the user of the light source about how object colors *should* look like under the light source by paying special attention to the vividness and/or the naturalness of the object colors. Naturalness should not be confused with color fidelity as object colors are often (but not always) considered "more natural" if their saturation is enhanced by the light source. Certain unsaturated objects (e.g., butter or Caucasian skin) can be considered more natural if they are less saturated than under the reference light source. According to the above, the distinction between color preference and color fidelity is very important for the spectral engineering of LED light sources, that is, for the question how to tailor the shape of the LED's SPDs across the visible wavelength range. This issue is further discussed in Section 5.5 where a more detailed definition of color preference is given. It is also important to mention already at this point that color preference strongly depends on the profession and cultural background of the subject and on the context of the illuminated scene, see also Section 5.5.

The impression of color harmony (see e.g., [18]) is the result of a similar aesthetic judgment about *combinations* of object colors with 2, 3, or more components. In Figure 5.2, the color harmony impression of pairs or triads of watercolors (e.g., yellow–orange or yellow–orange–red) changes if the SPD of the light source changes.

The concept of memory color rendering [19, 20] is based on the similarity of an object color under the current light source to its long-term memory color. The latter is a color perception of an often-seen object saved in the memory of the light source user and remembered later. For example, different observers tend to have a common memory color about Caucasian skin color perceptions which is presented in the bottom left corner of Figure 5.2. The long-term memory color of an object (e.g., Caucasian skin or green grass) differs from both actual color perceptions (e.g., under the daylight phase D65) and from the preferred color of that object [6].

To accentuate one color quality aspect (e.g., to provide preferred color appearance for reddish objects) or to accentuate more aspects of color quality at the same time with an LED light source (which exhibits very flexible SPDs hence there are huge possibilities for spectral design compared to non-LED light sources), numeric descriptors (or indices) were defined for each color quality aspect. These indices can be computed from instrumentally measured SPDs and optimized by changing the technological parameters (e.g., phosphor concentrations in the phosphor mixture, see Section 3.4) of the LED light source [14, 21].

For an effective spectral design, it is advantageous to obtain a set of *independent* visual aspects of color quality. In a real field test [8] using a colorful still life (tabletop) arrangement with real artificial objects (Figure 5.2), observers scaled nine color quality aspects under different light sources visually. Six more or less independent factors were extracted from the correlations among these visual color quality scales (CQSs): resemblance to long-term memory colors, color preference, brightness, color fidelity, color gamut, and the perceptibility of small color differences [8].

In a computational approach, correlations among existing numeric indices of color quality were investigated [22]. The relevance of such studies (see also [23]) is that if two indices correlate well for a broad set of light sources with different SPDs then a usable co-optimum can be obtained for a specific LED spectral design that optimizes both color quality aspects at the same time. The target aspect of color quality optimization and the set of important color objects depend on the application [14]. In practice, economic viability is also important where phosphor converted LEDs (PC-LEDs) provide an affordable solution following due to the use of only one blue pumping LED and a good combination of high quality phosphors [14], see Section 3.4. The optimization issue is discussed further in this section as well as in Section 7.1.

To turn back to the earlier-mentioned computation of the different indices [22], it included the CIE general color rendering index (CRI) (R_a, Section 5.2) [15], the so-called flattery index (a color preference index (CPI), R_f) [24], another type of CPI [25], a color discrimination index (CDI) that predicts the perception of small color differences under the light source being evaluated [26], the so-called color rendering capacity (CRC) that quantifies color gamut under the light source [27], a similar color gamut measure called cone surface area (CSA) [28], and another type of CRI (RP) [29]. From the correlation analysis among these measures, two principal components were extracted: (i) a color gamut based component explaining gamut metrics and a reference light source based component explaining such metrics that compare two color appearances of the same reflecting object, one appearance under the light source in question and one appearance under a reference light source (see Section 5.2).

A combined color fidelity–color preference computation method, the National Institute of Standards and Technology (NIST) CQS [30] is described in Section 5.5. Correlations between the indices resulting from the CIE CRI method (Section 5.2), its modern version (CRI2012, Section 5.2) and the CQS color preference indices are also shown in Section 5.5.

In today's standard lighting engineering practice, however, the measures used to predict and optimize LED lighting quality [31] or to co-optimize lighting quality and energy efficiency [32] are limited to the standard photometric quantities and some derived measures including luminance, illuminance, the contrast rendering factor as well as the different disability and discomfort glare measures, see Figure 5.4. It is pointed out below why this is a serious limitation impairing LED spectral design.

As can be seen from Figure 5.4, the measures to evaluate or to design a given lighting application are all based on the CIE (1924) photopic luminous efficiency function ($V(\lambda)$) that allows to consider – for the spectral engineering of light sources – only those rays whose wavelengths are in its limited spectral range by underweighting important wavelengths to which the chromatic and Circadian mechanisms of human vision are sensitive. The shape of the $V(\lambda)$ function is compared with the spectral sensitivities of the other human vision mechanisms and the ipRGC (intrinsically photosensitive retinal ganglion cell) mechanism in Figure 2.6.

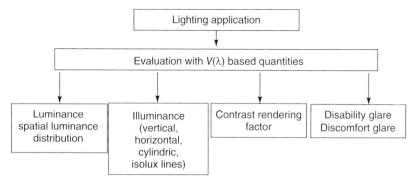

Figure 5.4 Today's prediction and optimization method of lighting quality for photopic (interior lighting) applications. (Ref. [33].)

Accordingly, $V(\lambda)$ based quantities are not suitable for a comprehensive, usable and profitable lighting quality design including color quality and light and health aspects. Important examples are brightness perception of the light source and lightness perception of colored objects which depend not only on their luminance or relative luminance but also on their chromaticity (see Section 5.6) and their scotopic luminance at lower (mesopic) luminance levels (see Section 6.1). Therefore, optimizing the luminous efficacy (lm W^{-1}) value of, for example, a phosphor LED by, for example, ray tracing [21] inside the light source by changing phosphor particle size, phosphor concentration, and LED light source geometry may yield a suboptimal brightness result because rays with wavelengths outside the central range of the $V(\lambda)$ curve are underweighted.

It is also well known that visual performance (e.g., reading speed) is an increasing function of both surround luminance level and luminance contrast (e.g., between characters and their background) but this function has a saturation plateau which means that over a wide range of task and lighting variables (especially at those levels that comply with international lighting standard specifications), "the change in relative visual performance is slight" [17]. This fact was also pointed out by Rea [34]. For general interior lighting at photopic levels, this optimum visual performance range can be ensured by providing a horizontal illuminance level of at least 500 lx. If this level has been reached by providing enough radiant flux at all relevant wavelengths of the LED light source then it is important to consider a more comprehensive scheme of lighting quality aspects to establish an advanced method of spectral engineering, see Figure 5.5.

As can be seen from Figure 5.5, there are two further groups of evaluation and design criteria in addition to the classic V(λ) based scheme of Figure 5.4: color quality indices and light and health measures. Color quality indices are the subject of this section and are summarized in Figure 5.5. Light and health (see Section 2.5) measures are not so well established at the time of writing as lighting quality indices. But a usable measure does exist [35] which is introduced in Section 5.10.

What is perhaps most interesting is the fact that, depending on the application of the LED light engine to be designed, the three groups of lighting quality measures can be given different weights during the spectral design or the spectral optimization of the LED light source, and subcategories can be weighted inside the optimization target function. Generally speaking, the target function T of spectral optimization can then be written in the form as shown in the schematic Eq. (5.1).

$$T = a\,\text{VL} + b\,\text{CQ} + c\,\text{LH} \tag{5.1}$$

In Eq. (5.1), VL means the set of $V(\lambda)$ based quality measures (e.g., illuminance), CQ corresponds to the descriptors of color quality (e.g., color fidelity index) and LH means a measure of light and health aspects (e.g., the Circadian stimulus (CS), see Section 5.10). The weights a, b, c can be changed according to what is relevant for a given application. For example, for general office lighting, color fidelity, visual clarity, and illuminance level may be relevant hence the weights a and b (or the corresponding sub-weights of the particular aspects) can be inserted with a high

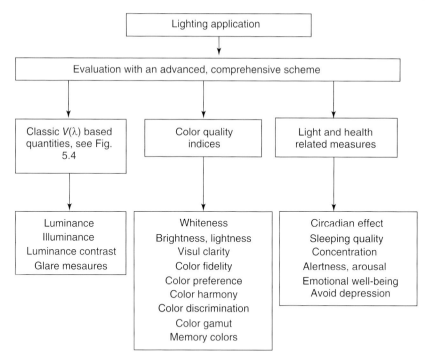

Figure 5.5 Comprehensive scheme of lighting quality aspects for an advanced spectral engineering for photopic (interior lighting) applications. (Ref. [33].)

value in Eq. (5.1). In some cases (e.g., in a retirement home or in a hospital) light and health aspects may play a crucial role hence the value of c will be high when maximizing the value of T by changing the technological parameters of the LED light source.

Combining VL and CQ measures with equally high weightings (e.g., $a = 1, b = 1, c = 0$) may, however, lead to a contradicting optimization criterion. For example, if luminous efficacy (in lm W^{-1} units) is co-optimized with a type of color fidelity index, for example, the CIE general CRI R_a [7] by simulating the changing of the physical parameters of a PC-LED, for example, by the use of a ray tracing program, the use of luminous efficacy as one of the target variables seriously underweights (or completely suppresses) important rays by the use of the $V(\lambda)$ function. But such wavelengths are necessary to maximize the value of the CRI R_a (or a special CRI), the other important target variable.

Note that for a *realistic* optimization usable to develop a real product, it is unrewarding to work with only *relative* SPDs and try to optimize the *luminous efficacy of radiation* (LED) [7], which is the conversion factor from optical power (watt) to luminous flux (lumen). What is relevant is luminous efficacy, the conversion efficiency from *electrical* power to luminous flux. Such a simulation, however, can be carried out only with a comprehensive model including LED models (Chapter 4),

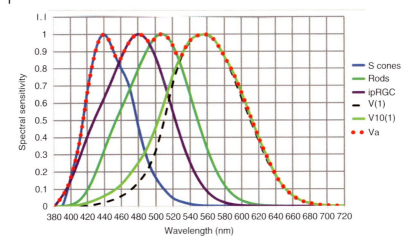

Figure 5.6 Red dots: shape of a possible alternative spectral weighting function, $V_a(\lambda)$. This can be a solution to the discrepancy of using the $V(\lambda)$ based luminous efficacy (lm W^{-1}) as a co-optimization criterion for the color quality optimization (or Circadian optimization) of an LED light source.

a phosphor model (Chapter 3), and an optical model of the LED light source with a ray tracing program, see for example [21].

The earlier mentioned contradiction may be solved by the use of an *alternative* luminous efficacy value in Eq. (5.1) in the future, by computing its value by an *alternative* spectral weighting function $V_a(\lambda)$ instead of $V(\lambda)$. This idea (according to the knowledge of the authors, no such solution exists in literature) was inspired by Loe [36]: $V_a(\lambda)$ should have a broader shape than $V(\lambda)$ in order not to underweight any wavelength that can be reflected from typical colored objects including red, blue, and purple objects hence contributing to the enhancement of the CRI (Section 5.4). A possible shape can be seen in Figure 5.6.

As can be seen from Figure 5.6, $V_a(\lambda)$ was tentatively defined here as the upper envelop function of the spectral sensitivities of human vision mechanisms that contribute to color quality and light and health aspects including rod and L-, M-, S-cone sensitivities as well as the ipRGC mechanism (compare with Figure 5.61). It may be interesting to consider the co-optimization of this alternative luminous efficacy like quantity and a color quality index and compute the standard $V(\lambda)$ based quantities *after* the optimization in order to check whether the spectral design complies with the standards.

5.2
Color Rendering Indices: CRI, CRI2012

The concept of color fidelity (the alternative – more widely used but less specific – term is color rendering) was defined and illustrated in Section 5.1. It was mentioned that for the color quality assessment of general lighting applications, the fidelity aspect of color quality is of vital importance. It is equally important to be able to compute from the SPD of the light source a numeric descriptor quantity (a numeric index) or a set of such quantities that correlate well with the visual perception of color fidelity resulting from the "conscious or subconscious comparison of the color appearance of the objects with their appearance under the reference illuminant" [15]. In this section, two methods to compute such indices are described: the CRI method recommended by the CIE [15] at the time of writing this book and its modernized version intended to avoid its deficiencies, the so-called CRI2012 method [37].

The CRIs of both methods are based on the computation of numeric color differences for a set of specific homogeneous colored objects, so-called test color samples (TCSs), between their test and reference color appearance: the greater the color difference, the less the resemblance to the color appearance under the reference light source hence the lower the value of the CRI (see Eq. (5.3)).

5.2.1
CIE CRI Color Rendering Index

The computational method of the so-called *general* CIE CRI (R_a) [15] has the following steps (its block diagram is shown in Figure 5.7):

Step 1. A reference illuminant is selected. This has the same CCT (T_{cp}) as the test light source. If the value of T_{cp} is less than 5000 K then a blackbody radiator (see Figures 2.26 and 2.27) of the same color temperature is the reference illuminant. If the value of T_{cp} is 5000 K or greater than 5000 K then a phase of daylight (see Figures 2.26 and 2.27) of the same T_{cp} value is used as reference illuminant. The Euclidean distance between the test light source and the reference illuminant in the u, v color diagram will not be greater than 5.4×10^{-3}. Otherwise, according to this method, the test light source cannot be considered a shade of white. As mentioned in Section 5.1, the quality of perceived whiteness is a very important lighting quality criterion. This is why a whole Section 5.7 is devoted to this issue.

Step 2. 14 TCS are selected from the Munsell color atlas. The first eight TCSs are used in average to compute the value of the general CRI (R_a). For every one of the 14 TCSs, 14 so-called special CRIs are computed. Figure 5.8 shows the spectral reflectance functions of the first eight unsaturated TCSs (TCS01–08). Figure 5.9 shows the spectral reflectance of the TCS09–14 (TCS09–12 are very saturated colors).

Step 3. The CIE 1931 tristimulus values X, Y, Z are computed for the 14 TCS (TCS01–14) under the test light source and the reference light source. These values are transformed into CIE 1960 UCS (Uniform Chromaticity Scale) co-ordinates (u, v) and into the today outdated CIE 1964 U^*, V^*, W^* color space [15, 39].

Step 4. In order to describe the chromatic adaptation between the white points of the test light source and the reference light source, the chromaticity of

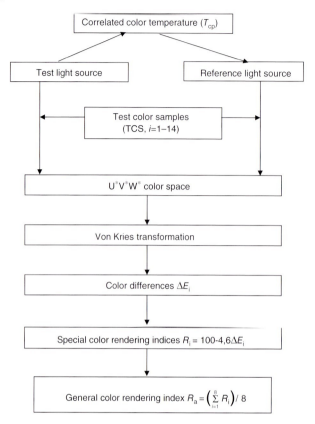

Figure 5.7 Block diagram of the computational method of the general color rendering index CIE CRI R_a [15]. (After Ref. [38].)

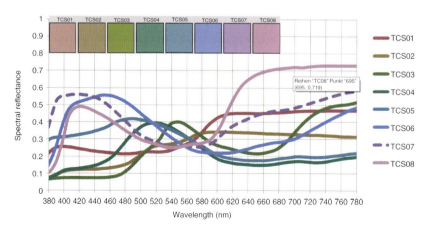

Figure 5.8 Spectral reflectance of the six test color samples TCS09–14. TCS09–12: saturated samples, TCS13: a type of human complexion color; TCS14: leaf green.

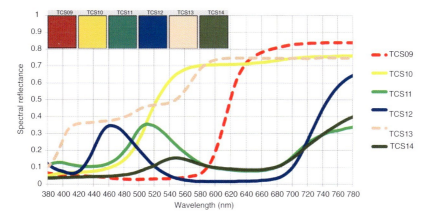

Figure 5.9 Spectral reflectance of the first eight unsaturated test color samples (TCS01–08).

the test light source is transformed into the chromaticity of the reference illuminant by the aid of a today outdated so-called *von Kries transformation* [15].

Step 5. 14 CIE 1964 color differences (ΔE_i) [39] are computed for each one of the 14 TCSs ($i = 1, \ldots, 14$) from the differences of the U^*, V^*, W^* values under the test light source and the reference light source (ΔU^*_i), (ΔV^*_i), (ΔW^*_i):

$$\Delta E_i = \sqrt{(\Delta U^*_i)^2 + (\Delta V^*_i)^2 + (\Delta W^*_i)^2} \tag{5.2}$$

Step 6. For every TCS (TCS01–14), a so-called *special* CRI is computed (R_i; $i = 1, \ldots, 14$) by Eq. (5.3).

$$R_i = 100 - 4.6\, \Delta E_i \tag{5.3}$$

The indices R_i are scaled according to Eq. (5.3) in the following way:
1. the value of $R_i = 100$ means a complete agreement between the test and reference appearances of the TCS; and
2. the CIE illuminant F4 [40] (a warm white fluorescent lamp) has the value of $R_a = 51$. This is established by the factor 4.6 in Eq. (5.3).

Step 7. The general color rendering index (R_a) is defined as the arithmetic mean of the first eight special color rendering indices (see Figure 5.8), see Eq. (5.4).

$$R_a = \left(\frac{1}{8}\right) \sum_{i=1}^{8} R_i \tag{5.4}$$

5.2.2
Deficiencies of the CIE CRI Color Rendering Method

The current CRI computation method [15] is known to exhibit several problems. It was shown in visual experiments and in computations that the general CRI R_a is unable to describe the perceived color rendering property of light sources correctly [9, 41]. The color rendering rank order of light sources obtained in visual experiments is often predicted incorrectly by their general CRI values (R_a). The reasons for these deficiencies are described below.

Choice of the TCSs. The first eight TCSs (TCS01–08, see Figure 5.8) are unsaturated and do not represent the variety of colored natural and artificial objects (including saturated objects, see Section 5.4), compare the first eight TCS (Figure 5.8) with Figures 5.22 and 5.24. An average value of the first eight special CRIs (Eq. (5.4)) cannot describe the variety of the spectral reflectances of colored objects in interaction with the often discontinuous emission spectra of the diverse test light sources. This interaction is analyzed in Section 5.4 in detail. An example of the interaction between emission and reflectance spectra is shown in Figure 5.10.

As can be seen from Figure 5.10, the red peak in the relative spectral radiance of the RGB LED light source interacts with the sharp slope of the spectral reflectance curve of the saturated red TCS09 in the wavelength range 620–660 nm. Because of this interaction the color perception of TCS09 is extremely saturated – very different from the reference color appearance. This effect cannot be predicted by the general CRI R_a which is based on the average of eight unsaturated TCSs (TCS01–08).

Nonuniform color space. One of the most serious deficiencies of the CIE CRI color rendering method represents the outdated color difference formula used to predict the perceived color difference between two color appearances of the same TCS: under the test light source and under the reference light source. This color

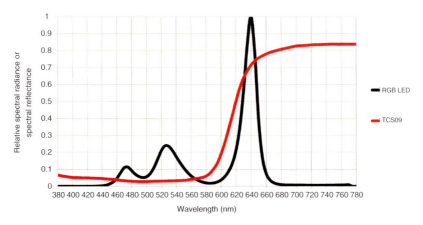

Figure 5.10 Interaction of the spectral reflection of TCS09 (saturated red) with the relative spectral radiance function of an RGB LED light source with a sharp red maximum.

difference should be predicted by a *visually relevant* color difference metric which is suitable to describe all kinds of color differences for all types of color stimuli and all magnitudes of color differences from visually just noticeable up to large differences. This outdated color difference formula is based on the obsolete and visually non-uniform CIE 1964 U^*, V^*, W^* color space [15]. Therefore, the color differences ΔE_i computed via Eq. (5.2) are perceptually incorrect. The correlation between the color differences computed in U^*, V^*, W^* color space (ΔE_i) and the perceived color differences (ΔE_{vis}) is low. Such a color space is called visually nonuniform. In addition, the U^*, V^*, W^* color space predicts the state of chromatic adaptation of the human visual system erroneously. These problems can be solved by using the perceptually uniform CAM02-UCS color space introduced in Section 2.3 that performed suitably both for small and large color differences [42]. Another advantage is that it also uses an advanced chromatic adaptation formula hence it is applicable to a wide range of the illuminant's CCTs. CIE2012, the updated CRI method, is based on this CAM02-UCS color space, see Section 5.2.3.

Interpretation of the values on the numeric CRI scale. It is not easy for the user of the light source and sometimes even for lighting engineering experts to interpret the values on the CRI scale as a benchmark of color quality. Is the color rendering property of the light source moderate, good or very good if $R_a = 83$, 87, 93, or 97, respectively? The interpretation of the differences on the same scale is also problematic, for example, how different is the visual color rendering judgment for $\Delta R_a = 0.5$, 3.0, or 25.0? This problem is dealt with in Section 5.3.

5.2.3
CRI2012 Color Rendering Index

To solve the problems discussed in Section 5.2.2, the following principles were considered when constructing the CRI2012 method [37], a highly developed update for the CIE CRI method:

1) A large set of representative TCSs was used;
2) Color stimuli of these TCSs were represented in the modern CAM02-UCS color space (see Section 2.3);
3) Every such color stimulus has two versions: one stimulus is the TCS illuminated by the reference illuminant and the other one is illuminated by the test light source. Both versions are represented in CAM02-UCS color space. The vector that originates from the reference stimulus and points towards the test stimulus is called *color shift*. Figure 5.13 illustrates such sets of color shift vectors for the TCSs of the CRI2012 method.

The CRI2012 method [37] was developed in the framework of CIE TC 1–69, "Color Rendition by White Light Sources." Here, its current version (not yet endorsed by the CIE at the time of writing of this book) is presented although we suggest checking recent CIE publications to get a further update. Starting from the relative spectral radiance distribution of the light source, the CRI2012 method computes a new general CRI, $R_{a,2012}$ and also special CRIs ($R_{i,2012}$).

The components of the CRI2012 method [37] are summarized below in comparison with the CIE CRI method (Figure 5.7):

1) The reference illuminant is the same illuminant as in the CIE CRI method.
2) Instead of the CIE TCS01–14 there are new TCSs.
3) Instead of the CIE 1964 color difference calculation (Eq. (5.2)), there is a new color difference calculation in the CAM02-UCS color space.
4) Instead of Eq. (5.4), there is a new way of representing the mean value (root mean square (RMS)) of the calculated color differences for the TCSs.
5) Instead of the linear scaling of the CRI according to Eq. (5.3) there is a new nonlinear scaling function to transform the CAM02-UCS color difference ($\Delta E'$) into the new index value $R_{i,2012}$.

The CRI2012 computation method [37] has the following steps, see Figure 5.11:

1) Selection of the reference illuminant for the current test light source, the same illuminant as in the CIE CRI method.

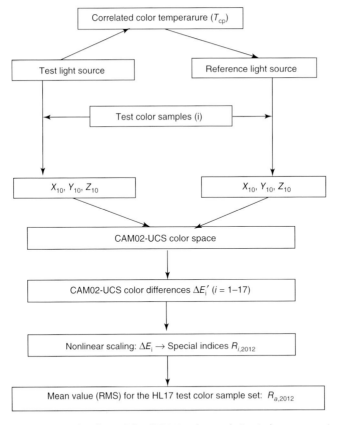

Figure 5.11 Flowchart of the CRI2012 color rendering index computation method to obtain the special color rendering indices $R_{i,2012}$ and the general color rendering index $R_{a,2012}$ [37]. Compare with Figure 5.7.

2) There are three different sets of new TCSs. They are defined with their spectral reflectance curves.
3) The 10° tristimulus values X_{10}, Y_{10}, and Z_{10} are calculated for every TCS both under the test light source and under the reference light source. These values are transformed into CAM02-UCS color space assuming the following CIECAM02 viewing condition parameters: $L_A = 100$ cd m^{-2}; $Y_b = 20$; $F = 1.0$; $D = 1.0$; $C = 0.69$; and $N_c = 1.0$.
4) CAM02-UCS color differences ($\Delta E'_i$) are computed for every TCS.
5) These color differences ($\Delta E'_i$) are scaled in a nonlinear manner to obtain the CRI values ($R_{i,2012}$, $R_{a,2012}$).
6) The value of the general CRI $R_{a,2012}$ is obtained as an RMS value calculated from the 17 $\Delta E'_i$ values of the 17 TCSs of one of the three different sets, the so-called HL17 set. It is not suggested to use these 17 TCSs as a basis of special color rendering indices.

5.2.3.1 Test Color Samples in the CRI2012 Method

The starting point to define the TCSs was the so-called *Leeds 100 000* set [37] with spectral reflectance curves of more than a hundred thousand natural and artificial objects measured at the University of Leeds (England). On the basis of this dataset, three sets of TCSs were generated:

1) By the aid of mathematical considerations, a virtual set of 1000 representative test colors was defined. This is the so-called *Leeds 1000 set*. This set is intended to represent the spectral reflectance curves of all Leeds 100 000 test colors.
2) A further representative set was also defined: the so-called *HL17 set* with only 17 virtual test colors [43] (see Figure 5.12) to compute the general CRI $R_{a,2012}$.

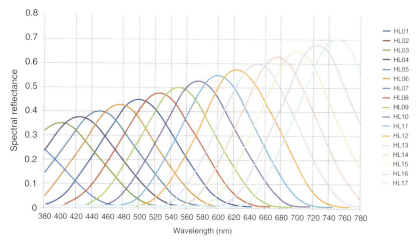

Figure 5.12 Spectral reflectance curves of the 17 virtual test color samples HL1–HL17 of the CRI2012 color rendering index method. (Data source: CRI2012 worksheet, version 9.0.)

3) A set of 210 *real test colors* was also selected. This set contains the following test colors that cover the whole hue circle at all saturation levels approximately uniformly:
   a. 90 test colors whose color stimuli change only a little if the spectral emission of the test light source changes (so-called *high color constancy samples* or HCC);
   b. 90 test colors whose color stimuli change a lot if the spectral emission of the test light source changes (so-called *low color constancy samples* or LCC); and
   c. 10 typical artist's colors and 4×5 different skin tones (African, Caucasian, Hispanic, Oriental, and South Asian). Every skin tone was taken four times in order to increase its weight within the set.

Figure 5.13 illustrates three sets of color shift vectors for the three sets of TCSs in the CIECAM02 a_M, b_M diagram if the reference light source (origin of the vectors) is replaced by the test light source (end of the vectors). The test light source is CIE illuminant F4 ($T_{cp} = 2938$ K; CRI $R_a = 51$) and the reference light source is a blackbody radiator at 2938 K.

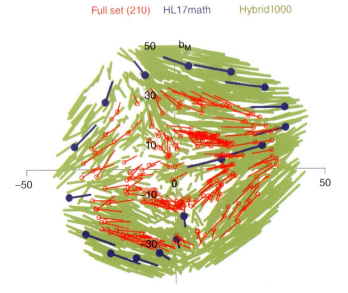

Figure 5.13 CIECAM02 a_M, b_M diagram with three sets of color shift vectors for the three sets of test color samples if the reference light source (origin of the vectors) is replaced by the test light source (end of the vectors). The test light source is CIE illuminant F4 ($T_{cp} = 2938$ K; CRI $R_a = 51$) and the reference light source is a blackbody radiator at 2938 K. The three sets: Hybrid1000 (green): Leeds 1000 set; HL17math (blue, filled blue circles: reference color appearance): HL17 set with the 17 virtual test colors HL1–HL17; full set (red; red circles = reference color appearance): the 210 real test colors. Image source: CRI2012 worksheet of CIE TC 1–69, version 9.0.

The following can be seen from Figure 5.13:

1) The direction and magnitude of the color shifts depend on the type of test color, on CIECAM02 hue angle h and on CIECAM02 colorfulness M (distance from the origin).
2) The Leeds 1000 set contains many saturated test colors that are absent from the other sets.
3) There are no HL17 test colors in the hue angle range $280-360°$.

It can also be seen from Figure 5.13 that it is not possible to compress the information of such large sets of color shift vectors reasonably into only one scalar number (e.g., the CRI2012 general CRI) although this would be important for simple understandable light source datasheets, specifications, and package labels. For example, if the composition of a phosphor mixture of an LED light source is optimized with the sole criterion that only one number – the general CRI – should be maximal then certain object colors can be rendered still poorly even at a very high value of the general CRI. This issue is further discussed in Section 5.4.

In literature, this kind of spectral engineering of a light source that concentrates on the optimization of the general CRI (R_a) with its known deficiencies is sometimes called *spectral gaming* [37, 43]. The HL17 set, however, has been defined with the mathematical criterion of being insensitive against spectral gaming [37, 43]. This is why the CRI2012 method defines the general CRI $R_{a,2012}$ on the basis of the HL17 set although it is not suggested to use the special CRIs based on the individual TCSs of the HL17 set because the HL17 set works correctly *only as a whole*.

The CRI2012 method calculates special CRIs for the 210 real test colors and, as an option, also for the *Leeds 1000* test color set. As an optional supplementary information, in addition to the value of $R_{a,2012}$, the *worst* special CRIs of the subsets of the 210 test colors can be specified (like the 90 LCC samples, the 90 HCC samples or the artist's colors and skin tones).

5.2.3.2 Root Mean Square and Nonlinear Scaling

In the CRI2012 method, the CAM02-UCS color difference is computed for every one of the 17 test colors of the HL17 set ($\Delta E'_i$, $i = 1-17$) between the two color appearances of the test color under the test and reference light sources. Then the RMS value of the set of these 17 CAM02-UCS color differences is calculated (Eq. (5.5))

$$\Delta E'_{rms} = \sqrt{\frac{\sum_{i=1}^{17} (\Delta E'_i)^2}{17}} \tag{5.5}$$

The advantage of the RMS formula (Eq. (5.5)) is that if a certain color difference value ($\Delta E'_i$) is high (i.e., if there is a very poorly rendered TCS) then it has more influence than in the case of the arithmetic mean (Eq. (5.4)). This corresponds to the visual effect that if one test color is rendered very poorly then this deteriorates the visual color rendering property significantly. This is why the RMS value is more suitable to condense the color rendering property of a light source than an arithmetic mean value.

The scale of the general CRI $R_{a,2012}$ [37] is defined by the aid of the nonlinear scaling of Eq. (5.6).

$$R_{a,2012} = 100 \left[\frac{2}{1 + e^{[(1/55)(\Delta E'_{rms})^{1.5}]}} \right]^2 \quad (5.6)$$

Figure 5.14 shows the function $R_{a,2012}$ ($\Delta E'_{rms}$) of Eq. (5.6).

The special indices $R_{i,2012}$ of the *Leeds 1000* set and the *210 real* set as functions of the $\Delta E'_i$ color difference values can be computed by using the same curve as in Figure 5.14 (substituting the symbol $\Delta E'_{rms}$ by $\Delta E'_i$ in Eq. (5.6)). The nonlinearity introduced by Eq. (5.6) describes signal compression, a general property of the human visual system at the extremes of perceptual scales.

Concerning the exponent (1/55) in Eq. (5.6) which was considered a preliminary value [37], the value (1/55) corresponds to the criterion that the arithmetic mean value of the $R_{a,2012}$ values for the 12 CIE illuminants F1–F12 should have the same value ($R_{a,2012,F1-F12,mean} = 75$) as the mean of the CIE CRI R_a values for the same 12 CIE illuminants F1-F12 ($R_{a,F1-F12,mean} = 75$). Other exponent values can also be considered. For example, the value of (2/45) corresponds to a *psychophysically relevant* criterion, a quasi-linear relationship with the so-called semantic interpretation scale (R) [44]. We see in Section 5.3 that semantic categories assigned to certain fixed values on the R scale help specify the meaning of any numeric value on the CRI scale in terms of natural (everyday) language (so-called *semantic interpretation of numeric values*), For example, "very good," "good," or "moderate" color rendering which is understandable for nonexpert light source users.

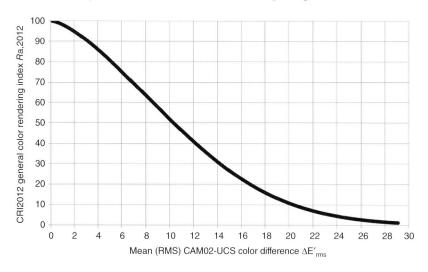

Figure 5.14 The function $R_{a,2012}$ ($\Delta E'_{rms}$) of Eq. (5.6).

5.3 Semantic Interpretation of Color Differences and Color Rendering Indices

Nonexpert users cannot understand the meaning of the numeric values of the CRI and the differences between them. It is not straightforward to figure out, even for experts, whether the value of $R_a = 87$ corresponds to "low," "moderate," or "good" color rendering for example, from the point of view of a demanding LED light source user of an interior application and if for example, the difference between $R_a = 84$ and $R_a = 87$ means a significant categorical change, for example, "good" instead of "moderate." Such categorical labels (e.g., "good" or "moderate") corresponding to certain numeric values are called semantic interpretations.

To derive the semantic interpretation of the values of the CRI in this sense, it is essential to deal with the semantic interpretation of the values of the underlying color difference metric as CRIs are computed from color difference values. As already seen, semantic interpretations provide important additional information to the continuous numeric scale of a metric about the meaning of a certain value of the metric in the following sense: the continuous scale of the color difference metric (or the scale of the CRI) is assigned a further continuous scale, the so-called semantic interpretation scale (or R scale).

Any user of the light source, even a nonexpert user, should be able to understand the values on the R scale immediately because a set of consecutive values on the R scale is labelled by the following categories taken from everyday language, for example, "very good" ($R = 1.0$), "good–very good" ($R = 1.5$), "good" ($R = 2.0$), "moderate–good" ($R = 2.5$), "moderate" ($R = 3.0$), "low-moderate" ($R = 3.5$), "low" ($R = 4.0$), "bad–low" ($R = 4.5$), and "bad" ($R = 5.0$) [44]. Because of this fixed labelling, values on the R scale can be understood and communicated among scientists, engineers, designers, manufacturers, retailers, and customers without difficulty.

To develop the semantic interpretation scale for the metric color difference values and the CRIs, a series of three visual experiments was carried out [44]. In these experiments, observers had to tell a semantic interpretation category corresponding to the perceived similarity of color appearance between two color stimuli: two versions of the same colored object illuminated by the reference light source and the test light source, respectively, for example, "good" if the similarity of color appearance was "good."

Here, the method of one of the visual experiments is summarized. The other two experiments also had a similar method. These experiments provided a basis for the mathematical modelling of the experimentally determined semantic interpretations. The model is simple. It consists of a single continuous function that computes for every instrumentally measured color difference (ΔE) between the two color stimuli a value R as the descriptor of the visual similarity of their color appearance. Every R value has a semantic interpretation in the above sense, for example, $R = 3.5$ means "low-moderate" similarity. This function, the so-called semantic interpretation function $R(\Delta E)$, is used to interpret the values of the CRIs of the two methods presented in Section 5.2 (CRI and CRI2012).

5.3.1
Experimental Method of the Semantic Interpretation of Color Differences

In the visual experiment, a three-chamber viewing booth was used [45]. Figure 5.15 shows the scheme of the viewing booth while Figure 5.16 visualizes the chambers furnished with three times the same collection of colored objects.

The three-chamber viewing booth (Figures 5.15 and 5.16) worked at two CCTs, at 2700 K or at 5400 K. The white points of the three chambers were adjusted to match visually at the middle of the white bottom in every chamber. The illuminance was equal $E = 1100\,\text{lx} \pm 50\,\text{lx}$ at the bottom of every one of the three chambers. In the left chamber (with test light source A), current market-based WLED light sources (WLEDs) and RGB chip LEDs were installed. Their combination was set to the following general CRI values: $R_{a,2700\,K} = 69$ and $R_{a,5400\,K} = 71$.

The middle chamber contained the reference light sources of the experiment, a tungsten halogen lamp at 2700 K and a fluorescent lamp at 5400 K with $R_{a,2700\,K} = 99$ and $R_{a,5400\,K} = 91$, respectively. In the right chamber (with test light source B), compact fluorescent lamps (CFLs) and RGB chip LEDs were installed ($R_{a,2700\,K} = 84$; $R_{a,5400\,K} = 81$). In every chamber, the same 30 real colored objects were arranged in the same manner: a MacBeth ColorChecker chart, textiles, artificial fruits, artificial flowers, and a Barbie doll®, see Figure 5.16. The spectral radiance at the bottom of each chamber and the spectral reflectance of the objects were measured with a calibrated spectroradiometer (Konica Minolta CS2000A).

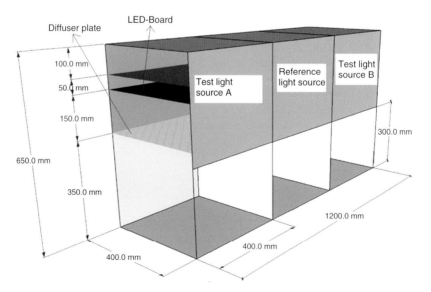

Figure 5.15 Scheme of the three-chamber viewing booth used in one of the experiments for the semantic interpretation of color differences [45].

Figure 5.16 Three-chamber viewing booth with the same three set of colored objects. Observers had to assess the similarity of color appearance of the same object (e.g., the red rose) between the left (test A) chamber and the middle (reference) chamber and between the right (test B) chamber and the middle (reference) chamber visually. The combination of 4 × 6 homogeneous test colors is the MacBeth ColorChecker® Chart.

Twenty observers with normal color vision had to assess the similarity of color appearance of the same colored objects between the left and middle chambers and in the same way, also between the right and middle chambers (one reason for using three chambers instead of only two chambers was to save time by carrying out two simultaneous assessments). Only 18 colored objects had to be assessed; the other objects were used as context elements. Similarity had to be judged for every object individually by using the following categories: "excellent," "very good," "good," "moderate," "bad," and "very bad."

Combinations of these categories were not allowed. Observers had to use only one of these categories. In the other two experiments (not described here) the category "low" was also used between "moderate" and "bad." Before the experiment, observers were instructed to act as very demanding users of interior lighting for example, in the context of the approval of a high-quality color reproduction.

Every judgment (e.g., "good" similarity of color appearance) was assigned an instrumentally measured CAM02-UCS (see Section 2.3) color difference ($\Delta E'$) between the two color stimuli (two versions of the same colored object under the test and the reference light sources). This dataset of 1440 semantic judgments (20 subjects × 18 objects × 2 CCTs × 2 test light sources) was merged with two similar datasets resulting from the other two experiments and the mean tendency of all data was used to derive the semantic interpretation function $R(\Delta E')$ [44]. The advantage of this function is that it *predicts* the semantic interpretation of color similarity from the *instrumentally* measured color differences between the two color stimuli.

5.3.2
Semantic Interpretation Function R(ΔE′) for CAM02-UCS Color Differences

Figure 5.17 shows the semantic interpretation function $R(\Delta E')$ [44] with the semantic category labels attached to the ordinate.

As can be seen from Figure 5.17, the semantic interpretation function $R(\Delta E')$ [44] never reaches the categories excellent (0.0) and very bad (6.0). This means that these categories were avoided by the observers. The function $R(\Delta E')$ is monotonically increasing, as expected: the greater the color difference, the worse the color similarity. For $\Delta E' \geq 11$, the value of R equals a constant (approximately 5.6). For $\Delta E' = 0.0$, the value of R equals 1.0 (very good). $R(\Delta E')$ is a cubic function with the polynomial coefficients $(a_0 - a_3)$ listed in Table 5.1.

If for example, the color difference between the two versions of the color stimuli of a colored object (i.e., under the test light source and the reference light source) equals $\Delta E' = 3.0$ then we obtain for the value of the semantic interpretation scale

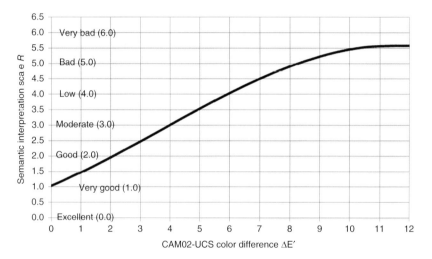

Figure 5.17 Semantic interpretation scale (ordinate) for the similarity of color appearance between two color stimuli (R) and CAM02-UCS color differences (ΔE′) between them (abscissa). Black curve: semantic interpretation function $R(\Delta E')$ [44] with the semantic category labels attached to the ordinate.

Table 5.1 Polynomial coefficients $(a_0 - a_3)$ of the cubic semantic interpretation function $R(\Delta E')$ for $\Delta E' < 11$.

a_0	a_1	a_2	a_3
1.03904809	0.4073793	0.03377417	−0.00302202

For $\Delta E' \geq 11$, $R = 5.585$.
Source: Reproduced with permission from *Color Research and Application* [44].

$R = 2.48$. This value corresponds to "good-moderate" similarity of color appearance.

5.3.3
Semantic Interpretation of the CRI2012 Color Rendering Indices

As the CRIs of the CRI2012 method are based on CAM02-UCS color differences ($\Delta E'$, Eq. (5.6)), the CRI2012 scale ($R_{a,2012}$ or $R_{i,2012}$) can be interpreted by the semantic interpretation function $R(\Delta E')$ directly. The question is whether a certain test light source with for example, $R_{a,2012} = 87$ provides "very good," "good," or "moderate" similarity to its reference light source concerning the color appearance of a collection of typical colored objects [46]. As mentioned before, the answer to such a question is especially relevant for nonexperts who would like to buy and install a light source but do not know what the value of the CRI means.

The function $R(R_{a,2012})$ was derived from the semantic interpretation function $R(\Delta E')$ in the following way [33]:

1) a set of different $\Delta E'$ values was generated ($\{\Delta E'_j\}$);
2) a $R_{a,2012,j}$ value was computed for every $\Delta E'_j$ value via Eq. (5.6);
3) a value R_j on the semantic interpretation scale was computed for every $\Delta E'_j$ value using Table 5.1;
4) a continuous function $R(R_{a,2012})$ was fitted on the dataset $\{R_j; R_{a,2012,j}\}$.

Figure 5.18 shows the function $R(R_{a,2012})$.

The semantic interpretation scale values of the special CRIs $R(R_{i,2012})$ can be computed in the same way by using the same fourth order polynomial function

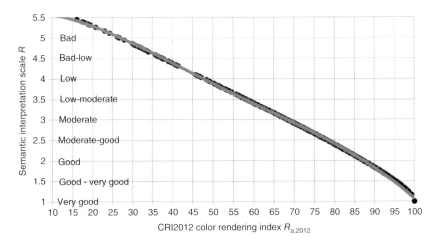

Figure 5.18 Semantic interpretation of the CRI2012 color rendering index (Section 5.2.3). Black dots: computed data points; Gray curve: a fourth order polynomial function fitted to the data points, see Table 5.2. (After Ref. [33].)

Table 5.2 Polynomial coefficients (a_0–a_4) of the fourth order polynomial semantic interpretation function $R(R_{a,2012})$ [33] shown as the grey curve in Figure 5.18.

a_0	a_1	a_2	a_3	a_4
5.5887800492912	0.0145577420367	−0.0019268842519	0.0000248325694	−0.0000001153459

defined in Table 5.2. Instead of the TCSs of the CRI2012 method (Section 5.2.3.1) any colored object can be used if its spectral reflectance function is known. The corresponding CRIs are called *object specific color rendering indices*, $R_{obj,2012}$ and the value of the semantic interpretation scale $R(R_{i,2012})$ can be calculated in the same way for them by using Table 5.2.

5.3.4
Semantic Interpretation of the CIE CRI Color Rendering Indices

Despite its deficiencies (Section 5.2.2), the CIE CRI color rendering method has been generally used worldwide at the time of writing this book. From the point of view of the semantic interpretation of the CIE CRI index values, the most important deficiency of the CIE CRI method is that it is based on the perceptually nonuniform, obsolete CIE 1964 $U^*V^*W^*$ color space. This non-uniformity results (unlike CRI2012 which is based on the uniform CAM02-UCS space), in dissimilar semantic interpretation functions for each CIE TCS (Figures 5.8 and 5.9) or every colored object when we try to derive a semantic interpretation function for special CIE CRIs (R_i) or object specific CIE CRIs (R_{obj}).

First, we compute the values R of the semantic interpretation scale for the first eight CIE TCSs (TCS01–08) as the basis of the general CRI (Eq. (5.4)). In this computation [33], several test light source SPDs including all types of light sources were taken and then, both the R values and the special CIE CRIs (R_i) were computed for every TCS (TCS01–08). Figure 5.19 shows the result.

As can be seen from Figure 5.19, R values scatter intensely as a function of the CIE CRI R_i values. The reason is (as it was indicated earlier) that the R values were computed from $\Delta E'$ values of the perceptually (almost) uniform CAM02-UCS color space while the R_i values were computed from the nonuniform, today outdated CIE 1964 U^*, V^*, W^* color space in which perceived color differences depend much on the chromaticity of the TCS (TCS01, TCS02, …, TCS08, see Figure 5.8).

Nevertheless, according to the prevailing (and unfortunate) use of the CIE general CRI R_a at the time of writing in lighting practice, a mean (cubic) semantic interpretation function was fitted to the $R(R_i)$ data points (black dots in Figure 5.19). This function [33] is represented by the gray dots in Figure 5.19 and it can be considered as the semantic interpretation function of the CIE general CRI, $R(R_a)$.

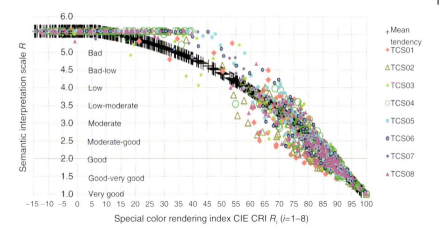

Figure 5.19 Semantic interpretation of the special CIE CRIs R_i ($i = 1-8$, the basis of the general color rendering index R_a). (After Ref. [33]). During the computation of the data points (colored symbols of different colors for TCS01–08) [33], several test light source spectra were taken and then, both the R values and the special CIE CRIs (R_i) were computed for every TCS (TCS01–08). Black plus signs: a cubic function intended to represent a mean tendency [33] which can be considered as the semantic interpretation function of the CIE general color rendering index, $R(R_a)$.

Table 5.3 Polynomial coefficients ($a_0 - a_3$) of the cubic polynomial mean semantic interpretation function $R(R_i)$ [33] shown as the grey curve in Figure 5.18.

a_0	a_1	a_2	a_3
5.6542214712829	−0.00868708870	−0.00025988179	−0.00000127635

This function can be considered as $R(R_a)$ for $R_a > 6.7$. If $R_a < 6.7$ then the constant value 5.58458851 can be used instead of the cubic function.

According to the function $R(R_a)$ defined in Table 5.3, the scale of the CIE CRI R_a obtains the following semantic interpretation: $R_a = 100$: very good; $R_a = 94$: good–very good; $R_a = 87$: good; $R_a = 82$: moderate–good; $R_a = 74$: moderate; $R_a = 58$: low; $R_a = 33$: bad. The current (at the time of writing) commonly used standard EN 12464 for interior lighting requires at least $R_a = 80$ for general interior lighting purposes which is a little bit worse than "moderate–good," apparently not enough for today's demanding light source users. At this point, the question is, how reliably the $R(R_a)$ function based on the mean tendency of the first eight unsaturated CIE TCSs can be used to interpret individual special or object specific CRIs. Here, the example of a saturated red object, a red rose, is shown. This was a real red rose whose spectral reflectance function was carefully measured in the Laboratory of the authors at the Technische Universität

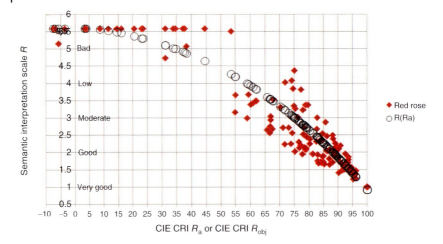

Figure 5.20 Red squares: semantic interpretation scale values (R) of the object specific CIE CRI color rendering index R_{obj} for a real red rose (whose spectral reflectance was measured in our laboratory) for a set of simulated light sources. Open black circles: prediction of the $R(R_a)$ function.

Darmstadt. Its semantic interpretation scale values (R) are compared with the prediction of the $R(R_a)$ function in Figure 5.20.

As can be seen from Figure 5.20, semantic interpretation scale values (R) of the object specific CRI of the red rose (R_{obj}) cannot be predicted from the $R(R_a)$ function which is based on the mean tendency of the first eight unsaturated TCSs (TCS01–TCS08). Also, there is a hug scatter of the semantic interpretation scale values. Depending on the relative SPD of the test light source, for example, the value of $R_{obj} = 75$ of the red rose can be associated with very different semantic interpretations in the range $R = 2.0$ (good) – $R = 4.5$ (bad-low).

The reason is that the *perceived* color difference between the test and reference color appearances of the red rose corresponding to the value $R_{obj} = 75$ depends strongly on the *direction* of the color shift (see Figure 5.13) if the color stimuli are represented in the obsolete, perceptually nonuniform $U^*V^*W^*$ color space underlying the CIE CRI method. The advantage of the CRI2012 method is that its CRIs are associated with unambiguous semantic interpretations [33] because of the perceptual uniformity of its underlying color space, CAM02-UCS.

To provide an overview at the end of this section, Table 5.4 summarizes and compares the CRI values ($R_{a,2012}$ or $R_{i,2012}$; and R_a) corresponding to the semantic categories.

In Table 5.4, $R_{a,2012}$ or $R_{i,2012}$ values were computed using the exponent (2/45) instead of (1/55) in Eq. (5.6). As mentioned earlier, the reason is that if the value of (2/45) is used then the relationship between R and $R_{a,2012}$ (or $R_{i,2012}$) becomes quasi-linear [46]. The advantage of specifying the color rendering property of a light source in terms of an R value (or a semantic category) is that the R value corresponds to a fixed, *psychophysically relevant* scale anchored by the semantic

Table 5.4 Comparison of the color rendering index values ($R_{a,2012}$ or $R_{i,2012}$; and R_a) corresponding to the semantic categories.

Semantic category	CRI2012 ($R_{a,2012}$ or $R_{i,2012}$)	R_a
Very good	100	100
Good	87	87
Moderate	67	74
Low	47	58
Bad	26	33

In this table, $R_{a,2012}$ or $R_{i,2012}$ values were computed using the exponent (2/45) instead of (1/55) in Eq. (5.6). R_a values reflect the mean tendency of Figure 5.19 computed by considering all types of light source spectra.

ratings of real subjects (e.g., 1: "very good" and 2: "good") who played the role of demanding light source users. On the contrary, the numeric scales of CRIs depend on scientific or industrial *agreements* and are subject to change if their agreed-on constant (4.6 in Eq. (5.3) or 1/55 in Eq. (5.6)) is changed.

5.4
Object Specific Color Rendering Indices of Current White LED Light Sources

In the computational analysis of this section, spectral reflectance characteristics of two selected colored object groups (human skin tones and reddish objects) are studied with the aim of deriving implications for the spectral engineering of WLED light sources. The general CRI $R_{a,2012}$ (or, more precisely, its semantic interpretation scale R) is compared with the R-scales of some sample object specific CRIs (for a limited set of representative colored objects) considering a representative sample set of 34 WLED light sources (resulting from a market-based choice plus a few LEDs with very high quality phosphor mixtures or composed of multiple phosphor-converted LED light sources). All of these WLEDs are intended for general interior lighting purposes.

The relative SPDs of the WLED light sources used in this Section were grouped after their CCT (warm white, neutral white, and cool white) and their R_a level (>93; 85–93, 78–85 and <78). Differences between the general CRI values and the object specific CRI values are analyzed in terms of the semantic interpretation scales of the CRI2012 CRIs (see Section 5.3) especially with respect to whether there are categorical changes (e.g., "moderate" instead of "good") if the general color rendering is used instead of object specific indices.

5.4.1
Spectral Reflectances of Real Colored Objects

As we have already learnt, in our natural and built environment a permanent interaction takes place between the emission spectra of the light sources and the reflection spectra of the colored objects (see Figure 5.10). This interaction influences the appearance of perceived colors in the human visual system very strongly, see Figure 2.19. In a real situation, for example, in a living room or in a museum, the impression of the color rendering property of the light source depends not only on the SPD of the light source itself but also on the type of colored objects that are currently viewed under the light source. The reason is that certain kinds of object combinations (e.g., the skin tones of a group of persons) can be rendered by the current light source better than another collection of objects (e.g., a collection of green–yellow objects).

The set of typical (prevailing) object colors in interior lighting is highly application dependent. Just consider how different the typical colored objects of for example, offices, kitchens, living rooms, butcheries, bakeries, green-grocer's, and fashion shops are. Even if the SPD of for example, a WLED light source is optimized with the general CRI or a predefined set of special CRIs as an optimization target, certain object colors that are important for a particular application may still become less optimally rendered. Therefore, it is necessary to measure and analyze the spectral reflectance functions of the typical natural and artificial objects in order to be able to select an application relevant set of object specific CRIs and to carry out an application relevant optimization of the WLED light source.

In some applications, certain colored objects are more important than other objects and these relevant objects should get a high weight in an *application oriented* color rendering optimization of the WLED's relative SPD. For example, at the blue jeans department of a fashion shop, the different types of blue jeans spectral reflectances should be rendered in an excellent way. In this section, we point out that for the excellent color rendering of special groups of colored objects, it is worth using their own set of object specific CRIs. For example, to illuminate blue jeans by a modern LED luminaire, it is essential to maximize the object specific CRIs of the blue textile class of goods.

To compute the object specific CRIs, it is essential to have a spectral reflectance dataset of the colored objects belonging to important interior lighting applications. To this end, spectral reflectance data of a set of real colored objects and human skin tones were measured that are representative for the different applications of interior lighting [47]. Colored objects included fruits, leaves, flowers, hair, textiles, skin tones, printed products, and food. The objects were grouped according to their prevailing color: skin tones, reddish, yellowish, greenish, and bluish-violet objects.

This kind of grouping is advantageous because in a given lighting application, often the colored objects belonging to the *same* color group can be rendered at the best possible level, for example, different skin tones to record a movie scene in a studio (see Figure 7.13), yellowish objects in a bakery (see Figure 7.16), or reddish objects at the butcher's (see Figure 7.17). The aim of this section is to understand the spectral reflectance behavior of these real objects to be able to fine-tune the SPD of a WLED light source until its interaction with the spectral reflectance curves yields excellent color rendering for the human visual system. Here (for illuminating practice the most relevant) object groups will be analyzed as

 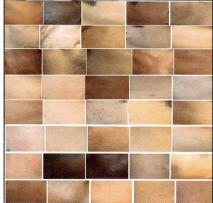

Figure 5.21 Demonstration of the visual effect of spectral optimization of a light source. Optimization target: maximize the object specific color rendering indices for 40 various skin tones [33]. Left: bad color rendering of skin tones under a sub-optimal spectral power distribution; right: high quality color rendering of skin tones under an optimized spectral power distribution (optimization target: maximize the object specific color rendering indices of all skin tones).

examples: skin tones and reddish objects. A more detailed analysis with all groups of colored objects can be found elsewhere [33].

As a first example, Figure 5.21 visualizes bad color rendering and high quality color rendering (a possible result of spectral optimization of the light source) for a collection of skin tones taken from a panel of 40 subjects covering all types of skin tones [33].

Figure 5.22 shows the measured spectral reflectance curves of the 40 skin tones from Figure 5.21 together with the spectral reflectance of CIE TCS 13 (Caucasian skin) from Figure 5.9.

As can be seen from Figure 5.22, there is a fundamental difference between the spectral reflectance of the CIE Caucasian skin color (TCS13) and the measured skin reflectances. For TCS13, the increase of reflectance begins already at 540 nm while for the measured Caucasian and Asian skin tones the increase begins only at 580 nm. Caucasian and Asian skin colors also exhibit some local spectral maxima at 510 and 560 nm and local minima at 420, 537, and 573 nm. These minima can be explained by hemoglobin absorption bands.

The skin tones of the observers 56M (thick black curve), 59W (orange dash curve), and 60W (brown dash curve) have a lower spectral reflectance without local minima or local maxima and with no increase at the above wavelengths, for example, compared to the skin spectral reflectance of subject 42W (thick orange curve). In general, measured curves increase from 430 until 510 nm slowly and then faster from 580 until 680 nm. The implication of this finding for the spectral engineering of WLED light sources is that if emission components in the spectral range 450–510 nm and from 625 nm on are missing then the color rendering of skin tones will be poor, see Figure 5.21 (left).

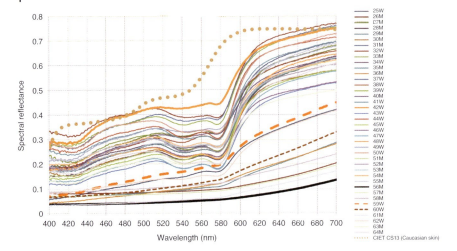

Figure 5.22 Measured spectral reflectance curves of the 40 skin tones from Figure 5.21 together with the spectral reflectance of CIE TCS 13 (Caucasian skin, dot curve) from Figure 5.8.

Figure 5.23 Demonstration of the visual effect of spectral optimization of a light source. Optimization target: maximize the object specific color rendering indices of reddish objects [33]. (a) bad color rendering of reddish objects under a suboptimal spectral power distribution; (b) high quality color rendering of reddish objects under an optimized spectral power distribution (optimization target: maximize the object specific color rendering indices of reddish objects).

As a second example, Figure 5.23 illustrates the diversity of reddish colored objects that tend to appear in a typical built indoor environment under two light sources, once with bad color rendering (left) and then also with high-quality color rendering optimized for reddish objects (right). Of course, Figure 5.23 can just provide a rough visualization for the purpose of this book but the authors could observe a similar, striking visual effect in reality when experimenting with a modern multi-LED engine in the laboratory.

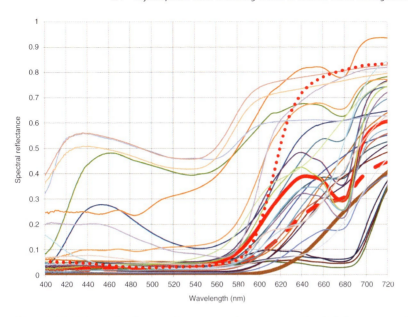

Figure 5.24 Measured [47] spectral reflectance curves of reddish objects (not the same objects as in Figure 5.23). Thick red dash curve: strawberry; continuous thick red curve: red apple; thick brown curve: dark red peony; dot curve: the deep red test color sample CIE TCS09.

Figure 5.24 shows the measured [47] spectral reflectance curves of reddish objects (not the same objects as in Figure 5.23).

As can be seen from Figure 5.24, the deep red CIE TCS (TCS09, red dot curve) has a reflectance value of about 5% from 400 nm (violet) up to 580 nm (yellow). The curve increases from about 580 nm until 640 nm rapidly. This spectral behavior means that this saturated red TCS has a high reflectance already in the yellow (580–590 nm) and orange (590–610 nm) range well before the beginning of the red range from about 610 nm on.

Figure 5.24 also shows that most of the measured typical reddish objects exhibit a different spectral reflectance behavior than TCS09. Certain reddish objects really challenge the light source engineer, for example, the dark red peony (thick brown curve) reflects below 605 nm barely any radiation but its reflectance increases rapidly toward the deep red spectral range. Therefore, illuminant wavelengths below 610 nm are hardly effective. In order to render this dark red peony correctly, as it appears under incandescent light or daylight, the WLED light source to be designed must emit enough radiation from 610 nm on. A similar consideration holds true for the strawberry (red dash curve), the red apple (continuous thick red curve) and several other reddish objects, too.

Therefore, for a high color rendering of reddish objects, it is not enough to use a red phosphor or a red semiconductor (chip) LED of a peak wavelength at 610 nm. Taking a closer look at Figure 5.24, we can conclude that the consequence of the spectral engineering of WLED light sources is that such a PC-LED and/or chip-LED combination can be used to cover the important spectral range between 620 and 680 nm. Longer wavelengths lose their visual importance because of the limited sensitivity of the L cones of the human visual system above about 680 nm, see Figure 2.6. The engineer can appraise the effect of this interaction between emission spectra (current test light source) and reflectance spectra (colored objects) spectra on the human light source user by taking a look at the values of the object specific CRIs (and their semantic interpretations) of the current test light source. Examples for such an analysis are presented in the next section.

5.4.2
Color Rendering Analysis of a Sample Set of White LEDs

5.4.2.1 Definition of the Sample Set of White LEDs

Figure 5.25 shows the relative SPDs of the selected 34 WLED light sources.

As can be seen from Figure 5.25, this sample set covers all types of WLEDs available on the market and some very high quality WLEDs including white PC-LEDs with one phosphor, a mixture of two phosphors or a combination of multiple PC-LED light source, as well as two LEDs with the characteristic peaks of red chip LEDs at 625–630 nm. The maxima of the blue pumping LEDs vary between 445 and 460 nm and they exhibit various relative intensities related to the phosphor intensities according to their CCT between 0.2 (warm WLEDs) and 1.0 (cool WLEDs).

For the wide range of possible interior lighting applications, it is essential to group the LEDs according to their CCT (e.g., cool or neutral white tones for offices and warm white for restaurants and home lighting) and color rendering level (some applications like museum lighting or the illumination of feature film studios require very good color rendering). The 34 LEDs of the sample set were grouped according to the scheme depicted in Figure 5.26.

As can be seen from Figure 5.26, WLEDs were grouped first according to their CCT, WW: warm white; NW: neutral white, and CW: cool white. Second, they were grouped according to their color rendering level: if the value of the semantic interpretation scale (R) of the LED light source is less than 1.5 (i.e., better than "good–very good") then it belongs to CR (color rendering) Group 1 and others, see the caption of Figure 5.26. This new grouping method (introduced here first) is straightforward and easy to understand. It is also *visually relevant* because the R values were computed from the $R_{a,2012}$ values of each light source using Table 5.2.

As the CIE general CRI R_a is still widely used at the time of writing, Figure 5.26 may offer another usable tool for the lighting specialist. As a rule of thumb, using a

---

Figure 5.25 Relative spectral power distributions of the 34 selected typical market-based white LED light sources (sample set). This set also includes a few LEDs with very high quality phosphor mixtures or those composed of multiple phosphor-converted LED light sources as well as two warm white LEDs containing an additional red semiconductor LED. Top: WW (warm white) with CCT < 3500 K; middle: NW (neutral white) with 3500 K ≤ CCT < 5000 K; bottom: CW (cool white) with CCT ≥ 5000 K.

5.4 Object Specific Color Rendering Indices of Current White LED Light Sources

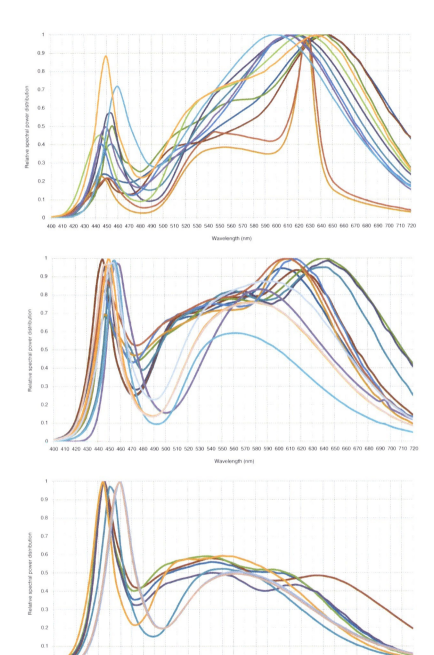

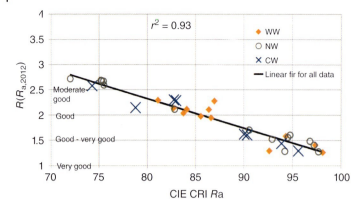

Figure 5.26 Grouping scheme of the sample set of 34 white LED light sources (typical market-based LEDs at the time of writing). Every symbol corresponds to one of the 34 LEDs. WW (warm white): CCT < 3500 K; NW (neutral white): 3500 K ≤ CCT < 5000 K; CW (cool white): CCT ≥ 5000 K. The semantic interpretation scale (R) was computed from the $R_{a,2012}$ value of every light source. Grouping after color rendering: CR Group 1: $R < 1.5$; CR Group 2: $1.5 \leq R < 2.0$; CR Group 3: $2.0 \leq R < 2.5$; CR Group 4: $R > 2.5$, see Table 5.5.

Table 5.5 CIE CRI R_a limits of the color rendering groups identified in Figure 5.26 for the sample set of 34 white LEDs.

CR group	Semantic interpretation scale (R)	CIE CRI R_a
1	$R < 1.5$	$R_a > 93$
2	$1.5 \leq R < 2.0$	$85 < R_a \leq 93$
3	$2.0 \leq R < 2.5$	$78 < R_a \leq 85$
4	$R > 2.5$	$R_a \leq 78$

linear fit equation to all 34 data points forced to fit the ($R_a = 100$; $R = 1$) point, one can estimate the color rendering group of a typical, commercially available WLED light source if the value of the CIE general CRI, CIE CRI R_a is known: $R = -0.0666 R_a + 7.683$. For example, the category label "good" corresponds to $R_a = 85$ which does not deviate much from the mean tendency of Figure 5.19 where it is $R_a = 87$ which corresponds to the category label "good." Note that Figure 5.19 was computed using all types of light source SPDs, not just WLEDs.

Table 5.5 contains the CIE CRI R_a limits of the color rendering groups (CR Group 1, CR Group 2, etc.) identified in Figure 5.26 for the sample set of 34 WLEDs.

5.4.2.2 Definition of the Sample Set of Object Reflectances

To provide an instructive color rendering analysis, the set of object reflectances chosen was reasonably limited to serve as a demonstrative example for all color groups (skin, red, yellow, green, and blue-purple). As the TCSs of the CIE color

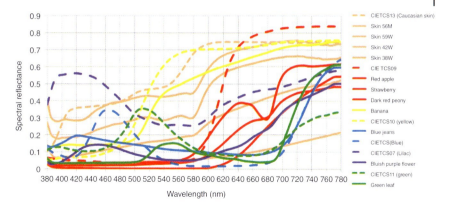

Figure 5.27 Spectral reflectance curves of the sample set of test colors (CIE TCS [15] and real objects [47]) for the color rendering analysis of the 34 sample LEDs. Color coding (skin tone, red, yellow, green, blue, lilac) corresponds to the grouping of the colored objects/TCSs. Dash curves: CIE TCSs. Continuous curves: real objects.

rendering method (TCS01–14) are in widespread use, five TCSs (Caucasian skin, red, yellow, green, blue, and violet) were also included as "standard" samples. These TCSs are compared in this section with the most typical colored objects. The spectral reflectance curves of the sample set are shown in Figure 5.27.

As can be seen from Figure 5.27, the spectral reflectances are very different within a certain group of colors. Now the question is how relevant these differences are from the point of view of the rendering of these color groups by the WLEDs and also whether the whole variety of spectral reflectances of all kinds of objects can be appreciated by a single average CRI ($R_{a,2012}$). This is analyzed in the coming section in terms of the semantic interpretation scale.

5.4.2.3 Examples for the Relationship among the Color Rendering Indices in Terms of the Semantic Interpretation Scale (R)

Figure 5.28 shows the semantic interpretation scale (R) values of the object specific CRI $R(R_{TCS12,2012})$ for the deep blue CIE TCS12 (ordinate) opposed to the R value of the general CRI, $R(R_{a,2012})$. For every one of the 34 WLED light sources in Figure 5.25, both indices (and then their R values) were computed and a symbol was plotted (Figure 5.25).

As can be seen from Figure 5.28, for a narrow range of general R ($R_{a,2012}$) values, a wide range of $R(R_{TCS12,2012})$ may come into play for the deep blue TCS. For example, for the WLED light sources for which R ($R_{a,2012}$)) ranges between 2.0–2.3 ("good"), the "deep blue" CRI ranges between 1.7 and 3.1. This implies that it is not possible to predict or optimize the color rendering of deep blue objects by means of the general CRI. This example shows the relevance of special or object specific CRIs.

Another question is the relationship between two bluish test colors, TCS012 and the blue jeans (real test object) which is shown in Figure 5.29.

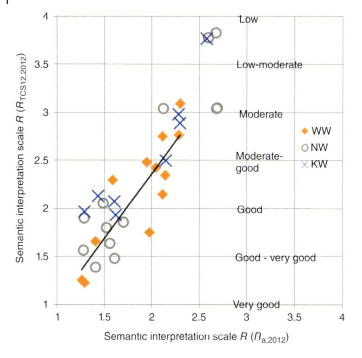

Figure 5.28 Semantic interpretation scale (R) value of the object specific color rendering index $R(R_{TCS12,2012})$ for the deep blue CIE TCS12 (ordinate) and the R value of the general color rendering index, $R(R_{a,2012})$. Every symbol corresponds to one of the 34 white LED light sources of the sample set in Figure 5.25.

Figure 5.29 shows that the color rendering of the less saturated blue jeans is generally better than the color rendering of the highly saturated TCS12 for the whole sample WLED light source set. This is an example of how misleading the use of a predefined "blue" colored object (TCS12) can be – instead of using the reflectance curve of the real object (blue jeans) to compute an object specific CRI – if the aim of spectral engineering is to illuminate the real object (e.g., at the blue jeans department of the fashion shop) at high color quality.

5.4.2.4 Object Specific Color Rendering Bar Charts with Semantic Interpretation Scales (R)

A solution to the problem described in Section 5.4.2.3 for the lighting practitioner may be the use of a bar chart displaying the semantic interpretation scale (R) of the CRIs of those objects that are relevant for a given application. Four examples with the 17 colored objects from Figure 5.27 and the general $R(R_{a,2012})$ value can be seen in Figure 5.30, for four WLEDs selected from the sample set of Figure 5.25.

As can be seen from Figure 5.30, the high quality WW LED light source with a broad red peak around 645 nm (2596 K, $R_a = 98$) in the top left corner exhibits very good color rendering for all objects (the main slope in its well-made relative SPD in

5.4 Object Specific Color Rendering Indices of Current White LED Light Sources

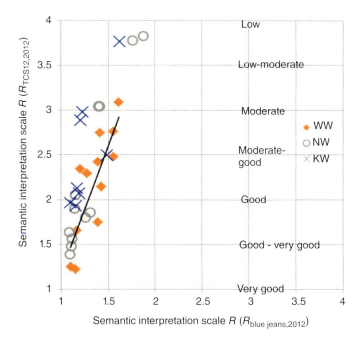

Figure 5.29 Semantic interpretation scale (R) value of the object specific color rendering index $R(R_{TCS12,2012})$ for the deep blue CIE TCS12 (ordinate) and the R value of the object specific color rendering index for the real test object blue jeans $R(R_{blue\ jeans,2012})$. Every symbol corresponds to one of the 34 white LED light sources of the sample set in Figure 5.25.

the inset approximates its reference light source, the Planckian radiator at 2596 K). It is very instructive to compare this example (top left) with its concurrent warm WLED (3008 K, $R_a = 86$) at the top right corner and to compare the two insets, top left and top right.

The emission spectrum in the top right corner exhibits serious deficiencies: the deep and broad local minimum around 485 nm and the high red peak at 630 nm which is too steep to cover the important red wavelength range from 635 nm on. The dark red peony just begins to reflect at this wavelength (Figure 5.24) so it is not surprising that its object specific color rendering is only "moderate–good" ($R = 2.6$). The broad red peak around 645 nm in the top left corner renders this flower "very well."

While the WW LED in the top right corner is able to render skin tones "very well," there is a further problem with CIE TCS12 (2.8, "moderate") because of the lack of radiation at the position of the broad minimum around 485 nm exactly in that spectral range in which the decreasing branch of TCS12's spectral reflectance curve still yields an important contribution compared to its relatively low reflectance maximum (only about 0.35).

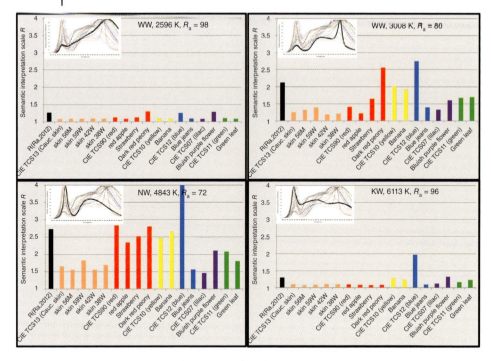

Figure 5.30 Examples of object specific color rendering bar charts with semantic interpretation scales (R) for four white LEDs from Figure 5.25 (see the thick black curves in the insets).

A neutral WLED of inferior quality is shown in the lower left corner of Figure 5.30 (NW, 4843 K, $R_a = 72$). Its relative SPD exhibits a very deep local minimum at about 490 nm (see the inset). Compared to this minimum, the blue peak (at about 455 nm) is too high and overlaps with TCS12's reflectance maximum (at about 460 nm) which causes a serious blue oversaturation and, consequently, "low" ($R = 4.0$) color rendering for TCS12. Because of the lack of enough radiant flux (for a color rendering analysis, this should be judged *relative* to the high blue peak of the pump LED's radiation) from 620 nm on, the color rendering of yellowish and reddish objects is only "moderate" – "moderate-good," see their spectral reflectance in Figure 5.27.

The last example (KW, 6113 K, $R_a = 96$) in the bottom right corner of Figure 5.30 is a high color quality WLED but this LED produces a cool white tone. This WLED exhibits "very good" color rendering for all objects except for the deep blue TCS12 ("good") because of the high blue maximum of the blue pump LED at 445 nm.

5.4.3
Summary

The spectral reflectance of object colors are often different from the spectral reflectance of TCSs of predefined sets, for example, of (TCS01 – TCS14).

Therefore, for high quality applications with specific object colors (e.g., museum or shop lighting), predefined samples are often not suitable, nor is an average CRI derived from them. Generally, it is not possible to specify or optimize the color rendering property of a light source by using a predefined set or a general CRI for a specific application requiring high color quality. The reason is the strong interaction between the colored objects' reflectance spectra and the emission spectra of the WLED light source.

To represent how the human visual system reacts to this interaction, a practical method with object specific bar charts (examples are shown in Figure 5.30) was proposed based on the spectral reflectance of all objects relevant to the application (as extensive spectral reflectance databases are available today). Every bar in the chart corresponds to the semantic interpretation scale value (R) of an object specific CRI, $R_{obj,2012}$. Taking a look at such a diagram of a particular LED light source, colored objects of "very good," "good," "moderate" color rendering can be identified at a first glance and the optimization target of spectral engineering can be for example, to force the color rendering property of every application relevant object into CR group 1 ($R < 1.5$, see Table 5.5).

5.5
Color Preference Assessment: Comparisons Between CRI, CRI2012, and CQS

As mentioned in Section 5.1, in addition to color rendering, there are several different criteria to evaluate the color quality of the environment lit by the WLED light source. Perhaps the most important additional criterion is color preference which corresponds to a cognitive, subjective, and emotional processing of the color perception of object colors in the human visual brain [33]. This processing leads to the color preference judgment of the observer.

The concept of color preference is defined here from the point of view of the lighting engineer (although there are also other definitions less related to engineering) as follows. Color preference is a subjective-emotional assessment of a subject about whether the subject likes or dislikes the current color appearance of a colored object under the current test light source [33]. This assessment depends on the object, the observer and the situation of observing the colored object. The extent of how the user appreciates the current color appearance of a colored object depends on the SPD of the illuminant very strongly. Hence color preference can be considered a property of the light source. Consequently, the aim is to develop a descriptor quantity of color preference from the SPD of the light source.

Because of the *subjective* nature of the color preference property, it depends (unlike color rendering which can be simply ascribed to perceived color differences) on a variety of subjective factors [33]:

- Most colored objects have a special (object specific) preferred color appearance that observers are keen to see (again and again) in the context of the object [48] illuminated by the test light source. For example for red roses, there is a

specific preferred shade of red and for the green leaves of a certain plant there is a specific shade of green or yellowish green. To define these specific shades of color, all three perceived attributes of color (hue, saturation, and even lightness) can be equally relevant.
- There are also more general (i.e., not object specific) *tendencies* of color preference, for every object or for every object within a group of objects like the preference of increased (or rarely decreased) chroma for certain object groups [49].
- Color preference depends on the region of origin or living and on culture [50]. In certain cultures, saturated colors are preferred while other cultures prefer rather unsaturated colors. For example in North American movies, the same objects are more saturated than in their Japanese counterparts to come close to the color preference of local cinema visitors.
- Certain well-known colored objects like skin, grass, flowers, or building materials can exhibit typical country specific or region specific varieties with typical spectral reflectance curves, hence with local color preferences [51]. The spectrum of the locally dominant phases of daylight, the clouds and the prevailing light source technology may vary from country to country. Local light source users become accustomed to their local viewing conditions and develop a regional color preference.
- Color preference depends on gender [50] and age [52].
- Color preference is individual: it depends on education and personality [53]: every profession, every occupational group may have its own color preference.

Because of the above influencing factors, it is difficult to establish a universally valid general-purpose metric (a numeric descriptor quantity to be computed from the SPD of the light source) to characterize the color preference property of light sources. Several metrics have been described in literature [24, 25, 54–58]. Here, only one method, the so-called CQS method [30] (so-called CQS, Version 7.5) of the NIST (United States) will be presented (Section 5.5.1) because it is already widely used at present. It has been validated in several visual color preference experiments.

In Section 5.5.2, the values of the descriptor quantities defined in the CQS method [30] are computed and analyzed for the same sample LED set of 34 typical, commercially available WLEDs intended for general lighting which was already analyzed in Section 5.4 plus another set of special purpose LED emission spectra. Descriptor quantities include the so-called general CQS Q_a and its alternative, CQS Q_p which emphasizes saturation enhancement. These CQS descriptor values are compared with the CRIs R_a and $R_{a,2012}$ in order to show their correlations. These correlations can be interesting to select or optimize a light source for a certain application in which either the color rendering or the color preference property may be more relevant.

However, the opinion of the present authors is that, for general interior lighting purposes, the primary aim of color quality optimization is color rendering (color fidelity) while color preference is only of secondary importance for a limited number of special applications. Color preference related indices of the CQS method should *not* be used as a *primary* criterion for general lighting to select or optimize the SPD of a WLED light source. Its important field of application is special (accent) lighting in which the colored objects should be rendered with increased chroma, that is, with higher CQS index values.

But for most general interior lighting applications, the increase of chroma is generally not desirable. In certain applications like medicine, color printing, cosmetics, film studios, textile industry, museums or food retail, color fidelity, that is, the resemblance of object colors to their appearance under a well-known reference illuminant is critical and optimizing for an increased chroma using CQS could result in misleading color decisions. Hence, in most general lighting applications, the CRI (CRI2012) will be the most important target parameter.

5.5.1
CQS: The Color Quality Scale

This section describes and discusses the computational method of the CQS [30].

5.5.1.1 Components of the CQS Method
The method has the following components [30]:

1) *Test colors*: the CQS method compares the color appearance of 15 test colors under a test light source and a reference light source. Figure 5.31 illustrates these test colors (so-called VS1–VS15) chosen from the Munsell color atlas. Figures 5.32 and 5.33 show the spectral reflectance curves of the 15 CQS test colors (VS1–VS15). It may be interesting to compare them with Figures 5.8 and 5.9 (CIE TCS 01–14) and Figure 5.12 (HL01–HL17).
2) *Reference light source*: It is the same reference light source as in the CRI method or in the CRI2012 method.
3) *CIE tristimulus values*: X, Y, Z tristimulus values are computed for every test color (VS1–VS15), both under the test light source and the reference light source.

Figure 5.31 Test colors of the CQS method (NIST, USA, version 7.5), VS1-VS15 [30]. Image source: CQS computational spreadsheet (NIST, USA). Not a calibrated reproduction (illustration only).

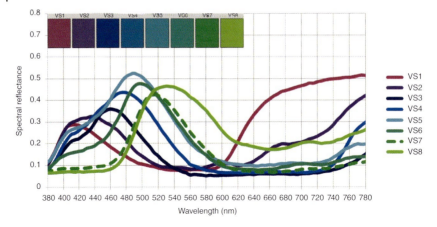

Figure 5.32 Spectral reflectance of the CQS test colors VS1–VS8.

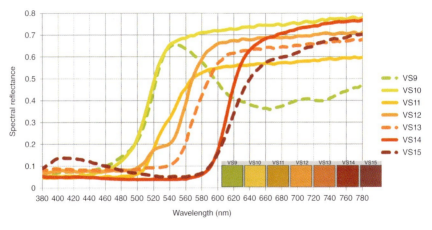

Figure 5.33 Spectral reflectance of the CQS test colors VS9–VS15.

4) *Chromatic adaptation*: X, Y, Z tristimulus values under the test light source are transformed into the corresponding X, Y, Z values under the reference light source using the chromatic adaptation transformation CMCCAT2000 [59].
5) *CIELAB values*: CIELAB L^*, a^*, b^*, and C^* values are computed for every test color, both under the test light source and the reference light source.
6) *CIELAB chroma difference*: For every test color ($i = 1–15$), the chroma difference is computed between the test light source and the reference light source: $\Delta C^*_{ab,i} = C^*_{ab,test,i} - C^*_{ab,ref,i}$.
7) *CIELAB color difference*: The value of $\Delta E^*_{ab,i}$ between the test light source and the reference light source is computed for every test color ($i = 1–15$).
8) *Chroma increment factor*: This is called *saturation factor* in the original article [30] but the present authors think that *chroma increment factor*

5.5 Color Preference Assessment: Comparisons Between CRI, CRI2012, and CQS

is a better term (while the original denotation $\Delta E_{ab,i,sat}^*$ was kept). In the CIE CRI and CRI2012 methods, those test light sources that increase the chroma of a test color are penalized. In the visual experiment [30] underlying the CQS method and several other visual studies, observers preferred the increase of chroma (while for some objects, the preference of decreased chroma cannot be excluded). Accordingly, the CQS method does not penalize the chroma increase of a test color when it is illuminated by the test light source instead of the reference light source, at least for the case of the main index (CQS Q_a) which is presented here. So if $\Delta C^*_{ab,i} > 0$ for a certain test color (i.e., if the CIELAB chroma of the test color is increased by the test light source compared to the situation when it is illuminated by the reference light source) then the CIELAB color difference $\Delta E_{ab,i}^*$ between the test and reference versions is transformed via Eq. (5.7) to obtain the quantity $\Delta E_{ab,i,sat}^*$. Otherwise the value of $\Delta E_{ab,i}^*$ remains unchanged.

$$\Delta E_{ab,i,sat}^* = \sqrt{(\Delta E_{ab,i}^*)^2 - (\Delta C_{ab,i}^*)^2} \quad \text{if } \Delta C*_{ab,i} > 0$$

$$\Delta E_{ab,i,sat}^* = \Delta E_{ab,i}^* \quad \text{if } \Delta C*_{ab,i} \leq 0 \tag{5.7}$$

Eq. (5.7) means that the transformed color difference ignores the chroma difference $\Delta C^*_{ab,i}$ provided that the chroma of the test color increases compared to the reference situation.

9) *RMS*: To calculate a mean value from the transformed CIELAB color differences $\Delta E_{ab,i,sat}^*$ of the 15 test colors, the RMS is used instead of the arithmetic mean, see Eq. (5.8).

$$\Delta E_{rms} = \sqrt{\frac{1}{15} \sum_{i=1}^{15} (\Delta E_{ab,i,sat}^*)^2} \tag{5.8}$$

10) *Scaling*: The ΔE_{rms} value of Eq. (5.8) is scaled according to Eq. (5.9) to obtain the preliminary variable CQS $Q_{a,rms}$.

$$Q_{a,rms} = 100 - 3.1 \Delta E_{rms} \tag{5.9}$$

The factor 3.1 of Eq. (5.9) corresponds to the traditional criterion (already introduced at the end of Section 5.2) that the arithmetic mean value of 12 specific values of the final CQS indices (the so-called CQS Q_a values from Eq. (5.12)) should have the same value as the arithmetic mean value of 12 specific CIE CRI R_a values. The 12 specific values are computed for the 12 CIE illuminants F1–F12 that represent the relative SPDs of typical fluorescent lamps. Note that the CRIs described in Section 5.2 were also scaled according to this criterion.

11) *Rescaling for the interval 0–100*: In order to avoid negative index values, the value of $Q_{a,rms}$ is transformed into the interval 0–100, see Eq. (5.10)

$$Q_{a,0-100} = 10 \ln \left\{ \exp\left(\frac{Q_{a,rms}}{10}\right) + 1 \right\} \tag{5.10}$$

12) *CCT factor:* Test light sources of lower CCTs (T_{cp} < 3500 K) obtain a lower index value via a multiplicative factor M_{CCT} according to Eq. (5.11).

If T_{cp} < 3500 K

then $M_{CCT} = T_{cp}{}^3(9.2672 \times 10^{-11}) - T_{cp}{}^2(8.3959 \times 10^{-7})$
$+ T_{cp}(0.00255) - 1.61$;

otherwise $M_{CCT} = 1.0$ \hfill (5.11)

13) *General CQS index (CQS Q_a):* the final index is computed via Eq. (5.12).

$$Q_a = Q_{a,0-100} \cdot M_{CCT} \hfill (5.12)$$

14) *Special CQS indices (CQS Q_i):* these indices are computed in a similar way as the general index but, from the point 10 on, instead of ΔE_{RMS} the quantity $\Delta E^*_{ab,i,sat}$ can be used for every individual test color.

Figure 5.34 summarizes the steps of the CQS computational method [30].

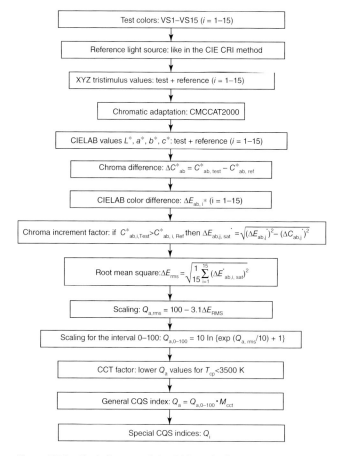

Figure 5.34 Block diagram of the CQS method.

5.5.1.2 Discussion of the CQS Method

The CQS method was validated in an independent study [60]. Not surprisingly, it was found that the CQS Q_a index correlated with the visual color preference assessments of the subjects better than the CRI CIE CRI R_a [60]. Nevertheless, according to the opinion of the present authors, the CQS method considers *only one* (generally important) tendency of color preference, that is, the increase of chroma. The other attributes of perceived color (hue and lightness) are not dealt with. Even for chroma, for some desaturated objects, a decrease of chroma may be preferred which should be subject of future investigations.

An unfavorable property of the CQS method is the use of the visually nonuniform CIELAB color space that represents an outdated model of perceived color differences. CIELAB is the underlying color space of today's widely used version of the CQS method (i.e., version 7.5) [30] but the reader is encouraged to check further developments of the method as its later versions already use the more uniform CAM02-UCS color space.

In addition to the above described CQS Q_a and Q_i indices, other color preference indices can also be found in literature. In the earlier-mentioned study that compared visual color preference results from several visual experiments of independent research groups [60], such color preference related indices [24, 25, 55–58] (different from CQS) were analyzed. Some of the latter indices correlated well with visual color preference results. It should be noted that today's color quality metrics (including CQS and the other indices) are not able to predict the effect of subjective factors (personality, gender, age, region, and culture) mentioned at the beginning of Section 5.5 which should also be the subject of future studies.

The CQS method defines two other indices in addition to the general CQS Q_a, the so-called gamut area scale Q_g and the so-called color preference scale Q_p. The gamut area scale Q_g is calculated as the relative gamut area formed by the CIELAB a^*, b^* coordinates of the 15 color samples illuminated by the test light source. Q_g is normalized by the gamut area of the CIE illuminant D65 and multiplied by 100 [30]. The color preference scale Q_p was intended to place additional weight on the preference of object color appearance. This metric emphasizes "the notion that increases in chroma are generally preferred and should be rewarded" [30]. Hence Q_p is calculated using the same procedure as CQS Q_a (see Figure 5.34) *except* that Q_p rewards light sources for increasing object chroma [30] by replacing Eq. (5.9) by the more complex Eq. (5.13).

$$Q_{a,\mathrm{rms}} = 100 - 3.78 \left[\Delta E_{\mathrm{rms}} - \frac{1}{15} \sum_{i=1}^{15} \Delta C^*_{ab,i} K(i) \right] \qquad (5.13)$$

In Eq. (5.13), if $C^*_{ab,i,\mathrm{test}} \geq C^*_{ab,i,\mathrm{ref}}$ then $K(i) = 1$ and if $C^*_{ab,i,\mathrm{test}} \leq C^*_{ab,i,\mathrm{ref}}$ then $K(i) = 0$. This means that if the chroma of the test color increases under the test light source compared to the reference light source then the value of the index increases in order to account for the higher color preference of chroma increment. The Q_p score is scaled with the scaling factor 3.78 so that the average value of Q_p is the same for the 12 fluorescent illuminants F1–F12 as for the CIE CRI R_a.

The color preference scale CQS Q_p was compared with visual results in an independent study [61]. In this study, 60 subjects made user preference judgments (concerning naturalness and colorfulness of colored objects) about a limited set of different light emission spectra generated by an LED luminaire built from 12 different types of LEDs. These emission spectra contained sharp local minima and maxima and had a color rendering level of $R_a = 80$ (except three of the spectra with $R_a = 96-98$).

This study [61] worked with 21 LED SPDs at three different CCTs (2700, 4000, and 6500 K) in a light booth. These 21 LED emission spectra were different from the set of 34 WLEDs of Figure 5.25. Three of these emission spectra were optimized for a high general CRI R_a value, $R_a = 96-98$ (for every CCTs), three for a high value of Q_g, three for a low and three for a high FCI (Feeling of Contrast Index) value, another three were designed for a low and three a for high Q_p value and the last three were designed to mimic fluorescent lamps.

Within this specific set of LED light sources, subjects preferred those SPDs under which the chroma and colorfulness values of the object colors were higher. It was found [61] that the CIE CRI R_a was not a good indicator of the observers' preference for these 21 LED SPDs. The metrics CQS Q_p (CQS version 7.5), Q_g (CQS version 7.5), and FCI [57] were good preference indicators for the limited set of LED emission spectra of this study [61], that is, LED emission spectra with higher Q_g, Q_p, and FCI values were preferred.

5.5.2
Relationship between the Color Quality Scale (CQS Q_a, Q_p) and the Color Rendering Indices (CRI, CRI2012)

For the analysis and optimization of the SPDs of WLED light sources, it is important to study the relationship between the CQS values (which correlate with color preference) and the CRI2012 index scales (which represent color fidelity). Although, as mentioned above, maximal color *fidelity* is the primary aim of color quality optimization for *general* interior lighting, there *are* certain special applications (e.g., accent lighting of colored objects like certain theatre or movie scenes, discotheques, or specific retail areas) in which an oversaturated color appearance of most of the object colors may be preferred by the limited user group that a WLED light source is intended for.

Therefore, it is interesting to investigate the sample set of 34 WLEDs of Figure 5.25 intended for general lighting, especially whether there are WLEDs of higher and lower CQS Q_a values within each CRI group. LEDs of higher Q_a values may also be used for the purpose of a special application with (slightly) higher object color saturation while maintaining the color rendering level required by the given CR group (see Table 5.5). The WLED set of Figure 5.25 is analyzed from this point of view. Figure 5.35 shows the CQS Q_a values of the sample set of 34 WLED light sources (Figure 5.25) grouped after CCT (warm white, neutral white, cool white, see Figure 5.26) and ordered according to the semantic interpretation scale $R(R_{a,2012})$ and the categories of color rendering (very good, good–very good, good, moderate–good) along the ordinate.

As can be seen from Figure 5.35, the correlation between $R(R_{a,2012})$ and CQS Q_a is high for every CCT group. It can also be seen that the LED light sources with CQS Q_a values ranging between 90 and 98 are located at the category "good–very good" on the ordinate. This finding mitigates the earlier mentioned possibility of unsafe color decisions because of object oversaturation at least for this representative group ($Q_a > 90$) of the 34 WLEDs. Specifically, these 34 LEDs are intended for general interior lighting without oversaturating colored objects. Therefore, even the WLEDs with the highest CQS Q_a values exhibit "good–very good" color rendering. The reason is that, because of the general lack of sharp local minima and maxima in the emission spectra of this set of 34 LEDs (see Figure 5.25), these WLEDs either do not fulfill the condition $\Delta C^*_{ab,i} > 0$ of Eq. (5.7) or the absolute value of $\Delta C^*_{ab,i}$ is small so that the CQS Q_a values correlate well with the semantic interpretation scale value of the CRI, $R(R_{a,2012})$.

In this respect, according to Eq. (5.7), CQS Q_a can be considered as a combined color rendering – color preference index. An implication of this statement is that the CQS Q_a ranges corresponding to the semantic categories of color rendering in Figure 5.35 also provide a *rough* semantic labelling for the CQS Q_a metric within this group of WLEDs: CQS Q_a values between 90 and 98 correspond to "good–very good," 76–86 to "moderate–good–good," and 69–76 to "moderate–good."

Concerning the CQS Q_p metric of Eq. (5.13), it is interesting to consider another computational example. This second example contains three RGB LEDs and five

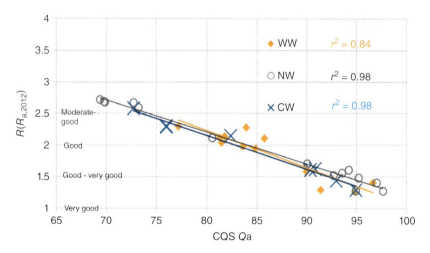

Figure 5.35 CQS Q_a values of the sample set of 34 white LED light sources (Figure 5.25) grouped regarding their CCT (warm white, neutral white, cool white, see Figure 5.26). Every symbol corresponds to one of the white LEDs ordered according to the semantic interpretation category labels at the ordinate which represents the semantic interpretation scale $R(R_{a,2012})$.

so called multi-LEDs. These multi-LEDs were constructed as a linear combination of six relative SPDs, one white PC-LED plus five colored chip-LEDs. This set of eight light sources is fundamentally different from the LED set of the color preference study [61]. The five multi-LED relative SPDs were constructed to obtain a constant CQS Q_p level ($Q_p = 104-105$) at different levels of the CRI2012 general CRI, $R_{a,2012}$. Relative SPDs are shown in Figure 5.36.

As can be seen from the example of Figure 5.36, the five different relative SPDs with different color rendering properties ranging between "very good" and "moderate" result in the *same* Q_p index level (104–105). The CIELAB chroma values of the 15 CQS test colors under the five test light sources of Figure 5.36 ($C^*_{ab,i,test}$, see Figure 5.31) can be seen in Figure 5.37.

As can be seen from Figure 5.37, a fixed CQS Q_p level ($Q_p = 104-105$) can result in a wide range of chroma values for the individual test colors at the five different color fidelity levels. For example, the CIELAB chroma of the CQS test color VS7 ($C^*_{ab,7,test}$, green) ranges between 55 and 86. This means that although the CQS Q_p level is fixed within the set of relative SPDs of a multi-LED light source, the individual test colors can exhibit a wide range of chroma values. The same is true for the three RGB LEDs. According to the general tendency of increased color preference, this means that the value of CQS Q_p *alone* cannot be used for color preference optimization. A better strategy for color preference optimization is to concentrate on an application specific group of colored objects (e.g., reddish objects, see Figure 5.23) and increase the chroma of these objects by optimizing the SPD of the WLED and taking care of the quality of perceived whiteness of its white point [62], see Section 5.7.

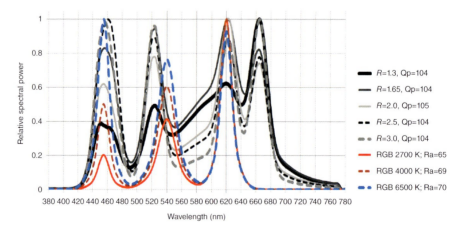

Figure 5.36 Black and gray curves: multi-LEDs, sample relative spectral power distributions of a computational example concerning the CQS Q_p metric of Eq. (5.13). The designation R in the legend means the semantic interpretation scale for the CRI2012 general color rendering index $R_{a,2012}$. The level of the CQS Qp index is constant ($Q_p = 104-105$) while the level of the CRI2012 general color rendering index $R(R_{a,2012})$ varies between "very good" and "moderate" ($R = 1.3-3.0$). Colored curves: three RGB LEDs.

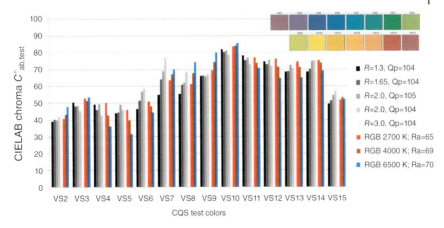

Figure 5.37 CIELAB chroma values of the 15 CQS test colors under the eight test light sources of Figure 5.36 ($C^*_{ab,i,test}$, see Figure 5.31).

Figure 5.35 shows the relationship between the Q_a scale and the semantic interpretation scale of $R_{a,2012}$. In the following, the semantic interpretation scale is disregarded for the moment, and, for the sake of practical applications, numeric values of CRIs are considered. Three CRIs are analyzed: the general CRI CIE CRI R_a, the average value of all 14 special CRIs ($R_1 - R_{14}$, Eq. (5.3)) and a modified CRI2012 general CRI (designated by $R_{a,2012(VS1-VS15)}$) which uses the color samples VS1–VS15 (Figure 5.31) instead of HL1–HL17 according to Eq. (5.14).

$$\Delta E'_{rms,VS1-VS15} = \sqrt{\frac{\sum_{i=1}^{15}(\Delta E'_i)^2}{15}} \qquad (5.14)$$

The reason of the choice of the CIE CRI R_a index (average of the $R_1 - R_8$ based on the unsaturated TCSs TCS01–TCS08, Figure 5.9) is that (despite its deficiencies, see Section 5.2.2) it is most widely used at the time of writing this book. The average of $\{R_1 - R_{14}\}$ is an interesting compromise to include the more saturated TCS09–12, a type of human complexion color (TCS13) and leaf green (TCS14), see Figure 5.8. Finally, $R_{a,2012(VS1-VS15)}$ was included to provide a fair comparison with CQS by using the same set of test colors (VS1–VS15).

Figure 5.38 shows the correlations between Q_a and the three CRIs for the WLED light sources of Figure 5.25 (a choice of market based WLEDs intended for general lighting). Figure 5.39 shows the same correlations for the choice of eight LED light sources in Figure 5.36, that is, multi-LEDs at a fixed Q_p level plus three RGB LEDs.

As can be seen from Figure 5.38, there is a good correlation ($r^2 = 0.94 - 0.96$) between all three types of CRIs and CQS Q_a. The reason is, as already mentioned before (when Figure 5.35 was analyzed) that these 34 LEDs are intended for general interior lighting. These LEDs (from Figure 5.25) do not cause object oversaturation: the absolute value of $\Delta C^*_{ab,i}$ in the CQS equations is small. Just the opposite is true for the other example of the eight LEDs (from Figure 5.36)

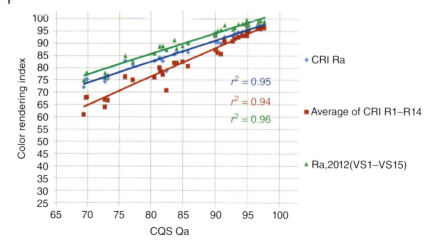

Figure 5.38 Correlations between CQS Q_a and three types of color rendering indices (see text) for the 34 white LED light sources of Figure 5.25.

whose emission spectra exhibit sharp local maxima. These LEDs cause more or less oversaturation for certain colored objects (see Figure 5.37) and the difference between the two types of indices is clearly visible from the lower ($r^2 = 0.69$–0.77) correlations except for $R_{a,2012(VS1-VS15)}$, see Figure 5.39. The high ($r^2 = 0.97$) correlation between CQS Q_a and $R_{a,2012(VS1-VS15)}$ shows the relevance of the choice of test colors. If both indices are based on the same set (VS1–VS15 in this case) then the two types of indices correlate well.

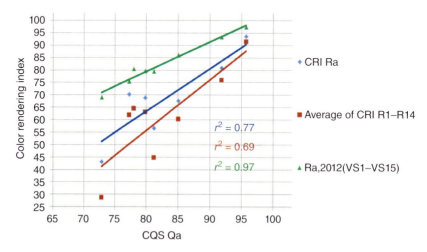

Figure 5.39 Correlations between CQS Q_a and the three types of color rendering indices (see text) for the eight LED light sources of Figure 5.36.

Based on the results of Section 5.5.2 we can conclude that

1) The relationship between the descriptor quantities of the CQS (Q_a, Q_p) and the CRIs (CRI2012) depends on the type of LED light source very strongly. Different correlation patterns arise when white PC-LED light sources for general lighting are used (without extensive local minima and maxima in their emission spectra, see Figure 5.25) or when special RGB LEDs or multi-LEDs with significant local maxima (spectral emission peaks, see Figure 5.36) are considered.
2) There is a good correlation between CQS Q_a and $R_{a,2012}$ for white PC-LEDs for general lighting (Figure 5.25).
3) There is a good correlation between CQS Q_a and CIE R_a for white PC-LEDs for general lighting (Figure 5.25) but not for the special RGB LEDs and multi-LEDs in Figure 5.36. Note that (i) such RGB LEDs (Figure 5.36) are not used in lighting practice and (ii) the SPD of such multi-LEDs can be varied very flexibly and Figure 5.36 presents only a very limited number of examples designed to obtain a narrow CQS Q_p range (Q_p = 104–105) with different color rendering groups.

5.6
Brightness, Chromatic Lightness, and Color Rendering of White LEDs

As mentioned in Section 5.1, the brightness impression of the room illuminated by the LED light source is an important aspect of lighting quality. The brightness perception of the light source itself and the lightness perception of the colored objects illuminated by the light source (refer to Section 2.3 for the concepts of brightness, lightness as well as related and unrelated color stimuli) depend not only on luminance (a $V(\lambda)$ based quantity) but also on chromaticity. The brightness impression of a white wall illuminated by a warm white and a cool WLED light source is different even if their luminance is the same. Also, the lightness impression of a saturated red object and an unsaturated yellow object of the same relative luminance is different. This is the so-called Helmholtz-Kohlrausch effect or, in other words, brightness-luminance discrepancy. An example is shown in Figure 5.40.

The reason for the brightness-luminance discrepancy explained in Figure 5.40 (in the photopic luminance range) is that, in the human visual system, it is not only the luminance channel that contributes to brightness perception but also the two chromatic channels (L−M) and (L+M−S) where the minus sign means a difference of two cone receptor signals in the human visual system (refer to Figure 2.6 for the overview of the relative spectral sensitivities of the human visual mechanisms). The signal of the luminance channel is approximately equivalent to the sum of the weighted L-cone and M-cone signals which can be assigned to the $V(\lambda)$ function, at least approximately. Figure 5.41 shows a scheme about how these mechanisms merge to provide a brightness signal in the human visual system.

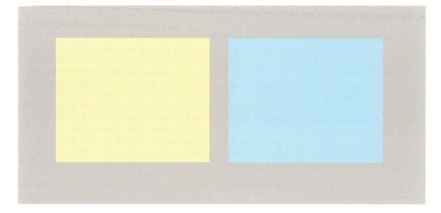

Figure 5.40 Heterochromatic brightness matching experiment. The bluish color stimulus on the right is being changed until the perceived brightness of the two color stimuli becomes the same. The color stimulus on the left (yellowish) remains constant. When the perceived brightness of the two circles matches visually, the luminance of the two color stimuli is still different: the bluish stimulus has a lower luminance than the yellowish stimulus [63].

The spectral design of LED light sources should take the chromatic brightness effect illustrated by Figure 5.41 into account. If this aspect is dealt with correctly, a significant increase of lighting quality and color quality can be achieved especially considering that there is an interesting correlation between chromatic lightness and color rendering for special groups of colored objects (this issue is shown below

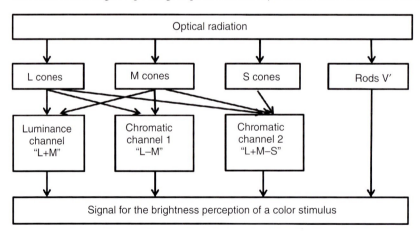

Figure 5.41 Scheme of human vision mechanisms that contribute to brightness perception. The sum of the L- and M-cone signals constitutes the luminance channel. The difference of the L- and M-cone signals represents chromatic channel 1 and the difference of the (L+M) and S-signals chromatic channel 2. In the mesopic (twilight) range of vision (Chapter 6), there is also a rod contribution with the spectral sensitivity $V'(\lambda)$. In the photopic range (generally speaking for $L > 10\,\text{cd}\,\text{m}^{-2}$ but the latter value depends on several factors), rods become inactive.

5.6 Brightness, Chromatic Lightness, and Color Rendering of White LEDs

for the example of red objects). However, the abandonment of this chromaticity dependence can be a devastating mistake for illuminating design. This mistake is sometimes made even today [64]: the success of a light source is sometimes incorrectly defined in terms of only luminous efficacy (lm W^{-1}) or only luminance (cd m^{-2}), see also Figure 5.4.

5.6.1 Modeling the Chromaticity Dependence of Brightness and Lightness

5.6.1.1 CIE Brightness Model

As seen before, brightness perception results from the sum of the luminance channel and two chromaticity channels in the brain hence it cannot be modeled by luminance alone [65]. One possible way of considering the chromatic contributions described earlier is the so-called CIE brightness model [66, 67]. The model's structure is shown in Figure 5.42. This model is intended to be valid for all luminance levels, even for mesopic luminance levels including scotopic (rod, $V'(\lambda)$) contribution. Mesopic levels, where both the rods and the cones are active range typically between 0.001 and 3.0 cd m^{-2} [68] but there are many parameters that influence the limits of this range [69] which is discussed in Chapter 6.

As can be seen from Figure 5.42, scotopic luminance (L') and photopic luminance (L) build the so-called achromatic channel which is modeled by the product $(L')^{1-a} L^a$. The weighting of scotopic luminance to photopic luminance is represented by the dynamic factor $a = L/(L+\alpha)$ with the parameter value $\alpha = 0.05$ cd m^{-2}. Note that the latter parameter value represents only one possibility for the transition from the mesopic range to the photopic range of vision. Especially for large viewing fields when subjects assess their so-called spatial

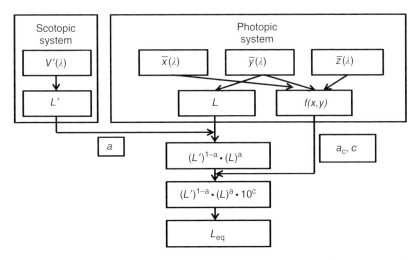

Figure 5.42 Structure of the CIE brightness model [66]. $\alpha = 0.05$ cd m^{-2}; $\beta = 2.24$ cd m^{-2}; $k = 1.3$ and $f(x,y)$ is a chromaticity weighting function [70].

mesopic brightness impression, the influence of the rods (discussed in Section 6 in more detail) can be more significant than predicted by this formula. This can be relevant for interior lighting – let us take a white wall covering a large viewing field illuminated by a WLED light source as an example.

The contribution of the two chromatic channels (R/G channel, red–green, and Y/B channel, yellow–blue) is modeled by the quantity c which appears in the exponent as 10^c. This quantity is combined with the achromatic channel in the scheme of Figure 5.42 to obtain the output of the model, a descriptor quantity of perceived brightness. This quantity is the so-called *equivalent luminance*, see Eq. (5.15).

$$L_{eq} = (L')^{1-a} \cdot L^a \cdot 10^c \text{ with } c = a_c \cdot f(x, y) \quad (5.15)$$

In Eq. (5.15), the exponent c is the product of the chromatic adaptation parameter a_c and the function $f(x,y)$ which represents a dependence on the chromaticity coordinates x and y. The chromatic adaptation parameter a_c is an increasing function of photopic luminance, see Eq. (5.16).

$$a_c = k\, L^{0.5}/(L^{0.5} + \beta) \text{ with } \beta = 2.24 \text{ cd m}^{-2} \text{ and } k = 1.3 \quad (5.16)$$

The function $f(x,y)$ [70] is defined by Eq. (5.17).

$$f(x, y) = \left(\frac{1}{2}\right) \cdot \log_{10}\{-0.0054 - 0.21x + 0.77y + 1.44x^2 - 2.97xy + 1.59x^2 - 2.11[(1 - x - y)y^2]\} - \log_{10}(y) \quad (5.17)$$

The concept of equivalent luminance is defined [66] so that it should predict a visual brightness match or a rank order of perceived brightness. Values of equivalent luminance (L_{eq}) predict whether two color stimuli result in the same brightness perception or whether the perceived brightness of one of them is greater than the perceived brightness of the other. But a series of color stimuli with uniformly ascending L_{eq} values is not necessarily perceived to be a uniform brightness scale.

5.6.1.2 Ware–Cowan Brightness Model

An alternative model of equivalent luminance (in the above defined sense) is the so-called Ware and Cowan formula [71]. It is intended for self-luminous photopic stimuli only. It can be applied for example, to compare the brightness perception of two white light sources, for example, a warm white and a cool white light source. In the first step of the computation of L_{eq} according to Ware and Cowan, the ratio (B/L) of brightness (B) to luminance (L) is computed, see Eq. (5.18). In the second step, this (B/L) ratio is multiplied by the luminance of the light source (in cd m^{-2}), see Eq. (5.19).

$$\log\left(\frac{B}{L}\right) = 0.256 - 0.184y - 2.527xy + 4.656x^3y + 4.657xy^4 \quad (5.18)$$

$$L_{eq} = \left(\frac{B}{L}\right) L \quad (5.19)$$

As an example, L_{eq} values were computed for the 34 WLEDs of Figure 5.25 assuming that all of them provide a fixed luminance ($L = 100$ cd m^{-2}), for example, on an achromatic white painted wall that the light source happens to illuminate. Figure 5.43 shows this result as a function of the CCT of the 34 WLEDs.

As can be seen from Figure 5.43, L_{eq} values increase with increasing CCT. For example, a LED light source with a yellowish white point (warm white) is predicted to be darker than a cool WLED. This is true for every color rendering group. This prediction agrees with the general tendency of several visual experiments at photopic luminance levels typical of interior lighting: a stimulus of a certain white tone of a higher CCT was generally perceived as brighter, at least *within* a certain color rendering group [72].

This review [72] of photopic brightness experiments with different white tones suggested that *both* descriptors, CCT *and* CRI R_a, shall be considered to predict the perceived brightness of an interior: "If lamp A has a higher CCT and CRI than lamp B, the interior will appear brighter under lamp A. If the lamps have similar CCT, the lamp of higher CRI will appear brighter, and similarly if they have similar CRI the lamp of higher CCT will appear brighter" [72]. This finding is important to predict the brightness impression if large achromatic interior surfaces is illuminated by a WLED light source within a given color rendering group.

The Ware–Cowan brightness model can also be applied to saturated (highly chromatic) stimuli. The formula [71] predicts that a saturated color stimulus of the same hue appears brighter than its unsaturated counterpart when both stimuli have the same luminance [63]. The reason is that the chromatic signals 1–2 in Figure 5.41 are more intensive for a more saturated stimulus. To explore hue characteristics, it is interesting to depict the dependence of the B/L ratio (Eq. (5.16)) on *wavelength* for quasi-monochromatic stimuli. This can be seen from Figure 5.44.

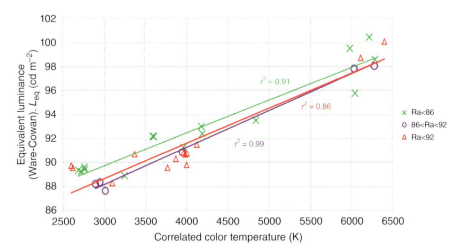

Figure 5.43 L_{eq} values (Ware and Cowan, ordinate) for the 34 white LEDs of Figure 5.25 assuming that all of them give rise to a fixed luminance of $L = 100$ cd m^{-2} at an achromatic white surface (e.g., a painted wall). Abscissa: correlated color temperature (CCT). Light sources are grouped according to their general color rendering index (CRI R_a) level.

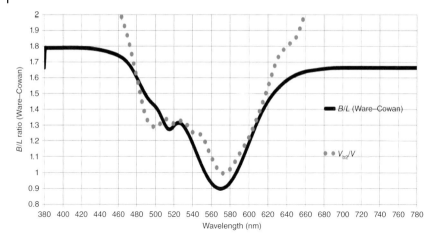

Figure 5.44 Wavelength dependence of the (B/L) ratio for quasi-monochromatic stimuli. For comparison: ratio of the CIE 2° brightness function ($V_{b,2}(\lambda)$ [73]) and the $V(\lambda)$ function ($V_{b,2}/V$).

As can be seen from Figure 5.44, the (B/L) ratio has a minimum at 570 nm (yellow) and a maximum in the blue wavelength range. This means that, at a certain fixed luminance level, yellow light evokes a weak brightness impression while blue light appears to be significantly brighter. There is also an increasing branch of the (B/L)(λ) function from yellow (570 nm) toward red (700 nm). The ratio ($V_{b,2}/V$) of the CIE 2° spectral brightness function ($V_{b,2}(\lambda)$ [73]) and the $V(\lambda)$ function exhibits a similar tendency. This finding corroborates the validity of the Ware–Cowan formula. Note that a similar tendency has already been described before for white tones: white tones of higher CCT (cool white, i.e., white tones with more blue content) exhibit a higher (B/L) ratio and more brightness than yellowish tones (low CCT, warm white).

5.6.1.3 A Chromatic Lightness Model (L^{**}) and Its Correlation with Color Rendering

If the light source illuminates colored *objects* in the interior then observers assess the perceived *lightness* of these objects as *related* color stimuli (i.e., the colored object is related to or compared with a reference to white stimulus for example, a white wall in the room). Similar to brightness, according to the scheme of Figure 5.41, perceived lightness also includes a chromatic component (Helmholtz–Kohlrausch effect). A possibility to model this effect is Fairchild and Pirrotta's L^{**} formula [68]. The formula is based on visual observations [74, 75]. Eq. (5.20) defines the L^{**} formula that corrects CIELAB lightness L^* (which is achromatic) by adding a chromatic component.

$$L^{**} = L^* + [f(L^*) \cdot g(h°) \cdot C^*] \text{ with}$$
$$f(L^*) = 2.5 - 0.025 L^* \text{ and}$$
$$g(h°) = 0.116|\sin((h° - 90)/2)| + 0.085 \tag{5.20}$$

5.6 Brightness, Chromatic Lightness, and Color Rendering of White LEDs

The symbols of Eq. (5.20) have the following meaning:

- $f(L^*)$: a correction function based on CIELAB lightness L^*;
- $g(h°)$: a correction function based on CIELAB hue angle h in degrees;
- C^*: CIELAB chroma; h: CIELAB hue angle in degrees.

As can be seen from Eq. (5.20), the correction term, that is, $[f(L^*) \cdot g(h°) \cdot C^*]$ increases linearly when the chroma of the stimulus (colored object) increases because of the fact that the contribution of the two chromatic channels in Figure 5.41 increases with chroma. The minimum of the hue correction function is located – not surprisingly – at the yellow hue angle $h = 90°$ compare with Figure 5.44 in which the minimum (B/L) ratio occurs at about 570 nm (yellow).

For lighting practice, it is interesting to investigate whether there is a correlation between the color rendering property and the perceived lightness of a colored object when the object is illuminated by different WLED light sources. In this section, the example of a red object (the CIE TCS09) is considered. The semantic interpretation scale value of the special CRI of the CIE TCS09, $R(R_{2012,TCS09})$ was computed for the set of 34 WLED light sources (from Figure 5.25). The value of L^{**} was also computed for TCS09 when it was illuminated by each one of these 34 LEDs. Figure 5.45 shows the correlation between the two quantities, $R(R_{2012,TCS09})$ and L^{**} (TCS09).

As can be seen from Figure 5.45, if the deep red TCS09 has a better semantic interpretation category which corresponds to a higher CRI within each CCT group then its chromatic lightness is also higher. A similar result was found for

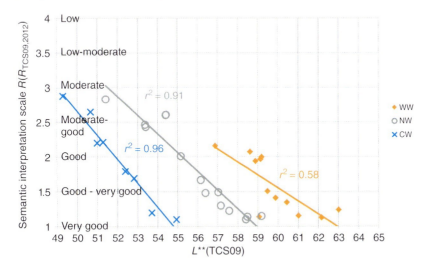

Figure 5.45 Ordinate: semantic interpretation scale value of the special color rendering index of the CIE test color sample TCS09, $R(R_{2012,TCS09})$; abscissa: chromatic lightness value L^{**} of TCS 09. Every symbol represents one of the 34 white LED light sources of Figure 5.25 grouped according to CCT; WW: warm white; NW: neutral white; CW: cool white.

other red objects, too [76]. The effect is demonstrated in Figure 5.23: on the left side of Figure 5.23 with poor color rendering, the chromatic lightness of reddish objects is low while on the right side of Figure 5.23 with good color rendering, the chromatic lightness of reddish objects is high. This implies that it is possible to optimize the SPD of the WLED light source so as to increase both the chromatic lightness of the visually important reddish objects and their color fidelity *simultaneously*. This means that a co-optimization of two important aspects of the color quality of the LED light source is possible in order to achieve a visually appealing lit interior environment.

5.7
White Point Characteristics of LED Lighting

As mentioned in Section 5.1, the first task of the spectral engineering of WLEDs is to assure an appropriate shade of white, a so-called target white tone chromaticity, by optimizing the SPD of the LED light source. The chromaticity of the white tone of the LED light source (ranging between warm white and cool white via neutral white) is an important aspect of lighting quality as it determines the chromaticity of achromatic or nearly achromatic objects (spectrally aselectively reflecting surfaces) in the lit environment. Such surfaces (e.g., white walls, window sills, or gray furniture) typically cover a large part of the room illuminated by the WLED light source hence their perceived white or gray tone contribute to the impression of lighting quality significantly.

The perceptual issue of white tone quality (the term *white point* is also used because the white tone of the light source is represented by a point in the chromaticity diagram) is twofold: perceived whiteness and white preference. Perceived whiteness is the degree to which a light source or a surface appears white, that is, without any chromatic shade, without for example a slight yellowish, greenish, or purplish tint (similarly, perceived achromaticness can be defined for grey surfaces). However, even if the perceived color of a light source or a surface contains a chromatic shade (usually a yellowish shade, e.g., for a warm WLED), it is often preferred by (at least European and North American) light source users against a light source of greater perceived whiteness (e.g., against a cool WLED) in the context of living rooms or restaurants (to provide a relaxing environment). In contrast, a cool WLED (which contains a slight bluish shade) is rather preferred for office lighting.

Thus, white tone preference (the most probable choice of a certain white tone of a certain user group for a given application) depends on the application (e.g., office lighting vs living room lighting) and the group of observers (e.g., Europeans vs Far Eastern people) the light source is designed for. Apart from white preference, the most important question for the engineer of the WLED light source is what light source chromaticities are accepted by the light source users as white illuminants for general lighting (while for special or accent lighting purposes like discotheques or theater scenes virtually any chromaticity is possible even saturated purples, greens, etc. is possible), see Figure 5.46.

5.7.1
Whiteness Perception, Correlated Color Temperature, and Target White Chromaticities for the Spectral Optimization of White LEDs

The question of what illuminant chromaticities are usually accepted or preferred for general lighting has already been answered in Section 5.2: according to the CIE CRI method [15], such white tones are located at the blackbody radiators' (or Planckian) locus (see Figure 2.26) for CCT < 5000 K and at the daylight locus (see Figure 2.26) for CCT ≥ 5000 K with the criterion that the Euclidean distance between the test light source and the reference illuminant in the u, v color diagram will not be greater than $\Delta uv = 0.0054$ (the quantity Δuv is denoted by DC [15] and CCT is often denoted by T_{cp}). These chromaticities along the Planckian locus and the daylight locus are generally considered as the primary *target chromaticities* in WLED spectral design.

The validity and meaning of this target chromaticity limit ($\Delta uv = 0.0054$) and other white chromaticity difference magnitudes are be analyzed in Section 5.8 as this issue is very relevant to the chromaticity binning of WLEDs. Also, note that if the luminance of the light source or of the brightest aselectively reflecting surface in the field of view of the observer is low (e.g., less than $1\,cd\,m^{-2}$) then there is no white perception in the scene, the surface (e.g., a wall) or the faint light source itself appears to be grey.

As mentioned earlier, the grouping of WLEDs according to their white tone (warm white, neutral white, and cool white) is very important from the point of view of selecting or optimizing a light source appropriately for a given lighting

Figure 5.46 Illustration of a room illuminated by different white tones. Left: a white tone acceptable for general lighting; right: not acceptable because of an unexpected greenish shade (e.g., because of the failure of LED thermal management). Note that light source users assess those surfaces visually that they expect to be achromatic (walls, grey furniture). (Image source: [77].)

application (e.g., cool white for an office). In Figure 5.26, a widely used (intuitive) grouping scheme was applied according to CCT (the concept was introduced in Chapter 2): CCT < 3500 K for warm white, CCT between 3500 and 5000 K for neutral white and CCT ≥ 5000 K for cool white. To validate this grouping scheme and to refine the grouping limits, the results of a previous visual study are summarized and further analyzed below.

In this study [78], eight observers had to judge the type of a white tone (either "warm" or "cool") on a white standard at 50 and 200 lx (note that these illuminance levels are acceptable to create a relaxing atmosphere in a living room or in a restaurant while for office lighting, illuminance levels of greater than 500 lx are necessary). The CCT of the (multi-LED) light source was varied between 2200 and 6300 K in 300 K steps. The white point of all light sources was on the blackbody locus with an error of $\Delta uv = 0.0002$ (less than the target chromaticity limit of $\Delta uv = 0.0054$). Note that (as specified before) the target chromaticity should lie on the daylight locus for CCT ≥ 5000 K and not on the blackbody locus [15]. The CIE CRI R_a values ranged between 89 and 94. Every observer repeated his/her judgment seven times. The frequency of judging "cool" was plotted as a function of CCT for every observer and a probit model was used to determine the 50% point which can be considered as the middle of the neutral white range.

Among the eight observers, the CCT of the neutral white point in this sense ranged between 3640 and 4265 K for 50 lx (average: 3953 K) and between 3751 and 4434 K for 200 lx (average: 4093 K). The middle point of the neutral range of the grouping scheme of Figure 5.26 equals 4250 K, which can be considered plausible in the view of this study [78]. A further analysis of the 200 lx experimental data [78] was carried out by the present authors: the last CCT with 0% frequency of "cool white" answer (an estimate of the upper limit of the warm white range) and the first CCT with 100% frequency of "cool white" answer (an estimate of the lower limit of the cool white range) was read from Figure 1b of [78] for each one of the eight observers. So the upper limit of warm white ranged between 3000 and 3700 K and the lower limit of cool white ranged between 4900 and 5400 K. The tendency of this visual result does not contradict the grouping limits of Figure 5.26.

As mentioned before, the white point of the light sources of this study [78] were always positioned on the blackbody locus, even for CCT ≥ 5000 K, for which a target white point on the daylight locus (and not on the blackbody) is most widely accepted [15]. In this context, the following two questions arise: (i) at a fixed CCT, which white point is perceived to be of maximum whiteness, in other words, which white tone is tinted neither yellow nor blue? (ii) What is the percentage of hue (green–yellow and purple–violet) for the different white tones at the same CCT along the so-called isotherm lines that are perpendicular to the blackbody locus in the u, v chromaticity diagram (see Eq. 2.30)? If we find the answer to these questions then we can tell the type and magnitude of chromatic tints of the usual target chromaticities on the blackbody locus and the daylight locus exhibit.

To answer the above questions, the method and the results of a previous study [79] are summarized here. In this experiment, seven different SPDs were generated (by using a multi-LED light source in combination with a halogen lamp) for each one of the following fixed CCTs: 2700, 3000, 3500, 4100, 5000, 6500 K along the isotherm lines perpendicular to the blackbody locus. One of the seven SPDs was always situated on the blackbody locus. Twenty observers evaluated the white tone of the interior of a viewing box (with an aselectively reflecting white paint) covering their whole field of view. Observers had to judge whether the white point was either purple–violet or green–yellow (hue-choice judgments) and then also to judge what percent of that hue was present in the illuminated cube (hue-magnitude judgments, 0% corresponded to pure white, and 100 % corresponded to maximum saturation of the given hue). Both tasks [79] were carried out both immediately after seeing the viewing box and after 45 s of exposure (to account for chromatic adaptation effects).

Figure 5.47 reproduces the contours of constant tint from the reference [79] in the x, y chromaticity diagram for the "after 45 s exposure" condition as this condition is more relevant for interior lighting.

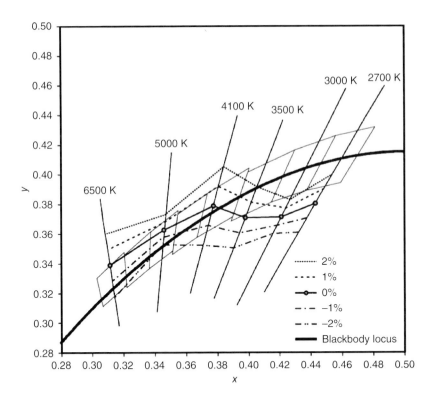

Figure 5.47 Reproduction of the contours of constant tint from Ref. [79] in the x, y chromaticity diagram. +%, percentage of green–yellow hue; −%, percentage of purple–violet hue. The u, v isotherm lines of the six different CCTs (thin lines), the blackbody locus (thick line), and the ANSI chromaticity quadrangles for solid-state lighting [80] (see Section 5.8) are included. (Reproduced with permission from *Color Research and Application*.)

The following can be seen from Figure 5.47:

1) Using white points on the blackbody locus as target white tone chromaticities above about 5000 K corresponds to about 1–2% of purple–violet hue.
2) Between 5000 and 6500 K, the daylight locus (unfortunately not plotted in the diagram) approximately corresponds to the dash-dot line of 1% of purple–violet.
3) Between 5000 and 6500 K, white tones of maximum whiteness (0%) are located above the blackbody locus and above the daylight locus.
4) Between 4100 and 5000 K, the blackbody locus exhibits a purple–violet hue and the phases of daylight are perceived to be approximately achromatic. Note that in a typical office with windows, often daylight like white points can be measured in this range. Thus, instead of using the blackbody locus as WLED target chromaticity, the daylight locus can also be considered as an alternative.
5) For CCT < 4100 K, white tones of zero hue (0%) are located well below the blackbody locus and the blackbody locus itself exhibits (as it is well known) an increasing amount of yellow hue as CCT decreases which is accepted and preferred for living room lighting in Western culture.

5.7.2
Analysis of the White Points of the Sample Set of 34 White LEDs

The white points of the sample set of 34 WLEDs of Figure 5.25 areanalyzed in this subsection. Figure 5.48 shows the chromaticity difference Δuv (denoted by DC [15]) between the white point of each LED of this set and the white point of its reference illuminant on the blackbody locus or the daylight locus (refer to Figure 2.26). Latter white point is considered as the target chromaticity of the WLED at its CCT which has been achieved more or less accurately. Figure 5.48 illustrates the relationship between these Δuv values and the semantic interpretation scale (R) value of $R_{a,2012}$ similar to the scheme of Figure 5.26.

As can be seen from Figure 5.48, 26 of the 34 WLEDs comply with the target chromaticity criterion $\Delta uv < 0.0054$ [15] (they are located to the left from the red line) while eight WLEDs do not fulfil this criterion (they are located to the right of the red line). Latter "white" LEDs exhibit disturbing chromatic shades. However, this chromaticity criterion is only a very rough approach to ensure the white tone quality of the LED light source. An advanced evaluation scheme of WLED chromaticities and a corresponding WLED binning strategy are presented in Section 5.8.

Another issue is the relationship of whiteness with color rendering. As can be seen from Figure 5.48, there is no significant correlation between Δuv and $R(R_{a,2012})$ for the WW and NW groups but there is a correlation of $r^2 = 0.57$ for the CW group of LEDs. Latter correlation means that if the white point of a CW LED is better then its color rendering property is also better.

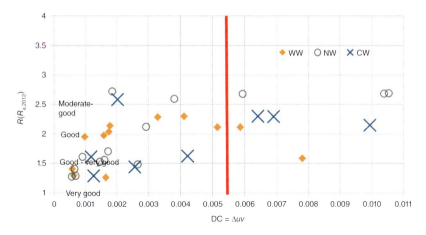

Figure 5.48 Abscissa: chromaticity difference Δuv (denoted by DC [15]) between the white point of each LED of the set of 34 white LEDs of Figure 5.25 and the white point of its reference illuminant. Ordinate: semantic interpretation scale (R) value of $R_{a,2012}$ similar to the scheme of Figure 5.26. Red line: target chromaticity limit ($\Delta uv = 0.0054$) according to [15].

5.7.3 Summary and Outlook

As it was pointed out in Section 5.7, the target chromaticities for the optimization of the white point of LED light sources do not necessarily coincide with those that exhibit maximum perceived whiteness. Target chromaticities of high CCTs (on the daylight locus, cool white) may show a slight bluish-violet tone while low CCTs (on the blackbody locus, warm white) show a characteristic yellowish tint. A rough estimate of white tone coloration was taken from the CIE CRI method [15]: if the WLED light source complies with the criterion $\Delta uv < 0.0054$ then it can be considered as usable. More advanced white tone criteria are described in Section 5.8.

Recent studies showed that CIE colorimetry may break down when narrow band RGB LEDs are used as WLED light sources and they should be matched with broad band SPDs [81].

RGB LED light sources whose instrumentally measured chromaticity coordinates match a broadband light source, for example, a high-quality white PC-LED light source or an incandescent lamp, can exhibit a different visual white point, for example, a disturbing shade and this effect depends much on the observer (high interobserver variability) and on the visual angle of the white surface it illuminates. Latter findings provide reasons not to use RGB LED light sources with pure semiconductor LEDs and sharp spectral peaks for general lighting purposes.

5.8
Chromaticity Binning of White LEDs

The concept of chromaticity binning means the sorting of LEDs immediately after production to different chromaticity groups, that is, different white tone categories for white PC-LEDs under a certain driving condition, for example, at an ambient temperature of 25 °C (or 85 °C, so-called hot binning condition), a current of 350 mA (or 700 mA) and a pulse duration of 20 ms. It is better to sort the LEDs under a realistic condition which is close to a real application, for example, hot binning helps avoid surprises because of higher temperatures inside the LED luminaire. Principally, chromaticity binning is necessary due to the technological fluctuations of semiconductor and phosphor materials during LED mass production. Without chromaticity binning, LEDs of different chromaticities would be used and homogeneous illumination would not be possible, see Figure 5.49.

As can be seen from Figure 5.49, homogeneity cannot be guaranteed without a visually fine enough chromaticity binning (an objective criterion is described below). The method of chromaticity binning influences not only LED production workflows and datasheet specifications but also the representation of the visual effect of temperature and current changes as well as the aging of LEDs as the white point of the LED changes according to these characteristics. The question is whether the LEDs of a luminaire or the LEDs in more luminaires in a room still belong to the same binning category despite more or less substantial white tone changes. For example, the LED luminaire in the left-hand side of Figure 5.49 has to be replaced after 6000 h by a new one which is now visible on the left, but the luminaire on the right remained to be the old one. The reader can see the result in Figure 5.49: the old luminaire on the right exhibits a disturbing yellow shade.

Figure 5.49 Without LED chromaticity binning or with a false LED binning strategy, the manufacturer of the LED luminaire uses LEDs of different chromaticities and a visually disturbing inhomogeneous illumination is produced. (Image source: Technische Universität Darmstadt.)

Besides chromaticity binning, there are many other possibilities for LED binning: the LEDs coming out from the production line can also be sorted into groups according to their luminous flux, forward voltage, and other parameters while a binning strategy according to a combination of parameters like luminous flux *and* chromaticity is also possible. But it is the chromaticity binning method that is of key relevance for lighting practice with WLEDs.

There are several different chromaticity binning methods for WLED light sources. All of them are based on specifying a mesh of neighboring chromaticity regions of parallelogram shape and different size in the x, y or u', v' chromaticity diagram depending on CCT and the distance from the blackbody locus or daylight locus. Take a look at Figure 5.50 as a standard example. A WLED gets to one of such chromaticity regions after its chromaticity is measured after production. There are binning methods with a fine mesh of regions (with many small regions) and coarse lattices (with only a few chromaticity regions). Fine lattices with warm, neutral, and cool white binning groups are intended for customers of different needs for interior lighting while coarse binning methods with predominantly cool white regions are rather intended for street lighting applications.

5.8.1
The ANSI Binning Standard

"To specify the range of chromaticities recommended for general lighting with solid state lighting (SSL) products, as well as to ensure that the white light chromaticities of the products can be communicated to consumers" [80], the ANSI chromaticity binning standard [80] has been developed. As it is widely used in industry, the main features of this standard binning method are described here briefly. Note that the chromaticity binning requirement of the standard is intended for general indoor lighting applications only.

There are eight nominal CCT categories in the ANSI binning standard [80]: 2700, 3000, 3500, 4000, 4500, 5000, 5700, and 6500 K. There is an additional category with flexible CCT between 2700 and 6500 K to account for products with other CCTs than these eight nominal CCTs. The centers of the chromaticity binning ranges gradually shift from the Planckian locus (at low CCTs) toward the daylight locus (high CCTs) and the size of each parallelogram shaped binning range corresponds to a seven-step MacAdam ellipse (see Figure 2.25) in the x, y or u', v' chromaticity diagrams, see Figure 5.50. The chromaticity distance in the x, y diagram between any point on the contour of a one-step (i.e., original) MacAdam ellipse [82] and the center of the ellipse represents 1 standard deviation of visual color matching (SDCM) between the center and the point.

As can be seen from Figure 5.50, MacAdam ellipses exhibit a less elliptical (i.e., more rounded) shape when they are drawn in the visually more uniform u', v' chromaticity diagram (right) instead of the x, y chromaticity diagram. The radius of an original (i.e., one-step) MacAdam ellipse roughly corresponds to the Euclidean distance of $\Delta u'v' = 0.001$ (10^{-3}) on the u', v' diagram. While the ANSI standard uses seven-step MacAdam ellipses, some companies offer finer binning grids

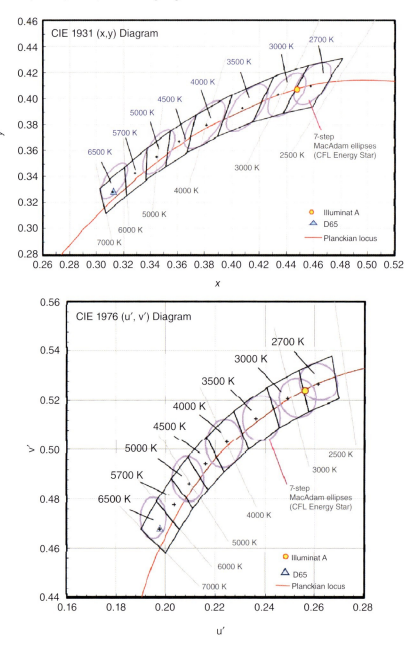

Figure 5.50 The ANSI chromaticity binning method for white LEDs of interior lighting. Upper diagram: x, y chromaticity diagram; Lower diagram: u′, v′ chromaticity diagram. The centers of the parallelogram shaped chromaticity binning regions gradually shift from the Planckian locus (at low CCTs) toward the daylight locus (high CCTs). The size of each region corresponds to a seven-step MacAdam ellipse. (Reproduced with permission from Ref. [80].)

based on four-step MacAdam ellipses or even two-step ellipses intended for interior lighting customers of highest demand.

5.8.2
A Visually Relevant Semantic Binning Strategy

As mentioned before, for the LED luminaires for interior lighting, in order to achieve a specific target white tone chromaticity and to realize and maintain color homogeneity on white surfaces (and on the illuminated colored objects), an efficient chromaticity binning strategy of the LED light sources is of importance. The ANSI binning method and the other industrial binning methods of individual companies are based on chromaticity specifications formulated in terms of a fine or coarse mesh of binning regions in the u', v' or x, y chromaticity diagrams often specified in terms of MacAdam ellipse steps. Such binning methods represent widespread tools to communicate the binning categories between manufacturers and customers today provided that chromaticity tolerances (expressed in terms of u', v' chromaticity differences) can be measured accurately enough by an affordable instrument.

But the question of the *visual relevance* of the above described binning agreements between customers and manufacturers remains open according to the following reasons:

1) MacAdam ellipses [82] are based only on a very limited number of observers.
2) MacAdam ellipses represent the SDCM, these are actually very small visual color differences generally less than a *just noticeable* chromaticity difference. It is known from literature that the linear magnification of MacAdam ellipses cannot predict supra-threshold (small, medium, and large) visual color differences [83]. The shape and orientation of supra-threshold ellipses is different from MacAdam ellipses. Therefore, the concept of two-step, four-step, or seven-step MacAdam ellipses is visually irrelevant for binning.
3) MacAdam ellipses were obtained with a chromatic adaptation to illuminant C (simulated daylight at 6774 K) only. Their validity for other chromaticities (e.g., 3000 K) is limited.
4) Neither manufacturers, nor their customers or end users can readily understand and interpret (refer to Section 5.3) the meaning of the numeric values intended to express white tone tolerances, neither the $\Delta u'v'$ values (i.e., Euclidean distances in the u', v' chromaticity diagram) nor the number of MacAdam steps (one-step, two-step, four-step, seven-step). The magnitude of *visually acceptable* chromaticity differences between two different white tones (depending on CCT and the distance from the Planckian or daylight loci) is unknown. What does a chromaticity shift of for example, $\Delta u'v' = 0.007$ (i.e., roughly a "seven-step MacAdam") mean visually? If there is such a change, for example, because of aging, can the user still accept it or, does the LED module have to be replaced? A semantic interpretation similar

to the one described in Section 5.3 (very good, good, moderate, low, bad visual similarity between two white tones) is missing.

After having recognized the deficiencies of current chromaticity binning methods for WLED light sources, a visually relevant semantic binning strategy [84] is described below. This semantic binning strategy assigns any instrumentally measured chromaticity difference expressed by the color difference metric CAM02-UCS (introduced in Section 2.3 and being the basis of the CRI2012 method described in Section 5.2.3) a semantic interpretation scale value (R, see Section 5.3) around any chromaticity center. This chromaticity center can be a target white tone chromaticity (see Section 5.7) for the WLED light source to be optimized. The chromaticity center can also be a reference white chromaticity to compute white tone deviations in order to characterize spatial color inhomogeneities on a white wall (see Figure 5.49) or during aging (e.g., to compare a WLED module at 0 hours of operation and after 6000 h of operation).

The computational method of this semantic binning strategy [84] transforms at a fixed luminance level every value of the CAM02-UCS color difference metric ($\Delta E'$) into chromaticity differences on the CIE 1931 x, y chromaticity diagram around any chromaticity center in any direction. Then, the contours of "very good," "good," "moderate," "low," or "bad" visual agreement with the chromaticity of the chromaticity center are computed in every direction around the chromaticity center by using the semantic interpretation function $R(\Delta E')$ [44] of Figure 5.17.

These so-called *semantic contours* resulting from the computation are shown in the CIE 1931 x, y chromaticity diagram of Figure 5.51 for the case of a warm white chromaticity center (the Planckian radiator at 3000 K, $x = 0.437; y = 0.404$) as an example. In Figure 5.51, semantic contours are compared with contours of constant chromaticity differences ($\Delta u'v' = 0.001 - \Delta u'v' = 0.007$) in the $u'-v'$ diagram measured from the chromaticity center. These constant $\Delta u'v'$ contours roughly approximate one-step to seven-step MacAdam ellipses. The computer program implementing the computational method [84] can calculate a semantic contour figure for any chromaticity center.

It is easy to understand the quintessence of the semantic binning strategy [84] from Figure 5.51: if the chromaticity point of a given WLED is located on a semantic contour then its binning category is the category of that semantic contour (e.g., very good). If the chromaticity point of a given WLED is located within the areas between two neighboring semantic contours then its binning category is the category corresponding to the nearest semantic contour. This binning strategy is visually relevant because it is based on the visual experiment described in Section 5.3.1 and on the semantic interpretation function (Figure 5.17) deduced from this experiment.

Turning back to the example of the warm white chromaticity center (target white tone chromaticity) of Figure 5.51, the following can be seen:

1) The orientation of the constant $\Delta u'v'$ contours and the orientation of the semantic contours is different.

5.8 Chromaticity Binning of White LEDs

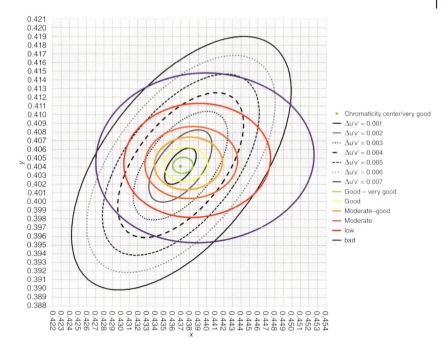

Figure 5.51 Semantic contours of the visually relevant semantic binning method [84] in the CIE 1931 x, y chromaticity diagram for a warm white chromaticity center (Planckian radiator at 3000 K, $x = 0.437$; $y = 0.404$). Contours of constant chromaticity differences ($\Delta u'v' = 0.001 – 0.007$) in the $u' - v'$ diagram, so-called constant $\Delta u'v'$ contours (that roughly approximate one- to seven-step MacAdam ellipses), are also shown.

2) At this chromaticity center ($x = 0.437$; $y = 0.404$), the different white tones along the $\Delta u'v' = 0.001$ ellipse (approximately one-step MacAdam) may correspond to "good – very good," "good," or "moderate-good" color agreement with the center (the light green point, i.e., the target white point) depending on direction, see the interSection points of the $\Delta u'v' = 0.001$ ellipse (black curve) with the green, yellow, and orange semantic contours in Figure 5.51.

Another example of semantic contours is shown in Figure 5.52 around the chromaticity center of the D65 illuminant (a phase of daylight at 6500 K; $x = 0.313$; $y = 0.329$). These semantic contours are compared with constant $\Delta u'v'$ contours again.

The following can be seen from Figure 5.52:

1) The orientation of the constant $\Delta u'v'$ contours and the orientation of the semantic contours is nearly similar.
2) At this chromaticity center ($x = 0.313$; $y = 0.329$), the different white tones along the $\Delta u'v' = 0.001$ ellipse (roughly a one-step MacAdam ellipse) correspond approximately to a "good – very good" agreement with the center as the green semantic contour exhibits a similar course as the black ellipse.

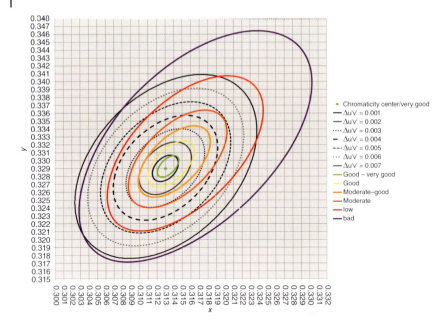

Figure 5.52 Semantic contours [84] around the chromaticity center of the D65 illuminant (a phase of daylight at 6500 K; $x = 0.313$; $y = 0.329$) compared with constant $\Delta u'v'$ contours.

3) The different white tones along the $\Delta u'v' = 0.002$ ellipse may correspond to "good" or "moderate–good" color agreement with the center (the light green point, i.e., the target white point), depending on direction, see the intersection points of the $\Delta u'v' = 0.002$ ellipse (grey) with the yellow and orange semantic contours in Figure 5.52.

5.8.3
Comparison of the Semantic Binning Method with a Visual Binning Experiment

In this section, the predictions of the semantic binning method described in Section 5.8.2 are compared with the results of an independent visual binning experiment [85]. This binning experiment used a portable color consistency test setup with six WLED modules of more or less varying white tones. The test setup illuminated a white wall at a vertical illuminance of 600 lx and a CCT of 3000 K with a target white tone chromaticity on the blackbody locus. Observers were asked to comment on the color differences they perceived on the wall because of the varying white tones, see Figure 5.53.

Observers assessed three different types of test setup called class 1, class 2, and class 6 setups [85] according to the different levels of chromaticity variations within the setup:

- The class 1 setup was the most homogeneous, with a CCT variation of ±60 K and blackbody locus distance (Δuv, see Section 5.7.1) variation of ±0.001 among

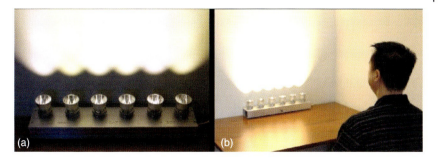

Figure 5.53 Visual chromaticity binning experiment using a portable color consistency test setup with six LED modules (a) that illuminated a white wall at a vertical illuminance of 600 lx and at a CCT of 3000 K (b). Observers were asked to comment on the color differences they perceived on the wall (b). (Reproduced with permission from the CIE from Ref. [85].)

the WLED modules of the setup. This corresponds in average to the category "good" at the 3000 K color center of Figure 5.51.
- The class 2 setup had a CCT variation of ±100 K and a blackbody locus distance (Δuv) variation of ±0.002 which covered about a four-step MacAdam ellipse. This corresponds in average to the category "moderate" at a warm white color center (see Figure 5.51).
- The class 6 setup had a CCT variation of ±175 K and a blackbody locus distance (Δuv) variation of ±0.006 which covered about a seven-step MacAdam ellipse. This setup is important because it corresponds to the ANSI binning standard [80]. This corresponds in average to the category "bad" at the 3000 K color center of Figure 5.51.

Despite the fact that the experiment was carried out "with hundreds of observers in Europe, United States, and Japan" [85], a consistent result was obtained. The class 6 set (roughly corresponding to the semantic category "bad") was "consistently rejected as having *unacceptable* chromaticity error." Observers did not want to specify, make or use fixtures with the "class 6" setup [85].

The "class 2" setup (roughly corresponding to the semantic category "moderate") "gave a more mixed response: all the observers immediately noticed a color difference … most [observers] concluded that they would not accept these types of color differences for architectural and accent lighting applications. Some of them would also not accept these color difference for other applications, such as general lighting, while others thought that it might be acceptable" [85].

The "class 1" setup (roughly corresponding to the semantic category "good") "was considered as acceptable by almost all observers. Color differences were still noticed, but it would require a bit more time and effort to see the actual color difference. A small number of observers concluded that they would like to see *even lower* color differences than this" [85], possibly a white tone agreement inside the binning category "very good," see the green semantic contour in Figure 5.51. This finding means that for the general user acceptability of the spatial homogeneity of white tones for interior lighting, the usage of WLED light sources at least within the semantic binning category "good" should be ensured.

5.8.4
Evaluation of the White Points of the Sample set of 34 White LEDs in Terms of the Visually Relevant Semantic Binning Strategy

Besides spatial homogeneity considerations, the semantic binning strategy can also be applied to indicate how well a certain test WLED light source comes close to its target white tone chromaticity (usually a chromaticity point of a certain CCT located at the blackbody locus or the daylight locus). If a test LED is located within the region of the "good" binning category in the x, y chromaticity diagram centered at a certain reference white point then the white point of this LED has a "good" visual similarity to the reference white point, that is, the target white tone chromaticity. Accordingly, the target of the spectral optimization of WLEDs would be a white point inside this "good" region around the target chromaticity point (e.g., 3000 K on the Planckian locus or 6500 K on the daylight locus).

In this respect, as an application of the semantic binning strategy, it is interesting to compute the similarity of the chromaticity of the white point of each LED from the set of 34 WLEDs (Figure 5.25) to their own reference white point in terms of the semantic interpretation scale (R). Figure 5.54 shows the semantic interpretation scale values (R) of the similarity of the white tone of each one of the set of 34 LEDs of Figure 5.25 to their own reference illuminant as a function of the quantity

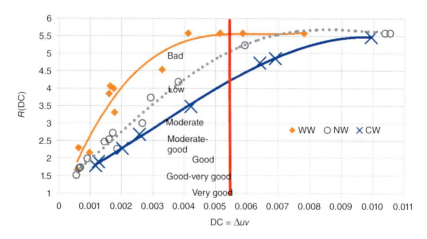

Figure 5.54 Ordinate: semantic interpretation scale of the similarity of the white tone of each one of the set of 34 LEDs of Figure 5.25 to their own reference illuminant. Abscissa: DC (Δuv), the distance from the blackbody locus (CCT < 5000 K) or the daylight locus (CCT ≥ 5000 K). Cubic polynomial trend lines are also shown for the three CCT groups (WW, NW, and CW) separately. The red line visualizes the $\Delta uv = 0.0054$ criterion value for white tones according to the CIE CRI method [15].

DC ($=\Delta uv$), the distance from the blackbody locus (CCT < 5000 K) or the daylight locus (CCT ≥ 5000 K).

As can be seen from Figure 5.54, only a few WLEDs of this set of 34 LEDs comply with the requirement of the semantic category "good" ($R=2$, compare with Figure 5.17). Following the polynomial trend lines of the three CCT groups, this requirement can be expressed in terms of Δuv values: the similarity of a cool WLED to its reference light source is "good" (or better than "good") if $\Delta uv < 0.0015$. The same criterion is $\Delta uv < 0.001$ for neutral white and $\Delta uv < 0.0006$ for warm white. To achieve a "better than moderate" similarity, $\Delta uv < 0.0033$ (for cool white), $\Delta uv < 0.0023$ (for neutral white), and $\Delta uv < 0.0012$ (for warm white) should be ensured.

The tendency is that, in terms of the u, v or u', v' metric, there is a more severe numeric limit for warm white than for cool white. This can also be seen when comparing the size of the ellipses of Figures 5.51 and 5.52. The $\Delta uv = 0.0054$ criterion value for white tones (see the red line in Figure 5.54) according to the CIE CRI method [15] corresponds to "low" to "bad" similarity to the reference white tone (i.e., the usual target white tone chromaticity). This criterion is intended to help decide whether a test light source can be considered as a more or less usable white tone *at all*. It should not be used as a spectral design criterion for WLEDs.

Another important issue is the relationship between color rendering and visual similarity to the target white tone. The combination of these two parameters represents the most often used target function (or constraint parameter) for the color quality optimization of WLED light sources. Similar to Figure 5.48, this relationship is shown for the example of the 34 WLEDs (taken from Figure 5.25) in Figure 5.55.

The most important findings in Figure 5.55 are:

1) If the WLED possesses a "good" or "better than good" white point then its general color rendering property is also "good" or "better than good."
2) There are numerous light sources with a "good" or "better than good" general color rendering property but with an inferior white point.

5.8.5
Summary

As seen in this section, white tone chromaticity binning means the sorting of LEDs into different chromaticity groups (different white tone categories). The main aims of chromaticity binning methods are: (i) to achieve a target white tone chromaticity with a reasonable accuracy; (ii) to maintain a target chromaticity among all LED light sources in a luminaire or in all luminaires illuminating an interior environment; and (iii) to be able check whether the LEDs are still in the same binning category after several thousand hours of operation, that is, to evaluate the effect of aging on the white tone chromaticity of the LED.

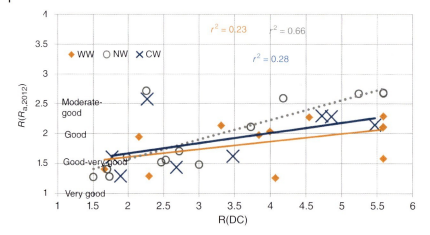

Figure 5.55 Abscissa: semantic interpretation scale value (R) of the chromaticity difference between the white point of each LED of the set of 34 white LEDs (from Figure 5.25) and the white point of its reference illuminant. Ordinate: semantic interpretation scale (R) value of the general color rendering index $R_{a,2012}$.

For practical applications, a visually relevant binning strategy is of vital importance. It was shown that binning methods using the magnification of MacAdam ellipses, for example, the use of seven-step ellipses in the ANSI binning standard, are visually irrelevant. For the effective communication of white tone chromaticity tolerances, a visually relevant binning strategy was described. It is based on the semantic interpretation categories ("very good," "good," "moderate," etc.) introduced in Section 5.3 and applied for the visual similarity of white tone chromaticities.

So-called semantic contours of the semantic binning categories "very good–good," "good," "good–moderate," "moderate," and so on were plotted in x, y chromaticity diagrams at a warm white and a cool white chromaticity center (Figures 5.51 and 5.52). In general, these visually relevant binning regions exhibit different shape and size than the magnified MacAdam ellipses while the seven-step MacAdam ellipses of the ANSI binning standard generally correspond to the category "bad." According to the above reasons, the visually relevant semantic binning strategy is recommended for the spectral design, selection, and characterization of WLED light sources.

The binning category limits were corroborated by an independent visual study [85] at least for warm white chromaticities. It can be concluded that, for the general user acceptability of the chromatic consistency of white tones for demanding interior lighting applications, the semantic binning category "good" can be ensured. This criterion can be translated into the language of the Δuv chromaticity difference metric via Figure 5.54 by considering the three CCT groups of WLEDs (WW, NW, CW) separately. The $\Delta uv = 0.0054$ criterion value of the CIE CRI method [15] corresponds to "low"–"bad" similarity to the reference white tone and it should not be used as a spectral design criterion for WLEDs.

For the sample set of 34 WLEDs of this chapter, it was also pointed out that a "good" or "very good" white point (close to the target white point) implies a "good" or "very good" general color rendering. But a "good" or "very good" general CRI does not necessarily imply a white point of high visual quality. Therefore, both requirements, white point and color rendering, have to be optimized simultaneously during the spectral design of WLED light sources.

5.9
Visual Experiments (Real Field Tests) on the Color Quality of White LEDs

As it was pointed out in Section 5.1, color quality has several visual aspects, see Figures 5.1 and 5.2. In the subsequent sections, some descriptor quantities were introduced for certain aspects, for example, the CRI, the CPI or the distance of the white point from the blackbody locus. It was also shown that the emission spectra of (single or combined) WLED light sources exhibit a previously unknown flexibility: a plethora of different shapes of WLED SPDs exists. By the aid of spectral engineering, the width and height of local minima and maxima and their slopes can be changed according to the engineer's intent within the visible wavelength range by changing phosphor compositions and concentrations and the wavelengths of pump LEDs and colored chip LEDs. Therefore, different aspects of color quality can be optimized, co-optimized, or emphasized.

To get a feeling about the visual experience of light source users in a real interior with real colored objects (see e.g., Figure 5.2) illuminated by different types of WLED light sources with very different emission spectra (also in comparison with conventional light sources) and in order to realize how relevant the spectral design of WLED light sources for the user acceptability of color appearance is, it is essential to get familiar with the methodology and the results of visual experiments (real field tests) with real subjects (observers) who assessed their impression on the color quality they experienced and whose answers were accumulated and analyzed. Two such real field studies are summarized in this Section to enrich the scope of this book concerning WLED spectral evaluation and design.

5.9.1
Color Quality Experiments in the Three-Chamber Viewing Booth

Two visual experiments on color quality are described in which the same three-chamber viewing booth [45] was used as shown in Figures 5.15 and 5.16. In the first experiment [86], the task of the observers was different from the one described in Section 5.3 but the light sources and the colored objects remained the same as described in Section 5.3. In the second experiment (results are first published here), the light sources and the colored objects were changed and observers had a different task from the one in Section 5.3.

5.9.1.1 The First Color Quality Experiment

To point out the effect of the different LED emission spectra, these spectra are shown in Figure 5.56 [86].

An interesting task for the reader is to analyze the deficiencies of the SPDs in Figure 5.56 (deep local minima with very low radiance in certain wavelength ranges) similar to the analysis of the emission spectra of Figure 5.30. Here, we would like to point out the effect of these spectra on visual color quality.

Observers had two tasks [86]: (i) assess the rank order of chromatic lightness (see Section 5.6.1.3) of three versions of each object placed in the middle of every chamber while only one object per chamber was visible at a time (e.g., left: darkest, middle: brightest); and (ii) establish a rank order of the three chambers according to the color preference of all colored objects viewed together (as seen in Figure 5.16). Twenty subjects carried out the experiment (16 male, 4 female, 32 years old in average).

Concerning the results of the chromatic lightness task, the rank order of the three chambers (rank 1, 2, or 3) was converted into a continuous visual chromatic lightness variable (1.0, 2.0, or 3.0). Then, in turn, the average values of this variable were computed for all colored objects. Results can be seen from Figure 5.57.

A tendency is clearly visible from Figure 5.57: those two LED light sources that are based on PC-LEDs and RGB LEDs in the left chamber trigger a stronger chromatic lightness impression on the colored objects they illuminate than the other four light sources. The reason is that both of them exhibit high and broad enough RGB peaks with shallow enough minima in between in their emission spectra (see Figure 5.56) to oversaturate the colored objects compared to the other two light sources at both CCTs. Note that as the illuminance level was approximately the

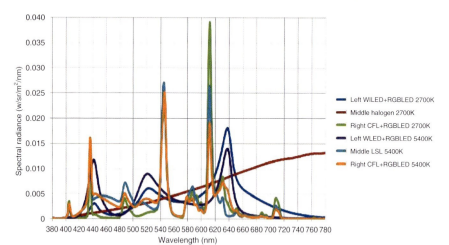

Figure 5.56 Spectral radiance distributions of the light sources of the first experiment in the three-chamber viewing booth [86], see Figures 5.15 and 5.16 and the subsequent text. $E = 1100\,\mathrm{lx}/50\,\mathrm{lx}$. Left chamber: $R_{a,2700\,K} = 69$; $R_{a,5400\,K} = 71$; middle chamber $R_{a,2700\,K} = 99$; $R_{a,5400\,K} = 91$; right chamber $R_{a,2700\,K} = 84$; $R_{a,5400\,K} = 81$.

5.9 Visual Experiments (Real Field Tests) on the Color Quality of White LEDs | 311

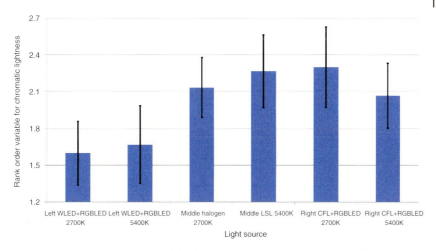

Figure 5.57 Rank order variable of the light sources in the three chambers of Figure 5.16 according to the visual chromatic lightness assessment of each colored object: average for all 30 colored objects and their 95% confidence intervals.

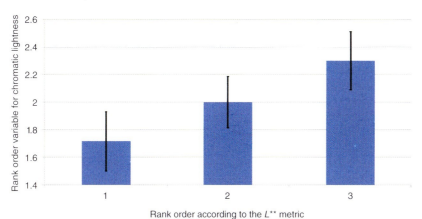

Figure 5.58 Rank order variable of the light sources in the three chambers of Figure 5.16 according to the visual chromatic lightness assessment of each colored object: average for all 30 colored objects and their 95% confidence intervals grouped according to the rank order as predicted by the chromatic lightness metric L^{**} from Eq. (5.20).

same ($E = 1100\,\text{lx} \pm 50\,\text{lx}$) this is predominantly a chromatic lightness effect (see Section 5.6.1.3). The chromatic lightness metric (L^{**}) of Eq. (5.20) is able to predict this tendency, see Figure 5.58.

Concerning the result of the color preference task, a visual color preference scale was computed in the following way: the number of observers who assigned the best color preference (first place) of all colored objects viewed together for a certain light source was multiplied by the factor 3.0 and the number of observers who assigned the same light source the second place was multiplied by the factor 1.0.

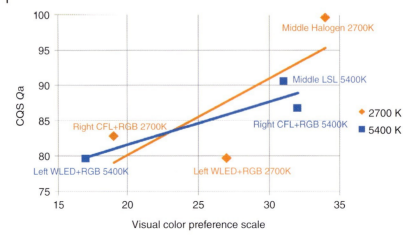

Figure 5.59 Visual color preference scale of the light sources by assessing all colored objects at the same time (abscissa) and CQS Q_a values (Section 5.5) (ordinate).

These two values were summed up and put on the abscissa of Figure 5.59 while the corresponding CQS Q_a values (Section 5.5) are shown on the ordinate.

As can be seen from Figure 5.59, the halogen light source (2700 K) in the middle chamber provided an illumination of highest color preference. This is an emission spectrum observers are used to and prefer as a reference illuminant. For the case of CCT = 5400 K, the CQS Q_a metric is able to predict visual findings with good correlation. But for 2700 K, there is an exception: the WLED+ RGB LED light source in the left chamber. Although it has a medium visual color preference value, its CQS Q_a value is placed in the lower range ($Q_a = 80$).

5.9.1.2 The Second Color Quality Experiment

As mentioned earlier, in this visual experiment, the same three-chamber viewing booth was used but the collection of 30 colored objects was replaced by another collection of only eight colored objects: a banana, an orange, a red pepper, a green pepper (these four objects were purchased in the morning of the test day and the experiment took only about 4 h), an artificial gentian flower, a piece of pink textile, a piece of turquoise textile, and a piece of green textile. The fresh objects were cut into three equal parts and positioned into the three chambers. Three equal copies of the same artificial objects/textiles were used. The white bottom of the three chambers was illuminated at the same illuminance level ($E = 1000 \text{ lx} \pm 20 \text{ lx}$) in all three chambers.

Twelve subjects (4 women and 8 men aged between 30 and 50) took part in this experiment and only three light sources were used to illuminate the chambers: a halogen light source (reference) in the middle chamber and two different high-end PC-LEDs with high quality phosphor mixtures in the left and right chambers. Their relative SPDs can be seen in Figure 5.60. Table 5.6 summarizes their colorimetric properties.

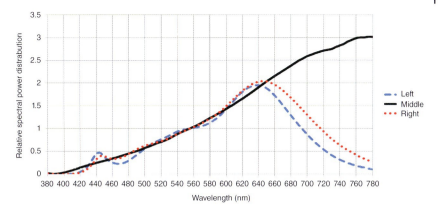

Figure 5.60 Relative spectral power distributions of the three light sources used in the three-chamber viewing booth during the second color quality experiment: Left and right chambers: two high-end PC-LEDs with high quality phosphor mixtures. Middle: halogen reference, see also Table 5.6.

Table 5.6 Colorimetric properties of the three light sources used in the three-chamber viewing booth during the second color quality experiment.

	Left	Middle	Right
CCT	2610	2585	2554
CRI R_a	97	99	97
CRI R_9	97	94	99
$\Delta u'v'$	0.0002	0.0011	0.0022

As can be seen from Figure 5.60 and from Table 5.6, the two high-quality PC-LEDs with composite phosphor mixtures exhibit similar, well balanced emission spectra, good–very good white points and very good color rendering properties. Because of the red phosphor component in both PC-LEDs (see the two decreasing spectral emission branches from 640 nm on in Figure 5.60), the red R_9 indices are high but slightly higher in the right chamber than in the left chamber (99 vs 97) because of the even more accentuated long-wavelength emission of the red phosphor in the right chamber. Note that the white point of the right PC-LED is further away from the reference white point than the white point of the left PC-LED ($\Delta u'v' = 0.0022$ vs 0.0002), which means an inferior but still "good" white point. Observe that the left PC-LED exhibits a local emission minimum at about 470 nm.

As all three light sources of this experiment equally represent *high* color quality in opposition to the first experiment discussed in Section 5.9.1.2, it is very interesting to explore whether observers can perceive significant visual difference even in this case when all three light sources are expected to yield high color quality in view of the knowledge we learnt in the previous sections. Observers assessed the following aspects of color quality: color preference (the naturalness aspect), color rendering (perceived visual color differences related to the reference light source in the middle), brightness, and color harmony.

To assess the naturalness aspect of color preference, each chamber was viewed separately and each one of the eight objects was judged individually after each other by using the following attributes: "perfect," "very good," "good," "moderate," "bad," or "very bad" naturalness. To establish a numeric naturalness score, six points per object were added to the score for a "perfect" judgment and one point was added to the score for a "very bad" judgment with continuously decreasing points in between. These points were summed up for all eight objects which yielded a maximum naturalness score of 48 for a "perfect" light source. The mean naturalness score of the 12 observers amounted to 32.1 (left), 34.3 (middle), and 36.1 (right). A significant difference (T test, $\alpha = 5\%$) resulted only between the left and the right chambers, that is, between the two PC-LED, the right PC-LED with the improved red phosphor being superior. The analysis of the naturalness scores of the individual objects showed that the PC-LED in the right-hand chamber was especially advantageous for banana, orange, and green pepper.

Concerning the visual color rendering property, observers had to answer the following question: where is the color appearance difference of an object higher: left to middle (halogen reference) or right to middle? Wherever a subject indicated a higher color difference, the score increased by 1 but it did not increase otherwise. Observers viewed all three chambers simultaneously and assessed each object separately. The PC-LED on the left generally exhibited smaller color differences (thus a better visual color rendering property) than the PC-LED on the right. But for the purple textile and the red pepper the right PC-LED caused less color difference (hence better reddish CRIs) than the left PC-LED for purple textile and red pepper because of its improved red phosphor emission branch emitting at longer wavelengths.

Concerning the color harmony among the eight objects, observers had to use both an interval scale (0–100) and a categorical scale ("perfect," "very good," "good," "moderate," "bad," and "very bad") to quantify their harmony impression by viewing all three chambers simultaneously and all eight objects simultaneously. The PC-LED on the left obtained the lowest interval scale value (average of 12 subjects and 95% confidence interval: 61 ± 11). The same score was equal, 75 ± 8, for the halogen reference and 71 ± 9 for the PC-LED on the right. The visual harmony difference between the left and the middle chambers was significant at the 6% level (t test, $p = 5.3\%$). The categorical scale was transformed into a continuous variable ("perfect" = 6.0, ..., "very bad" = 1.0) and this resulted in the following mean values: 3.7 ± 0.6 (left), 4.7 ± 0.5 (middle), and 4.2 ± 0.5 (right). Only the harmony scale difference between the left and the middle chambers was significantly different at the 5% level (t test, $p = 0.04$).

Brightness was judged without the colored objects by viewing the white bottom and the walls of the three chambers ($E = 1000\,\text{lx} \pm 20\,\text{lx}$ at the bottom). Observers had to establish a brightness rank order, for example, "left" (brightest), "right" (medium), "middle" (darkest). To get a numeric brightness score, the first choice added four points and the second choice added two points to the score while the third choice did not increase the score (0 point). The final score of a light source was equal the sum of the points for its first, second, and third choice for all 12 subjects (max. 48). The left pcLED obtained a visual brightness score of 30, the middle chamber (halogen reference) 32 (brightest), and the right pcLED only 10 (darkest). The last finding may be associated to the slightly poorer white point of the right PC-LED ($\Delta u'v' = 0.0022$). Eight of the 12 observers noticed that there was a slight red tone in the white point of the PC-LED illuminating the right chamber.

5.9.2 Concluding Remarks

Visual field tests with subjects observing arrangements of color objects are useful tools to ultimately validate the color quality performance of a WLED illumination. Important components of such studies include thermal stabilization of the LED light source, accurate spectral measurements, and the choice of possibly naive subjects who answer carefully formulated questions about color quality or assess it on categorical or continuous scales.

The three selected experiments of this section showed that there are significant visual differences among the different WLED spectral emission curves even in case of high-end to high-end comparisons. A further improvement of the emission spectrum of a high-end pcLED (e.g., covering longer wavelength with an improved red phosphor) can yield a visually significant improvement of color naturalness, color harmony, and the special color rendering property for reddish objects. Finally note that the literature on color quality field tests is abundant. Let some interesting studies be mentioned here: [8, 11, 19, 87–90].

5.10 Circadian Stimulus, Color Temperature, and Color Rendering of White LEDs

As mentioned in Section 2.5, visible electromagnetic radiation affects not only the human visual system but also the Circadian clock [91] responsible for the timing of all biological functions according to daily rhythm. This effect depends on the intensity, the spatial and SPD and the timing and duration of visible electromagnetic radiation reaching the human eye. Experimental data on the effect of light source SPDs on *nocturnal melatonin suppression* were described in Section 2.5 together with an additive weighting function model of the extent of melatonin suppression. As described in Section 2.5, melatonin is a sleep hormone, an important component of the Circadian clock as it fosters sleep-in. The exposure to light suppresses the production of melatonin and, consequently, alertness increases.

As mentioned in Section 2.5, the additive weighting function model tries to account for the spectral effect of light on the human Circadian system by using a hypothetical single-opsin type spectral sensitivity curve similar to but not identical with the spectral sensitivity of melanopsin, the photopigment in the ipRGCs (see Section 2.5). Deficiencies of modeling in terms of a single spectral weighting function were described in Section 2.5.

5.10.1
The Rea et al. Model

A more thorough analysis of the experimental data on nocturnal melatonin suppression showed that a complex modeling going beyond the spectral weighting function (as described in Section 2.5) was necessary [91]. Such a model is the so-called Rea et al. model which uses the signals of more retinal mechanisms in addition to the ipRGC signal to compute an output quantity called Circadian stimulus (CS). Experimental melatonin suppression data (responsible for the extent of alertness) were found to correlate well with the logarithm of CS [92] which can be computed if the absolute spectral irradiance distribution from a light source at the human eye is known.

The Rea et al. model [91, 92] is described in this section. The spectral sensitivity of its contributing retinal mechanisms is depicted in Figure 5.61.

As can be seen from Figure 5.61, the signals of the S cones, the rods, the ipRGCs as well as the L and M cones (see Section 2.1) represented by $V_{10}(\lambda)$ contribute to the CS. $V_{10}(\lambda)$ is the large-field spectral luminous efficiency function [93] which is shown in comparison with $V(\lambda)$ in Figure 5.61 while $V(\lambda)$ is not used in the Rea et al. model. Figure 5.62 illustrates the workflow of the Rea et al. model.

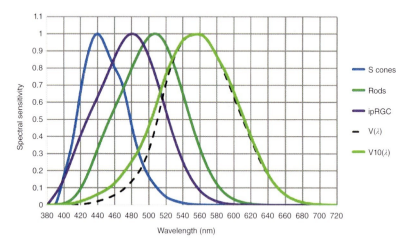

Figure 5.61 Relative spectral sensitivity of the retinal mechanisms contributing to the Circadian stimulus (CS), the output of the Rea et al. model [91]. The additive combination of the L- and M-cone mechanisms is represented by $V_{10}(\lambda)$, the large-field spectral luminous efficiency function [93] shown in comparison with $V(\lambda)$. $V(\lambda)$ is not used in the Rea et al. model.

5.10 Circadian Stimulus, Color Temperature, and Color Rendering of White LEDs

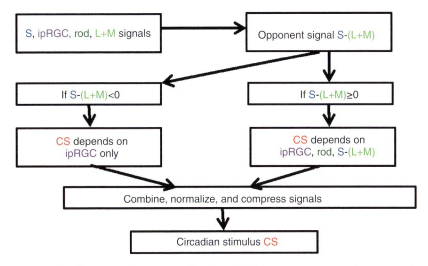

Figure 5.62 Workflow of the Rea *et al.* model [91]. Melatonin suppression data responsible for alertness were found to correlate well with the logarithm of CS [92], see also Figure 5.61.

As can be seen from Figure 5.62, the Rea *et al.* model [91] consists of the following steps:

- S-cone, ipRGC, rod and (L+M) signals are computed by multiplying the spectral irradiance at the human eye by the spectral sensitivity of the corresponding mechanism and integrating between 380 and 780 nm (using $V_{10}(\lambda)$ to represent L+M).
- The opponent signal S−(L+M) is computed from the S and (L+M) signals.
- If S−(L+M) is negative then the CS depends only on the ipRGC signal.
- If S−(L+M) is non-negative then CS depends on the ipRGC, the S−(L+M) and the rod signals.
- For high irradiance levels, the model predicts rod saturation, that is, no rod contribution.
- The contributing signals are combined and the resulting value is normalized to CIE illuminant A.
- Signal compression to fit experimental nocturnal human melatonin suppression data results in the CS value.

5.10.2
Application of the Rea *et al.* Model to White LED Light Sources

In this section, the complex equations resulting in the CS were implemented in a computer program [91]. CS values were computed for the set of 34 WLED SPDs (Figure 5.25) analyzed earlier in several cases in Chapter 5. The aim of this computation was to investigate the dependence of CS on the CCT of the WLEDs in the warm white, neutral white, and cool WLED groups, or, if possible, to find an easy-to-compute or easy-to-measure quantity to predict CS at for these typical WLEDs to provide an easy-to-use tool for the lighting practitioner.

Figure 5.63 shows the dependence of the CS on CCT for the 34 WLED SPDs (Figure 5.25) assuming that every one of the 34 WLEDs provides a constant vertical illuminance of 700 lx at the eye in the plane of the cornea (this is a typical value in a well-illuminated office).

As can be seen from Figure 5.63, although the illuminance is fixed (700 lx), different CS values arise because of the fact that illuminance is computed by weighting the spectral irradiance distributions of the WLEDs by $V(\lambda)$, which is based on a combination of L and M, while CS has three further contributing retinal signals, S cones, rods, and ipRGC (see Figure 5.62).

It can also be seen from Figure 5.63 that the correlation between CCT and CS is low especially in the cool white (CW) group. It is interesting to take a look at the SPDs of the two WLEDs with the lowest and highest CS values within the WW group, WLED-6 (min. CS), and WLED-12 (max. CS). As can be seen in the two inset diagrams with the schematic emission spectra of these two LEDs, although both LEDs belong to the WW group, WLED-6 has a blue maximum of lower intensity while WLED-12 has a high blue intensity. To get a warm white point, the high

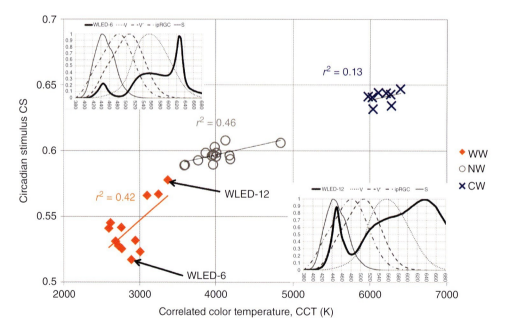

Figure 5.63 Circadian stimulus (CS) versus correlated color temperature (CCT) for the 34 white LED spectral power distributions from Figure 5.25. Every one of the 34 white LEDs provides a constant vertical illuminance of 700 lx at the eye. Inset diagrams: schematic spectral power distributions; left: WLED-6, right: WLED-12.

blue peak of WLED-12 is counterbalanced by a relatively high amount of radiation from 520 nm when compared to WLED-6. But the high blue peak yields a high ipRGC signal hence a greater CS value for the case of WLED-12.

It is interesting to compute the so-called ipRGC to photopic ratio (α) defined by Eq. (5.21).

$$\alpha = \frac{\int_{380\text{ nm}}^{780\text{ nm}} ipRGC(\lambda) S(\lambda) d\lambda}{\int_{380\text{ nm}}^{780\text{ nm}} V(\lambda) S(\lambda) d\lambda} \tag{5.21}$$

In Eq. (5.21), $S(\lambda)$ is the SPD of the LED light source, $ipRGC(\lambda)$ is ipRGC spectral sensitivity taken from Figure 5.61 and $V(\lambda)$ is the CIE (1924) photopic luminous efficiency function. As can be seen from Eq. (5.21), the quantity α represents the ratio of two weighted integrals of $S(\lambda)$ in the visible spectral range, one with ipRGC and the other one with $V(\lambda)$. Similar to Figure 5.63, Figure 5.64 shows the dependence of the CS on the quantity α for the 34 WLED SPDs (Figure 5.25) with the same assumption of constant vertical illuminance of 700 lx.

As can be seen from Figure 5.64, the correlation between the CS and the ipRGC to photopic ratio (α) is considerably better than between CS and CCT (compare with Figure 5.63) within the WW, NW, and CW groups and also over the whole set of 34 WLEDs. A quadratic curve can be used with reasonable accuracy ($r^2 = 0.99$) as a general fit curve for this illuminance value (700 lx), see Eq. (5.22).

$$CS = -0.035 \alpha^2 + 0.1958 \alpha + 0.386 \tag{5.22}$$

The importance of Eq. (5.22) for the lighting practitioner is that it is very easy to compute the value of α when the SPD of the LED light source is known (or using

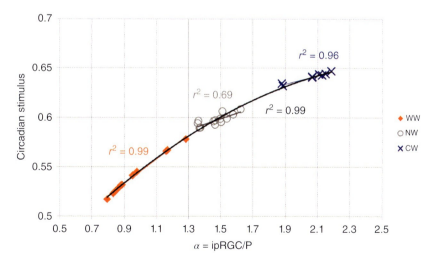

Figure 5.64 Circadian stimulus (CS) versus ipRGC to photopic ratio (α) defined by Eq. (5.21). Every one of the 34 white LEDs provides a constant vertical illuminance of 700 lx at the eye. Black curve: quadratic fit curve to all data points, see Eq. (5.22).

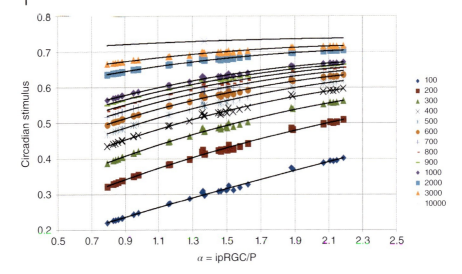

Figure 5.65 Circadian *stimulus* (CS) versus ipRGC to photopic ratio (α) defined by Eq. (5.21). Every one of the 34 white LEDs is assumed to provide a constant vertical illuminance between 100 and 10 000 lx (see legend, lx values). Black curves: quadratic fit curves to the data points at fixed illuminance levels.

a filter-detector combination that approximates the ipRGC(λ) sensitivity curve). Figure 5.65 shows the dependence of CS on the value of α for different illuminance values for the same set of WLEDs.

The CS of the different WLEDs can be approximately determined by using Figure 5.65 if the illuminance level is specified and the value of α is known.

5.10.3
Relationship between the Color Rendering Index $R_{a,2012}$ and the Circadian Stimulus for the White LED Light Sources

Figure 5.66 shows the relationship between the CRI $R_{a,2012}$ and the CS for the set of 34 WLED light sources from Figure 5.25.

As can be seen from Figure 5.66, there is no correlation between the CRI $R_{a,2012}$ and CS. At a fixed illuminance level, a WLED with a high CS value provides high melatonin suppression during evening or night hours that boosts alertness and helps maintain or increase work performance. By day, as there is hardly any melatonin to suppress, the color rendering property of the light source becomes more important. The best choice for office work is a combination of the two, that is, the upper right WLED in the CS versus $R_{a,2012}$ diagram. A WLED with a low CS value is, in turn, better for a calming environment for example, home lighting in the evening in order not to suppress melatonin to foster relaxation. But such an LED should also provide excellent color rendering. Therefore, to choose or optimize WLED light source SPDs, a parallel view of both properties is necessary.

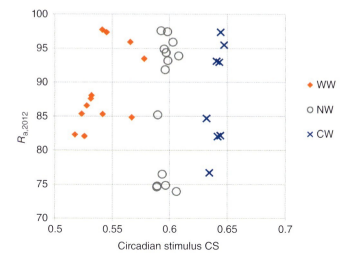

Figure 5.66 Color rendering index $R_{a,2012}$ (ordinate) and circadian stimulus CS (abscissa) for the set of 34 white LED light sources at 700 lx from Figure 5.25.

5.10.4
Summary

To predict the relaxing or alerting effect of WLED lighting in the evening or during night hours, we need a good model to predict the CS. The aim of the optimization of the CS can be either maximizing for better alertness and work activity or minimizing to provide a relaxing environment. Circadian optimization as a part of spectral engineering is efficient during the dark hours only. Otherwise there is hardly any effect because during the daytime, melatonin is suppressed. The CS of the Rea *et al.* model was computed for the representative set of 34 WLED light sources at fixed illuminance levels. It turned out that for these WLEDs, the value of CS can be computed from the so-called ipRGC to P ratio (α), which is easy to determine, at every fixed illuminance level.

5.11
Flicker and Stroboscopic Perception of White LEDs under Photopic Conditions

A huge advantage of LED lighting is the easy implementation of dimming. Because of the very short rise time and falling time of the light output of WLEDs and their fast electronics, it is possible to develop adaptive lighting systems for interior and outdoor lighting applications. One of the dimming methods is the constant current reduction (CCR) method. Because of its higher cost and the undesirable effect that the current change has an impact on the form and on the absolute intensity of the emitted spectrum (see Sections 4.3 and 4.6), in most applications, the pulse-width-modulation (PWM) method is preferred to regulate the amount of the LED's optical power. Turning the LED on and off at frequencies beyond the visual flicker fusion threshold prevents the perception of single light pulses and allows to control the brightness by varying the on-time of the LED.

The flicker effect is the rapid modulation of the optical radiation in a periodical manner. Within a short time, observers can perceive a periodic series of brighter and darker light stimuli falling on a certain area inside their viewing field. Human sensitivity to the flicker depends on the following parameters:

- Modulation frequency and modulation depth;
- Duty cycle of the modulated light at a defined frequency;
- Wavelength or CCT;
- Adaptation luminance; and
- Location of the flickering light stimulus in the visual field.

The stroboscopic effect is the indirect perception of the flicker effect if an object is moving at a certain speed (longitudinally or circularly) inside the visual field. The object is illuminated with periodically modulated optical radiation, for example, by a PWM controlled WLED. It is then observed as a series of multiple images, see Figure 5.67.

In this section, studies on the flicker and stroboscopic effects including Bullough *et al.*'s study [94] in comparison with the authors' experiments in a real office room are described. These studies had the goal of pointing out the effect of frequency, modulation depth, duty cycle, and color temperature on the perception of flicker and the stroboscopic effect.

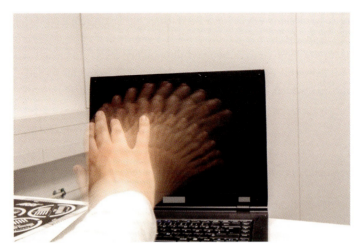

Figure 5.67 Stroboscopic effect of a waved hand in a room illuminated by PWM controlled white LEDs. (Image source: Technische Universität Darmstadt.)

5.11.1
Flicker Research Results in the Past

The first fundamental research studies on the flicker effects were performed by Kelly [95] who illuminated a large visual field of a radius of 30° with different luminance levels from 0.03 to 5000 cd m^{-2} and found that with a modulation of more than 100 Hz, the flicker perception at the fovea and the periphery is nearly negligible even with a modulation of 100%.

Hershberger et al. [96] reported that 97% of the observers could detect a stroboscopic phantom array when they changed the visual fixation point in the illumination field of LEDs driven with a frequency of 200 Hz and with 100% modulation. In a further study [97], a phantom array was perceived by most observers at a frequency of 500 Hz.

5.11.2
The Bullough et al. Study on Flicker and the Stroboscopic Effect with a LED-Luminaire

The study was conducted in a dark-painted windowless room with the following aim [94]:

- Investigate flicker perception and the stroboscopic effect;
- Point out the influence of modulation frequency, modulation depth, duty cycle, CCT, and wave form on the acceptability, detection, and comfort behavior of the subjects.

Ten volunteer subjects (three female and seven male, aged between 23 and 55 years, mean 34 years) were tested by the aid of the apparatus shown in Figure 5.68.

In the context of flicker and stroboscopic effect, *modulation depth* is defined by Eq. (5.23)

$$MD = \frac{I_{max} - I_{min}}{I_{max}} \tag{5.23}$$

In Eq. (5.23), I_{max} is the maximal light intensity and I_{min} is the minimal light intensity. Subjects sat in front of a table with a laptop and with an LED lamp unit on the right side producing a horizontal illuminance of 400 lx at the center of the desktop. The lamp unit was driven by LED current sources with the test conditions listed in Table 5.7.

Subjects were asked to rate perceived flicker by using the computer, looking at the LED luminaire or at the point A and to assess the stroboscopic effect by shifting their gaze between the points A and B and also by waving their hand underneath the luminaire. In additional, they were requested to judge the acceptability of flicker or stroboscopic perception on the following rating scale:

- +2: very acceptable;
- +1: somewhat acceptable;
- 0: neither acceptable nor unacceptable;
- −1: somewhat unacceptable; and

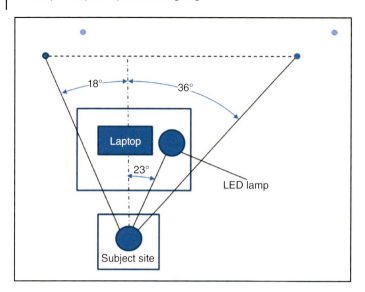

Figure 5.68 Approximate angular positions of objects in the subject's field of view [94], reproduced with permission from Lighting Research and Technology.

Table 5.7 Experimental conditions [94].

Condition	Flicker frequency (Hz)	Modulation (%)	Flicker index	Duty cycle (%)	Waveform	CCT (K)
1	50	100	0.5	50	Rectangular	4000
2	60	100	0.5	50	Rectangular	4000
3	100	100	0.5	50	Rectangular	4000
4	120	100	0.5	50	Rectangular	4000
5	300	100	0.5	50	Rectangular	4000
6	120	100	0.9	10	Rectangular	4000
7	120	33	0.17	100	Rectangular	4000
8	120	100	0.41	60	Chopped sine wave	6000
9	120	100	0.9	10	Rectangular	2700

Source: Reproduced with permission from Lighting Research and Technology.

- −2: very unacceptable.

The following results were obtained:

- Flicker perception percentage at 50 and 60 Hz was very high (≥90%).
- Flicker perception percentage at 100 Hz and higher was very low (≤10%) with an acceptability near +2.
- Flicker perception and stroboscopic effect averaged 80% at 50 and 60 Hz, 50% at 100 Hz, and 30% even at 300 Hz when subjects were asked to shift their gaze between the points A and B.

- For the analysis of the impact of modulation depth (at 120 Hz, conditions 4 and 7 in Table 5.7), flicker or stroboscopic effects were detected about one third of the time when modulation was high (100%, condition 4) if gaze was shifted between A and B. When modulation was lower (condition 7), the effects were never detected.
- There were no significant differences in flicker or stroboscopic perception between the different waveforms (rectangular or sine wave) or for the change of CCT from 2700 to 4000 K.

Limitations of this experiment [94] include the small number of subjects (only 10 persons), the use of frequencies only up to 300 Hz and the context of the tasks subjects had to carry out (they were not really typical office tasks).

5.11.3 The Study of the Present Authors on Flicker Perception and the Stroboscopic Effect with an LED Luminaire

5.11.3.1 Experimental Setup

The LED luminaire was set up in a white painted windowless room at the Laboratory of Lighting Technology of the Technische Universität Darmstadt (Germany). The desktop contained a laptop (switched off during the experiment) on the left and some paper and a computer mouse on the right side. It was placed centrally under the LED luminaire as illustrated in Figure 5.69.

Horizontal illuminance was kept at a constant level of 250 lx for all tested PWM signals driving the LED luminaire so that the subjects had a photopic viewing condition. The luminaire consisted of eight WLED light sources with a CCT of about 4000 K and with a general CRI of 85. Figure 5.70 shows its relative SPD.

A custom current source was designed to generate the PWM signals. This driver enabled the generation of user defined PWM signals with a variable frequency, duty cycle and maximal current. The rise time and fall time of the LED light output were about 3×10^{-6} s. In the experiment, 11 different PWM conditions were used:

- Five frequencies: 100, 200, 300, 400, and 1000 with the duty cycles (dc's) of 20%;
- Five frequencies: 100, 200, 300, 400, and 1000 Hz with the duty cycles (dc's) of 50%.
- In additional, one flicker-free condition (DC-operation) was also investigated. All conditions were verified using an oscilloscope and an amplified photo silicon detector. The conditions were presented in a randomized order but in the same order for each subject.

5.11.3.2 The Subjects' Tasks

Before the experiment began, all subjects were instructed about their tasks under the same lighting conditions. In order to explain the stroboscopic effect, two PWM signals with the duty cycle of 20% at 50 and 80 Hz were presented. Then the subject was asked to perform five different tasks in the same order for all 11 conditions.

326 | *5 Photopic Perceptual Aspects of LED Lighting*

Figure 5.69 Experimental room to test flicker perception. (Image source: Technische Universität Darmstadt.)

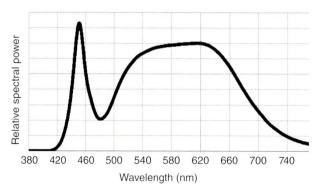

Figure 5.70 Relative spectral power distribution of the LED luminaire.

5.11 Flicker and Stroboscopic Perception of White LEDs under Photopic Conditions

The first task was to take place at the desktop. After each task the subject had to make two assessments about the noticeability and disturbance of the stroboscopic effect on a scale from −2 to 2. For the assessment of noticeability, the value of −2 means that the stroboscopic effect is not noticeable. The value of +2 means that the stroboscopic effect is clearly noticeable. The disturbance value of −2 means that

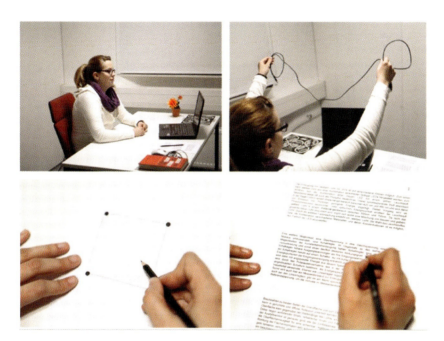

Figure 5.71 Task Nos. 2, 3, 4, and 5 of the flicker and stroboscopic experiment (see text).

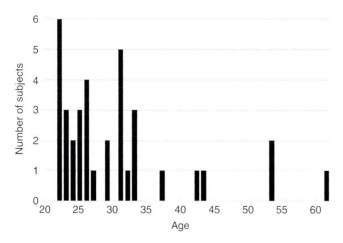

Figure 5.72 Age histogram of the subjects.

the lighting is not disturbing independent of whether the stroboscopic effect was noticed. The assessment of disturbance with +2 expressed a strong disturbance. Intermediate scale values were possible but not predefined.

Figure 5.71 illustrates four additional tasks. In the second task, the subject was asked to connect a computer mouse to the laptop in such a way that the cable should go behind the laptop. This requires a fast movement of the cable causing stroboscopic effect. The next task was to take a text with three paragraphs which was prepared on the desktop. The experiment instructor called an integral number between 10 and 30 for each paragraph and the subject had to count words and underline the word corresponding to the number. The fourth task was

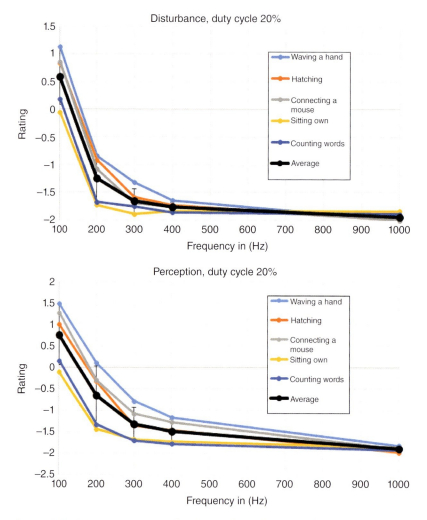

Figure 5.73 Mean assessments of all subjects, duty cycle of 20%. Lower diagram: assessment of noticeability of flicker perception; upper diagram: flicker disturbance rating.

to turn the paper and hatch a square. Finally, "waving of a hand" was performed. In order to save the assessments and to set the test conditions, a software program was developed.

Thirty-seven subjects aged from 22 to 61 took part in this experiment (23 male, 14 female: 24 subjects without lens correction, 10 glass carriers, and 3 with contact lenses). Figure 5.72 illustrates their age histogram.

5.11.3.3 Experimental Results

The average assessment over all subjects for each task and the duty cycle of 20% as a function of PWM frequency is illustrated in Figure 5.73. The lower diagram shows the assessment of noticeability while the upper diagram shows the results on the flicker disturbance level.

As can be seen from Figure 5.73, the ratings exhibit similar trends when they are depicted as functions of PWM frequency in both diagrams (lower and upper). In lighting practice, the duty cycle of 20% is of relevance because, from the physiological and psychophysical point of view, a dimming of the luminous flux of the luminaires down to 18% should cause a brightness appearance reduction of 50% so that a dimming factor of 18% to −0% is the most widely used value in a typical office room with adaptive lighting systems. It can be seen from Figure 5.73 that for the tasks "waving the hand" or "connecting the mouse," the frequency for a weak flicker perception is 400 Hz yielding a scale value near −1.

Figure 5.74 shows the cumulative histogram (percentage of the subject responses) for the task "connecting a mouse" at 200 Hz and 20% duty cycle. Only 20% of the subjects of 30 years of age or older (they were lighting experts) gave a rating of −2 (weak flicker perception) and about 80% of them could still detect the stroboscopic effect. Therefore, a frequency of 200 Hz should not be used for office lighting applications.

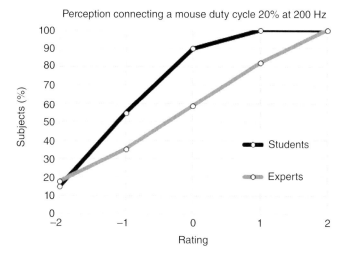

Figure 5.74 Cumulative histogram of the subjects' response at 200 Hz and 20%.

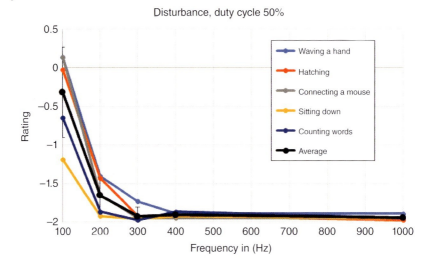

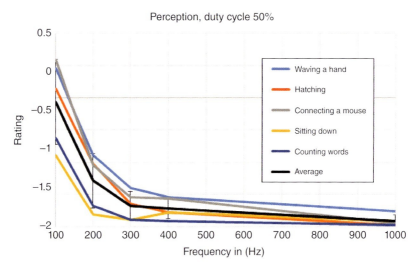

Figure 5.75 Mean assessments of all subjects, duty cycle of 50%. Lower diagram: assessment of noticeability of flicker perception; upper diagram: flicker disturbance rating.

Results with the duty cycle of 50 % are shown in Figure 5.75.

As can be seen from Figure 5.75, the subjects' ratings are lower than in the case of the duty cycle of 20%. This was to be expected, as the low duty cycle leads to long dark phases which are responsible for irritations in case of fast movement such as waving a hand or connecting a mouse.

In order to check the reproducibility and the relevance of the test results, additional tests were performed with a male subject (age: 25 years) on one single day. Five experiments started at 9:30 o'clock in the morning and then four times again, in consecutive 1.5 h intervals. In a second experiment, a male subject (age: 33

years) was tested on five days at the same hour, 7:30 in the morning. In both experiments, good reproducibility was found.

In order to prove if the subjects' ratings exhibit significant differences, a Student's t-test was performed. Ratings among most experimental tasks turned out to be significantly different both for perception and disturbance. Even at high PWM frequencies, the differences among the "perception" ratings of the subjects were statistically significant. On the other hand, there were no significant differences in "disturbance" ratings between dc = 50% and 300 Hz, dc = 50% at 400 Hz, both duty cycles at 1000 Hz and the constant lighting conditions.

5.11.4
Conclusions

Under the lighting conditions of the experiment on flicker perception and the stroboscopic effect presented in Section 5.11.3, the stroboscopic effect turned out to be perceptible even at 1000 Hz. Disturbance ratings showed a decreasing tendency with rising duty cycle. Under the given conditions, duty cycles from 50% on were not critical. To design an LED luminaires on the PWM base, a PWM frequency of at least 400 Hz is recommended. In additional, the "waving of a hand" task is a suitable one to assess the lighting quality concerning the perception of the stroboscopic effect.

References

1. Kronqvist, A. (2013) Review of office lighting research. Proceedings of the CIE Centenary Conference Towards a New Century of Light, Paris, France, April 15–16, 2013.
2. Tralau, B., Wellmann, T.U., Dehoff, P., and Schierz, Ch. (2010) Objective measurable criteria of lighting quality – transformation of the ergonomic lighting indicator into a measurable system. Proceedings of the CIE 2010 "Lighting Quality and Energy Efficiency", Vienna, Austria, March 14–17, 2010.
3. Dehoff, P. (2007) ELI: lighting criteria at one glance. Proceedings of the 26th Session of the CIE, Beijing, 2007, CIE Publication 178:2007.
4. U.S. Department of Energy (2013) Energy Efficiency and Renewable Energy, SSL Fact Sheets: Optical Safety of LEDs, June 2013.
5. Rautkylä, E., Puolakka, M., and Halonen, L. (2012) Alerting effects of daytime light exposure – a proposed link between light exposure and brain mechanisms. *Light. Res. Technol.*, **44** (2), 238–252.
6. Bodrogi, P. and Khanh, T.Q. (2012) *Illumination, Color and Imaging: Evaluation and Optimization of Visual Displays*, Wiley-VCH Verlag GmbH & Co. KGaA.
7. Ohno, Y. (2005) Spectral design considerations for white LED color rendering. *Opt. Eng.*, **44** (11), 111302.
8. Bodrogi, P., Brückner, S., Khanh, T.Q., and Winkler, H. (2013) Visual assessment of light source color quality. *Color Res. Appl.*, **38** (1), 4–13.
9. Commission Internationale de l'Eclairage (2007) Color Rendering of White LED Light Sources. Publication CIE 177:2007.
10. Halstead, M.B. (1977) *Proceedings of the AIC Color 77*, Adam Hilger, Bristol, pp. 97–127.
11. Viénot, F., Mahler, E., Ezrati, J.J., Boust, C., Rambaud, A., and Bricoune, A.

(2008) Color appearance under LED illumination: the visual judgment of observers. *J. Light Visual Environ.*, **32** (2), 208–213.

12. Ayama, M., Tashiroi, T., Kawanobe, S., Kimura-Minoda, T., Kohko, S., and Ishikawa, T. (2013) Discomfort glare of white led sources of different spatial arrangements. Proceedings of the CIE Centenary Conference Towards a New Century of Light, Paris, France, April 15–16, 2013.

13. Higashi, H., Koga, S., and Kotani, T. (2012) The development of evaluation for discomfort glare in LED lighting of indoor work place – relationship between UGR and subjective evaluation, CIE x037:2012. Proceedings of the CIE 2012 Lighting Quality and Energy Efficiency, Hangzhou, China, September 19–21, 2012.

14. Merck Patent GmbH (2013) Verfahren zur Optimierung der Farbqualität von Lichtquellen (Procedure to optimize light source color quality). Patent Application, No. DE 102012021223 A1 and WO 2014063791 A1.

15. Commission Internationale de l'Éclairage (1995) Method of measuring and specifying color rendering properties of light sources. CIE Tutorial and Expert Symposium on Measurement Uncertainties in Photometry and Radiometry for Industry, Vienna, Austria, September 11–12, 2014, Publication CIE 13.3-1995.

16. Hashimoto, K. and Nayatani, Y. (1994) Visual clarity and feeling of contrast. *Color Res. Appl.*, **19**, 171–185.

17. Boyce, P.R. (2011) On measuring task performance. *Color. Technol.*, **127**, 101–113.

18. Szabó, F., Bodrogi, P., and Schanda, J. (2009) A color harmony rendering index based on predictions of color harmony impression. *Light. Res. Technol.*, **41**, 165–182.

19. Smet, K., Ryckaert, W.R., Pointer, M.R., Deconinck, G., and Hanselaer, P. (2011) Colour appearance rating of familiar real objects. *Color Res. Appl.*, **36** (3), 192–200.

20. Smet, K.A.G., Ryckaert, W.R., Pointer, M.R., Deconinck, G., and Hanselaer, P. (2010) Memory colors and color quality evaluation of conventional and solid-state lamps. *Opt. Express*, **18**, 26229–26244.

21. Bois, C., Bodrogi, P., Khanh, T.Q., and Winkler, H. (2012) White LED light characteristics as a function of phosphor particle size. *ECS J. Solid State Sci. Technol.*, **1** (5), R131–R135.

22. Guo, X. and Houser, K.W. (2004) A review of color rendering indices and their application to commercial light sources. *Light. Res. Technol.*, **36**, 183–199.

23. European Association of National Metrology Institutes (EURAMET) (2012) EMRP Joint Research Project ENG05 "Metrology for Solid State Lightning". Report on color quality indices, 2012.

24. Judd, D.B. (1967) A flattery index for artificial illuminants. *J. Illum. Eng. Soc.*, **62**, 592–598.

25. Thornton, W.A. (1974) A validation of the color preference index. *J. Illumination Eng. Soc.*, **4**, 48–52.

26. Thornton, W.A. (1972) Color-discrimination index. *J. Opt. Soc. Am.*, **62**, 191–194.

27. Xu, H. (1993) Colour rendering capacity and luminous efficiency of a spectrum. *Light. Res. Technol.*, **25**, 131–132.

28. Fotios, S.A. (1997) The perception of light sources of different color properties. PhD thesis. UMIST, Manchester.

29. Pointer, M.R. (1986) Measuring color rendering – a new approach. *Light. Res. Technol.*, **18**, 175–184.

30. Davis, W. and Ohno, Y. (2010) Color quality scale. *Opt. Eng.*, **49** (3), 033602.

31. Commission International de l'Eclairage (2013) Review of Lighting Quality Measures for Interior Lighting with LED Lighting Systems, 2013, CIE Publication 205:2013.

32. Kirsch, R. and Völker, S. (2013) Lighting quality versus energy efficiency. Proceedings of the CIE Centenary Conference Towards a New Century of Light, Paris, France, April 15–16, 2013.

33. Khanh, T.Q., Bodrogi, P., Vinh, T.Q., and Brückner, S. (2013) *Farbwiedergabe von konventionellen und Halbleiter-Lichtquellen: Theorie, Bewertung, Praxis* (*Color Rendering of Conventional and Solid-State Light Sources: Theory, Evaluation and Practice*), Richard Pflaum Verlag GmbH & Co. KG, Munich (in German).
34. Rea, M.S. (2012) The trotter Paterson lecture 2012: whatever happened to visual performance? *Light. Res. Technol.*, **44** (2), 95–108.
35. Rea, M.S., Figueiro, M.G., Bierman, A., and Bullough, J.D. (2010) Circadian light. *J. Circadian Rhythms*, **8**, 2. doi: 10.1186/1740-3391-8-2
36. Loe D, Correspondence: lighting metrics, *Light. Res. Technol.* **44**(1), . 85–86, 2012.
37. Smet, K.A.G., Schanda, J., Whitehead, L., and Luo, M.R. (2013) CRI2012: a proposal for updating the CIE color rendering index. *Light. Res. Technol.*, **45**, 689–709.
38. LiTG (2012) Deutsche Lichttechnische Gesellschaft e.V., LiTG-Schrift, Farbwiedergabe für moderne Lichtquellen, in German: "Color rendering for modern light sources", LiTG Publication No. 28, 2012.
39. Commission Internationale de l'Eclairage (CIE) (1971) Colorimetry, CIE Publication 15:1971.
40. Commission Internationale de l'Eclairage (CIE) (2004) Colorimetry, 3rd edn, CIE Publ. 015:2004.
41. Bodrogi, P., Csuti, P., Horváth, P., and Schanda, J. (2004) Why does the CIE colour rendering index fail for white RGB LED light sources? Proceedings of the CIE Expert Symposium on LED Light Sources: Physical Measurement and Visual and Photobiological Assessment, Tokyo, Japan, 2004.
42. Wang, H., Cui, G., Luo, M.R., and Xu, H. (2012) Evaluation of colour-difference formulae for different colour-difference magnitudes. *Color Res. Appl.*, **37** (5), 316–325.
43. Smet, K. and Whitehead, L. (2011) Meta-standards for color rendering metrics and implications for sample spectral sets. CIC19: Color Imaging Conference, San Jose, CA, November 7–11, 2011.
44. Bodrogi, P., Brückner, S., Krause, N., and Khanh, T.Q. (2013) Semantic interpretation of color differences and color-rendering indices. *Color Res. Appl.*, **39**, (3), June 2014, 252–262.
45. Bodrogi, P., Krause, N., Brückner, S., Khanh, T.Q., and Winkler, H. (2012) The psychometry of color quality: a three-chamber viewing booth method. Predicting Perceptions: The 3rd International Conference on Appearance Edinburgh, Scotland, April 17–19, 2012.
46. Bodrogi, P., Krause, N., Brückner, S., and Khanh, T.Q. (2013) Semantic interpretation of color rendering indices: a comparison of CRI and CRI2012. CIE Centenary Conference, Paris, France, April 15–16, 2013.
47. Krause, N., Bodrogi, P., and Khanh, T.Q. (2012) Bewertung der Farbwiedergabe: Reflexionsspektren von Objekten unter verschiedenen weißen Lichtquellen. *Licht*, **5**, 62–69.
48. Schloss, K.B., Strauss, E.D., and Palmer, S.E. (2013) Object color preferences. *Color Res. Appl.*, **38** (6), 393–411.
49. Pitchford, N.J. and Mullen, K.T. (2005) Influence of saturation on color preference and color naming. *Perception*, **34S**, 25.
50. Ling, Y., Robinson, L., and Hurlbert, A. (2004) Colour preference: sex and culture. *Perception*, **33S**, 45.
51. Quellman, E.M., Boyce, P.R., and Berman, S. (2002) The light source color preferences of people of different skin tones. Discussions. Authors' reply. *J. Illum. Eng. Soc.*, **31** (1), 109–118.
52. Ou, L.C., Luo, M.R., Sun, P.L., Hu, N.C., and Chen, H.S. (2004) Age effects on color emotion, preference, and harmony. *Color Res. Appl.*, **29** (5), 381–389.
53. Ling, Y., Johnson, N., and Hurlbert, A. (2007) Colour preference and personality. *Perception*, **36S**, 195.
54. Ou, L.C., Luo, M.R., Woodcock, A., and Wright, A. (2004) A study of color emotion and color preference. Part III:

54. colour preference modeling. *Color Res. Appl.*, **29** (5), 381–389.
55. Schanda, J. (1985) A combined color preference – color rendering index. *Light. Res. Technol.*, **17**, 31–34.
56. Smet, K.A.G., Ryckaert, W.R., Pointer, M.R., Deconinck, G., and Hanselaer, P. (2010) Memory colors and color quality evaluation of conventional and solid-state lamps. *Opt. Express*, **18** (25), 26229–26244.
57. Hashimoto, K., Yano, T., Shimizu, M., and Nayatani, Y. (2007) New method for specifying color-rendering properties of light sources based on feeling of contrast. *Color Res. Appl.*, **32** (5), 361–371.
58. Freyssinier-Nova, J.P. and Rea, M.S. (2010) A two-metric proposal to specify the color-rendering properties of light sources for retail lighting. Tenth International Conference of Solid-State Lighting, Proceedings of SPIE, San Diego, CA, 2010, p. 77840V.
59. Li, C.J., Luo, M.R., Rigg, B., and Hunt, R.W.G. (2002) CMC 2000 chromatic adaptation transform: CMCCAT2000. *Color Res. Appl.*, **27**, 49–58.
60. Smet, K.A.G., Ryckaert, W.R., Pointer, M.R., Deconinck, G., and Hanselaer, P. (2011) Correlation between color quality metric predictions and visual appreciation of light sources. *Opt. Express*, **19** (9), 8151–8166.
61. Islam, M.S., Dangol, R., Hyvärinen, M., Bhusal, P., Puolakka, M., and Halonen, L. (2013) User preferences for LED lighting in terms of light spectrum. *Light. Res. Technol.*, **45**, 641–665.
62. Khanh, T.Q., Kirsten, M., Trinh, Q.V., and Bodrogi, P (2014) Verkaufs-fördernde LED – Platinen – Leuchte, Utility Patent (Gebrauchsmuster) Application, 2014., No. 20 2014 100 673.7
63. Khanh, T.Q. and Bodrogi, P. (2013) Farbqualität von weißen LEDs und von konventionellen Lichtquellen-systematische Zusammenhänge und praktische Bedeutungen, in *Handbuch für Beleuchtung*, Lange.
64. Khanh, T.Q. (2013) V(λ)–Lichttechnik: Entstehung, Wesen der Wahrnehmung, Defizite und neue Aspekte für eine wahrnehmungsgerechte Lichttechnik. Eine Publikation zum 100jährigen Jubiläum der CIE (1913–2013), in *Handbuch für Beleuchtung*, Lange.
65. Commission International de l'Eclairage (2001) Testing of Supplementary Systems of photometry, CIE Technical Report Publication 141–2001, CIE, Wien.
66. Sagawa, K. (2006) Toward a CIE supplementary system of photometry: brightness at any level including mesopic vision. *Ophthal. Physiol. Opt.*, **26**, 240–245.
67. Commission International de l'Eclairage (2011) CIE Supplementary System of Photometry, CIE Publication 200:2011.
68. Fairchild, M.D. and Pirrotta, E. (1991) Predicting the lightness of chromatic object colours using CIELAB. *Colour Res. Appl.*, **16** (6), 385–393.
69. Stockman, A. and Sharpe, L.T. (2006) Into the twilight zone: the complexities of mesopic vision and luminous efficiency. *Ophthal. Physiol. Opt.*, **26**, 225–239.
70. Nakano, Y., Yamada, K., Suehara, K., and Yano, T. (1999) A simple formula to calculate brightness equivalent luminance. Proceedings of the CIE 24th Session 1, pp. 33–37.
71. Commission Internationale de l'Éclairage (1995) CIE Collection in Colour and Vision, 118/2: Models of Heterochromatic Brightness Matching, 1995, Publication CIE 118–1995.
72. Fotios, S.A. (2001) Lamp color properties and apparent brightness: a review. *Light. Res. Technol.*, **33** (3), 163–181.
73. Commission Internationale de l'Eclairage (1988) Spectral Luminous Efficiency Functions Based upon Brightness Matching for Monochromatic Point Sources 2° and 10° Fields, Wien, CIE Publication 075–1988.
74. Wyszecki, G. and Stiles, W.S. (2000) *Color Science: Concepts and Methods, Quantitative Data and Formulae*, Wiley Series in Pure and Applied Optics, 2nd edn, Wiley-Interscience.

75. Wyszecki, G. (1967) Correlate for lightness in terms of CIE chromaticity coordinates and luminous reflectance. *J. Opt. Soc. Am.*, **57**, 254–257.
76. Khanh, T.Q. and Bodrogi, P. (2012) Farbwiedergabe- und Helligkeitswahrnehmung von weißen leuchtstoffkonvertierten Hochleistungs-LEDs in der Innenraumbeleuchtung. Teil I- 6.13.12, in *Handbuch für Beleuchtung*, Lange.
77. Pepler, W., Böll, M., Bodrogi, P., and Khanh, T.Q. (2012) Einfluss unterschiedlicher Beleuchtungskonzepte und Lampenspektren in der Innenraumbeleuchtung. Licht 2012, Berlin.
78. Davis, W., Weintraub, S., and Anson, G. (2013) Perception of correlated color temperature: the color of white, Proceedings of the 27th Session of the CIE, Sun City, 2011, pp. 197–202.
79. Rea, M.S. and Freyssinier, J.P. (2013) White lighting. *Color Res. Appl.*, **38** (2), 82–92.
80. ANSI_ANSLG C78.377-2011, (2011) American National Standard for electric lamps: Specifications for the Chromaticity of Solid State Lighting Products, National Electrical Manufacturers Association.
81. Csuti, P. and Schanda, J. (2008) Colour matching experiments with RGB-LEDs. *Color Res. Appl.*, **33** (2), 108–112.
82. MacAdam, D.L. (1942) Visual sensitivities to color differences in daylight. *J. Opt. Soc. Am.*, **32** (5), 247–274.
83. Kuehni, R.G. (2000) Threshold color differences compared to suprathreshold color differences. *Color Res. Appl.*, **25** (3), 226–229.
84. Bodrogi, P. and Khanh, T.Q. (2013) Semantic interpretation of the color binning of white and colored LEDs for automotive lighting products. Proceeedings of the 10th International Symposium on Automotive Lighting, ISAL 2013, Darmstadt, Germany, 2013.
85. Harbers, G., McGroddy, K., Petluri, R., Tseng, P.K., and Yriberri, J. (2010) Visual color matching of LED and tungsten-halogen light sources. Proceedings of CIE 2010 Lighting Quality and Energy Efficiency, Vienna, Austria, March, 2010, CIE Publications x035:2010.
86. Krause, N., Brückner, S., Bodrogi, P., and Khanh, T.Q. (2012) Neue 3-Kammer-Methode für die Psychometrie der Farbqualität für weiße LEDs. Licht 2012, Berlin.
87. Schanda, J. and Németh-Vidovszky, A. (2013) Colour rendering: color fidelity or color preference, AIC 2013. The 12th International AIC Colour Congress: Bringing Colour to Life, July8–12, 2013.
88. Jost-Boissard, S., Fontoynont, M., and Blanc-Gonnet, J. (2009) Perceived lighting quality of LED sources for the presentation of fruit and vegetables. *J. Mod. Opt.*, **56**, 1420–1432.
89. Schanda, J., Madár, G., Sándor, N., and Szabó, F. (2006) Colour rendering – color acceptability. 6th International Lighting Research Symposium on Light and Color, Florida, 2006.
90. Nakano, Y., Tahara, H., Suehara, K., Kohda, J., and Yano, T. (2005) Application of multispectral camera to color rendering simulator. Proceedings of 10th Congress of the International Color Association, Granada, Spain, 2005, pp. 1625–1628.
91. Rea, M.S., Figueiro, M.G., Bierman, A., and Bullough, J.D. (2010) Circadian light. *J. Circadian Rhythms*, **8** (2), 1–10.
92. Rea, M.S., Figueiro, M.G., Bullough, J.D., and Bierman, A. (2005) A model of phototransduction by the human circadian system. *Brain Res. Rev.*, **50**, 213–228.
93. Schanda, J., Morren, L., Rea, M., Rositani-Ronchi, L., and Walraven, P. (2002) Does lighting need more photopic luminous efficiency functions? *Light. Res. Technol.*, **34** (1), 69–78.
94. Bullough, J.D., Sewater Hickcox, K., Klein, T.R., and Narendran, N. (2012) Effects of flicker characteristics from solid-state lighting on detection, acceptability and comfort. *Light. Res. Technol.*, **43**, 337–348.
95. Kelly, D.H. (1961) Visual response to time-dependent stimuli: I.amplitude

sensitivity measurements. *J. Opt. Soc. Am.*, **51**, 422–429.

96. Hershberger, W.A. and Jordan, J.S. (1998) The phantom array: a perisaccadic illusion of visual direction. *Psychol. Rec.*, **48**, 21–32.

97. Hershberger, W.A., Jordan, J.S., and Lucas, D.R. (1998) Visualizing the perisaccadic shift of spatiotopic coordinates. *Percept. Psychophys.*, **60**, 82–88.

6
Mesopic Perceptual Aspects of LED Lighting

Tran Quoc Khanh, Peter Bodrogi, Stefan Brückner, Nils Haferkemper, and Christoph Schiller

6.1
Foundations and Models of Mesopic Brightness and Visual Performance

In exterior lighting (road or street lighting, vehicle lighting) visual tasks of road users occur in the mesopic (twilight) range of vision, typically between 0.05 and about $5-10\,\text{cd}\,\text{m}^{-2}$. This mesopic luminance range is very important for the spectral design of LED (light emitting diode) light sources for exterior lighting. The upper luminance limit depends on the position (eccentricity in the field of view, for example, 5° or 20° off the visual axis of the observer) and chromaticity (e.g., blue or red) of the visual target (object) that the visual performance of the user of the light source is related to.

6.1.1
Visual Tasks in the Mesopic Range

Typical visual tasks in the mesopic range include [1]:

- visual assessment of brightness impression;
- visual search;
- detection of (usually dangerous) visual targets (objects);
- quick reaction to these visual targets; and
- recognition of these visual targets.

The most relevant task for nighttime traffic is object detection (without recognition) because this is the basis for a quick reaction to it, for example, in order to avoid a traffic accident. In the following section, these visual tasks are discussed in detail.

The assessment of the perceived *brightness* of the whole street is a long process typically lasting for more than 3 s. During this time, the whole amount of visible radiation reflected from the street and the objects lying or standing on it are being captured and processed visually. From these impressions, the subject derives her or his own feeling of security and an aesthetic judgment of the scene being viewed.

LED Lighting: Technology and Perception, First Edition.
Edited by Tran Quoc Khanh, Peter Bodrogi, Quang Trinh Vinh and Holger Winkler.
© 2015 Wiley-VCH Verlag GmbH & Co. KGaA. Published 2015 by Wiley-VCH Verlag GmbH & Co. KGaA.

During the *visual search* process, the scene is being analyzed and interpreted to find certain objects of known shape, contrast or color that interest the observer, for example, certain shop windows or traffic signs. To determine the efficiency of visual search, the conspicuity of the search target and the number and conspicuity of its distractor objects (that distract visual attention) are important.

The visual task of *detection* means to discover a (dangerous) visual target (object) that is present (visible) in the scene and to react to it without recognizing that object, that is, without seeing its details. The reason is that, for a quick reaction, it is not necessary to identify a visual target. For example, to avoid a collision on a country road during the night, it is irrelevant what kind of object the visual target is (e.g., a deer or a cyclist), the aim is just to avoid collision and it is not necessary to see its fine details.

For the visual task of object recognition, however, the fine visual details of the object (of high spatial frequency) have to be seen and evaluated. To do so, the image of the object at the observer's retina has to be resolved spatially fine enough. The recognition process takes place by comparing the just seen object with an object *template* stored in the visual brain. An example is the recognition of a red stop sign. The object properties "octagonal" and "red" and "STOP with white letters" are then being compared with the observer's long-term memory template of a stop sign and then a previously learned behavior (breaking) is initiated.

6.1.2
Mesopic Vision and Its Modeling

Within the mesopic range, there is a continuous and significant change of the characteristics of human perceptual mechanisms, depending on the parameters of the visual scene that the observers experience. The most important parameters influencing human mesopic visual perception are:

- the nature of the visual task, for example, brightness assessment versus detection;
- mesopic luminance level, for example, 0.05 cd m^{-2} versus 0.5 cd m^{-2};
- the chromaticity of the light source used for street lighting or vehicle lighting, for example, a yellowish high pressure sodium lamp (HPS) versus a cool white (CW) LED street lamp or a car's LED headlamp with phosphor conversion optimized to enhance mesopic vision;
- the position of the object to be detected for example, a central object position in the middle of the viewing field versus an eccentric appearance in the periphery at about 3° up to 30° (these are typical values of eccentricity in the viewing field for a road with bends if the driver is fixating to the middle of the street) [2].

The reason for the significant dependence of the characteristics of mesopic visual perception on these parameters is that the type and extent of the *interactions* among the different types of retinal photoreceptors, L, M, and S cones and rods (R) (see Figure 5.41) depend on these parameters. Therefore, the four receptor signals L, M, S, and R (LMSR signals) are being combined in a different way for different parameters to yield a relevant neural signal for a certain task (e.g., a brightness signal vs a detection signal). The way of combining these LMSR signals is explained below.

Conventional photometry, that is, the quantities based on the $V(\lambda)$ function (luminance and illuminance) can account for the visual task of object recognition in the central visual field (involving L and M cone signals) only. Therefore, the concepts of (conventional photopic) luminance and luminance contrast can be used only to predict the *recognition* of small foveal visual targets subtending max. 2°. But this visual task and this viewing condition have no relevance to everyday mesopic (traffic) scenarios the LED light sources are designed for.

To describe the more relevant other mesopic tasks (excluding foveal object recognition), *mesopic* models of human vision can be used like the different mesopic brightness models [2] or the brightness model of the International Commission on Illumination (CIE) [3] which was described in Section 5.6.1.1. To model mesopic *reaction times*, the so-called X model or unified system of photometry (USP) was published [4]. A further model, the so-called MOVE model [5] unified the results of three types of *mesopic visual performance* experiments (reaction time, recognition, and detection) and arrived at a similar mathematical form as the X model but with different coefficients.

Mesopic brightness models are based on the observation procedure of heterochromatic brightness matching [2], see Figure 5.40. Note that the task of the subject in a heterochromatic brightness matching experiment is significantly different from the visual task of detecting a dangerous object of low contrast, for example, on the roadside. Latter task can be described by a mesopic visual performance model like the earlier-mentioned X model [4] or the MOVE model [5] which are more relevant for traffic safety and the design of road, street, or vehicle lighting (including automotive lighting).

The difference between the detection task and the brightness assessment task is that the important visual targets to be detected appear *with a low contrast* on a mesopic background (they tend to appear at the visual *detection threshold* defined below) while brightness evaluation is a suprathreshold task, that is, extended surfaces like a street or building façades are being viewed and assessed, see Figure 6.1. Because of this difference, LMSR signals are combined in a different way and, in turn, there is a difference between mesopic visual performance models (for detection and reaction time) and mesopic brightness models (which yield a numeric correlate of perceived brightness).

In the example of Figure 6.1 (left), the artificial gray visual target (marked by the red arrow) appears at 5° peripherally with a certain radiance difference (contrast) in comparison to its background. This typical mesopic detection target (object) is perceived at the detection threshold which can be defined as follows: the contrast at which the visual target is detected at the probability of 75%. Thus this is a threshold observation with weak LMSR signal differences between the target and the background. The target appears in the periphery of the retina (i.e., the viewing

Figure 6.1 Comparison of two mesopic visual tasks: 1. detection and 2. brightness assessment. Filled red circle: the central viewing field at the fixation point of the observer with a diameter of 2°. Red circle: a viewing field with a diameter of 10°. Left: detection of a typical mesopic visual target, here an artificial gray visual target in a real field test at 5° in the periphery, at the red arrow. Right: mesopic brightness impression: sustained viewing of the whole street scene. The red arrow shows the possible position of (sudden) appearance of a target to be detected (e.g., a pedestrian) between the two parking cars. (Reproduced from [1], copyright © 2011 permission from HJR Verlag.)

field), off the fixation point, typically between 3° and 30°. The neural signal responsible for detection arises as a combination of the spatial LMSR signal differences (at low spatial frequencies) between the target and the background.

In the example of Figure 6.1 (right), the human subject (the LED based exterior lighting system is designed for) observes and assesses the brightness of the whole scene. The central viewing field at the fixation point (2°) and a part of the whole viewing field of a diameter of 10° are marked in the figure. A typical position of appearance (5°) of a visual target to be detected (e.g., a pedestrian) between the two parking cars is marked by a red arrow. The brightness of the neighborhood of the target (extending typically about 10°) provides an "operating point" for the detection mechanism: the higher the mesopic brightness, the lower contrasts can be detected at a high probability. The scheme of the combination of human vision mechanisms (LMSR) that contribute to brightness perception has already been shown in Figure 5.41 while the CIE brightness model that uses the concept of equivalent luminance has been described in Section 5.6.1.1.

To *recognize* an object (e.g., letters), the human subject moves her or his central 2° viewing field (red filled circles in Figure 6.1) that possesses the highest visual acuity via eye movements toward the object. Then, the subject scans the fine spatial details of the object. The efficiency of object recognition in the central field of view can be described by the concept of conventional (photopic) *luminance contrast* both in the photopic and in the mesopic range. But to account for the other (more relevant) mesopic visual tasks, the earlier-mentioned models of *mesopic vision* can be applied.

Mesopic *reaction time* decreases with increasing contrast between the target and the background. Increasing the contrast in the suprathreshold contrast range, reaction time as a function of contrast converges to an asymptotic reaction time value [6] which is lower for higher background luminance levels. But to describe the latter effect in the mesopic range, instead of conventional (photopic) luminance, another quantity, the so-called *mesopic luminance*, can be used to characterize the background on which the target appears. The using of photopic and scotopic luminance results in an *erroneous* prediction of reaction time (and detection threshold contrast).

To compute the value of mesopic luminance from instrumentally measured spectral radiance distributions, the CIE recommends a mesopic visual performance model [7] which is described in Section 6.1.3. It was shown experimentally that (except for the lowest mesopic levels below 0.01 cd m^{-2}) the signals of the two chromatic channels (see Figure 5.41) influence reaction time [6] and detection thresholds [8]. Anyway, for the sake of simplicity, the CIE mesopic visual performance model [7] works with an achromatic approximation excluding chromatic signals. Currently, research efforts focus on the development of an advanced, practicable mesopic visual performance model that includes chromatic signals.

Contrast perception is a further widely used concept which merges two different visual tasks, object detection and object recognition. For object *detection*, the perception of low spatial frequency contrasts between the target and the background (blurred targets without identification) are enough. For object *recognition*, high spatial frequencies (fine object details, e.g., letters to be identified) have to be perceived. In the mesopic range, the visual mechanisms underlying these two different types of tasks are very different.

For the detection task, the observer has to discover and react to a visual target without necessarily recognizing it. The observer has to make a decision *only* about whether *any* object has been seen. In this context, object details of higher spatial frequencies are irrelevant as only the receptor signals of lower spatial frequencies are being processed. To avoid an accident, all dangerous objects have to be detected, even low contrast objects, at all positions in the field of view. For traffic lighting and vehicle lighting, the most important positions of target appearance range between 3° and 30°.

Concerning *object recognition*, observers have to recognize the shapes formed by the spatial details of the object, for example, the letters on the number plate of the preceding car. The question is what kind of object has been seen – and to answer this question, the contrast perception ability of high spatial frequencies (i.e., visual acuity) is important. This visual task takes places in the central 1° region of the retina (in the fovea) as receptor density (and consequently, spatial resolution ability) is at its maximum there. But it is only the L+M mechanism (accounted for by the $V(\lambda)$ function) that contributes to this visual task.

To carry out the mesopic *detection* task, however, rod and chromatic signals are also important. The contribution of the chromatic mechanisms of human vision can be seen from the shape of the so-called spectral sensitivity curves ($V_{\text{mes,det}}$) obtained in psychophysical experiments in which human subjects were asked to establish their mesopic detection threshold contrasts of low contrast visual targets of different colors (using quasi-monochromatic (QM) visual targets of different wavelength across the visible spectrum) appearing at different retinal positions on different mesopic backgrounds.

These $V_{mes,det}(\lambda)$ curves exhibit more local maxima [9–11] in the visible spectral range in opposition to the $V(\lambda)$ and $V'(\lambda)$ functions which are simply bell-shaped. The presence and the wavelengths of these (local) maxima depend on mesopic luminance level and the position of the detection target in the field of view [8]. These empirical spectral sensitivity curves of mesopic detection could be approximated by a linear combination ($V_{mes,lin}(\lambda)$) of the (known) spectral sensitivities of the visual mechanisms that were shown for the case of brightness perception in Figure 5.41. The combination of the mechanisms of Figure 5.41 yields a detection signal in the visual brain in this case, instead of the brightness signal indicated at the bottom of Figure 5.41. Note that the signal combination for detection has different weights (different from brightness) and includes spatial receptor signal *differences* between target and background.

The relative weights of the mechanisms that contribute to the mesopic detection signal depend on mesopic luminance level and the angular position of the object in the field of view [8]. Figure 6.2 shows the fit functions $V_{mes,lin}(\lambda)$ for different mesopic luminance levels and detection target positions. Note that instead of the chromatic channel 2 of Figure 5.41 (L+M−S), the standalone $S(\lambda)$ (i.e., S cone) sensitivity function was used in this case as a fit template (from mathematical reasons).

The different courses of the spectral sensitivity functions in Figure 6.2 reveal that the contribution (weighting) of the different mechanisms to the detection of low contrast quasi-monochromatic (QM) mesopic objects of different wavelength depends on target position and mesopic background luminance level. For example, at the $10°/1.0\,cd\,m^{-2}$ condition, there is significant S cone contribution and L−M chromatic contribution to the detection signal of an object, just in opposition to $30°/0.5\,cd\,m^{-2}$. Comparing the $10°/1.0\,cd\,m^{-2}$ condition (dark orange curve) and the $10°/0.1\,cd\,m^{-2}$ condition (red curve), it can be seen that rod contribution is significantly stronger for $0.1\,cd\,m^{-2}$ than for $1.0\,cd\,m^{-2}$. This is the so-called Purkinje effect, the sensitivity shift toward blue wavelengths for lower luminance levels.

It is interesting to compare the spectral detection sensitivity curves shown in Figure 6.2 with the spectral sensitivity curves resulting from Kokoschka's heterochromatic brightness matching experiment [12] (the principle is shown in Figure 5.40) in which observers had to match the brightness of QM stimuli of different wavelengths and the brightness of a constant reference stimulus (QM, 530 nm) at different mesopic luminance levels (0.001, 0.01, 0.1, 1.0, and $10\,cd\,m^{-2}$) and for different stimulus sizes (3°, 9.5°, and 64°). The results taken from the numeric tables [12] at 1.0 and $0.1\,cd\,m^{-2}$ are depicted in Figure 6.3.

As can be seen from Figure 6.3, for decreasing mesopic luminance levels ($0.1\,cd\,m^{-2}$, black curves, instead of $1.0\,cd\,m^{-2}$, gray curves) and for increasing stimulus size (64° or 9.5° instead of 3°), the spectral brightness sensitivity curve is

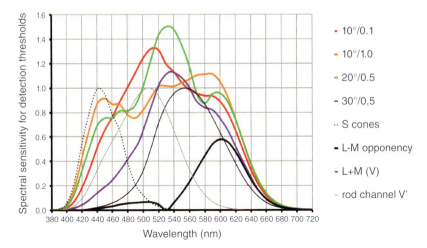

Figure 6.2 Colored curves: modeled mesopic spectral sensitivity functions $V_{mes,lin}(\lambda)$ for the detection of low contrast quasi-monochromatic mesopic objects of different wavelength, depending on target position (eccentricity from the fixation point in degrees) and mesopic background luminance level. For example, 10°/1.0 corresponds to the target appearance at the retinal position of 10° off the visual axis and the adaptation to the background luminance of 1.0 cd m^{-2}. Black and gray curves: spectral sensitivities of the individual mechanisms of the visual system (spectral templates) used to interpret the colored $V_{mes,lin}(\lambda)$ curves, compare with Figures 2.6 and 5.41.

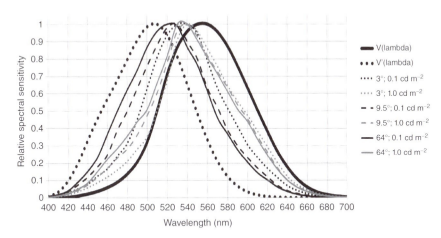

Figure 6.3 Spectral sensitivity curves resulting from Kokoschka's heterochromatic brightness matching experiment [12]. Observers had to match the brightness of quasi-monochromatic stimuli of different wavelengths and the brightness of a constant reference stimulus for different stimulus sizes (3°, 9.5°, and 64°). Results at 1.0 cd m^{-2} and 0.1 cd m^{-2} are shown. Abscissa: wavelength in nm, ordinate: relative spectral sensitivity of brightness (logarithmic).

shifted toward shorter wavelengths, that is, toward $V'(\lambda)$, the spectral sensitivity of the rod mechanism according to the fact that rods (that become more active with decreasing luminance levels, Purkinje effect) are more sensitive to shorter wavelengths than the L+M mechanism ($V(\lambda)$) and that there are more rods in the periphery of the retina than in its central region (while rods are absent in the center of the fovea).

6.1.3
The Mesopic Visual Performance Model of the CIE

The mesopic visual performance model of the CIE [7] is an intermediate model between the two earlier-mentioned mesopic visual performance models, the X model [4] and the MOVE model [5, 7]. Illuminating systems and their components for exterior lighting (e.g., those used to ensure nighttime traffic safety) must guarantee excellent mesopic visual performance. This is why the most important application of the model is road lighting, street lighting and vehicle lighting including automotive lighting. In these applications, mesopic viewing conditions between 0.05 and 1.5 cd m^{-2} predominate. The CIE model [7] is intended to yield a practical method that accounts for all three aspects of mesopic visual performance, detection, reaction time, and object recognition.

As mentioned in Section 6.1.2, the detection task is most relevant for the design of exterior lighting systems especially to avoid nighttime traffic accidents. In order to increase nighttime traffic safety, the detection probability of typical detection targets (dangerous objects with low contrasts around the detection contrast threshold, see Figure 6.1, left) can be increased by illuminating them and their background correctly: road lighting, street lighting, and vehicle lighting systems can be designed *mesopically*. The mesopic model of the CIE (see below) offers a possibility to do so. Applying conventional (photopic) photometry, however, may result in an erroneous design – as a typical target object is being detected by a *mesopic* visual process. A typical target appears between 3° and 30° off the visual axis (i.e., off the fixation point, the center of the visual field) where both the cones and the rods are active at the typical mesopic luminance levels. A typical target object has, as mentioned earlier, a low contrast, that is, a low radiance difference related to the background.

To detect such an object safely, an object-background contrast *above* the detection *threshold contrast* is necessary. If the value of the threshold contrast itself is low then even a low contrast object can be detected safely. So, the aim is to lower the detection contrast *threshold* (above which the visual system can detect safely) by using an appropriate illuminating system.

In the mesopic range, the value of the threshold contrast can be reduced correctly by increasing the *mesopic* (and not the photopic) luminance of the background on which the target appears. The background is the "operating point" of the detection mechanisms of the human visual system. The background can be, for example, the road surface illuminated by a cool white (CW) LED providing a mesopic luminance. The application of photopic or scotopic luminance as an "operating point" for the detection mechanism is incorrect [7]. Therefore, the aim of exterior lighting design is to increase the *mesopic* luminance of the background in order to increase the detection probability of low contrast targets that tend to appear on it.

Although current standards (at the time of writing) provide maintenance values for exterior lighting in terms of photopic luminance values, lighting engineers have the choice of light sources of different correlated color temperatures (CCTs) or chromaticities (e.g., choose a CW LED street luminaire instead of a low pressure sodium lamp, both providing 1.5 cd m^{-2} on the road surface to comply with the standard). Different light sources of different CCTs or chromaticities provide different mesopic luminance values *at the same photopic luminance* hence they provide different levels of mesopic visual performance (including detection performance).

In the mesopic visual performance model of the CIE [7], mesopic luminance values are computed from photopic luminance values (roughly the L+M signal) and scotopic luminance values (the rod signal). The relative weighting of the two types of luminance values depends on the luminance level and the so-called S/P ratio (scotopic luminance divided by photopic luminance) of the stimulus. For example, at the same photopic luminance value, the mesopic luminance of a road surface illuminated by a CW LED with a high S/P ratio is higher than its mesopic luminance if it is illuminated by a yellowish HPS lamp with a low S/P ratio while the road surface is assumed to be gray (spectrally aselective). But the extent of this effect depends on the luminance level. This way, detection performance can be increased by using light sources with high S/P ratios.

The iterative computational method of the mesopic visual performance model of the CIE [7] can be summarized as follows.

1) Input: photopic luminance L in cd m^{-2} as well as S/P ratio;
2) Compute the scotopic luminance L' (rod signal);
3) $m_0 := 0.5$ (initial value for the adaptation coefficient m to combine L' and L);
4) Compute an initial value for the mesopic luminance: $L_{mes,1}$;
5) Compute m_1 (first approximation for m) from the value of $L_{mes,1}$;
6) Compute the next value of mesopic luminance ($L_{mes,2}$) from L, L', and m_1;
7) Compute the next value of the adaptation coefficient (m_2) from the value of $L_{mes,2}$.

The steps Nos. 6 and 7 can be carried out iteratively until the value of mesopic luminance (L_{mes}) remains constant up to a desired accuracy. In most cases, 4–7 iterations yield a good result. The model is purely photopic above 5.0 cd m^{-2}. The following equations show the exact mathematical form of the above steps.

$$L' = (S/P)L \tag{6.1}$$

$$L_{mes,n} = \frac{m_{n-1}L + (1 - m_{n-1})L'(683/1699)}{m_{n-1} + (1 - m_{n-1})(683/1699)} \tag{6.2}$$

$$m_n = 0.767 + 0.3334 \log_{10}(L_{mes,n}); \ 0 \leq m_n \leq 1 \tag{6.3}$$

Table 6.1 shows the dependence of mesopic luminance (L_{mes}) on photopic luminance (L, in the first row of Table 6.1) and on the S/P ratio (in the first column of Table 6.1) in the mesopic visual performance model of the CIE [7] for a set of application relevant selected values.

As can be seen from Table 6.1, a higher S/P ratio yields a higher mesopic luminance L_{mes} at the same photopic luminance L. The relevance of the mesopic visual performance model of the CIE [7] for the lighting engineer is that the value of L_{mes} depends not only on the photopic luminance of the stimulus (in practice, the mesopic background on which the detection target, e.g., a pedestrian appears) but also on its S/P ratio.

Table 6.1 Dependence of mesopic luminance (L_{mes}) on photopic luminance (L, in the first row) and on the S/P ratio (in the first column) in the mesopic visual performance model of the CIE [7]. All L_{mes} values were computed by Eqs. (6.1)–(6.3). The model is photopic above 5.0 cd m^{-2}.

S/P	Photopic luminance L (cd m^{-2})						
	0.01	0.03	0.10	0.30	1.00	3.00	4.50
0.25	0.0025	0.0145	0.0705	0.2467	0.9130	2.9265	4.4782
0.35	0.0035	0.0174	0.0750	0.2545	0.9253	2.9367	4.4812
0.45	0.0045	0.0198	0.0793	0.2620	0.9373	2.9468	4.4842
0.55	0.0057	0.0220	0.0834	0.2693	0.9492	2.9568	4.4872
0.65	0.0069	0.0239	0.0873	0.2764	0.9608	2.9666	4.4901
0.75	0.0079	0.0258	0.0911	0.2833	0.9722	2.9763	4.4929
0.85	0.0088	0.0275	0.0947	0.2901	0.9835	2.9859	4.4958
0.95	0.0096	0.0292	0.0983	0.2967	0.9945	2.9953	4.4986
1.05	0.0104	0.0308	0.1017	0.3032	1.0054	3.0046	4.5014
1.15	0.0111	0.0323	0.1051	0.3096	1.0161	3.0139	4.5041
1.25	0.0118	0.0338	0.1083	0.3158	1.0267	3.0230	4.5068
1.35	0.0125	0.0353	0.1115	0.3220	1.0371	3.0319	4.5095
1.45	0.0132	0.0367	0.1147	0.3280	1.0473	3.0408	4.5122
1.55	0.0138	0.0381	0.1178	0.3339	1.0575	3.0496	4.5148
1.65	0.0145	0.0395	0.1208	0.3398	1.0674	3.0582	4.5174
1.75	0.0151	0.0408	0.1238	0.3455	1.0773	3.0668	4.5200
1.85	0.0157	0.0421	0.1267	0.3512	1.0870	3.0753	4.5225
1.95	0.0163	0.0434	0.1295	0.3568	1.0966	3.0836	4.5250
2.05	0.0169	0.0446	0.1324	0.3623	1.1060	3.0919	4.5275
2.15	0.0174	0.0459	0.1352	0.3677	1.1154	3.1001	4.5299
2.25	0.0180	0.0471	0.1379	0.3731	1.1246	3.1082	4.5323
2.35	0.0185	0.0483	0.1406	0.3784	1.1338	3.1162	4.5347
2.45	0.0191	0.0495	0.1433	0.3836	1.1428	3.1241	4.5371
2.55	0.0196	0.0506	0.1459	0.3888	1.1517	3.1319	4.5395
2.65	0.0201	0.0518	0.1485	0.3939	1.1605	3.1396	4.5418
2.75	0.0207	0.0529	0.1511	0.3989	1.1693	3.1473	4.5441

Source: Reproduced with permission from the CIE Ref. [7].

Table 6.1 shows a wide range of S/P ratios, for example, the yellowish HPS lamp exhibits a low S/P value (0.422) because it has a low blue content in its emission spectrum hence it causes low mesopic luminance values when its light is reflected from a gray road surface. On the other hand, the CW LED lamp has a high S/P value according to the high blue content in its emission spectrum (2.240) which results in a high mesopic luminance. Compare the mesopic luminance values with the corresponding photopic luminance values in the first row of Table 6.1.

6.2 Mesopic Brightness under LED Based and Conventional Automotive Front Lighting Light Sources

Visual experiments on the mesopic brightness perception of typical automotive headlamp spectra (H7 tungsten halogen, Xenon, and LED) [13] and on street lamp spectra [14] showed that "white" light sources (those without yellowish tones, e.g., certain fluorescent lamps (FLs) and CW LEDs) produced a significantly brighter lit environment than yellowish light sources, for example, HPS lamps.

The ratio (R) of the photopic luminance values of the two types of light sources after matching their brightness visually (see Figure 5.40) was significantly different from 1.00 indicating that photopic luminance cannot describe mesopic brightness. For example, the visually obtained mean ratio for a white FL and a HPS lamp was $R(\text{FL/HPS}) = 0.72$ [14]. This means that less photopic luminance was necessary for the same perceived brightness in the case of the FL.

The aim of this section is to present the method and the result of a visual brightness study comparing an LED car headlamp with its incandescent and Xenon counterparts [15]. The design of automotive front lighting light sources for safe nighttime driving represents a special application of the knowledge on the mesopic luminance range. In this application, the adaptation luminance ranges between $0.1\,\text{cd}\,\text{m}^{-2}$ (at 60–70 m from the car) and $5-7\,\text{cd}\,\text{m}^{-2}$ (not far away in front of the car).

Therefore, car drivers have to adapt to mesopic conditions but ECE and SAE regulations are based on photopic, that is, $V(\lambda)$ based photometry independent of the type of light source used [15]. As besides mesopic luminance, mesopic brightness is a possible quantity to represent the "operating point" of the object detection and object recognition mechanisms, a high mesopic brightness results in high visual performance in nighttime driving including high object detection probability, short reaction time, and good visual acuity.

As pointed out in Sections 5.6 and 6.1.2, this photometric system (especially the quantity photopic luminance) is unable to assess lights in terms of perceived brightness (neither in the photopic nor in the mesopic range) because of the number of visual mechanisms (including the chromatic mechanisms and the rod mechanism, see Figure 5.41) that are not incorporated in $V(\lambda)$.

Latter effect is sometimes referred to as brightness-luminance discrepancy. Its two reasons were pointed out above (chromatic contribution to brightness – the

Helmholtz–Kohlrausch effect – and the contribution of the rod photoreceptors – the Purkinje effect) and a model, the CIE brightness model [3], which results in the value of equivalent luminance L_{eq} to account for brightness was described (see Figure 5.42).

In the automobile lighting industry, we can observe the application of a number of light sources for front lighting. Since the beginning of the 1960s, we have had the halogen incandescent lamps ((ILBs, incandescent light bulbs)), since about 20 years the HID lamps with Xenon-Technology and now, we can see more and more LED headlamps on the roads. One aim of this chapter is to evaluate the earlier-mentioned brightness-luminance discrepancy and the CIE brightness model [3] for automotive front lighting light sources to bring the consequences of mesopic brightness evaluation into the focus of attention of lighting engineers for a correct evaluation of front lighting light sources [13].

The values of photopic luminance L_P and equivalent luminance L_{eq} are compared in the present section for those light stimuli whose brightness perceptions match when we analyze the results of the earlier-mentioned experiment [15]. For the design of car headlamps, a further interesting issue is interobserver variability and the identification of groups of observers of different brightness perception tendencies.

Today, it is known that there exist at least two types of observers according to their brightness perception a chromatic and an achromatic type [16]. The chromatic type has more contribution to perceived brightness from her/his chromatic visual mechanisms. Thus, the other aim of the present section is to describe this interindividual variability of mesopic brightness perception by establishing two groups (so-called clusters) of observers regarding how well the CIE supplementary system of photometry [3] works for different observers.

6.2.1
Experimental Method

During the earlier-mentioned experiment [15], subjects were sitting in front of a double-chamber light box in a dark room. The inner walls of the two separate chambers were coated with barium sulfate to produce homogeneous illumination. The light of the reference lamp, that is, the ILB often used as a car headlamp light source (H7: $x = 0.43$; $y = 0.41$; $T_{cp} = 3100$ K) was coupled into the right chamber. The reference light level was varied to achieve three luminance levels relevant for nighttime driving (0.1, 0.5, and 1.5 cd m^{-2}).

In the left chamber, two further widely used types of automotive light sources were used as test lamps: a HID Xenon lamp ($x = 0.40$; $y = 0.41$; $T_C = 3500$ K) and a white LED lamp ($x = 0.30$; $y = 0.28$; $T_{cp} = 6500$ K). Each of the 18 observers of normal color vision (10 young adults and 8 elderly observers) had to vary (spectrally aselectively) the photopic luminance of the left side (test) chamber by means of a mechanical iris stop to get the same mesopic brightness appearance as in the right chamber illuminated by the ILB H7 at the above three luminance levels. Each

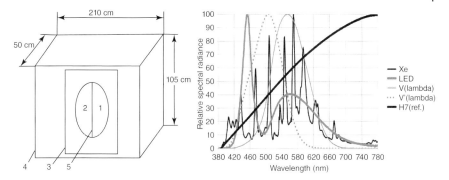

Figure 6.4 Left: setup for the mesopic visual brightness matching experiment. 1, Reference field (H7); 2, test field with Xenon or LED; 3, circular bipartite field (diameter: 55 cm, corresponding to 30.8° of viewing angle) with a reference stimulus (fix) and a test stimulus (to be adjusted by observers); 4, light box; 5, separator plate. The observer sat in front of the light box at a viewing distance of 100 cm. Right: relative spectral power distributions of the light sources together with the silhouettes of the $V'(\lambda)$ and $V(\lambda)$ curves. (Reproduced from Ref. [15].)

observation was repeated three times. The setup and the relative spectra of the two test light sources are shown in Figure 6.4.

As can be seen from Figure 6.4, for the case of the LED light source spectrum (at 6500 K), there is more overlap with $V'(\lambda)$, that is, with the rod (scotopic) sensitivity than for H7 (at 3100 K). This implies that the LED light source exhibits a higher scotopic to photopic (S/P) ratio than H7 which has a yellowish tone and a lower CCT.

6.2.2
Mean Results of Brightness Matching

First, the mean matching values of *photopic* luminance were analyzed for the two test light sources (i.e., L_p for Xenon and L_p for LED) and the three luminance levels of the H7 incandescent reference light source ($L_{ref} = 0.1$, 0.5, and 1.5 cd m^{-2}). For these conditions, the mean values of the ratios $R_{LP} = (L_p/L_{ref})$ were computed for all the 18 observers' three repetitions. Figure 6.5 shows the result.

As can be seen from Figure 6.5, the general tendency is that, to obtain the brightness match, less photopic luminance is required for both test light sources (LED and Xenon) than for the H7 reference light source, at all three luminance levels. The reason is the earlier-described brightness-luminance discrepancy. It can also be seen from Figure 6.5 that the LED light source needs less photopic luminance than the Xenon light source to be matched in perceived brightness with the incandescent reference lamp (H7). T tests showed the significance ($p < 0.05$) of brightness-luminance discrepancy. This can also be seen from the nonoverlapping or only slightly overlapping confidence intervals in Figure 6.5 (right).

The next step of the analysis concerned the values of equivalent luminance L_{eq} [3] intended to account for perceived brightness. The ratios $R_{Leq} = (L_{eq}/L_{ref,eq})$

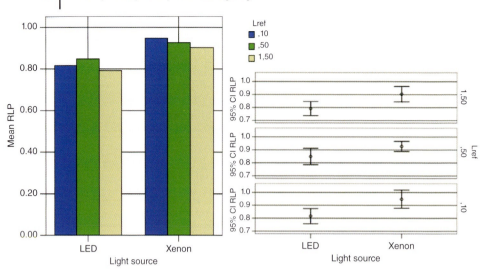

Figure 6.5 Left: Mean values (18 observers × 3 repetitions) of $R_{LP} = (L_P/L_{ref})$ for the two test light sources (LED and Xenon) and for the three H7 reference luminance levels ($L_{ref} = 0.1$, 0.5, and 1.5 cd m^{-2}). Right: 95% confidence intervals (CI) for R_{LP}. (Reproduced from Ref. [15])

were computed for all conditions similar to the previous analysis shown in Figure 6.5. L_{eq} denotes the values of equivalent luminance found by the subjects at their visual brightness match for the two test sources and $L_{ref,eq}$ denotes the values of equivalent luminance for the incandescent reference light source ($L_{ref,eq} = 0.12$, 0.53, and 1.54 cd m^{-2}). All L_{eq} calculations were made by using the CIE brightness model [3]. Figure 6.6 shows the mean matching values of equivalent luminance, similar to Figure 6.5.

In Figure 6.6, all values of R_{Leq} should be equal to 1.00 because the equivalent luminance computation method was introduced to predict perceptual brightness matches. As can be seen from Figure 6.6 (right), this goal was fulfilled for comparing LED and Xenon brightness. The confidence intervals of LED and Xenon overlap. This finding was also corroborated by a T test. But all values of R_{Leq} differ from 1.00 significantly. This indicates that a further improvement of the equivalent luminance computation method [3] is necessary, especially for higher mesopic light levels.

6.2.3
Interobserver Variability of Mesopic Brightness Matching

Figure 6.7 shows the interobserver variability of the mesopic brightness matching experiment in terms of the computed R_{Leq} values.

As can be seen from Figure 6.7, the CIE supplementary system of photometry [3] does work to some extent for some observers (Cluster 1 with $0.9 < R_{Leq} < 1.1$, observers Nos. 1, 6–10, 12–13, and 15–16), but the model predicts too low values of equivalent luminance for some other observers (Cluster 2 with $R_{Leq} < 0.9$,

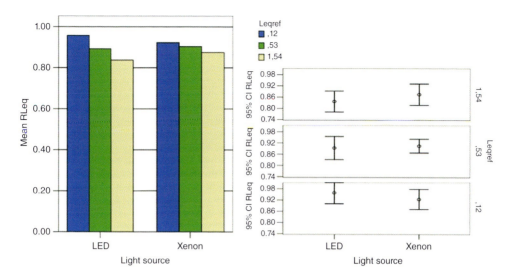

Figure 6.6 Left: mean values (18 observers × 3 repetitions) of $R_{Leq} = (L_{eq}/L_{ref,eq})$ for the two test light sources (LED and Xenon) and for the three H7 reference luminance levels ($L_{ref,eq} = 0.12$, 0.53, and 1.54 cd m^{-2}). Right: 95% confidence intervals (CI) for R_{Leq}. (Reproduced from Ref. [15])

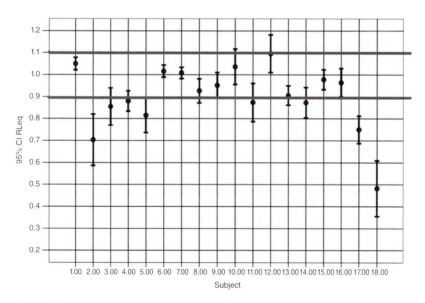

Figure 6.7 Mean values (2 test light sources × 3 mesopic luminance levels) of $R_{Leq} = (L_{eq}/L_{ref,eq})$ for the 18 subjects. (Reproduced from Ref. [15].)

352 6 *Mesopic Perceptual Aspects of LED Lighting*

Nos. 2–5, 11, 14, and 17–18). According to the definition of $R_{\text{Leq}} = (L_{\text{eq}}/L_{\text{ref,eq}})$, a value of $R_{\text{Leq}} < 0.9$ means that the equivalent luminance of the two light sources Xenon and LED is less than the equivalent luminance of the ILB H7. The reason is the *different* relative intensities of the photopic luminance, chromatic, and rod mechanisms involved in mesopic brightness perception in the visual systems of the *different* observers. But, of course, the model uses the *same* coefficients for *all* observers.

It is interesting to examine the interobserver variabilities of the three component mechanisms of the CIE brightness model [3] (see Figure 5.42 and Eqs. (5.15)–(5.17)): the rod mechanism $R = (L_S)^{1-a}$ ($L_S = L'$, the index S refers to "scotopic") the photopic luminance mechanism $P = (L_P)^a$ and the chromatic mechanism $C = 10^{(faC)}$. The relative intensities of R, P, and C computed from the brightness matching answers of the observers for the reference light source are depicted in Figure 6.8, as a function of the type of test light source and luminance level, for each observer separately.

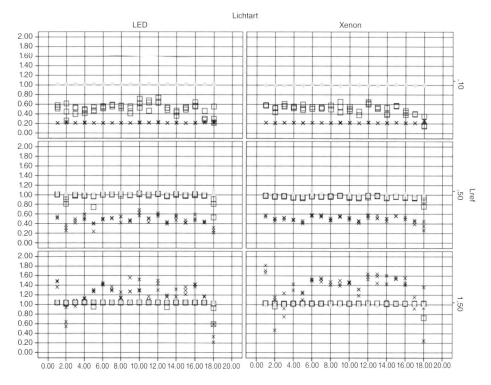

Figure 6.8 Ordinate: relative intensity of the rod (squares), photopic luminance (crosses) or chromatic (filled gray circles) mechanism computed from the brightness matching answers of the observers for the reference light source (R, P, and C, see text). All three repetitions of each observer are shown. Abscissa: observer Nos. (1–18). (Reproduced from Ref. [15].)

As can be seen from Figure 6.8, the rod mechanism $R = (L_S)^{1-a}$ is very changeable at the lowest luminance level (0.1 cd m^{-2}) but it is almost constant at the highest luminance level (1.5 cd m^{-2}). Just the opposite is true for the photopic luminance mechanism $P = (L_P)^a$. The chromatic mechanism $C = 10^{(f \cdot aC)}$ is the least variable mechanism according to the fact that the chromaticity of the test light source was fixed in this experiment. The interobserver variabilities of Figures 6.7 and 6.8 point toward using more than one (possibly two) "standard" observer for mesopic brightness prediction in the future. With different parameter values of the R, P, and C mechanisms for each "standard" observer it may become possible to reduce the interobserver variability of L_{eq} for the visual brightness matching condition.

6.2.4
Conclusion

At mesopic luminance levels between 0.1 and 1.5 cd m^{-2}, a HID Xenon lamp ($x = 0.40$; $y = 0.41$; $T_C = 3500$ K) required about 5–10% less photopic luminance than an ILB (H7: $x = 0.43$; $y = 0.41$; $T_C = 3100$ K) to match in perceived brightness (depending on the luminance level). The white LED lamp ($x = 0.30$; $y = 0.28$; $T_C = 6500$ K) required about 20% less photopic luminance. Thus the tendency is that if the CCT is higher (i.e., the light source provides more bluish light) then the light source looks brighter than predicted by its value of photopic luminance. The perceived brightness of the illuminated mesopic scene is an important parameter for nighttime traffic because it provides the "operating point" for the visual detection (reaction time), visual acuity and visual search tasks. Therefore, it is important to consider this result for the design of automotive front lighting light sources. As seen from the interobserver variability analysis, there exist two clusters of observers whose mesopic brightness perception is different. The CIE brightness model [3] works only for the observers of Cluster 1 reasonably.

6.3
Mesopic Visual Performance under LED Lighting Conditions

The aim of this section is to show application examples of the mesopic visual performance model of the CIE [7] described in Section 6.1.3 (see Eqs. (6.1)–(6.3)) for typical LED light sources. Mesopic luminance values were computed for the sample set of 34 selected typical market-based white LED light sources in Chapter 5 (Figure 5.25) at different luminance levels. Figure 6.9 shows the mesopic luminance values computed for the relative spectral power distributions of the 34 LEDs at 0.1, 0.5, and 1.5 cd m^{-2}, as a function of the S/P ratio of the LEDs (grouped after CCT: WW, NW, and CW).

As can be seen from Figure 6.9, mesopic luminance (L_{mes}) increases with increasing S/P ratio at all three luminance levels, $L = L_P = 0.1$, 0.5, and 1.5 cd m^{-2}.

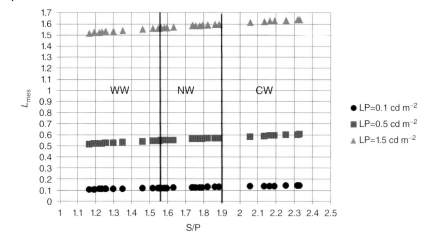

Figure 6.9 Mesopic luminance (L_{mes}) [7] computed for the relative spectral power distributions of the 34 LEDs of Figure 5.25 at $L = L_p = 0.1\,cd\,m^{-2}$, $0.5\,cd\,m^{-2}$, and $1.5\,cd\,m^{-2}$, as a function of the S/P ratio of the LEDs (grouped after CCT: WW, NW, and CW).

In lighting practice, it can be imaged that the light of the LED light source is reflected from the gray road surface and then these three (photopic) luminance values are measured using a conventional luminance meter. If a CW LED light source (e.g., S/P = 2.3) is used instead of a warm white LED light source (with e.g., S/P = 1.3) at the same photopic luminance value (e.g., $L_p = 0.1\,cd\,m^{-2}$) then higher L_{mes} values are obtained implying an enhanced mesopic visual performance under CW LEDs.

To get a better impression about the improvement of visual performance, mesopic luminance (L_{mes}) is divided by the corresponding photopic luminance (L_p) on the ordinate of Figure 6.10.

As can be seen from Figure 6.10, at $L_p = 0.1\,cd\,m^{-2}$ the gain of mesopic visual performance in terms of the L_{mes} value can be up to about 30% if a CW LED is used instead of a warm white LED, up to about 17% at $L_p = 0.5\,cd\,m^{-2}$ and up to about 10% at $L_p = 1.5\,cd\,m^{-2}$. As pointed out in Section 6.1, a gain of visual performance means that the detection contrast threshold decreases. Therefore, it becomes possible for the user of the LED light source (e.g., the car driver) to detect (dangerous) objects of lower contrasts (e.g., a pedestrian wearing a gray coat with gray trousers and walking on the roadside). Note that the exact *magnitude* of the decrease of mesopic detection contrast thresholds by increasing the value of the mesopic luminance of the background is currently unknown. Such an internationally endorsed and validated experimental dataset does not exist although experiments are underway at the time of writing.

Another question is how this L_{mes}/L_p ratio (a measure of the gain of mesopic visual performance by using white LEDs of different S/P values) depends on photopic luminance (L_p) which is measurable by conventional equipment at the time of writing. This dependence is shown in the sample computation of

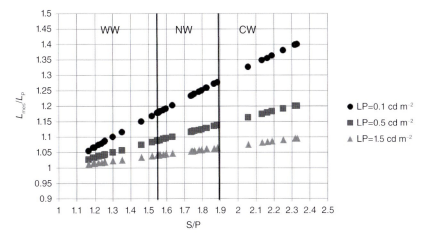

Figure 6.10 Mesopic luminance (L_{mes}) [7] divided by the corresponding conventional photopic luminance (L_p) for the 34 LEDs of Figure 5.25 at $L = L_p = 0.1$ cd m^{-2}, 0.5 cd m^{-2}, and 1.5 cd m^{-2}, as a function of the S/P ratio of the LEDs (grouped after CCT: WW, NW, and CW).

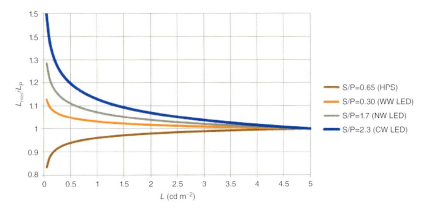

Figure 6.11 Dependence of the L_{mes}/L_p ratio on the photopic luminance (L_p) of four types of light sources: a typical warm white LED (S/P = 1.3), a typical neutral white LED (S/P = 1.7), a typical cool white LED (S/P = 2.3), and a type of high pressure sodium (HPS) lamp (S/P = 0.65).

Figure 6.11 where a typical warm white LED (S/P = 1.3), a typical neutral white LED (S/P = 1.7) and a typical CW LED (S/P = 2.3) are compared with a common light source in road lighting, a type of HPS lamp (S/P = 0.65).

As can be seen from Figure 6.11, the HPS lamp causes a loss of mesopic visual performance compared to the LED light sources and this loss increases at low mesopic luminance levels. The *improvement* of mesopic visual performance introduced by the LED light sources at each fixed photopic luminance level compared to HPS is more expressed for lower luminance levels, and this can be read from Figure 6.11 at least in terms of the L_{mes}/L_p ratio.

6 Mesopic Perceptual Aspects of LED Lighting

But this graph shows only the *improvement* and not the absolute visual performance in terms of numeric values of the mesopic detection threshold contrast (as mentioned before, such an internationally endorsed and validated model does not exist at the time of writing according to the best knowledge of the authors). Therefore, it is necessary to follow the minimum luminance values required by standards. For example, the EN 13201 standard requires the maintenance value of $L_p = 0.3$ cd m^{-2} for the lighting class ME6. According to Figure 6.11, if this value is ensured by illuminating the road by a CW LED (S/P = 2.3) then an enhanced mesopic visual performance can be obtained compared to a HPS street luminaire (S/P = 0.65).

Because a common conventional light source for road lighting is the HPS lamp, for today's lighting practitioners it may be of interest to use a type of HPS lamp with S/P = 0.65 as a standard lamp and compute the ratio of two photopic luminance values: (i) L of a test light source of a given S/P value and (ii) L_{HPS} with S/P = 0.65 for every relevant value of L_{HPS} (between e.g., 0.1 and 5.0 cd m^{-2}). The value of L is computed by using the criterion of *equal mesopic* visual performance with L_{HPS} in terms of L_{mes} [17], that is, $L_{mes}(L) = L_{mes}(L_{HPS})$. Figure 6.12 shows the result of such an analysis for the three typical S/P ratios of warm white, neutral white, and CW LEDs.

As can be seen from Figure 6.12, for example, while using a CW LED (S/P = 2.3), less photopic luminance is enough to ensure the same value of L_{mes}. For example, with a luminance level of 0.5 cd m^{-2} of a HPS-type relative spectral power distribution can be ensured by 23% less luminance if a CW LED light source is used instead of HPS. Another question is how much electric energy is saved: this is *strongly* influenced by the luminous efficacy of the light source, for example, a

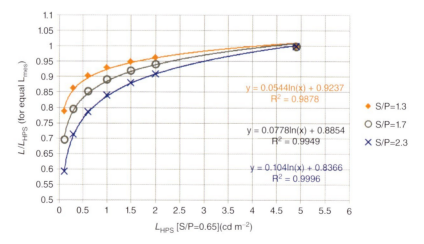

Figure 6.12 L/L_{HPS} for an equal value of mesopic luminance (L_{mes}). L: photopic luminance of a test light source (S/P = 1.3, 1.7, and 2.3) that yields the same L_{mes} value as a type of HPS lamp with S/P = 0.65 at different photopic luminance values (L_{HPS}) [17]. Logarithmic fit curves are also shown for the three typical S/P values of WW (S/P = 1.3), NW (S/P = 1.7), and CW (S/P = 2.3) LEDs.

certain type of HPS or a certain type of CW LED. Latter aspect is a complex technological issue.

As a closing remark, note that, as mentioned before, it is unknown how the change of the L_{mes} value of a mesopic background affects the detection contrast threshold for a low contrast target that happens to appear on it. Therefore, it is suggested to comply with current illumination standards for different situations (i.e., ensure the required luminance values) and try to illuminate with a (white) light source of a high S/P value in order to increase L_{mes} and, in turn, to increase mesopic visual performance.

6.4
Visual Acuity in the Mesopic Range with Conventional Light Sources and White LEDs

6.4.1
Introduction

The last decade of mesopic research was dominated by visual performance, reaction time, detection as well as brightness considerations. Several interesting effects besides the Purkinje effect could be identified that significantly influence the mesopic perception of light and thus the evaluation of illuminated scenes perceived by night. Even though there are clear tendencies, there is still no real progress in the area of standardization (except for the CIE publications [3, 7]) which would help transfer this scientific knowledge to applications like the evaluation of different headlamp and street light source spectra to efficiently use these findings in order to design better vehicle and street lighting. This is because of some still missing knowledge about the general understanding of human vision and the uncertainties that come along with this.

As mentioned in Section 6.1 already, visual acuity is the ability to recognize relatively small objects with high contrast. These small objects can be a group of so-called Landolt rings; observers have to recognize the position of the opening of the rings. Landolt rings exhibit abrupt light–dark edges with high spatial frequencies. If these rings are imaged on the retina then the ring openings can be recognized only by the foveal zone around 2° with a very high density of L cones and M cones. In other words, at the central fovea, the image can be scanned with a very high spatial sampling frequency. A Landolt ring is shown in Figure 6.13.

In automotive lighting, visual acuity plays an important role. During nighttime driving on the road, visual acuity is responsible for recognition, a component of visual performance, as described in Section 6.1 – as opposed to the detection task. In this task, objects (e.g., an animal, a stone, a tree, or a person) that are fixated (brought into the central fovea) are identified after being detected. For street lighting, visual acuity (at least in the sense as described below) is the characteristic parameter for face recognition of persons seen by night. According to the above reasons, a better visual acuity improves both objective traffic safety and the feeling of safety (in the sense of well-being of the subject) by night.

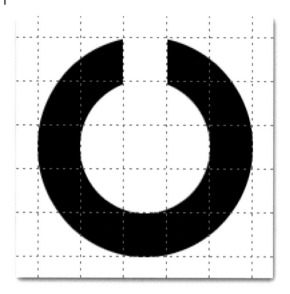

Figure 6.13 A Landolt ring used to characterize the mesopic recognition task.

Having the duty to improve visual performance in driving, several attempts were made from universities, headlamp manufacturers and LED companies to make use of the actual research and optimize the LEDs' spectral distributions towards higher "mesopic efficiency". To minimize the error of this optimization and contribute to a more complete understanding of mesopic vision, a visual experiment was conducted where an aspect of visual acuity (in a sense that is relevant for exterior lighting design, as described below) was determined under a set of five different light sources at mesopic luminances (LEDs, automotive HID lamps and halogen tungsten lamps).

Physiologically, the assumption was that, in terms of visual acuity, spectral sensitivity should not change because of the missing rods in the fovea. As there is actually no validated model devoted solely to visual acuity in the mesopic range depending on the light spectrum, a visual experiment [18] is described in this section to show the necessity of such a model.

6.4.2
Test Method

The idea behind this experiment [18] on the dependence of visual acuity on the type of light source spectral power distribution was to have a visual task which enables to compare visual performance under different light sources at several mesopic light levels. Typically, a so-called Snellen chart is used for visual acuity testing but this is the usual case for photopic luminances only. The specialty of mesopic vision is the transition and thus the parallel activity of cones and rods. The different densities and sizes of these photoreceptors on the human retina suggest that visual acuity will vary across different light levels. This would imply a test chart with a higher resolution than the one the Snellen chart offers to be able to distinguish the effect of luminance level from chromatic (spectral) effects.

Therefore, instead of marching on the way of conventional visual acuity investigations that use variable letter sizes going down to very small sizes (according to the classic definition of visual acuity), it was decided [18] to choose a fixed (and not very small) size (which corresponds better to the typical recognition tasks in exterior lighting) and vary the contrast between the object and the background. This is sometimes called a "letter contrast acuity experiment". Such an experiment is more easily applicable to the visual tasks that demand acuity in typical night-time driving or face recognition situations. This technique is common in neurophysiological investigations: it evaluates contrast discrimination [19].

6.4.2.1 Test Chart

The Pelli-Robson Contrast Sensitivity Chart was chosen to be the visual task as it is well known (there are several studies using this test). So it is possible to compare the results from the present study with previous studies. Using the recommended viewing distance of 3 m (see Figure 6.14) with the letter size of 48.5 mm, this condition results in an angle 2α which is slightly above 1°. This is equivalent to an object size of 0.8 m at a viewing distance of 50 m. This corresponds to a visual acuity of 0.1 being about one-fifth of typical photopic visual acuity. The chart consists of predefined letters in groups of three and decreasing contrast from group to group. During the test, a group is considered as recognized when two of three letters were named correctly.

As mentioned in Section 6.1, mesopic luminances are defined by the CIE to range between 0.001 and "some" $cd\,m^{-2}$. For nighttime driving the relevant luminances range from 0.1 up to $10\,cd\,m^{-2}$ [20], so the luminances were varied

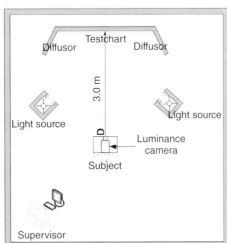

Figure 6.14 Left: Pelli Robson test chart used in the letter contrast acuity experiment [18]; right: experimental setup.

from 0.1 to 2 cd m^{-2} in this study [18]. A luminance of 0.1 cd m^{-2} corresponds to the luminance of a dry road surface at a distance of 50 m in front of the car headlamp with a tungsten halogen light source. For this test, a standard tungsten halogen lamp (H7), a neutral white LED with 4693 K often used as a neutral white LED for street lighting and as car headlamps as well as a HID (D2S) car headlamp were selected.

To have a greater variation in color temperatures, modified HID lamps were also tested, see Table 6.2 and Figure 6.15. As pointed out previously, there is a correlation between CCT (in K) and the S/P ratio (scotopic to photopic or, roughly speaking, rod signal to L+M signal ratio which is often used in mesopic models) and therefore it is interesting to study a wide range of CCTs. Homogeneity is achieved by using a 45°-symmetric spatial radiance distribution of illumination. Fifteen subjects took part in the study: 12 students (20–29 years) and three elderly observers (60–69 years).

Table 6.2 Colorimetric properties of the tested light sources on the Pelli Robson chart surface [18]. Note that the values in the light source labels (e.g. HID 6000 K) are just nominal CCTs. The real values are listed in the row CCT (K).

	Tungsten	HID 4300 K	HID 5000 K	HID 6000 K	LED
Chrom. x	0.4447	0.3888	0.3663	0.3578	0.3544
Chrom. y	0.4071	0.3968	0.3751	0.3506	0.3603
CCT (K)	2894	3928	4392	4523	4693
S/P ratio	1.437	1.585	1.803	2.044	1.753

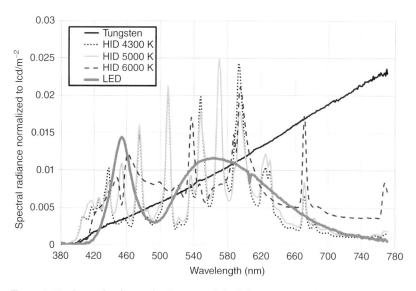

Figure 6.15 Spectral radiance distributions of the light sources used in the mesopic letter contrast acuity study [18] (normalized to 1 cd m^{-2}), see also Table 6.2.

Figure 6.16 Luminance image with the positions where contrasts were quantified.

The contrasts on the Pelli Robson test chart were measured by the aid of a luminance camera (see Figure 6.16) and then contrast values were calculated for each letter. There were only slight differences compared to Pelli Robson test chart specifications.

6.4.3
Letter Contrast Acuity Results

Results [18] showed increasing letter contrast acuity with increasing chart luminance as expected. Elderly subjects did not answer significantly different from younger observers (Figure 6.17).

Changing the light source (i.e., the scotopic to photopic ratio) by increasing CCT did not have a significant effect ($p = 0.41$) on the letter contrast acuity of the 15 observers. This leads to the important result that under these mesopic viewing conditions, letter contrast acuity was highly correlated with *photopic* luminance contrast, independent of illuminant chromaticity. These results do not imply a signficant rod effect (but it cannot be excluded based on these results). This finding matches the theoretical expectations because, during foveal observation ($2\alpha = 1°$) there are no rods involved and this produces an L+M cone like, i.e. $V(\lambda)$ like sensitivity function.

Concerning the application of mesopic research in automotive lighting and street lighting, this is a kind of bad news. This result (independence of illuminant

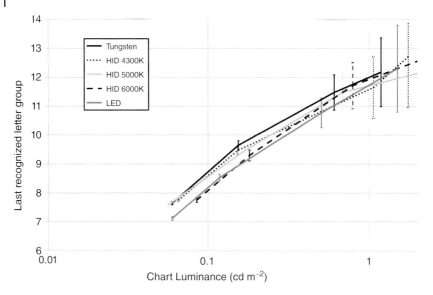

Figure 6.17 Mean values and standard deviations of the last recognized letter group (Figure 6.14) for the tested lamp types (Figure 6.15).

chromaticity in the mesopic range) suggests that mesopic models described in Sections 6.1–6.3 of this book do not apply for letter contrast sensitivity and cannot be applied to object recognition at foveal vision. This inevitably complicates the process of mesopically optimized spectra for automotive applications but shows that even in the deep mesopic regions, $V(\lambda)$ based light source judgments will never be obsolete, as far as the object recognition task is concerned.

6.5
Detection and Conspicuity of Road Markings in the Mesopic Range

6.5.1
Introduction

Road markings are a part of street equipment and are defined as colored markings on the road surface which have – depending on their application – different types, colors, and forms. The overall aim of road markings is to increase traffic safety for road users both by day and by night [21]. This effect is mainly based on a continuous visual information flow about the course of the street caused by a luminance and chromaticity contrast between the road marking and the usually darker road surface. During the day, the sun illuminates the road markings as well as the road surface with diffuse radiation.

At night, car headlamps and, in cities, street luminaires are the dominating light sources which emit directed light toward the street and its surroundings. Because of the retroreflective feature of road markings, light is mainly reflected back to the direction of the headlamps and the driver's eyes. In urban areas this effect is supplemented by the optical radiation of street lights. The part of reflected light which reaches the eyes causes a luminance contrast between the object (markings) and background (road surface).

Road markings were introduced in Germany in the early twentieth century and reached a high stage of development at the end of that century. In 2006 more than 70% of German roads were equipped with road markings of Type II with a higher nighttime visibility on wet roads [22]. Important standards for quality control of road markings in Germany include

- For materials: DIN EN 1423/DIN EN 1424;
- For the requirements on road markings: DIN EN 1436;
- For additional terms of contract: ZTV M 02.

These standards deal with luminance values and not with the spectral power distributions of the illuminating light sources. Road marking research of the last decades dealt mainly with the luminance contrast between road markings and the road surface. But the impact of different light spectra on visual conspicuity of road markings was not considered [23, 24].

As mentioned earlier in Chapter 6, for automotive front lighting, typical luminance values vary between about 2 and 10 cd m^{-2} at the front of the car. Most of the time, the driver focuses attention to a distance of 60–70 m in front of the car with an average luminance between 0.1 and 0.2 cd m^{-2}. The German DIN EN 13201 for street lighting requires that main roads with a high volume of traffic should have an average photopic luminance of max. 2.0 cd m^{-2}. Therefore, as pointed out earlier, viewing conditions in the automotive and street lighting belong to the mesopic range of vision.

Up to 1990, only tungsten and tungsten halogen lamps were available for automotive front lights. Therefore, there was no need to think about the influence of a light source's spectrum (in the sense of its chromaticity) on the *conspicuity* of road markings. In 1991, the first car with high intensity Xenon discharge headlamps was introduced, having a previously unexperienced spectral power distribution. In 2007, the first cars with LED headlamps were presented and a third light source, possibly with even higher S/P values, entered the market. This technological development and the better understanding of mesopic vision in the past 10–15 years (see e.g., [25, 26]) lead to the following question: What is the impact of spectral power distributions (in the sense of varying illuminant chromaticities) under mesopic vision conditions on the contrast perception and the conspicuity of road markings? (Refer to the fundamentals of mesopic vision in Section 6.1).

6.5.2
Spectral Characterization of Light Sources and Road Markings

This section describes the spectral characteristics of the different light sources and road markings which were used for the contrast calculations [21].

6.5.2.1 Automotive Light Sources

To take a look at automotive headlamps, three types of light sources are available. Their typical spectral power distributions were shown in Figure 6.4: a tungsten halogen lamp (CCT = 3100 K), a high intensity discharge lamp (CCT = 3500 K), and a white LED (CCT = 6500 K). The spectrum of the tungsten halogen lamp has a relatively low contribution in the wavelengths from 380 until 480 nm and a higher part in the orange and red wavelength region from 580 until 780 nm.

The Xenon HID lamps have a spectrum with many strong peaks and gaps in the visible range. Although the HID lamp has a relatively high luminous efficacy if weighting with the $V(\lambda)$ function, these gaps in the important regions dominating vision in the mesopic range indicate that this lamp type may not be efficient from the point of view of mesopic visual performance. But concerning the CW LED spectrum, especially in the range between 430 and 480 nm, this spectrum fits to the maximum of the mesopic detection mechanisms (see Figure 6.2).

6.5.2.2 Light Sources for Street Lighting

In Figure 6.18 the relative spectral radiance distributions of the light sources mainly used in current street lighting practice in Europe are illustrated.

As can be seen from Figure 6.18,

- High pressure mercury lamps have a distinct emission peak at 435 nm, while there is nearly no emission between 440 and 540 nm;
- HPS lamps have a high photopic efficacy because the major part of emission is between 560 and 660 nm (yellow–red);
- Lamps with fluorescent tubes have a spectral emission comparable to high pressure mercury lamps;

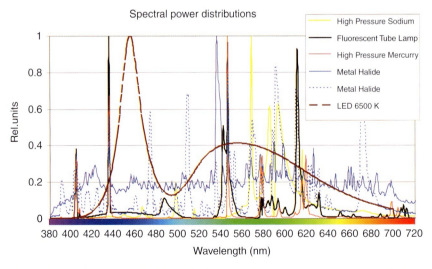

Figure 6.18 Relative spectral radiance distributions of typical light sources of street lighting.

- Cold white LEDs commonly have a local maximum between 450 and 460 nm and a second one at about 560 nm, coming from the yellow YAG-phosphor.

6.5.2.3 Spectral Characteristics of Road Markings

To characterize road markings, DIN EN 1436 requires special geometries and a measuring distance of 30 m. The geometries for light source, road marking, and observer are fundamentally different for automotive lighting and for street lighting. The aim of this section is a relative comparison among the different light sources assuming that the spectral characteristics of road markings are nearly constant for each viewing angle. Therefore the spectral characteristics of a set of road markings were measured as follows:

- Spectral calibration of the measurement setup consisting of a spectroradiometer for the absolute spectral radiance (in $W\,(m^2\,nm\,sr)^{-1}$ units), a tungsten halogen headlamp with a continuous spectrum and an ideal white diffuse reflecting material (barium sulfate) at a distance of 4 m (see Figure 6.19);
- Repetitions of the measurement using just road marking samples instead of the barium sulfate;
- Calculation of the relative spectral radiance factors $\beta(\lambda)$ for each road marking sample.

The following white road marking samples available on the market and used mainly in road technology were characterized in an uncorrupted new state:

- 3M A 340;
- 3M A 380 SD;
- 3M A 540;
- Berlack MF 30;
- Berlack MF 30 (with additional glass beads); and
- Premark 662180.

Figure 6.20 shows the spectral radiance factors $\beta(\lambda)$ for four selected retroreflective road marking samples.

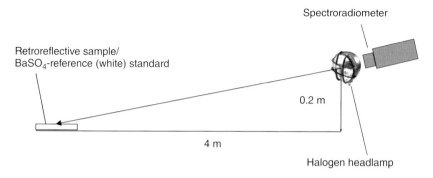

Figure 6.19 Scheme of spectral characterization of road marking samples.

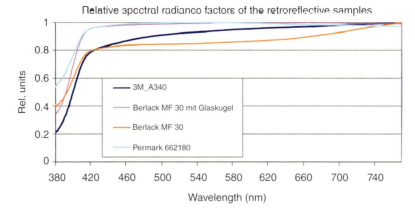

Figure 6.20 Spectral radiance factors $\beta(\lambda)$ of four selected road marking samples.

6.5.3
Contrast Calculations

The following measures were computed to characterize the contrast between the road markings and their background.

Contrast effective radiance: When radiation from a headlamp or a street light hits the road surface and the road marking, a part of the light is reflected towards the observer's eyes. A possible quantity to characterize this contrast is the so-called contrast effective radiance, $L(\text{contrast}, i)$ which can be calculated as shown in Eq. (6.4).

$$L(\text{contrast}, i) = \int L(\lambda, i) \cdot \beta(\lambda) \cdot K(\lambda, \alpha) \cdot d\lambda \qquad (6.4)$$

with:
$L(\lambda,i)$: spectral power distribution of the lamp i
$\beta(\lambda)$: spectral radiance factor of the road marking sample
$K(\lambda,\alpha)$: mesopic spectral sensitivity function for detection, for eccentricity α, see Figure 6.2

Photopic luminance: In order to have a correct comparison among the different light sources, each calculation has to be normalized to photopic luminance. The photopic luminance $L(v, i)$ of the road marking is calculated by using the $V(\lambda)$ function, see Eq. (6.5).

$$L(v, i) = 683 \ (\text{lm W}^{-1}) \int L(\lambda, i)\beta(\lambda)V(\lambda) \cdot d\lambda \qquad (6.5)$$

with:
$L(v,i)$: photopic luminance;
$L(\lambda,i)$: spectral power distribution of lamp i;
$\beta(\lambda)$: spectral radiance factor of the road marking samples;
$V(\lambda)$: spectral luminous efficiency function for daytime vision.

Luminance based contrast effective radiance: Eq. (6.4) divided by Eq. (6.5) results in the quantity called *luminance based contrast effective radiance* which is denoted by CL(i, α), see Eq. (6.6). CL(i, α) represents the ability of a light source to create a contrast perception for a constant luminance level.

$$CL(i, \alpha) = \frac{[683(\text{lm W}^{-1}) \cdot L(\text{contrast}, i)]}{L(v, i)} \quad (6.6)$$

with:
L(v,i): photopic luminance

6.5.4
Results and Evaluation

Tables 6.3 and 6.4 show the calculated results of luminance based contrast effective radiance CL(i, α) for two road marking samples, as examples, for 3M A 340 (with automotive headlamp light sources) and for Premark 662180 (with street light sources). In each cell of Tables 6.3 and 6.4, two values are shown. The first value is the calculated absolute CL(i, α) value. The second value is a relative value traced back to the reference light source which was the tungsten halogen H7 (for the case of automotive headlamps) and the HPS lamp (for the case of street lights), with a value of 1 corresponding to 100%.

Table 6.3 shows that there is nearly no difference between the tungsten halogen and the high intensity discharge headlamp concerning the prediction of contrast perception. But the tested headlamp with cold white LEDs is able to gain a contrast perception which is nearly 19% higher compared to both tungsten halogen and HID headlamps. This is of special interest because several dangerous objects (obstacles) in the visual field of the car driver can be seen under eccentricities of 10–20° (e.g., at a bend, at intersections or objects and persons being in the gap between two parking vehicles at the left or right roadside).

Because of their high photopic luminous efficacy, HPS lamps are one of the favored light sources for street lighting in major parts of America, Europe and Asia. The calculations show that – by far – all other tested light sources generate a higher contrast perception, see Table 6.4. While the contrast of road markings

Table 6.3 Luminance based contrast effective radiance CL(i, α) for automotive headlamps and the road marking sample type 3M A 340; reference: tungsten halogen lamp H7 (CL = 1,0). First value: CL(i, α); second value: relative to the reference.

Light source	Eccentricity		
	5°	10°	20°
Tungsten halogen H7	1.17/*1.0*	1.07/*1.0*	1.16/*1.0*
Discharge lamp (D1)	1.14/*0.98*	1.07/*1.0*	1.12/*0.97*
LED 6500 K	1.23/*1.05*	1.27/*1.19*	1.30/*1.12*

Table 6.4 Luminance based contrast effective radiance CL(i, α) for street lamps and the road marking sample type Premark 662180; reference: high pressure sodium lamp (CL = 1.0). First value: CL(i, α); second value: relative to the reference.

Light source	Eccentricity		
	5°	10°	20°
High pressure sodium	1.05/*1.0*	0.80/*1.0*	0.89/*1.0*
Fluorescent tube	1.19/*1.13*	1.10/*1.38*	1.19/*1.33*
High pressure mercury	1.15/*1.09*	0.99/*1.24*	1.08/*1.21*
Metal halide 1	1.21/*1.15*	1.20/*1.5*	1.24/*1.39*
Metal halide 2	1.16/*1.10*	1.05/*1.31*	1.12/*1.26*
LED 6500 K	1.26/*1.2*	1.30/*1.63*	1.33/*1.49*

illuminated with high pressure mercury lamps is 9–21% higher compared to HPS lamps, cold white LEDs are even up to 63% better.

Regardless of other aspects (e.g., glare), the results of this section show that white LEDs, especially those with higher CCTs, have the potential to gain a significant benefit in aspects of contrast perception for road markings. Driving tests in real traffic with LED headlamps showed comparable results for road markings [27]. Considering the fast technological development and the increasing market share of LEDs in traffic and street lighting, standards, and regulations should take the impact of the spectral power distributions (in terms of light source chromaticities) more into account.

6.6
Glare under Mesopic Conditions

6.6.1
Introduction: Categories of Glare Effects

The human visual system consists of the eye's optical system, the pupil aperture, the photoreceptors on the retina and the signal processing units at different stages along the pathway to the brain. After adaptation to a certain predominating luminance level, it is able to process visible radiation within a wide range of intensities: over twelve decades – from darkness to sunlight and skylight. But if radiation sources of very different intensities are present in the field of view *at the same time*, the human visual system can process luminance values within approximately three decades only. In Figure 6.21, luminance ranges for adaptation, for glare and for black shadow are shown.

As can be seen from the example of Figure 6.21, the car driver adapts to the average luminance of the road at about 25 m in front of the car with $L_A = 1.0 \, \text{cd m}^{-2}$ and cannot detect the objects with an object luminance less than

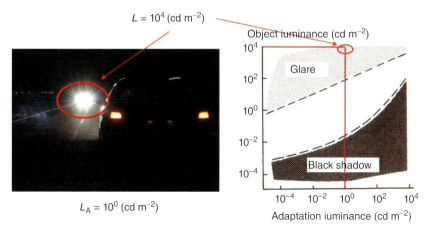

Figure 6.21 Luminance ranges for adaptation, for glare and for black shadow. (According to Hopkinson and Collins [28].)

$L_O = 0.01 = 10^{-2}$ cd m^{-2}. If a light source in the visual field of the driver with more than about 70–100 cd m^{-2} occurs, this light source can cause a glare effect.

Vos categorized glare effects into eight different phenomena [29] from which only four forms are relevant for street lighting and automotive lighting applications:

1) *Saturation glare* takes place in a very bright large visual field and to avoid this, low transmitting glasses (sun glasses) can be worn. This phenomenon occurs during longtime driving in sunny regions or making winter sport in a snow hill region.
2) *Adaptation glare* is the phenomenon if a human subject adapts for a long time to a certain adaption level, and, hereafter, the visual field changes luminance in a sudden manner. This is the case if for example, car drivers pass for a long time in a long tunnel and are glared at leaving the tunnel exit by experiencing suddenly bright skylight in front of their car.
3) *Disability glare* is the glare form that leads to the loss or the reduction of contrast between the surroundings and the object or object group to be detected. This glare is the subject of the next sections.
4) *Discomfort glare* does not disable the reduction of contrast of the objects in a direct way. When being glared by light sources within a short time repeatedly or over being glared continuously over a long period, for example, during driving along a country road with many opposing traffic cars that glare the driver again and again, concentration is reduced, distraction is increased and, finally, the ability to process the optical signals in the visual field in front of the car is diminished. This results, in turn, to an increased probability of causing a traffic accident. From this consideration, the two forms of glare, disability glare and discomfort glare are two results of the same visual stimulus, the same

visual phenomenon. Hence, the two forms of glare must not be separated in a practical analysis.

Glare is the main topic in the context of discussions of experts and traffic participants in street and automotive lighting applications. In the time between September 2009 and February 2010, a field study was conducted to evaluate the benefit of headlamp cleaning systems on detection ability and glare [30]. In this framework, Söllner et al. interviewed 419 drivers and asked two main questions about glare perception:

1) Do you feel glared by opposing traffic?
   a. (a) If so, how often?
   b. (b) When you feel glared, how intense is this feeling?
2) Do you feel glared by subsequent traffic?
   a. If so, how often?
   b. When you feel glared, how intense is this feeling?

Results of this glare interview are shown in Table 6.5.

As can be seen from Table 6.5, 419 drivers gave valid answers, 133 of them had a car with Xenon HID headlamps (bold numbers in the table) and 286 of them had a car with tungsten halogen headlamps (normal numbers). For the case of the HID headlamp, fewer drivers had a feeling of being glared by opposing traffic and also by subsequent traffic. This fact can be partly explained by the higher luminous flux on the road in case of HID headlamps which enables a higher state of adaptation. Gender differences among the drivers were also investigated. Women had a stronger glare perception, see Table 6.6.

6.6.2
Causes and Models of the Glare Phenomenon

6.6.2.1 Disability Glare

Optical radiation from different light sources in the viewing field passes the cornea and the other skin layers of the eye and is scattered by light scattering particles in the cornea, the eye lens and in the liquid chambers causing a veil, a more or less homogenous stray light distribution on the retina, see Figure 6.22. At and near the position of the images of the glare light sources on the retina, there is a dense Gaussian distribution of light intensity.

Table 6.5 Results of a glare interview [30]: 419 drivers gave valid answers, 133 of them had a car with Xenon HID headlamps (bold numbers); 286 of them had a car with tungsten halogen headlamps (normal numbers).

Traffic situation	Yes	No
Glared by opposing traffic?	**51%**/66%	**49%**/34%
Glared by subsequent traffic?	**27%**/54%	**73%**/46%

Table 6.6 Glare perception differences between men and women.

Traffic situation	Men (%)	Women (%)
Glared by opposing traffic?	61	78
Glared by subsequent traffic?	41	63

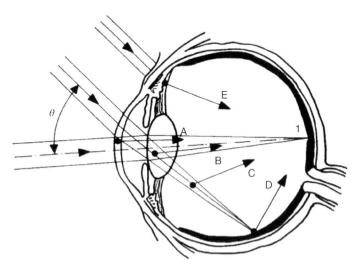

Figure 6.22 Stray light in the eye media that cause a reduction of contrast [31].

If the background of the light sources is relatively dark and there is no stray light then the retina should have an illuminance that causes all the receptors to be in a low mesopic or even scotopic adaptation state, except for the locations of the light source images with strong retinal illuminance so that the receptors at those retinal locations are forced to be in a high photopic local adaptation. With the additional contribution of stray light veil, the adaptation state of the receptors on the retina can be shifted toward different mesopic levels depending on the amount of stray light.

The effect of glare on contrast perception was investigated from 1926 until 2005 both in static and dynamic experiments in the laboratory and in real road conditions (test tracks) from 2007 on at the Technische Universität Darmstadt. Experimental setups were similar in all studies. Figure 6.23 shows a typical arrangement with the definitions of the most important glare parameters while Figure 6.24 demonstrates the impact of stray light on object detection.

As can be seen from Figure 6.24, if the glare source is absent or switched off, the object of viewing angle α can be detected at the background luminance L_{u1} if it exhibits a luminance difference ΔL_1 in comparison to the background luminance. When the glare source appears, the stray light content in the eye causes a retinal veiling illuminance being equivalent to the background luminance L_{u2} so

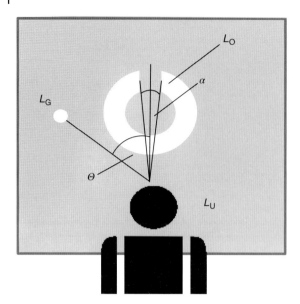

Figure 6.23 Geometry of a typical experimental setup used in glare studies [32]. 1: test object subtending the viewing angle α and with an object luminance L_o; 2: glare source with a viewing angle of β and with an eccentricity angle (glare angle) from the line of sight (Θ); L_U: background luminance and $\Delta L'$ luminance difference between the object and the background $\Delta L' = L_o - L_u$; E_B: glare illuminance (in lx) measured at the eye of the observer.

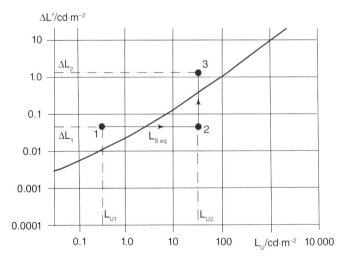

Figure 6.24 Glare (i.e., stray light in the eye) causes an increase of L', that is, an increase of luminance difference necessary to detect an object [31]. Continuous black curve: detection threshold luminance difference. L_U: background luminance on which the object to be detected appears.

6.6 Glare under Mesopic Conditions

Table 6.7 Observer age factor K in Eq. (6.8).

Years of age	Age factor K
20	6.3
30	9.2
40	12.1
50	15.0
60	17.8

that the former luminance difference ΔL_1 is below the detection threshold luminance difference curve (continuous black curve in Figure 6.24) and the object cannot be detected any more. In this case, the luminance of the object must be adjusted and increased to a new value so that the luminance difference is above the necessary threshold along the detection threshold curve. The equivalent veiling luminance $L_{s,eq} = L_v$ is the consequence of stray light from the glare source. This can be expressed in the form as shown in Eqs. (6.7)–(6.8).

$$L_{seq} = L_{u2} - L_{u1} \tag{6.7}$$

$$L_{S,eq} = K \cdot E_B \cdot \Theta^{-n} \tag{6.8}$$

In Eq. (6.8), E_B is the illuminance caused by the glare source, Θ is the viewing angle of the glare source and the factor K takes observer age into account, see Table 6.7.

Equation (6.8) was established by Holladay in 1926 and is currently used in most street lighting standards. It is valid in the range 1–30° of glare source angles measured from the line of sight. The K factor considers the fact that the elderly subjects have more disordered optical media in their eyes that cause more scatter than in case of young subjects. Until 1984, several studies were conducted and the form of the Holladay formula as well as the magnitude of its parameters did not change. In 1984, Carraro [33] found a new formula as shown in Eq. (6.9).

$$L_V = K \cdot \sum E_n^{0.58} \cdot \Theta_n^{-0.75} \tag{6.9}$$

The exponential function form in Eq. (6.9) is different from the Holladay formula (Eq. (6.8)). In 2002, the CIE published the so-called CIE equations for disability glare [34], see Eq. (6.10).

$$L_v = \sum \frac{10 \cdot E_n}{\Theta^3} + E_n \left(1 + \left(\frac{A}{62.5}\right)^4\right) \cdot \left(5\frac{E_n}{\Theta^2} + 0.1\frac{p}{\Theta}\right) + 0.025p \tag{6.10}$$

The parameters of Eq. (6.10) have the following meaning:

p: degree of eye pigmentation;
L_v: equivalent veiling luminance (cd m^{-2});
A: observer age;
E_n: illuminance at the eye from the nth glare source (lx);
Θ: angle of the nth glare source measured from the line of sight (in degrees).

Applying the three glare formulae (i.e., the Holladay, Carraro and CIE formulae) to a real traffic situation with two opposing cars driven along two parallel lanes with constant headlamp luminous intensity as depicted in Figures 6.25 and 6.26 illustrates the dependence of veiling luminance on the distance of the two cars.

As can be seen from Figure 6.26, veiling luminance as a function of car distance is constant according to the Carraro formula while it is increasing according to the CIE formula. This contradiction leads to certain considerations. Besides the glare parameters age factor, illuminance at the eye and glare angle, some other parameters can also be of decisive influence:

- Luminance of the glare source;
- Glare dose as a product of glare illuminance exposure time;
- Spectral distribution of the glare source (see Sections 6.6.3 and 6.6.4).

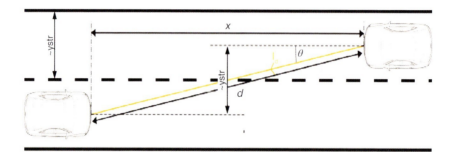

Figure 6.25 Geometry of a typical glare situation in nighttime driving: two cars approaching from opposite directions [35, 36].

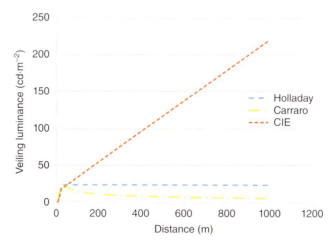

Figure 6.26 Veiling luminance as a function of car distance in a real traffic situation with two opposing cars [32].

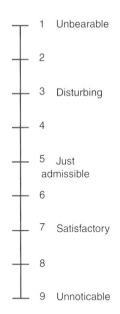

Figure 6.27 Discomfort glare rating scale according to de Boer [37].

6.6.2.2 Discomfort Glare

While disability glare can be characterized by the luminance difference between the object and its background in the presence of intensive, that is, glaring light sources and therefore by the loss of contrast between object and background, discomfort glare cannot be measured in such a direct way. However, the feeling of glare, the loss of concentration and the distraction from traffic events can be expressed as a psychophysical scale. In the past, several such scales were constructed but in the past 40 years, the so-called de Boer rating scale has mostly been used [37], see Figure 6.27.

In a comprehensive study in 1974, Schmidt-Clausen and Bindels [38] found a relationship between discomfort glare and the parameters of disability glare, see Eq. (6.11).

$$W = 5 - 2\lg \frac{E_{BL}}{\frac{0.003\,lx}{\min^{-0.46}} \cdot \left(1 + \sqrt{\frac{L_A}{0.04\,\frac{cd}{m^2}}}\right) \cdot \theta^{0.46}} \qquad (6.11)$$

The parameters of Eq. (6.11) are defined as follows:

W: de Boer discomfort glare rating scale;
$E_{BL} = E_b$: glare source illuminance at the eye;
$L_A = L_U$: adaptation luminance;
θ: glare angle in minutes of arc.

According to the best knowledge of the authors, Eq. (6.11) is the only formula in this context of glare science. But in a dynamic field test with modern measuring systems and with cars equipped with modern LED headlamps carried out at the Griesheim Airfield (Hessen, Germany), this formula could not be confirmed [35] so that further experimental studies are needed to clarify this aspect. From the above considerations, it can be concluded that the glare research of the past 90 years leaves many important questions open. In the next sections (Sections 6.6.3 and 6.6.4), the spectral sensitivity of discomfort and disability glare (determined in recent experiments) are presented. This is an important aspect for the spectral engineering of LED light sources.

6.6.3
Discomfort Glare under Mesopic Conditions – Spectral Behavior and Mechanisms

6.6.3.1 Introduction

As mentioned in Section 6.6.2, discomfort glare plays an important role in many practical applications including nighttime driving as it causes visual discomfort and visual fatigue on a long-term basis. Sometimes, this effect takes place without compromising visual performance immediately. Thus the aim of Section 6.6.3 is to analyze the spectral sensitivity of discomfort glare and interpret these results in terms of retinal mechanisms, compare with Figure 6.2. A typical angular size of 0.3° was chosen in this experiment that represents a typical value for a reflector headlamp of an oncoming car observed from 50 m distance on a typical background luminance of $L_B = 0.1$ cd m^{-2} and at a retinal eccentricity of 4.3° to the left of the fixation point where oncoming car headlamps tend to appear. This angular size of 0.3° also corresponds to a size of 16 cm of an LED module of an LED street lighting luminaire observed from 30 m distance.

Fekete et al. [39] also investigated spectral discomfort glare sensitivity. In their experiment, QM glare sources of 2° angular subtense at a background luminance level of less than 0.01 cd m^{-2} were viewed. They used the de Boer value of 3 which means "disturbing." But the present authors think that the setting of the de Boer value of 5 (just admissible) by the observer is easier to obtain experimentally. Also, the value of 5 is more relevant for automotive and street lighting applications than the value of 3. The value of 5 produces noticeable discomfort but it is just tolerable. For real automotive lighting applications, an adaptation of 0.1 cd m^{-2} is more relevant than a dark environment with less than 0.01 cd m^{-2} because of the luminance of a standard halogen tungsten headlamp on the road at a distance of 50–60 m from the car.

Bullough [40] analyzed several discomfort glare datasets by fitting a linear combination of $V_{10}(\lambda)$ and $S(\lambda)$ to the experimentally obtained spectral sensitivity function. In the present section, however, a more extended set of mechanisms $V(\lambda)$, $V'(\lambda)$, $S(\lambda)$ and a chromatic mechanism, $|L(\lambda)-M(\lambda)|$, were used. Here, $L(\lambda)$, $M(\lambda)$, and $S(\lambda)$ represent the spectral sensitivity functions of the long, middle, and short wavelength sensitive human cone photoreceptors, respectively (see Figure 6.2).

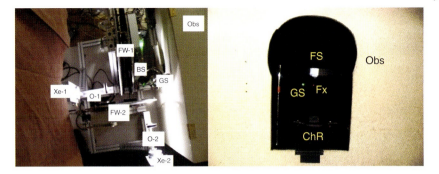

Figure 6.28 Experimental apparatus of the discomfort glare experiment, see text. (Reproduced with permission from the CIE Ref. [41].)

6.6.3.2 Experimental Method of a Discomfort Glare Experiment

In the discomfort glare experiment to be described here in detail [41], the apparatus consisted of the following parts (see Figure 6.28): two filter wheels (FW-1 and FW-2, only FW-1 was used) with 10 nm half-band width (HBW) interference filters, two optical channels (O-1 and O-2, only O-1 was used) each producing a QM beam from the two Xenon short arc high pressure lamps (Xe-1 and Xe-2, only Xe-1 was used), a beam splitter (BS, not used in the present experiment), a 0.3° aperture serving as the glare source (GS) of the experiment, a glare observation booth (Obs, interior: black) illuminated by four dimmable white LEDs mounted to the inside of the front cover to set adaptation luminance (not shown in Figure 6.28.), a forehead support (FS) and a chin rest (ChR) for the subject.

The spectral radiance of the glare source was measured by the 0.1° measuring field of a Konica Minolta CS2000A spectroradiometer looking into the booth and measuring in the middle of the 0.3° glare source at the position of the subject's left eye. The radiance of the glare source could be adjusted by the observer controlling a motorized continuous neutral wedge filter rotating before the glare source. Digital values d of the wedge were read by an electronic rotation encoder and converted into transmittance. Then, the spectral radiance of any of the 23 QM glare sources (with central wavelengths ranging between 430 and 650 nm) could be computed at any value of d obtained by the subject.

The subject was looking into the booth by fixating at the red blinking fixation LED (Fx in Figure 6.28) at 4.3° to the right of the glare source. Blinking was used to maintain the direction of fixation while adjusting glare source radiance until the value of 5 (just acceptable) was obtained on the de Boer scale. Adjustment was made by the aid of the "plus" or "minus" buttons in an iterative procedure starting from a low radiance. Adaptation to the mesopic background ($L_B = 0.1$ cd m^{-2}, S/P = 2.1) was carried out binocularly, at least 20 min at the beginning and at least 5 s between consecutive central wavelengths. Six subjects of normal color vision took part in the experiment. All of them were experienced discomfort glare observers. Observers completed three repetitions of all 23 glare sources except SB who completed two repetitions.

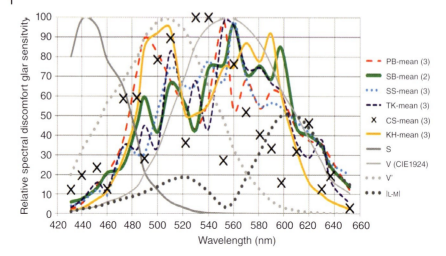

Figure 6.29 Spectral discomfort glare sensitivities of the individual observers. S, V, V', and |L−M|: S cone fundamental with a maximum at about 440 nm, the V (luminance mechanism) and V' (rods) functions as well as the |L−M| chromatic (opponent) mechanism. (Reproduced with permission from the CIE Ref. [41].)

6.6.3.3 Results and Discussion of the Discomfort Glare Experiment

Spectral discomfort glare sensitivities of the individual observers are shown in Figure 6.29.

As can be seen from Figure 6.29, minor local maxima are present at 450 and 473 nm for subjects CS, TK, and SS, possibly caused by the S (short wavelength sensitive) cones, see the S-curve of Figure 6.29. There is a local maximum at 490 nm for PB, SB, and TK and there is a local maximum at 510 nm for KH, TK, SS, and SB. The last maximum is a rod maximum, see the V' (rod sensitivity) curve in Figure 6.29. There is a peak at 550–560 nm for the observers PB, TK, SB, and SS. This is the luminance mechanism, see the V curve of Figure 6.29. There is a local maximum at 590–598 nm for KH, SB, PB, TK, SS interpreted by the chromatic |L−M| mechanism, see the |L−M| curve in Figure 6.29. Finally, there is a minimum at 520–540 nm for all subjects. In Figure 6.30, the average spectral sensitivity curve for all subjects and the $V(\lambda)$-function are illustrated.

Summarizing the above findings, it can be seen from Figures 6.29 and 6.30 that – in addition to the luminance mechanism characterized by $V(\lambda)$, the standard luminous efficiency function – three other retinal mechanisms (S, V', and |L−M|) contribute to discomfort glare perception whose contribution cannot be neglected. A linear combination of the spectral sensitivity functions of the contributing mechanisms seems to be a better descriptor than $V(\lambda)$. Furthermore, some subjects showed in the spectral range from 440 until 520 nm (blue–cyan–green color range) a higher spectral sensitivity than the $V(\lambda)$-function which also means that light sources with a high blue–cyan content can cause higher discomfort glare sensitivity at the same luminance or illuminance of the light sources.

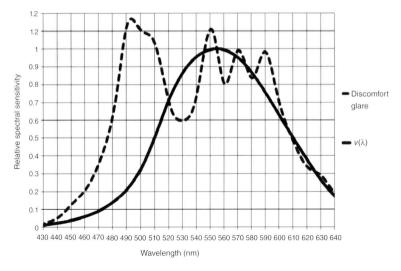

Figure 6.30 Average discomfort glare spectral sensitivity curve for all subjects and the $V(\lambda)$-function.

6.6.4
Experiments to Determine the Spectral Sensitivity of Disability Glare in the Mesopic Range under Traffic Lighting Conditions

Results on the spectral sensitivity of discomfort glare from Fekete, Bullough and Bodrogi *et al.* mentioned in Section 6.6.3 indicated a higher spectral sensitivity to discomfort glare in the range between 440 and 520 nm dissimilar to the $V(\lambda)$-function. Disability glare, by nature, is related to the reduction or loss of visual performance in the mesopic range. As described in Section 6.1, visual performance can be expressed in terms of contrast perception, or, more specifically, object detection probability at $p = 50\%$, 70%, or 75%.

In previous studies on visual performance, visual acuity was also regarded as a relevant criterion, for example, the recognition or identification of the opening of a Landolt ring, as described in Section 6.4. But, as pointed out in Section 6.1, from the point of view of traffic safety, the object detection task (without recognition) is more important because it initiates a quick reaction of the traffic participant to a (dangerous) object that appears rapidly, typically in the periphery of the viewing field.

According to this idea, a new, detection based disability glare paradigm was implemented in a recent experiment carried out at the Technische Universität Darmstadt [42]. This experiment used only seven QM (quasi-monochromatic) glare sources of different wavelengths. So it is considered as a pilot study while currently (at the time of writing) a further, more extended dataset is being accumulated at the Technische Universität Darmstadt. The setup of this pilot study can be seen in Figure 6.31.

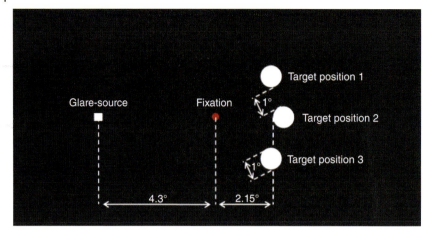

Figure 6.31 Setup of a pilot study [42] on disability glare perception with a new detection based experimental paradigm. The glare source emitted quasi-monochromatic radiation.

In the experimental setup depicted in Figure 6.31, subjects sat in front of a box illuminated by four cold white LEDs set to provide a background luminance of about $0.1\,\text{cd}\,\text{m}^{-2}$ to simulate the luminance of the road at 50–60 m in front of a car. The subject looked at the fixation point which was a dimmed red LED. At an eccentricity angle of about $2.15°$, three similar circular detection targets of the size of $1°$ appeared. They represented a typical object in a mesopic road scenario with a size of 0.87 m as observed from 50 m.

The angle $2.15°$ corresponds to a typical position of this object (e.g., a person or an animal) at the right side of the road being at a distance of 1.87 m measured from the center line of the road (the road width is assumed to be 3.75 m). The choice among the three targets (only one of them appeared at a time) was foreseen for the observer to prevent guessing. The background and the three targets were of the same CCT at 6170 K.

The glare source in Figure 6.31 had an angular size corresponding to the geometry of $0.2\,\text{m} \times 0.2\,\text{m}$ with a glare angle of $4.3°$ simulating the headlamp of an opposing car as seen from 50 m distance. It emitted QM radiation selected by the driving software and realized by an LED light source and a liquid crystal tunable filter (LCTF) with a FWHM of 7 nm.

Six subjects of 28–31 years of age and one person of 41 years of age took part in the experiment with seven glare source peak wavelengths $\lambda_{DA} = 455, 506, 519, 553, 568, 597$, and 627 nm, respectively. At each wavelength, subjects experienced different glare source radiances. They had to tell the position of the target (top, middle, bottom) once detected. The higher the glare source radiance, the lower the probability of a correct statement about its position is. In Figure 6.32, an example of fitting tanh functions to 12 detection probabilities for three different QM glare stimuli (as a function of glare source radiance) is illustrated, as an example.

As can be seen from Figure 6.32, for each test subject and each QM stimulus, a tanh function describing detection probability as a function of glare source

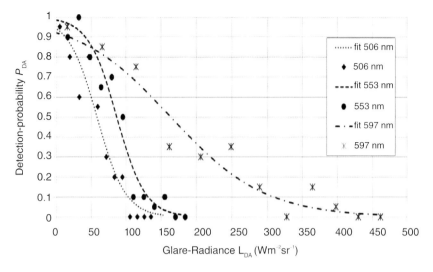

Figure 6.32 Example of fitting tanh functions to 12 detection probabilities for three different quasi-monochromatic glare sources (see Figure 6.31), as a function of the radiance of the glare source.

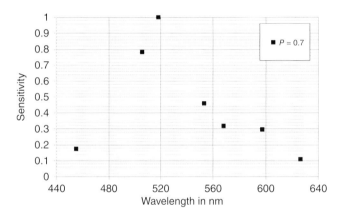

Figure 6.33 Spectral sensitivity to disability glare for the application relevant detection paradigm shown in Figure 6.31. Example: subject NH and $P_{DA} = 0.7$.

radiance can be fitted with reasonable accuracy. From these detection probability curves, the glare source radiance corresponding to a predefined detection probability, for example, $P_{DA} = 0.7$, can be determined (the index DA means disability). Because the monochromatic glare source stimuli of different wavelengths cause the same detection probability at different glare source radiances, a spectral sensitivity function can be derived for each individual subject. An example can be seen in Figure 6.33.

Figure 6.34 shows the average spectral sensitivity for all seven subjects in comparison to the $V(\lambda)$-function and the scotopic $V'(\lambda)$ function.

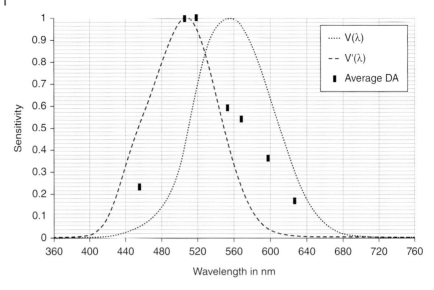

Figure 6.34 Mean spectral sensitivity to disability glare for the application relevant detection paradigm shown in Figure 6.31. Mean of seven observers; $P_{DA} = 0.7$. DA: disability.

The result in Figure 6.34. can be interpreted in the context of LED technology and the perception of disability glare under mesopic conditions in the following way.

- The experimentally determined spectral disability glare function in Figure 6.34 according to the novel application relevant detection based paradigm of Zydek et al. [42] means that the spectral disability glare sensitivity at a background luminance of $0.1\,\mathrm{cd\,m^{-2}}$ and for the background chromaticity with CCT = 6170 K (i.e., with a higher blue LED content that triggers the scotopic i.e., rod mechanism to be active) exhibits a maximum between about 507 and 519 nm. This function is not similar to the $V(\lambda)$ function. Therefore, disability glare evaluation with glare sources located at the periphery with an eccentricity of 4.3° and object size of 2.15° cannot be conducted by using $V(\lambda)$ weighted photometric quantities like illuminance in lx or luminance in $\mathrm{cd\,m^{-2}}$.
- A higher sensitivity at 507 nm or at 519 nm in comparison to 455 nm implies that a relatively small glare source radiance causes already disability glare while at 455 nm, a factor of 4–5 higher radiance can be used to ensure the same detection probability ($p_{DA} = 0.7$). Blue radiation in the range of 440–465 nm, which is characteristic of most blue LED chips, is not substantial for disability glare sensitivity, and there is more sensitivity in the cyan–green wavelength range around 507 and 520 nm.
- Under the test conditions of Zydek et al. [42] corresponding to real traffic conditions, the image of the glare source on the retina appears at 4.3° on one side of the fovea saturating the LMS photoreceptors (and bleaching the rods) *locally* at a *very high* photopic adaptation level. So the object detection process is locally

disabled, at the locations of the glare source image. The detection targets, however, occurred on the other side of the retina under 2.15° and had to be detected rather by the rods and partially by the LMS cones. At the other side of the retina, the veil of the glare source caused a more or less higher state of adaptation, depending on glare source radiance.

6.6.5
Glare in the Street Lighting with White LED and Conventional Light Sources

Figure 6.35 illustrates a pixel resolved absolute luminance image of an LED street lighting installation. The road luminance on the left side is between 0.05 and 0.5 cd m^{-2} and the pedestrian zone on the right side is at about 0.5 and 1.0 cd m^{-2}. These luminance levels define the adaptation field for car drivers. The luminance of the LED luminaire is above 2000 cd m^{-2} so that disability glare can be expected according to the characteristics in Figure 6.21. This implies that certain low contrast objects on the left side of the image with a luminance of only about 0.01 and 0.02 cd m^{-2} cannot be detected.

A parameter to express the reduction of visibility caused by the disability glare of the luminaires of a road lighting installation is the so-called *threshold increment* designated by TI [43]. The quantity TI is defined in Eq. (6.12).

$$\text{TI } (\%) = \frac{(\Delta L_{\text{thg}} - \Delta L_{\text{th}}) * 100\%}{\Delta L_{\text{th}}} \tag{6.12}$$

The symbols of Eq. (6.12) have the following meaning:

Figure 6.35 Pixel resolved luminance image of an LED street lighting installation. (Image source: Technische Universität Darmstadt.)

ΔL_{thg}: threshold luminance difference between object and background in the presence of glare;

ΔL_{th}: threshold luminance difference between object and background without glare.

The quantity TI describes the amount of additional contrast required to make the object just visible again after a glare source appeared in the field of view. TI can be predicted by using Eqs. (6.13) and (6.14).

$$L_{seq} = K \cdot \frac{E_B}{\Theta^2} = K \cdot \sum_{i=1}^{n} \frac{E_{Bi}}{\Theta_i^2} \quad (6.13)$$

$$TI = 65 \cdot \frac{L_{seq}}{L_U^{0.8}} \% \quad (6.14)$$

Note that Eq. (6.13) is a special form of Eq. (6.8) with the exponent value of $n = 2$ and $L_v = L_{seq}$ is the equivalent veiling luminance, E_{Bi} is the illuminance produced by the ith luminaire in the new state when it is on and causes glare on a plane normal to the line of sight and at the height of the observer's eye. Θ_i is the angle (in arc units) between the line of sight and the line from the observer to the center of the ith luminaire [44].

The observer is positioned at the center of each lane so that his eyes are at the height of 1.5 m above the road level. The longitudinal distance in meters in front of the calculation field is 2.75 (H-1.5) where H is the mounting height of the luminaire. The observation angle is 1° below the horizontal and in the vertical plane in the longitudinal direction passing through the eye of the observer.

The summation in Eq. (6.13) is performed from the first luminaire on for each luminaire row in the direction of observation [45] and is finished when a luminaire in that row contributes to the total veiling luminance less than 2%. Luminaires that are 20° above the horizontal passing through the observer's line of sight and intersect the road in the transverse direction are excluded from the calculation.

The calculation is started at the position described above and repeated with the observer moved forward in increments that are the same in number and distance as are used for the spacing of the luminance calculation points. The procedure is repeated with the observer positioned in the center of each lane. In each case, the initial average road surface luminance corresponding to the observer position is used. The operative TI value is the maximum value found [45]. In Table 6.8, the lighting criteria for the ME street classes according to the EN standards 13201 for street lighting [43, 44, 46] with TI values are listed.

The TI values for the ME street classes are allowed to be only in the range between 10% and 15% as computed by Eqs. (6.13)–(6.14). In the DIN EN standards for street lighting [43, 44, 46] in Europe which are also used in similar form in America and Asia, also so-called S class streets, are considered as smaller streets in residential areas for which the illuminance (in lx unit) is decisive. In this case, there are no TI value requirements in the standard. However, according to the authors' practical experience, many S class streets in Germany equipped with new

Table 6.8 Lighting criteria for the ME street classes according to the EN standards 13201 for street lighting [43, 44, 46]. Values in the table are road luminance values under dry condition required by the standard in cd m^{-2} unit.

Class	$\bar{L}$ (cd m^{-2}) [maintenance value]	U_0 [minimal value]	U_1 [minimal value]	TI (%) [maximal value]	SR [minimal value]
ME1	2.0	0.4	0.7	10	0.5
ME2	1.5	0.4	0.7	10	0.5
ME3a	1.0	0.4	0.7	15	0.5
ME3b	1.0	0.4	0.6	15	0.5
ME3c	1.0	0.4	0.5	15	0.5
ME4a	0.75	0.4	0.6	15	0.5
ME4b	0.75	0.4	0.5	15	0.5
ME5	0.5	0.35	0.4	15	0.5

Table 6.9 TI value reduction after increasing pole height. Measurement result for a real new LED street luminaire installation.

Pole height (m)	Pole distance (m)	Observer's eye height (m)	Observing distance (m)	Glare angle	Relative TI value
4	30	1.5	30	4.8°	1.0 (TI unchanged)
5	30	1.5	30	**6.65°**	**0.52**

LED street lighting luminaires can exhibit TI values up to TI = 30% or even 40% which causes strong glare perception for the users of street lighting during the dark hours.

One reason for this glare is the pole height of the old conventional luminaires of only 4.0–4.5 m at a pole distance of more than 30 m. This glare can be drastically reduced if technicians sent out to the premises from the local street administration authorities succeed in increasing pole height from this value (about 4 m) to about 5 m by means of an additional mechanical tube mounted on the top of the old pole. The effect of this pole height enlargement can be seen in Table 6.9.

As can be seen from Table 6.9, for the same LED street light luminaire, the TI value at 30 m pole height can be reduced by a factor of nearly 2 if pole height can be increased to 5 m instead of 4 m. Note that in Eqs. (6.13) and (6.14), TI values are described in terms of a photometric, that is, $V(\lambda)$ weighted parameter only. This parameter is glare illuminance at the eye of the observer (E_b). This means that two light sources with the same angular position that cause the same glare illuminance but with different chromaticities (very different forms of their relative spectral power distributions, for example, a white LED vs a yellow light sodium lamp, see Figure 6.18) should have the same impact on disability glare and on TI values.

This hypothesis has to be verified by well-designed glare experiments which is described herewith.

Niedling *et al.* [47] performed an experiment to point out the influence of glare light source chromaticity (relative spectral power distributions) on discomfort glare and disability glare under mesopic conditions in outdoor lighting. A half integrating sphere (a hemisphere) with a diameter of 1.5 m illuminated homogeneously by warm white LEDs (CCT = 3400 K) was used as a stimulus background at 0.05 cd m^{-2} (this is a very low luminance level not really typical for street lighting).

Subjects sat in front of the hemisphere and had to recognize Landolt rings of different gray contrasts, that is, different Landolt ring luminances compared to the background luminance and with an opening angle of 0.38° (see also Section 6.4). The glare source was the output of a light guide with a diameter of 3 mm positioned 4° above the line of sight. This was also the position of the center of the Landolt ring. Glare sources had an output luminance of about 500 000 cd m^{-2} corresponding to a luminous intensity of 3.53 cd. Therefore, a glare illuminance of about 5 lx was kept constant for all spectral power distributions at the eye of their [47] 28 subjects aged between 21 and 35 (the median age was 26 years).

Subjects were asked to look at the Landolt ring exhibiting different gray contrasts that were presented in a randomized order. If the subject was able to discern the opening of the ring then she or he had to turn off the glare source and tell the direction of the ring opening as well as to rate her or his feeling of discomfort glare on an (inverted) de Boer scale (refer to Figure 6.27). The following eight light sources with different spectra were used in the experiment as glare sources:

- White phosphor-converted LEDs (CCT = 7800, 4300, and 3300 K);
- Incandescent lamp (ILB);
- High pressure sodium lamp (HPS);
- Fluorescent lamp (FL, 6800 K);
- Metal halide lamp (MH); and
- High pressure mercury vapor lamp (HML).

Authors of the present book have to make three critical remarks at this point concerning the above described design of this experiment [47]:

1) The background luminance of only 0.05 cd m^{-2} is low. An S5 class small street in Germany with a mean illuminance of 3 lx will have a luminance of 0.21 cd m^{-2} on the road and the main adaptation area for car drivers on a dark country road has (at 30 m in front of the car) a luminance of 1–4 cd m^{-2} or at 50–60 m about 0.1 cd m^{-2}. Note that, at 0.05 cd m^{-2}, the activity of the LMS cones is weak.
2) The glare illuminance of 5 lx is very high. The Söllner *et al.* [30] study (see Table 6.5) reported on a typical illuminance being in the range between 0.5 and 2 lx at the point B50L according to the automotive lighting standard ECE

Table 6.10 Mean results of the discomfort and disability experiment (Niedling et al. study [47]).

Spectra	LED 7800 K	LED 4300 K	LED 3300 K	ILB	HPS	FL	MH	HML
De Boer (inverted!)	8	8	8	7	6.5	8	7.5	7.5
Grayscale contrast	31.0	32.0	31.5	31.0	32.5	32.0	31.5	31.0

48 and therefore glaring the participants of opposing traffic. At the eye position of the opposing traffic drivers at 50 m distance between the two cars, only an illuminance of max. 0.25 lx is allowed.

3) Many studies on mesopic visual performance and mesopic glare used Landolt rings as a criterion for object visibility. Visibility tests with Landolt rings represent a mesopic recognition experiment (see Section 6.1) in which foveal photoreceptors at a viewing angle of max. $2\alpha = 2°$ (rather 1.3°) are activated. In case of the Niedling et al. study [47], the ring opening had an angle of 0.38° so that the neural *recognition* signal was contributed by the L cones and the M cones and not by the S cones or by the rods. As pointed out in Section 6.1, for traffic safety analysis, the *detection* of (dangerous) objects at the roadside takes place in application relevant situations at eccentricity angles (measured from the line of sight) between 2.2° and up to 10°. At these locations, there is S cone and rod contribution to carry out the detection task, see Figure 6.2.

Results of the Niedling et al. study [47] are summarized in Table 6.10.
As can be seen from Table 6.10,

1) The grayscale contrast at which the opening of the Landolt rings was recognized is on the same level for all lamp spectra. This means that disability glare is independent of lamp chromaticity (relative spectral power distribution).
2) Discomfort glare rating is equal for the three white LEDs that exhibit different chromaticities (7800, 4300, and 3300 K). These discomfort glare ratings were not significantly different from those of the FL, MH, and HML lamps. For the ILB and the HPS lamp, glare rating was reduced with statistical significance.

The outcome of the Niedling et al. experiment [47] can be explained by the following two reasons:

1) The optical radiation of their six light sources with six different spectral power distributions enters the eye apparatus and is scattered delivering a veil on the retina. The three LED lamp types and the lamps FL, MH, and HML have a higher veil from blue radiation distributed on the retina and activating the whole retinal receptors, predominantly the rods, to a mesopic activity level. A higher *blue* radiation content reaching the retina at the same *photopic* luminance of 0.05 cd m^{-2} causes a significantly higher rod signal (observe the ordinate of Figure 6.11 at 0.05 cd m^{-2} for different S/P values and compare it with Figure 6.30). Thus, the six lamps with higher blue content result in a higher

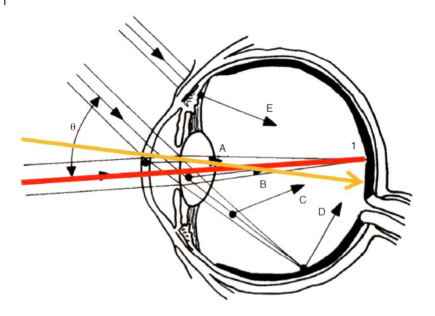

Figure 6.36 Explanation of stray light and the effect of glare on Landolt ring identification under illuminants of different chromaticities (spectral power distributions), see text.

discomfort glare rating. Note that the study used an *inverted* de Boer scale (see Table 6.10).

2) Now consider disability glare as evaluated with the Landolt rings. The Landolt rings are imaged along the optical axis of the eye lens onto the fovea (see the red line in Figure 6.36) with an angle of only 0.38° and the L cones and M cones work together to identify the opening of the ring. The glare source at 4° above the line of sight is imaged by the eye lens to a position at 4° below the center of the fovea (see the orange line in Figure 6.36). This glare source image saturates all photoreceptors at that location at a very high photopic adaptation level. There, rods are not active because of the high local retinal illuminance level. At the fovea, at 4° above this location, stray light level is high following light scattering and the imperfect optical imaging quality of the eye lens. Therefore, L cones and M cones experience a photopic or high mesopic adaptation level for all eight lamp spectra that enables the identification of the ring opening with the same efficiency.

6.7
Bead String Artifact of PWM Controlled LED Rear Lights at Different Frequencies

This section describes a serious visual artifact (the bead string artifact) that occurs at mesopic adaptation levels in traffic when LED light sources are applied as automotive rear lights and operated by pulse width modulation (PWM).

6.7.1
Introduction

LED are often used in today's car rear lights. In order to implement two functions (stop light and rear position light or tail light), two different LED dimming levels are necessary. A common way for dimming is the use of PWM. For stop light, simple DC operation is used. For tail light, the LEDs are typically operated at a duty cycle of approximately 13% (1 : 8 dimming ratio). A commonly used PWM frequency is 100 Hz (until 2008) or 200 Hz (as of writing in 2014).

The bead string artifact can be observed when the PWM driven LED rear light or the driver's eye moves very fast (see Figure 6.37).

Under the viewing condition depicted in Figure 6.37, temporally discrete light pulses are mapped onto different places on the observer's retina because of the fast movement (see Figure 6.38, right). If the rear light is being operated in DC mode then a continuous bright line appears on the retina.

Figure 6.37 Illustration of the bead string artifact on a mesopic background. (Image source: Technische Universität Darmstadt.)

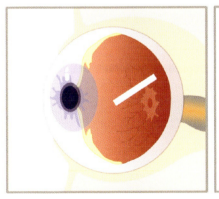 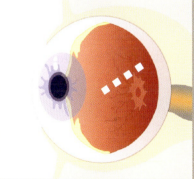

Figure 6.38 Image of the car's rear light on the human eye retina in DC operation (left picture) and in PWM operation at a certain PWM frequency (right picture).

Because the retinal image looks like a line of beads (see Figure 6.37), the effect is referred to as "bead string artifact." It should be noted that the human subjects' verbal descriptions of the effect vary in a broad range. For example, some observers describe a temporal "flickering." In difference to flickering that appears in case of *static* light sources and temporally discrete light pulses that are mapped to the *same* point onto the retina, the experiment reported in this section [48] was carried out under dynamic conditions with light source and/or eye moving.

The beads effect experienced by healthy subjects leads to some irritations and a loss of concentration during driving the car for a certain time. It leads, however, for photosensitive subjects (i.e., persons with epilepsy) probably to a loss of control of the car they are driving and possibly to a traffic accident. Therefore, the research on the impact of PWM controlled LED light output on flicker perception and on the bead string artifact has a high practical importance.

Analyzing the structure of the line of beads in Figure 6.38 (right), we can conclude that the bead string artifact can be theoretically reduced, if:

- PWM is operated at a very high frequency so that dark gaps between the individual beads are very short. But this high frequency can cause EMC problems including disturbance of the car's keyless entry system or the car's stereo receiver.
- PWM is operated with a trapezoidal or a cosine like pulse form or with a rectangular pulse form with a modulation depth being lower than 100%. But, in order to implement these requirements, complex electronic circuits need to be designed and installed.

6.7.2
A visual experiment [48] on the bead string artifact

6.7.2.1 Key Question
The key question of the experiment [48] was: Does the bead string artifact depend on the following issues?:

- Viewing conditions: foveal versus peripheral observation;
- Observer's age or gender;
- Observer's wearing eyeglasses or not;
- Observer's position in the car: driver or passenger seat; and
- Geometrical size of the lit area.

Car makers and LED luminaire manufacturers face a trade-off when changing PWM frequency: suppressing the bead string artifact by using a higher frequency on one hand and electromagnetic compliance (EMC) problems or more manufacturing costs for cables or electronics on the other hand. So the question is which PWM frequency can be recommended for practical use under real world conditions.

6.7.2.2 Experimental Setup

In order to test the correlation between PWM frequency and the geometrical size of the lamp's lit area, all tests were done with the rear lamps of the following three types of test cars:

- Audi A6 Avant;
- Mercedes SLK; and
- Volkswagen EOS.

Table 6.11 summarizes the characteristics of the experimental setup.

As can be seen from Table 6.11, the rear lamps were driven by a 13.5 V power supply. A power MOSFET in the path between the LED rear lamp and the power supply modulated the supply voltage. The MOSFET was driven by a microcontroller which created the PWM pulses. By the use of the PC software, the operator could adjust PWM frequency. The experiment was carried out under nighttime driving conditions at the Griesheim Airfield (Hessen, Germany). The low beam of the test cars was switched on and there were no street lamps.

Two subjects were sitting in a stationary car on the right lane while the test car with pulsed LED rear lamps was passing on the left lane at a speed of 50 km h^{-1}. This was repeated in several runs with different LED PWM frequencies. The run and the frequency assignment parameters were randomized and unknown by the subjects and by the test operator who was the test car driver. After each run, subjects had to place a cross on a questionnaire page about whether the bead string artifact was "clearly seen," "barely seen," or "not seen." So this experimental setup reproduced a typical highway traffic situation: the subject's car on the right lane was overtaken by the LED car at a speed difference of 50 km h^{-1}.

6.7.2.3 Peripheral Observation

In this part of the experiment, the subject's task was to fixate a target, a red lamp at about 100 m on the own lane all the time. Thus, while being overtaken by the pulsed LED test car, beads occurred at a peripheral position on the retina (see Figure 6.39).

Fifty-six observers took part in this first experiment (depicted in Figure 6.39) aged between 23 and 72 years. After this first experiment, authors found that the

Table 6.11 Characteristics of the experimental setup in the visual experiment [48] on the bead string artifact.

Condition	Fixed/var.	Value
Supply voltage	Fixed	13.5 V
PWM duty cycle	Fixed	13%
Speed of test car	Fixed	50 km h^{-1}
PWM frequency	Variable	45/60/100/130/170/220/280 Hz
Observing condition	Variable	Foveal, peripheral

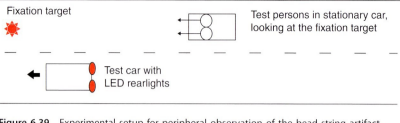

Figure 6.39 Experimental setup for peripheral observation of the bead string artifact.

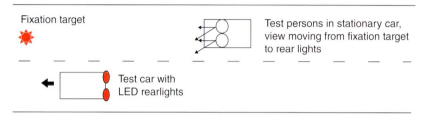

Figure 6.40 Experimental setup for foveal observation of the bead string artifact.

persons older than 45 years (there were 11 such subjects) have not seen the bead string artifact at all. In order to test the "worst case," which means young observers in this case, only subjects aged between 22 and 31 years of age were invited for the second experiment, which is described below.

6.7.2.4 Foveal Observation

In this second experiment, subjects were instructed to view the fixation target for the first moment. As the test car came into their field of view, subjects had to move their line of sight toward the LED rear lights immediately by the aid of a fast eye movement. Then their fixation point had to follow the rear lights by moving their eyes in saccades. Under this viewing condition, beads occurred in a foveal position (see Figure 6.40).

In comparison to the first experiment (depicted in Figure 6.39), in this case (i.e., in case of Figure 6.40), a higher sensitivity to PWM frequency was expected because of the fast eye movement from the fixation point to the rear lights. Therefore it was expected that higher PWM frequencies were needed to suppress the bead string artifact. 23 subjects participated in this second part of the experiment. All of them looked at the three rear lamps as in the first part of the experiment (Audi, Mercedes, and VW) with eight test runs each.

6.7.3
Results of the Visual Experiment [48] on the Bead String Artifact

The bead string artifact was found to depend on the following factors:

- Viewing condition: foveal versus peripheral observation;
- Observer's age;

- Geometrical size of the lit area.

This result corroborated the results of a similar experiment [49].
It turned out that the artifact does not depend on the following factors:

- Observer's gender;
- Observer's wearing eyeglasses or not;
- Observer's position in the car: driver or passenger seat.

6.7.3.1 Effect of Viewing Condition (Peripheral vs Foveal)

In Figure 6.41, the cumulative frequency of recognizing the bead string artifact is shown over the tested PWM frequency range. A cumulative frequency of 100 % means that all subjects could see the effect clearly. A value of 50% can be reached in different ways: either 100% of the subjects could (at least) just notice the effect, or 50% of the subjects could see the effect clearly while the other 50% could not see it.

As can be seen from Figure 6.41, for peripheral viewing (first experiment), the bead string artifact is recognized at relatively low PWM frequencies only (5% of the subjects could recognize the effect at about 130 Hz). For foveal viewing (second experiment), a clearly higher PWM frequency was needed to suppress the effect. At 280 Hz, more than 20% of the subjects could still recognize the effect. A mathematical extrapolation of the foveal tendency (see the blue line in Figure 6.41) and assuming 5% as an acceptance threshold criterion then a PWM frequency of about 370 Hz should be applied. This result was corroborated by the results of another current experiment which is being performed while writing this book.

6.7.3.2 Effect of the Observer's Age

From the results of the first experiment (peripheral viewing), elderly observers can be compared with young subjects: there were 36 test runs (each run consists of all 8 tested PWM frequencies) with subjects of less than 36 years of age and 20

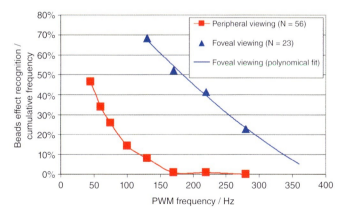

Figure 6.41 Cumulative frequency of recognizing the bead string artifact under the two viewing conditions, peripheral viewing and foveal viewing [48].

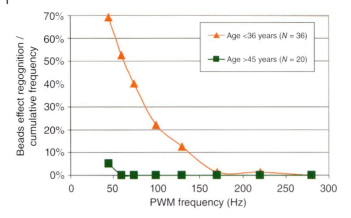

Figure 6.42 Bead string artifact recognition within the two age groups.

such test runs with subjects of more than 45 years of age. Figure 6.42 compares the tendencies of bead string artifact recognition within these two age groups.

As can be seen from Figure 6.42, bead string artifact recognition depends strongly on the observer's age. Elderly subjects tend to be insensitive to the effect while by the use of the PWM frequency of 100 Hz, the effect is clearly recognizable by young people.

6.7.4
Conclusion

The today widely used PWM frequency of 200 Hz is generally not sufficient to suppress the bead string artifact under most viewing conditions. There is no doubt that the number of discomfort reports caused by the effect would be dramatically reduced if car manufacturers and LED car headlamp suppliers would use a PWM frequency in the range of 350–400 Hz. Of course the result described here allows a statement for these specific viewing conditions only. The second experiment was carefully designed to represent a "worst case" in order to select a high PWM frequency which is needed to suppress the effect. Nevertheless, it cannot be excluded that there are some other viewing conditions for which even higher PWM frequencies are necessary to suppress the artifact.

6.8
Summarizing Remarks to Chapter 6

As we learnt from the considerations of Chapter 6,

1) Mesopic brightness [50] sets the adaptation level of mesopic vision and defines an "operating point" for the human visual system. It also defines the

level of mesopic visual comfort experienced by the users of mesopic lighting installations.
2) Object identification with Landolt rings without glare under mesopic conditions (Section 6.4) or with glare (i.e., disability glare for object identification, Section 6.6) is independent of light source chromaticity.
3) Discomfort glare is very important for automotive lighting applications but somewhat less important for street lighting. Discomfort glare depends on light source chromaticity.
4) Detection under mesopic conditions (including the effect of disability glare on object detection) for objects that tend to appear in the eccentricity angle range between 2.2° and 20° is a very important visual task for traffic lighting, for example, if a person suddenly occurs under 20° from the gap between two parking cars in the darkness on the right-hand side of a not well-illuminated street.

References

1. Bodrogi, P. and Khanh, T.Q. (2011) Mesopische Sehleistung – mesopische Photometrie. Teil III- 1.1.1, in *Handbuch für Beleuchtung (Lange)*, Ergänzungslieferung, vol. 51, Ecomed Sicherheit.
2. Eloholma, M., Liesiö, M., Halonen, L., Walkey, H., Goodman, T., Alferdinck, J., Freiding, A., Bodrogi, P., and Várady, G. (2005) Mesopic models – from brightness matching to visual performance of night-time driving: a review. *Light. Res. Technol.*, **37** (2), 155–175.
3. Commission Internationale de l'Eclairage (2011) CIE Supplementary System of Photometry, CIE Publication 200-2011.
4. Rea, M.S., Bullough, J.D., Freyssinier-Nova, J.P., and Bierman, A. (2004) A proposed unified system of photometry. *Light. Res. Technol.*, **36** (2), 85–111.
5. Goodman, T., Forbes, A., Walkey, H., Eloholma, M., Halonen, L., Alferdinck, J., Freiding, A., Bodrogi, P., Varady, G., and Szalmas, A. (2007) Mesopic visual efficiency IV: a model with relevance to night-time driving and other applications. *Light. Res. Technol.*, **39**, 357–383.
6. Walkey, H.C., Harlow, J.A., and Barbur, J.L. (2006) Characterising mesopic spectral sensitivity from reaction times. *Vision Res.*, **46**, 4232–4243.
7. Commission Internationale de l'Eclairage (2010) *Recommended System for Mesopic Photometry Based on Visual Performance*, CIE Publication 191-2010, CIE, Vienna.
8. Bodrogi, P., Vas, Z., Haferkemper, N., Varady, G., Schiller, C., Khanh, T.Q., and Schanda, J. (2010) Effect of chromatic mechanisms on the detection of mesopic incremental targets at different eccentricities. *Ophthal. Physiol. Opt.*, **30**, 85–94.
9. Haferkemper, N., Frohnapfel, A., Paramei, G., and Khanh, T.Q. (2007) A mesopic experiment series at automotive visual conditions. Proceedings of the 7th International Symposium on Automotive Lighting ISAL 2007, Technische Universitaet Darmstadt, pp. 402–409.
10. Vas, Z. and Bodrogi, P. (2007) Additivity of mesopic photometry. Proceedings of the 7th International Symposium on Automotive Lighting ISAL 2007, Technische Universitaet Darmstadt, pp. 187–194.
11. Freiding, A., Eloholma, M., Ketomäki, J., Halonen, L., Walkey, H., Goodman, T., Alferdinck, J., Várady, G., and Bodrogi, P. (2007) Mesopic visual efficiency I: detection threshold measurements. *Light. Res. Technol.*, **39**, 319–334.
12. Kokoschka, S. (1972) Untersuchungen zur mesopischen Strahlungsbewertung. PhD thesis. Universität Karlsruhe.

13. Khanh, T.Q., Böll, M., Schiller, Ch., and Haferkemper, N. (2008) Helligkeits- und Kontrastwahrnehmung im mesopischen Bereich. Licht Ausgabe 3, pp. 214–219.
14. Fotios, S.A. and Cheal, C. (2007) Lighting for subsidiary streets: investigation of lamps of different SPD, part 2 – Brightness. *Light. Res. Technol.*, **39** (3), 233–252.
15. Bodrogi, P., Böll, M., Schiller, Ch., and Khanh, T.Q. (2009) Brightness appearance of automotive front lighting light sources – a series of visual experiments. Proceedings of the 8th International Symposium on Automotive Lighting ISAL 2009, Technische Universitaet Darmstadt, 2009.
16. Kimura-Minoda, T., Kojima, Sh., Fujita, Y., and Ayama, M. (2007) Study on glare of LED lamp and individual variations of brightness perception. Proceedings of the ISAL 2007, 7th International Symposium on Automotive Lighting, Technische Universität Darmstadt, September 25–26, 2007.
17. Kostic, M.B. and Djokic, L.S. (2012) A modified CIE mesopic table and the effectiveness of white light sources. *Light. Res. Technol.*, **44**, 416–426.
18. Haferkemper, N. and Khanh, T.Q. (2013) Visual acuity in the mesopic range does not depend on the light spectrum. Proceedings of the ISAL 2013, 10th International Symposium on Automotive Lighting, Technische Universität Darmstadt, September 23–25, 2013.
19. Wender, M. (2007) Value of Pelli-Robson contrast sensitivity chart for evaluation of visual system in multiple sclerosis patients. *Neurol. Neurochir. Pol.*, **41** (2), 141–143.
20. Schmidt-Clausen, H.-J. and Freiding, A. (2004) Sehvermögen von Kraftfahrern und Lichtbedingungen im nächtlichen Straßenverkehr. Berichte der Bundesanstalt für Straßenwesen, Mensch und Sicherheit Heft M 158, 2004.
21. Schiller, C., Böll, M., Bodrogi, P., and Khanh, T.Q. (2010) Detection and conspicuity of road markings - impact of spectral power distribution. Lecture at the Conference V.I.S.I.O.N. 2010, Versailles, France, 2010.
22. Deutsche Studiengesellschaft für Straßenmarkierungen e.V (2007) DSGS Jahrbuch Straßenmarkierungen 2006/2007, Seite 75.
23. Meseberg, H. (1992) *Optimierte Fahrbahnmarkierungen und nächtliche Verkehrssicherheit*, Straßenverkehrstechnik, Heft 4/92, Kirschbaum Verlag, Bonn, p. S.197.
24. Schmidt-Clausen, H.-J. (1990) *Lichttechnische Anforderungen an Fahrbahnmarkierungen, Fahrbahnmarkierungen '90, Schriftenreihe der DSGM*, Heft 9/90, Kirschbaum Verlag, Bonn.
25. Khanh, T.Q. (2004) Zum optimalen Lampenspektrum zur Straßen- und Außenbeleuchtung – Diskussion, visuelle Tests. *Straße und Autobahn*, Heft 6/04, S. 333–S. 336.
26. Bodrogi, P., Vas, Z., Schiller, C., and Khanh, T.Q. (2008) Psycho-physical evaluation of a chromatic model of mesopic visiual performance. Proceedings of CGIV 2008, IS&T's Fourth European Conference on Color in Graphics, Imaging and Vision, 2008.
27. Schiller, C. and Khanh, T.Q. (2007) First field tests of cars with completely built-in LED headlamps under realistic driving conditions. Proceedings of the 7th International Symposium of Automotive Lighting, ISAL 2007, Darmstadt, Germany, pp. 135–142.
28. Hopkinson, R.G. and Collins, J.B. (1970) *The Ergonomics of Lighting*, McDonald & Co., (Publishers) Ltd, London.
29. Vos, J.J. (1999) Glare today in historical perspective: towards a new CIE Glare observer and a new glare nomenclature, in *Proceedings of the CIE 24th Session*, Warsaw, CIE, Vienna.
30. Söllner, S., Sprute, H., Polin D., and Khanh, T.Q. (2011) Field study: driver data on visual topics. ISAL 2011, The 9th International Symposium on Automotive Lighting, Darmstadt, Germany, September 26–28, 2011.
31. Khanh, T.Q. (2014) Student Lectures on Lighting Technology, Winter Semester 2013–2014, Technische Universität Darmstadt.

32. Sprute, H. (2012) Entwicklung lichttechnischer Kriterien zur Blendungsminimierung von adaptiven Fernlichtsystemen. PhD thesis. Technische Universität Darmstadt.
33. Carraro, U. (1984) Die Adaptationsleuchtdichte bei inhomogenen Leuchtdichtefeldern unter besonderer Berücksichtigung einer dynamischen Sehaufgabe. PhD thesis. Fakultät für Technische Wissenschaften, Technische Universität Ilmenau, Ilmenau.
34. Commission Internationale de l'Éclairage (2002) CIE Equations for Ddisability Glare, CIE TC 1–50, Wien.
35. Totzauer, A. and Khanh, T.Q. (2012) *Blendungsforschung für die Verkehrslichttechnik Dynamische Blendung, Schleierleuchtdichte, Blendbeleuchtungsstärke, Zeitschrift Licht*, Pflaum-Verlag, H.10/2012.
36. Totzauer, A. (2013) Kalibrierung und Wahrnehmung von blendfreiem LED-Fernlicht. PhD thesis. Technische Universität, Darmstadt.
37. de Boer, J.B. (1967) Visual perception in road traffic and the field of vision of the motorist, in *Public Lighting* (ed J.B. de Boer), Eindhoven, Philips Technical Library, pp. 11–96.
38. Schmidt-Clausen, H.-J. and Bindels, J.H. (1974) Assessment of discomfort glare in motor vehicle lighting. *Light. Res. Technol.*, **6** (2), 79–88.
39. Fekete, J., Sik-Lányi, C., and Schanda, S. (2010) Spectral discomfort glare sensitivity investigations. *Ophthal. Physiol. Opt.*, **30**, 182–187.
40. Bullough, J.D. (2009) Spectral sensitivity for extrafoveal discomfort glare. *J. Mod. Opt.*, **56**, 1518–1522.
41. Bodrogi, P., Wolf, N., and Khanh, T.Q. (2011) Spectral sensitivity and additivity of discomfort glare under street and automotive lighting conditions. Proceedings of the 27th Session of the CIE, South Africa, Publication CIE 197:2011, pp. 338 341.
42. Zydek, D., Bodrogi, P., Khanh, T.Q., and Haferkemper, N. (2011) A new concept of disability glare under traffic lighting conditions: experimental set-up, results and analysis of spectral sensitivity. ISAL2011, The 9th International Symposium on Automotive Lighting, Darmstadt, Germany, September 26–28, 2011.
43. European Committee for Standardization (2003) *Road Lighting, Part 2: Performance Requirements, Standard 13201–2*, European Committee for Standardization, Brussels, 16 p.
44. European Committee for Standardization (2003) *Road Lighting, Part 3: Calculation of Performance, Standard 13201–3*, European Committee for Standardization, Brussels, 41 p.
45. Ylinen, A.-M. *et al* (2011) Road lighting quality, energy efficiency, and mesopic design – LED street lighting case study. *Leukos*, **08** (1), 9–24.
46. European Committee for Standardization (2003) *Road Lighting, Part 4: Methods of Measuring Lighting Performance, Standard 13201–4*, European Committee for Standardization, Brussels, 14 p.
47. Niedling, M., Kierdorf, D., and Völker, S. (2013) Influence of a glare sources spectrum on discomfort and disability glare under mesopic conditions. Proceedings of the CIE Centenary Conference "Towards a New Century of Light", CIE Publication x038:2013, OP 47, pp. 341–347.
48. Brückner, S. and Khanh, T.Q. (2007) A field experiment on the perception of automotive rear lights using pulsed LEDs with different frequencies. Proceedings of the 7th International Symposium of Automotive Lighting, ISAL 2007, Darmstadt, Germany, 2007.
49. Wernicke, A. and Strauß, S. (2007) The analytical and experimental study of the effects of pulsewidth-modulated light sources on visual perception. Contribution to ISAL 2007 Symposium, Darmstadt, Germany, September 25–26, 2007.
50. Bodrogi, P., Schiller, Ch., Haferkemper, N., Böll, M., and Khanh, T.Q. (2010) Jüngste Labortestergebnisse der mesopischen Helligkeitswahrnehmung verschiedener Straßenlampenspektren und weißer LEDs. Proceedings of the Licht 2010, Vienna, Austria, October 17–20, 2010, pp. 43–46.

7
Optimization and Characterization of LED Luminaires for Indoor Lighting

Quang Trinh Vinh and Tran Quoc Khanh

In this chapter, the application fields and requirements of energy-efficient and long-term-stable LED luminaires for indoor lighting are formulated. Fundamentals for their development and design with high color and lighting quality are established. In order to achieve high color quality and lighting quality, suitable LED-components and modules have to be selected according to technological and lighting criteria. Following the recent trend of modern LED luminaires with dynamic lighting, variable correlated color temperatures, tunable spectra, and high-end color quality, the focus of this chapter is to explain how a LED luminaire can be designed for high efficiency and excellent color quality and what criteria can be used in this development process. The basic aspects of general indoor lighting are described in [1].

7.1
Indoor Lighting – Application Fields and Requirements

At the first stage of thinking, the terminology *"indoor lighting"* means lighting applications inside a building so that the operation conditions are relatively independent of weather and ambient influences like wind, rain, fog, and traffic situations so that the lighting systems and their users shall not be extensively protected. Going back to the history of human and industrial development, it can be concluded that indoor spaces have been the main working and dwelling areas since the beginning of the first industrial revolution at the end of the nineteenth century and throughout the revolutions of engineering and automobile industry in the period from 1960 until 2000 and of telecommunication (internet and mobile phone technology) in the past two decades. From the point of view of indoor lighting technology, the following applications are considered:

- Industrial lighting (e.g., manufacturing and mounting halls, stores, humidity controlled rooms in chemical industry);

LED Lighting: Technology and Perception, First Edition.
Edited by Tran Quoc Khanh, Peter Bodrogi, Quang Trinh Vinh and Holger Winkler.
© 2015 Wiley-VCH Verlag GmbH & Co. KGaA. Published 2015 by Wiley-VCH Verlag GmbH & Co. KGaA.

- Lighting for education (e.g., schools, lecture halls in university buildings);
- Health area lighting (e.g., hospitals, rooms for chirurgical operations and for general treatments, living houses for elderly people);
- Office lighting (e.g., offices in the industry, governmental buildings, buildings of finance, and the insurance sector);
- Hotel, restaurant, and wellness area lighting;
- Museum and gallery lighting;
- Shop lighting, lighting of representative areas (trade fairs, car shows, and fashion shows);
- Theater, film, and TV studio lighting.

All these applications represent different philosophies, aims and requirements, and, consequently, different lighting concepts. Currently, the planning process for indoor lighting is based on international and national standards containing the lighting quality parameters described in Figure 5.4 in Chapter 5. All these parameters are defined on the basis of the $V(\lambda)$ function (CIE 1924) and are intended to fulfill the two primary tasks of lighting:

1) Maximize task performance (e.g., text reading, assembling machines or their parts by using tools);
2) Minimize the frequency and severity of errors at the workplace, at home, or other places (e.g., error-free handling of machines, moving inside working halls without accidents).

The most important lighting parameters in Figure 5.4 are glare (in terms of UGR), horizontal illuminance E_m (in lx), and homogeneity U_o. With the introduction of LED luminaire technology, vertical illuminance E_v is regarded to be meaningful in order to avoid dark areas inside the room. In Table 7.1, some lighting parameter values for different working tasks [2] are listed.

In the past two decades of lighting and color research and with the introduction of LED technology, including color and white LED components, a rethinking process has taken place resulting in extensive and fundamental new knowledge. This can be described as follows:

1) The values of lighting parameters shall not be the same for all subjects (light source users) working in the same area. Novel luminaire design shall allow individual set-ups and lighting levels like direct/indirect illumination ratios and individually set color temperatures according to the preference of different ages and genders.
2) Dynamic lighting: correlated color temperature and illuminance levels shall not be fixed for all weather and daytime conditions. They shall be adjustable depending on the daytime, season, sunlight contribution during the day for the improvement of concentration of users during the work time and to foster relaxation after the working hours.
3) The chromatic component of optical radiation is analyzed and should play a more important role among modern lighting concepts especially in shop and hotel lighting, museum lighting, and film/TV studio lighting to improve

Table 7.1 Lighting parameters and their minimal allowed limits [2].

No.	Application	Task room	Minimal allowed average illuminance E_m in lx	Glare UGR	Homogeneity U_o	R_a
1.	Store and cooling rooms	Store rooms	100	25	0.4	60
		Shipping, and packaging rooms	300	25	0.6	60
2.	Office	Writing, reading, data processing rooms	500	19	0.6	80
		Technical design, drawing rooms	750	16	0.7	80
		Conference, meeting rooms	500	19	0.6	80
3.	School	Classrooms in ground schools	300	19	0.6	80
		Classrooms for evening school, adult classroom	500	19	0.6	80
4.	Art school	Drawing classrooms	750 5000 K < CCT < 6500 K	19	0.7	90
5.	Hospitals	Waiting rooms	200	22	0.4	80
		Analysis and treatment rooms	1000	19	0.7	90
		Operation rooms	1000	19	0.6	90

the visual quality of illuminated areas and objects (e.g., car show lighting, food lighting). These additional color quality parameters are illustrated in Figure 5.5 in Chapter 5.

Besides the well-known color rendering index (CRI) (see Section 5.2), color temperature preference depending on cultural and regional individuality and on the context of the colored objects to be illuminated is also important for user acceptance. Some studies in the past have shown that for food lighting a color temperature around 3300–3500 K is preferred which is also true for museum lighting. For office lighting purposes, a color temperature range around 4000–4500 K is preferred in Germany (see [3]). According to Table 7.1, a CRI for classrooms with drawing tasks in the art schools of 90 with a correlated color temperature between 5000 and 6500 K is required.

Analyzing the requirements presented in Table 7.1, some further conclusions can be drawn: the glare index UGR has to be 16 for working tasks with higher concentration requirements and 19 for standard requirements. Regarding color quality, the general CRI for office, industrial and school lighting has to be 80. It shall be better than 90 only for complicated applications in medical treatment and operation rooms. From the analysis in Section 5.3 (see Figure 5.19), we have learnt that a general CRI of 80 only means a visual semantic interpretation between "*moderate*" and "*good*" and of 90 a semantic interpretation only slightly better than "*good.*" A LED illumination and a LED lighting system can deliver a color rendering quality of "*just good*" only if the general CRI equals at least 86.

For typical white LEDs (see the example of 34 white LED in Chapter 5), this requirement ($R_a = 86$, based on the first eight unsaturated test color samples (TCSs) in Figure 5.8) is equivalent to the following criterion (which is well usable in practice for the design of white LED light sources): a mean CRI for *all 14* CIE test color samples (TCS1–TCS14) shall be greater than 80 and the special CRI R_9 for TCS09 (saturated red) shall be greater than 40. The basis of this criterion is illustrated in Figure 7.1: this is the relationship among the general CRI (R_a), the mean of all 14 special color rendering indices designated by mean ($R_1 - R_{14}$) in Figure 7.1 as well as R_9, for the sample LED set with the 34 typical white LEDs from Figure 5.25 plus 19 further similar white LEDs.

In the American product regulation Energy Star Eligibility Criteria version 2.1 [4], which is also widely used in many countries in the field of solid state lighting (SSL), the following requirement is formulated: "*the luminaire (directional luminaires) or replaceable LED light engine or GU24 based integrated lamp (non-directional luminaires) shall meet or exceed $R_a \geq 80$.*" But, according to Figure 7.1, for LED luminaires for indoor lighting with a "just good" semantic interpretation of its color rendering quality, a minimal general CRI value of $R_a = 86$ is required.

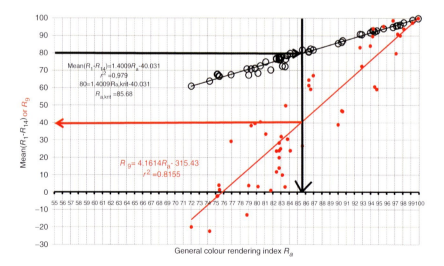

Figure 7.1 Relationship among the general CRI (R_a), the mean CRI for all 14 CIE test color samples designated by mean ($R_1 - R_{14}$) and R_9 for the 34 white LEDs from Figure 5.25 plus 19 further similar white LEDs.

7.2
Basic Aspects of LED-Indoor Luminaire Design

LED-indoor luminaires are technical products for lighting applications. Therefore, there are two methods to consider the design and development process of these luminaires:

- An electro-technical physical method structuring the whole luminaire as a composition of different functional groups (optical, electronic, thermal, and mechanical groups);
- A lighting engineering method analyzing the luminaire from the point of view of lighting aspects.

Both of these two methods are necessary and they are discussed separately here. Later, these two methods are considered together. Generally, the LED-indoor luminaire consists of the following functional units with different tasks and influencing parameters (see Figure 7.2.).

7.2.1
LED, Printed Circuit Board, Electronics

The LEDs are selected (see Section 7.3) and arranged in parallel and serial strings. The number of the LEDs are chosen according to the required luminous flux of the luminaire and by calculating the luminaire's optical efficiency. Three LED-packaging systems can be considered:

- LEDs are SMD-components and soldered on the circuit board with a predefined pitch. The luminous flux of the board is the sum of the luminous fluxes of all LEDs. This arrangement is preferred for big sized indoor luminaires for office, hospital, school lighting, and as soft light with diffuse lighting for gallery and photography studios. The secondary optics is micro-optics (micro-prisms plate) or a lens for each SMD-LED (see Figure 7.3) or diffuse plastic plates (PMMA, polycarbonate, see Figure 7.4).
- LED unit as chip-on-board (COB) with molded phosphor mixture and high luminous flux from a relatively small light emitting surface. This concept is rather used for down light luminaires with reflector optics for hotels, shops, restaurants, and in the high LED power class between 50 and 300 W for industrial lighting inside big factory halls.
- LED unit with remote phosphor plate rather with reflector optics for down light luminaires.

7 Optimization and Characterization of LED Luminaires for Indoor Lighting

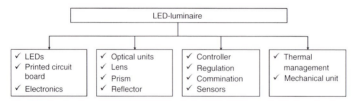

Figure 7.2 Principal structure of a LED luminaire.

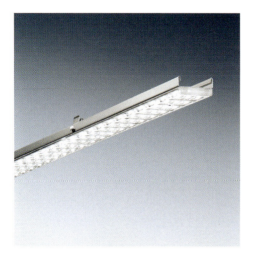

Figure 7.3 LED luminaire with secondary lenses. (Reproduced with permission from Trilux (Arnsberg, Germany).)

Figure 7.4 LED luminaire with diffuse plastic plate. (Reproduced with permission from Trilux (Arnsberg, Germany).)

The printed circuit board can be made of the materials FR4 (as standard), FR4 with thermal vias for an improved thermal conduction, metal core board (MCB) or ceramic material. For most applications, MCB is often used if the pitch of the LEDs on the board is not too small. For COB configurations with a very high thermal density, the ceramic material AlN with its high thermal conductivity of about 170 W (m K)$^{-1}$ is recommended, see Figure 3.70.

The driving electronics is selected according to the number of the LEDs for a serial string and their current. At 350 mA, most white LEDs have about 2.8–3.0 V and at 700 mA, the forward voltage can be 3.0–3.2 V so that the driving electronics can be chosen based on the allowed maximal voltage and electrical power. The selection of suitable electronics is also a question of the lifetime of the electronic components under the given ambient and system operation conditions (maximal load, maximal temperature, humidity). In many cases, the lifetime of the electronics is much shorter than the lifetime of the LED components so that reliability and lifetime shall be considered with system thinking and not only as a question of the lifetime of the LED light sources.

With respect to the electronics of the LEDs, the other question is how the LEDs can be dimmed by DC dimming or with PWM dimming. In the beginning time of color LEDs, that is, in the decades 1970–2000, PWM technology was preferred because the color coordinates were relatively constant during dimming (see Section 4.12). But this method caused stroboscopic effects (see Section 5.11) and it was not efficient under certain circumstances. Meanwhile, many manufacturers of the LED electronics have designed a mixed circuit topology between PWM and current dimming depending on the application. For example, for the first stage of dimming from 100% to 20%, current dimming can be used. From 20% to 1%, the PWM method shall be taken into consideration.

7.2.2
Optical Systems

While using conventional incandescent and discharge lamps, optical radiation was emitted in all solid angle directions (4π) so that reflector optics has been often used. With the introduction of white and color LEDs which emit optical radiation rather into a half-sphere solid angle (2π), lens optics is most widely used. This optical technology has the advantages of scalability, low cost, and a better manufacturing control. However, it has some disadvantages too and can be summarized as follows:

- The lens optics based LED arrangement leads to a periodic pattern of the light emitting area with dark-bright patterns which can increase glare perception for indoor and outdoor lighting applications under certain circumstances (see Figure 7.5).

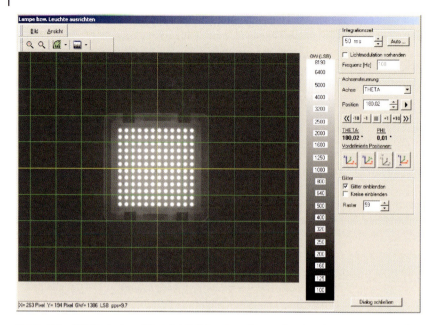

Figure 7.5 LED luminaire with a dark-light pattern. (Reproduced with permission from Ilexa (Ilmenau, Germany).)

- The plastic lenses (PMMA, PC) undergo an aging and yellowing process depending on ultraviolet and blue radiation energy, humidity, and temperature inside the material. This degradation is more important for outdoor lighting applications.

From these two aspects, reflector optics can be meaningfully used for LED tubes as an option for tubular fluorescent lamps and for COB LED configurations.

7.2.3
Controller-Regulation Electronics

The control and regulation electronics have different tasks that are important for current and future smart lighting systems:

1) With temperature sensors on the board or in heat sinks, the control electronics can dim the LED current or reduce PWM duty cycle in order to temporarily avoid the maximum temperature of the LEDs.
2) With optical sensors (as CCD or CMOS camera, infrared sensors) or ultrasonic sensors, the movement or presence of light source users can be detected so that the LEDs can be dimmed accordingly.
3) With a microcontroller (µC hardware), the color temperature, the luminous flux, and the ratio of direct and indirect lighting from the LED luminaire or luminous intensity distribution (so-called LID curve) can be selected or

changed individually or it can be programmed as a function of season, daytime, or orientation of the building relative to the sunlight direction.

4) Communication with the building management and with other building service systems (e.g., heating or warm water supply), with other luminaire systems in the building via DALI or DMX systems, with internet or other telecommunication platforms (e.g., Bluetooth).

5) With a time counter built on the board, lifetime, electrical current, and temperature profiles of the LEDs (data logging scheme) over a long time of 50 000 or 100 000 h can be recorded. It contributes to the decision of replacing certain LED modules.

7.2.4
Thermal Management

For conceptualizing the thermal management, the following steps shall be taken into account:

1) Measurement of LEDs to be used at the predefined currents so that the electrical power, optical power, luminous flux, chromaticity coordinates, and thermal power can be calculated.

2) Measurement or estimation of the thermal resistance for the whole chain from pn junction to the heat sink including the luminaire housing (see Figure 4.46 in Chapter 4).

3) From the thermal power of the LEDs, the thermal resistance of the system and, by assuming maximal ambient temperature for the luminaire, the maximal junction temperature and the maximal temperature for the electronics can be calculated. The maximal junction temperature at a predefined current has a great impact on LED lifetime so that four optimizing options can be used to reduce this temperature:
   a. Reduce the predefined current, which means more LEDs have to be used on the board;
   b. Select appropriate materials for the circuit board and for the adhesive layers with better thermal conductivity;
   c. Select better heat sinks (materials, forms, geometry) with lower thermal resistance;
   d. Apply active forced cooling systems (fan, water cooling, heat pipe, or a Peltier module).

Generally, all steps of thermal management optimization can be simulated by using modern thermal simulation software which is available on the market. With correct input data on geometry and material characteristics, luminaire designers can change and select the best configuration of the luminaire system from the thermal point of view. The deviation between the prognosis of very good thermal simulation programs and measurements with luminaire prototypes is about 3–5 K.

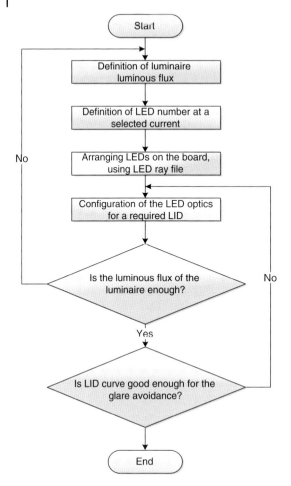

Figure 7.6 Flow chart of the optical and lighting design process. LID: luminous intensity distribution.

The procedure of lighting design, simulation, and optimization can be systematized according to the flowchart in Figure 7.6.

The specification of the luminaire's required luminous flux is either customer specific or it can be traced back to the luminous flux of conventional lamp types and mirror reflector luminaires of T5 fluorescent lamps. Generally, a down light luminaire for shop lighting should deliver 3000 lm and an office luminaire with a tubular fluorescent lamp with 80% luminaire optical efficiency and with a lamp luminous flux of 5800 lm (with 58 W T5 lamps) should have a luminous flux of 4100 lm.

The requirements on LID result from the following questions of design:

1) Is the intended luminaire a task luminaire, a free-standing luminaire for an office, a hanged, or a built-in luminaire?

Figure 7.7 An office LED luminaire with direct and indirect light contributions on the ceiling. (Reproduced with permission from Trilux (Arnsberg, Germany).)

2) Are there direct and indirect parts of the light from the luminaire? In the direct case, the light is emitted from the luminaire down to the workplaces. In the indirect case, light emanates from the luminaire to the ceiling and to the wall surfaces and from there, it is diffusely reflected to the workplaces and to their vicinity (see Figure 7.7).

3) How can the luminaire be assigned a LID curve to avoid glare perception? Glare limits are defined in Table 7.1 and, generally, UGR values of 16 or 19 are applied for most sitting locations in an office.

The starting point for optical simulation and optimization is the LED's ray file measured with a near-field goniophotometer. The ray file contains the absolute pixel-resolved angular luminance distribution of the LED light source, see Figure 7.8. In order to achieve the required LID curve, the optical characteristics of the lens or reflector optics as curvature, radius, focal length, material refractive index, surface roughness, the size of the optical element, and positions of the LEDs inside the optics or in the LED arrangement must be changed systematically and simulated.

7.3
Selection Criteria for LED Components and Units

Besides the design and layout of LED electronics, the selection of suitable LEDs is very important because it has a high impact on the lighting quality and color quality, the lifetime of the LED luminaire and on the user acceptance, and the economic aspects of the LED installation. The selection of the LEDs is guided by the criteria listed in Table 7.2.

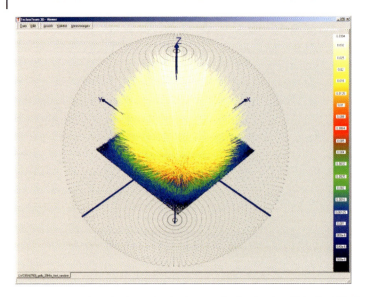

Figure 7.8 Pixel-resolved angular luminance distribution (ray file) of a LED device. (Reproduced with permission from Ilexa (Ilmenau, Germany).)

Table 7.2 Selection criteria for LED components.

No.	Characteristics			
1.	Optics	Geometry of LED package	Angular luminance distribution	Luminous efficacy
2.	Colorimetry	Spectral and colorimetric angular distribution	Warm/cold white LED	Color shift
3.	Electronic	ESD protection	Forward voltage	Maximal allowed current
4.	Thermal	Thermal resistance	Lifetime	

In the current section, selection criteria is analyzed in detail.

7.3.1
Geometry

Generally, the optics (lens, reflector) is designed for a certain LED chip geometry including the width and length of the active chip area, the dome height of the primary optics, see Figure 4.58 in Chapter 4. The LED chip is mounted at a defined distance relative to the focal point of the optics. Because the manufacturing cost of the designed optics is quite high and many LED manufacturers tend to change their LED products within a short time and also from the point of view of flexibility for the choice of the LEDs, it is advisable to select LEDs with a similar geometry and of similar dimensions. These LEDs shall be purchased from a leading LED light source manufacturer that ensures LED supply on a long-term basis so that the LED luminaire manufacturer can use their own optical concept and its software tool over a long time. It should also be possible to change the LED manufacturer without changing the optical and lighting characteristics of the LED luminaires remarkably. It is difficult to have the SMD LEDs as point light sources with a similar geometry from different LED manufacturers. It should be easier to use COB LED configurations from different manufacturers or customer-specific configurations with the same dimensions and same angular luminance characteristics together with the reflector optics.

7.3.2
Spectral and Colorimetric Angular Distribution

If the angular distribution of spectral radiance and the colorimetric properties of the LED to be selected are uniform (i.e., independent from emitting angle) then the chromaticity distribution on the illuminated area can be homogenous by using a simple optics. In case of LEDs with inhomogeneous angular chromaticity distributions, big color differences (color fringes) can be observed on the illuminated wall or table surfaces, see Figure 5.49. In these cases, complicated and expensive LED optics must be designed and applied.

The reason for inhomogeneous colorimetric distributions is the quality of molding of the phosphor mixtures on the blue LED chip with different phosphor thicknesses in different emitting angles. Therefore, blue radiation is absorbed differently along the different optical pathways. In Figure 7.9, the angular distribution of correlated color temperature is illustrated for three different LED types (0° corresponds to the direction perpendicular to the LED chip surface). As can be seen from Figure 7.9, the correlated color temperature of the LED type represented by

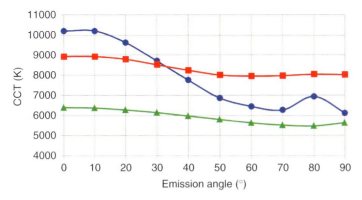

Figure 7.9 Angular distribution of the correlated color temperatures of three different LED types.

the blue curve diminishes by about 4000 K (from 10 000 K down to 6000 K) when the emission angle is changed from 0° to 90°.

7.3.3
Warm/Cold White LED

At the first sight, the selection of LEDs regarding the criterion warm white or neutral white or cool white is an intercultural/interregional question because of the earlier mentioned culture or region dependent color temperature preference of the users. Technologically, it is a question of luminous efficiency, heat development and lifetime. In Figure 7.10, the relationship between junction temperature and board temperature is shown for a warm white LED ($CCT = 2700$ K (correlated color temperature)) and a cool white LED ($CCT = 5000$ K) for different electrical currents.

It is obvious from Figure 7.10 that, at a board temperature of $T_s = 60\,°C$, the junction temperature of the warm white LED at 1000 mA is 97 °C and of the cool white LED is about 89 °C. This difference of 8 K means that the warm white phosphor absorbs more radiation producing more thermal energy. This, in turn, results in a shorter lifetime in comparison to the cool white LED.

7.3.4
Color Shift

Figure 4.66 shows that two different LED types of similar color temperature ($CCT = 4000$ K) from two leading LED manufacturers exhibit very different color

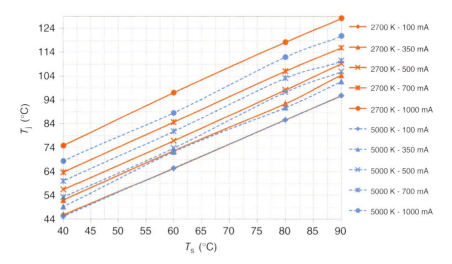

Figure 7.10 Relationship between the junction temperature and board temperature for a warm white and a cool white LED [5].

shift behaviors after 7000 h of degradation under the same electrical and thermal conditions.

7.3.5
Forward Voltage

In Figure 4.67, the LED type 1 showed after 7000 h of degradation a constant and relatively low forward voltage of 2.85 V if the current was 700 mA and the board temperature did not exceed 85 °C. The LED type 2, however, showed in Figure 4.68 a forward voltage of 3.15 V at the beginning and about 2.95 V after 7000 h. A lower voltage at the same current means a lower electrical resistance of the pn-semiconductor layers and the electrical contacts in the package. If LEDs with lower forward voltage are arranged in a long serial string, the total voltage for the driving electronics is even lower.

7.3.6
Choice of the Optimal Current for LEDs

The selection of the operation current is one of the main tasks of LED luminaire development. A higher current results in a higher luminous flux of the LED components and a lower number of LEDs necessary to achieve the predefined luminous flux. However, it is important to know that a higher current of the LEDs shortens their lifetime (see Section 4.10 and Figure 4.72) and reduces their luminous efficiency. Therefore, the optimal current is the current with the best compromise between luminous flux and luminous efficiency (or radiant efficiency). Figure 7.11 shows the relationship between these two quantities with varying LED current.

As can be seen from Figure 7.11, the intersection (showing the optimum current) between the radiant efficiency curves (η_e, dashed curves) and luminous flux curves (ϕ_V, continuous curves) is in the range between 350 and 550 mA. Each individual LED type has a specific internal structure with different technological parameters and therefore a different optimum current at the intersection point.

7.3.7
Thermal Resistance

The thermal resistance between the pn junction and a designated temperature measurement point is specified on the datasheet of LED manufacturers. A low thermal resistance means good thermal conductivity of the phosphor mixture, the semiconductor layers, the substrate, the submount, and the adhesive and, as a consequence, a lower pn junction temperature at constant board or ambient temperature. Modern white LEDs have a thermal resistance in the range of $4-8\,\text{K}\,\text{W}^{-1}$. With two LED types operated under the same electrical conditions, for example, 700 mA and 3.2 V, an electrical power of 2.24 W is being consumed. This causes a thermal power of about 1.568 W (at an optical efficiency of 0.7% or 70%).

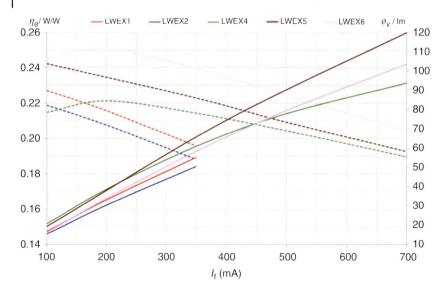

Figure 7.11 Relationship between luminous flux ϕ_v and radiant efficiency η_e for a warm white LED at 80 °C.

If the LED type A has 5 K W^{-1} and the other type B has 8 K W^{-1}, a difference of the thermal resistance of 3 K W^{-1} will result in a junction temperature difference of 3 K W$^{-1} \cdot 1.568$ W $= 4.7$ K. This temperature difference has substantial consequences for LED lifetime [5].

7.4
Application Fields with Higher Color and Lighting Requirements

In this section, some important application fields requiring high quality indoor lighting are described. Such applications include museum and gallery lighting, film and TV lighting, and shop lighting. Requirements for optimal lighting design with best color quality and luminous efficiency are formulated.

7.4.1
Museum and Gallery Lighting

Museums and galleries require high quality lighting as they are important cultural institutions in which art objects like paintings and sculptures from different eras are exhibited. These exhibited objects are the products of creativity and imagination of the artists through their forms, geometry, color, materials, and object sizes. The artists and sculptors of the past eras and the present time are makers, thinkers and observers of their natural, cultural, and social environment and have expressed their ideas and feelings on the social context, on philosophy and thinking streams, on the political and cultural developments via art works. From this

Figure 7.12 Lighting in a museum room. (Photo Source: Technische Universität Darmstadt.)

point of view, the art works in the museums and galleries are cultural transport tools with a high-level educational, aesthetic, cultural, and historical meaning for the next generations. In most cases, museums have a long history and tradition with a high expression power for modern society.

In this context, lighting guidelines can be formulated according to the architecture of the building: the structure of the museum rooms depending on the art epochs, the content, and the atmosphere in which the visitors learn and experience the art works. One important component of this atmosphere in the museum room is lighting, see Figure 7.12.

Museum lighting has two general tasks:

1) *Lighting task*: lighting systems with their illuminance levels, light source spectra and LIDs enable the appearance of color and light and the perception of forms, color contrasts, luminance contrasts, and color textures. Illumination helps visitors experience the value of the art work and the intention of the artist.
2) *Conservation task*: the illuminance of the art works and the spectra of the illuminating light sources have to be optimized in order to avoid or to minimize object damage caused by optical radiation (ultraviolet, visible, and infrared radiation).

Besides the well-known museums and galleries with art works from famous artists, there are also temporary exhibitions on different topics for which certain art works of outstanding artists can be loaned. At each such exhibition, the content, intention, philosophy, atmosphere, and historical epochs of the art works are changed. This content change requires a new temporary (time-limited) lighting concept with different illumination levels, white tones (warm white at

2500–3000 K or daylight white at 5500–6000 K), lighting directions, and local light distributions on the surfaces of the art works. For museum lighting with LED systems, high color rendering quality with variable correlated color temperatures can be achieved.

7.4.2
Film and TV Lighting

There are many applications of LEDs in the film and TV lighting domain. In studio lighting, warm white light sources (tungsten halogen lamps, fluorescent lamps) and for outdoor shooting applications, light sources with simulated daylight spectra have been used in order to illuminate high quality decorations, time-consuming makeups, and high-value costumes by appropriate lighting, see Figure 7.13. Therefore, the correlated color temperature for film and TV lighting should be variable between 3000 and 6500 K.

It has been the intention of film and TV makers for many decades to create film images of high quality and with a high level of color representation. Therefore, light sources and lighting concepts with a high CRI and brilliance for objects in the film scenes are required. For diffuse and shadow-free lighting, high-end luminaires with fluorescent lamps (see Figure 7.14) and for focused illumination spotlights, high-end luminaires with HMI discharge lamps and halogen tungsten lamps with relatively small half-width angles (see Figure 7.15) have been developed.

In Table 7.3, the general CRI (R_a) and the specific color rendering indices for the 14 CIE test color samples (TCS1 – 14, see Figures 5.8 and 5.9) of the tungsten halogen lamps ($CCT = 3200$ K) and of the fluorescent lamp KinoFlo 3200 K used

Figure 7.13 Studio lighting with focused spot lights. (Reproduced with permission from Arnold & Richter Cine Technik (Munich, Germany).)

7.4 Application Fields with Higher Color and Lighting Requirements | 417

Figure 7.14 A luminaire with long tube fluorescent lamps for diffuse and shadow-free lighting. (Reproduced with permission from Arnold & Richter Cine Technik (Munich, Germany).)

Figure 7.15 A studio spotlight with a HMI discharge lamp. (Reproduced with permission from Arnold & Richter Cine Technik (Munich, Germany).)

Table 7.3 Color rendering index of the tungsten halogen lamps and the fluorescent lamp KinoFlo 3200 K used in film industry.

No.	Color rendering index	Halogen tungsten lamp 3200 K	KinoFlo 3200 K
1.	R_1	99.7	92.0
2.	R_2	99.7	90.5
3.	R_3	**99.3**	**73.2**
4.	R_4	99.7	89.3
5.	R_5	99.7	88.3
6.	R_6	99.6	**79.2**
7.	R_7	99.9	87.1
8.	R_8	99.5	90.0
9.	R_9	98.1	97.1
10.	R_{10}	**99.0**	**66.1**
11.	R_{11}	99.7	82.4
12.	R_{12}	**98.8**	**68.1**
13.	R_{13}	99.7	92.4
14.	R_{14}	99.6	83.2
15.	R_a	99.6	86.2

in film industry are shown. From the table, it is obvious that the fluorescent lamps KinoFlo have lower color rendering indices that are subjects to be improved for the following TCSs: yellow-green (R_3), light blue (R_6), yellow (R_{10}), and blue (R_{12}).

In Table 7.4, the color rendering indices of the HMI discharge lamp and the fluorescent lamp for daylight-simulating lighting at 5600 K (KinoFlo 5600 K) are listed.

Table 7.4 Color rendering indices of the HMI discharge lamp and a fluorescent lamp (KinoFlo 5600 K).

No.	Color rendering index	HMI discharge lamp	Kinoflo 5600 K
1.	R_1	92.6	96.9
2.	R_2	94.4	97.3
3.	R_3	95.5	**89.7**
4.	R_4	92.5	96.2
5.	R_5	94.7	96.2
6.	R_6	92.7	93.1
7.	R_7	90.5	96.1
8.	R_8	**85.3**	96.3
9.	R_9	**60.7**	95.9
10.	R_{10}	87.7	87.3
11.	R_{11}	91.2	92.5
12.	R_{12}	91.5	91.9
13.	R_{13}	91.5	97.7
14.	R_{14}	97.5	93.5
15.	R_a	92.3	95.2

As can be seen from Table 7.4, the CRI of the HMI discharge lamp equals a relatively low value for saturated red (R_9), purple (R_8), and yellow (R_{10}). The CRI of the fluorescent lamp KinoFlo 5600 K has to be improved for the TCSs R_3 (yellow-green) and R_{10} (yellow). A general requirement is that LED luminaires for film and TV lighting have to be able to deliver high color rendering indices for all objects to be illuminated in film industry and also to exhibit a variable correlated color temperature.

7.4.3
Shop Lighting

Shop lighting has the task of delivering an appropriate brightness level by general illumination and also to direct the attention of the customers toward the value of an object for sale. Latter task is carried out by an object oriented and focused illumination system. The lighting concept contains a diffuse illumination component combined with a concentrated illumination component with imaging optics (imaging lens or reflector optics). During the design of the latter component, glare perception of the customers and the damaging impact of ultraviolet and infrared radiation have to be minimized.

The spectra of the usable light sources include warm white tones with correlated color temperatures between 2700 and 3200 K for the illumination of bakery products and cakes (see Figure 7.16) or for meat and sausage (see Figure 7.17). The neutral white or daylight-similar white with a color temperature in the range between 4000 and 5000 K is often applied for the illumination of car motor show, fashion show, and jewelry shops, where the higher part of the bluish optical radiation together with a focused lighting are intended to increase the brilliance of the objects (car metal, watch).

From Figures 7.16 and 7.17 it can be recognized that, for successful shop lighting, light sources with high color rendering quality especially in the red and orange

Figure 7.16 Illumination of a shop for bakery products. (Reproduced with permission from Bäro (Leichlingen, Germany).)

Figure 7.17 Illumination of a butcher's shop. (Reproduced with permission from Bäro (Leichlingen, Germany).)

color range should be developed and used in order to represent the originality and value of the exhibited objects.

7.4.4
Requirements for LED Luminaires with High Color Quality

From the previous discussions on film and TV lighting, museum lighting and shop lighting and in the view of the practical lighting conceptions for high-end hotels, restaurants, hospitals, and advertisement agencies, requirements for high quality LED luminaires are described in Table 7.5.

The requirements for high quality LED luminaires of Table 7.5 can be explained as follows:

Table 7.5 Requirements for high quality LED luminaires.

No.	Criterion	Value range	Remarks
1.	CCT	Variable 2700–6500 K	Taking into account intercultural color temperature preference
2.	R_a	Better than 93	See Figure 5.19
3.	R_9 (Red TCS 9)	Better than 85	See Figure 5.20
4.	R_{12} (Blue TCS 12)	Better than 85	
5.	Color difference from the reference white point	$\Delta u'v' < 0.002$	See Section 5.8.2
6.	Luminous efficacy	Better than 90 lm W^{-1}	Luminous efficacy of the luminaire

1) The adjustment of correlated color temperatures shall not only take the types of the illuminated objects and the illumination atmosphere into account but also the intercultural/interregional (for Latin America, Asia, North America, and Europe) color temperature preference.
2) The minimal limits for the color rendering indices shall follow the results of the semantic interpretation illustrated in Figures 5.19 and 5.20.
3) The chromaticity difference between the white point of the luminaire and the reference white point (with equal correlated color temperature) shall be small to avoid a slight but disturbing color tones in the white point of the luminaire when its light is directed onto a white surface. The white point will not be purplish, greenish, or bluish.
4) With the best conventional light sources for indoor lighting providing high quality illumination like the T5 fluorescent lamps, HMI or HTI (high pressure discharge lamps), a lamp luminous efficacy of $100-110\,\text{lm W}^{-1}$ shall be achieved. With an optical efficiency of 80%, a luminaire luminous efficiency of $80\,\text{lm W}^{-1}$ can be really expected. Nowadays, with modern LED types and intelligent conceptions, a luminous efficiency of $90\,\text{lm W}^{-1}$ or higher are therefore achievable.

7.5
Principles of LED Radiation Generation with Higher Color Quality and One Correlated Color Temperature

In the lighting laboratory of the authors, several types of white LEDs available on the market were analyzed and divided into three ranges according to the general CRI R_a:

1) $R_a = 80-86$ for school lighting, offices, rooms for basic treatment in hospitals, industries;
2) $R_a = 86-92$ for shop lighting, food lighting, general car shows, and hotels;
3) $R_a > 92$ for museum lighting, for film and TV lighting, fashion shows, and high-end shop lighting.

For this comprehensive analysis, a number of commercial white LEDs from leading LED manufacturers has been measured concerning their spectral radiant flux in W nm^{-1} between 380 and 780 nm at a current of 350 mA and a sensor temperature of 80 °C (see Figure 4.48). In this section, only some representative white LEDs are analyzed.

In Figure 7.18, the relative emission spectra of the warm white LEDs with $R_a = 80-86$ are shown.

Figure 7.18 shows that the position of the maximal wavelength of the blue pumping LED chip is between 440 and 460 nm, in the range of the best excitation for the phosphors in order to achieve the maximal conversion efficiency. A choice of the wavelength between 460 and 465 nm can yield better color quality but the

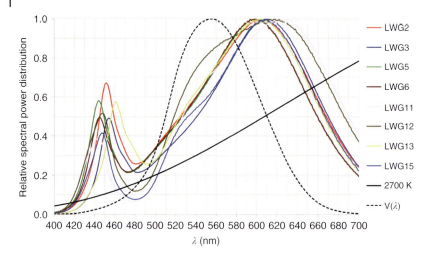

Figure 7.18 Relative emission spectra of the warm white LEDs with $R_a = 80-86$ (CCT = 2700–3000 K, 2700 K: black body radiator) [6].

conversion efficiency can then be reduced. Hereby it is obvious that the measured phosphor systems have a maximal emission wavelength between 600 and 620 nm.

If the maximal wavelength of the phosphor system comes closer to 620 nm then the CRI for red colors and skin tones is higher and the gap between the blue chip spectrum and the phosphor spectrum between 480 and 490 nm is lower which can be clearly observed for the case of the spectrum of the white LED type LWG11 (the purple curve in Figure 7.18). Although the general CRI R_a for the eight saturated CIE test colors is better than 80, the average CRI Mean $(R_1 - R_{14})$ for the 14 CIE test colors including the six saturated colors (TCS9–TCS14) is lower than 80 (see Figure 7.19).

Figure 7.19 shows that the special CRI R_9 for the saturated red color is low for the case of the white LEDs analyzed here. Only the white LED types LWG11 and LWG12 exhibit an R_9 value better than 35 and can be used in LED luminaires for just good color rendering quality, see Figure 7.1. These two LED types have a maximal phosphor wavelength of 615 and 620 nm. Generally, all these warm white LEDs can be used only in private household rooms with moderate color quality requirements. The luminous efficacy is not the focus of the analysis in this chapter because the current development of LED efficiency is dynamic. However, with a luminous efficacy between 57 and 72 lm W^{-1}, these LEDs can achieve the value of conventional compact fluorescent lamps (CFLs).

In Figure 7.20, the relative emission spectra of the warm white LEDs with $R_a = 86-92$ and $CCT = 2700-3000$ K are illustrated.

Analyzing the spectra in Figure 7.20, it can be concluded that the maximal wavelength of the orange-red phosphor systems is between 620 and 630 nm with the exception of the LED type LWVG1. The wavelength shift of 5–7 nm compared with the spectra in Figure 7.18 is the reason for the improvement of the special CRI R_9 for the red test color TCS9 being better than 50 with the exception

7.5 Principles of LED Radiation Generation

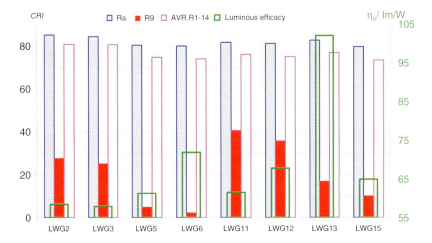

Figure 7.19 Luminous efficacy (η_v) and color rendering index of white LEDs from Figure 7.18 (2700–3000 K, $R_a = 80$–86). R_a: General color rendering index R_a; AVR.$_{R1-14}$: average color rendering index for all 14 CIE TCS, R_9: color rendering index for TCS9 (red test color) [6].

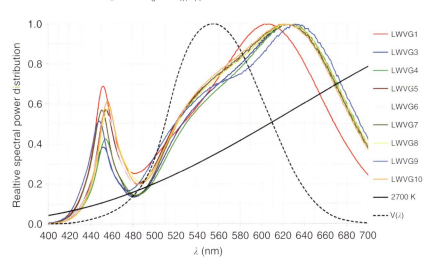

Figure 7.20 Relative emission spectra of the warm white LEDs with $R_a = 86$–92 (CCT = 2700–3000 K, 2700 K: black body radiator) [6].

of LWVG1 (see Figure 7.21). On the other side, the white LED LWVG1 has the less serious gap (local minimum with low spectral radiance) at 480 nm from all regarded white LEDs that results in the improved color rendering indices for blue, cyan, and light-green colors. The average CRI for the 14 CIE TCSs is better than 80 for all white LEDs with a luminous efficacy between 57 and 69 lm W^{-1}. This result indicates that a better color rendering does not automatically mean a lower luminous efficacy.

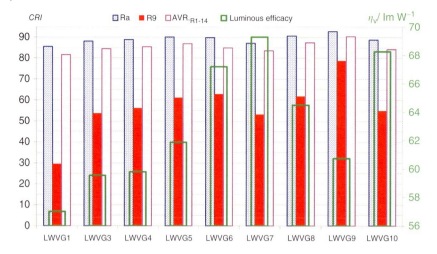

Figure 7.21 luminous efficacy (η_v) and color rendering index of white LEDs (2700–3000 K, $R_a = 86$–92). $AVR_{\cdot R1-14}$: average color rendering index for the 14 CIE TCS, R_9: color rendering index for TCS9 (red color) [6].

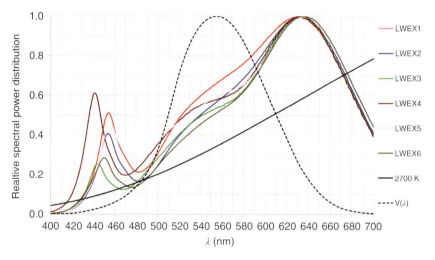

Figure 7.22 Relative emission spectra of the warm white LEDs with R_a better than 93 (CCT = 2700–3000 K, 2700 K: black body radiator) [6].

Considering the relative emission spectra of the warm white LEDs illustrated in Figure 7.22 with $CCT = 2700$–3000 K and with R_a being better than 93, the conclusion is that improved red phosphor systems with a maximal wavelength at 635 nm have been used for the case of these white LEDs.

These new red phosphor systems result from intensive phosphor development in the last years and allow an R_9 value of better than 70 for the white LEDs analyzed here. The innovation of phosphor technology is visible from these spectra: there is a second phosphor system between 520 (green) and 540 nm in Figure 7.23

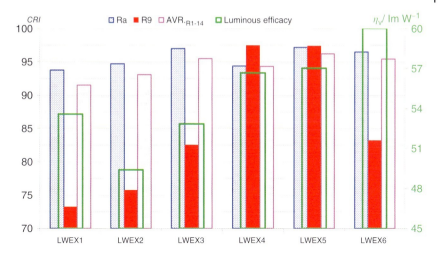

Figure 7.23 Luminous efficacy and color rendering index of white LEDs (2700–3000 K, $R_a > 93$) [6].

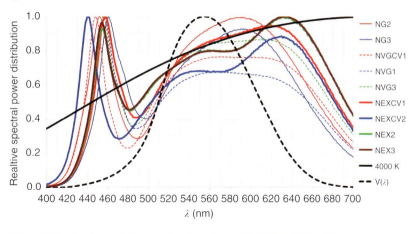

Figure 7.24 Relative emission spectra of neutral white LEDs with $R_a > 80$ (CCT = 4000–4800 K, 4000 K: black body radiator 4000 K) [6].

for the case of the three LED types LWEX3, LWEX4 and LWEX5. The positive consequence of this innovation is a high R_9 value and a high average CRI for all 14 CIE test colors.

Figure 7.24 shows the relative emission spectra of neutral white LEDs with color temperatures between 4000 and 4800 K and with a general CRI better than 80.

As can be seen from Figures 7.24 and 7.25,

- The maximal wavelength of the phosphor systems in white LED types NG2 and NG3 is around 590 nm and hence the R_9 value (for red test colors) equals 25. The average color rendering indices for the 14 CIE TCSs are below 80 (see Figure 7.25).

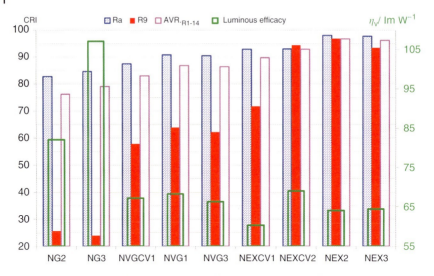

Figure 7.25 Luminous efficacy and color rendering index of white LEDs (4000–4800 K, $R_a > 80$) [6].

- The LED types NVGCV1, NVG1, and NVG3 have a CRI value between 86 and 92 (see Figure 7.25). Their relative emission spectra show a flat curve between 560 and 620 nm (see Figure 7.24) so that the R_9 value is between 55 and 65 and the average CRI of the 14 CIE TCSs is between 80 and 86.
- The LED types NEXCV2, NEX2, and NEX3 have a green-yellow phosphor system with a maximal wavelength around 540–560 nm and of a red phosphor system with a maximal wavelength at 630 nm. The indices R_a and R_9 are greater than 90 so that these white LED types can be used in high-end applications like fashion shows, museum and shop lighting, or in film and TV lighting.

All of these white LED light sources work at a fixed color temperature and many of them exhibit a CRI higher than 90. In order to combine both features, that is, a high CRI and variable color temperatures according to Table 7.5, a new method of LED luminaire design has to be found which is the subject of Section 7.6.

7.6
Optimization and Stabilization of Hybrid LED Luminaires with High Color Rendering Index and Variable Correlated Color Temperature

7.6.1
Motivation and General Consideration

The demand of high quality SSL applications worldwide such as film and television lighting, shop lighting, architectural lighting, and museum lighting has increased in the past years. Indeed, as mentioned already, a high quality SSL system is necessary to expose the merit and the beauty of famous paintings drawn by oil colors or by water colors in museums or in churches. Similarly, color quality of lighting is very important to make the clothes in fashion shops or fruits in food stores more attractive for the customers while the true color of their own skin, hair, textiles, or clothes must also be ensured. On the other hand, the shop or museum management intends to reduce the energy consumption of the lighting system although it still has to ensure both a high lighting quality and good temperature stability. Thus, in this section these constitute the main motivations of the authors of this book. To implement this, a general consideration of the optimization and stabilization of the most important lighting quality aspects of so-called hybrid LED lamps (those that mix the light from several chip-LEDs and phosphor converted LEDs or PC-LEDs, see below) in relation to specific color objects in high quality SSL applications is very important. As a first step, the essential concepts are defined and explained here:

7.6.1.1 Hybrid LED lamp and spectral LED models

Hybrid LED lamps consist of color semiconductor LEDs and white PC-LEDs that are combined together in the most appropriate form that optimizes lighting quality and color quality. The name *"hybrid LED lamp"* reflects this kind of combination. In order to optimize and stabilize their lighting quality aspects despite temperature changes, the LED light sources must be characterized accurately. Spectral LED models discussed in Section 4.6 are necessary to achieve the high accuracy in order to characterize the LED spectra.

7.6.1.2 Lighting Quality Parameters, Their Limits and Proposals for the Most Appropriate LED-Combination for Hybrid LED Lamps

In this section, essential lighting quality aspects of hybrid LED lamps are considered including chromaticity, correlated color temperature, whiteness, color rendering indices, luminous flux, luminous efficacy, temperature dependence, and current dependence. The aim is to describe examples for the optimization and stabilization of hybrid LED lamps in some particular lighting applications. The following criteria are fixed according to Table 7.5: color rendering indices should be higher than 90 and luminous efficacy should be higher than 90 lm W^{-1}. Thermal behavior must be stabilized and current dependence must be taken into account in the spectral LED models in order to ensure highest accuracy in all calculations for an available combination of color semiconductor LEDs and white PC-LEDs in these hybrid LED lamps.

7.6.1.3 Two Methodical Demonstration Examples and Their Tasks

First the example of museum lighting was chosen to demonstrate the optimization for the average value of a set of new object specific color rendering indices for oil colors which is called oil color index. The aim of this example is to optimize the hybrid LED lamp to provide an excellent illumination of high color quality for these oil color objects in a museum. As a second example, a hybrid LED lamp for shop lighting is shown to be optimized for the red objects of shop lighting

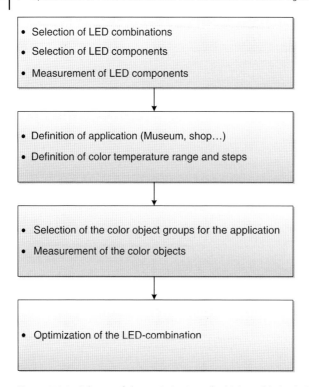

Figure 7.26 Scheme of the optimization of a high-end hybrid LED luminaire.

applications. The process of optimization for a high-end hybrid LED luminaire with variable color temperature consists of the steps shown in Figure 7.26.

According to Figure 7.26, the specific spectral reflectance of a representative set of color objects of the given application has to be measured and then the optimization should be carried out. For museum lighting, shop lighting and film lighting, hundreds of color objects were measured in the Laboratory of Lighting Technology of the Technische Universität Darmstadt. They are shown in a systematized manner (in addition to the set of color objects shown in Section 5.4) in Section 7.6.2.

7.6.2
Spectral Reflectance of Color Objects in Museum, Shop and Film Lighting

In the museums oil paintings from different historic eras are exhibited so that the optimization on color quality has to take the spectral reflectance of the real oil colors into account. In addition, the visitors with their skin tones, hairs, and clothes have to be illuminated with high color quality which is also the aim of shop and film lighting. Therefore, the spectral reflectance of different types of hair, skin tones, clothes, and oil colors are shown here.

7.6.2.1 Analysis of the Spectral Reflectance Curves of the Color Objects

On the basis of the measured results in Figure 7.27, it can be recognized that, although the spectral reflectance of the different skin types is different depending on gender, age, and country, their rising edges always appear at about 600 nm. Hence, in order to enhance the color quality for these human skin types, the peak wavelength of the spectral component of the PC-LED or of the color semiconductor LEDs in the hybrid LED lamp should appear at about 600 nm. Similarly, Figure 7.28 shows the very low spectral reflectance values of different hair types (below 0.5 in the range of 400–740 nm). These spectral reflectance curves are

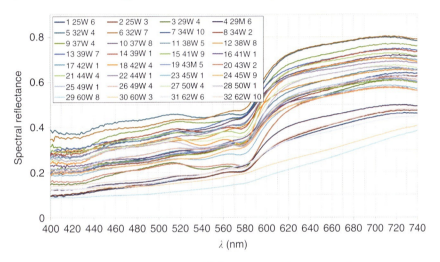

Figure 7.27 Spectral reflectance curves of different types of human skin [7].

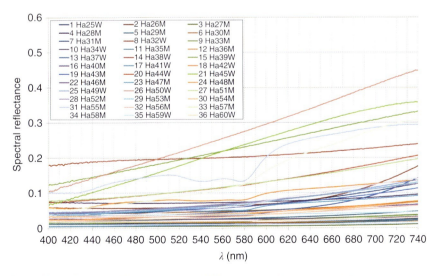

Figure 7.28 Spectral reflectance curves of different types of human hair [7].

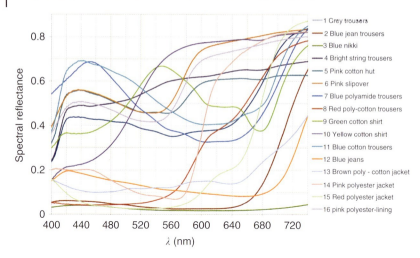

Figure 7.29 Spectral reflectance curves of different clothes [7].

often flat and some of them have their rising edges at about 680 nm in the deep red spectral region or about 600 nm in the amber/orange spectral region. The measured clothes exhibit higher spectral reflectance values and the distribution of their rising edges is fairly wide in the range of 400–740 nm such as 420, 530, 560, 590, or 680 nm (see Figure 7.29) in the blue, green, yellow, orange, and red spectral regions. This requires a hybrid LED lamp with a full spectrum with all spectral components.

7.6.2.2 Color Objects and Their Spectral Reflectance Curves in the Museum (Oil Paintings)

Figure 7.30 shows 79 original oil colors often used by painters. These colors are denoted as RO1–RO79.

In Figure 7.31, the spectral reflectance of a group of saturated red oil colors is illustrated. Red oil colors played an important role in painting art between the fifteenth and nineteenth centuries. Most red colors have only a reflectance of 0.1 (10%) until 580 nm and then exhibit a sharp edge with high spectral reflectance in the wavelength range between 610 and 740 nm. Therefore, red semiconductor LEDs or red phosphor systems with a high emission in the range between 610 and 680 nm should be selected.

In Figure 7.32, the spectral reflectance of a group of green oil colors is shown containing maximal reflectance values between 520 and 540 nm. Therefore, green phosphor systems and green semiconductor LEDs have to emit in this wavelength range reasonably.

Figure 7.33 shows the spectral reflectance of saturated blue oil colors. It is obvious, especially in case of the color ROG15, that spectral reflectance is high even in the wavelength region between 480 and 500 nm requiring a less serious (i.e., not so deep) gap between the spectral emission of the blue LED and the phosphor in

Figure 7.30 Seventy-nine well-known oil colors RO1–RO79 (oil paintings) [8].

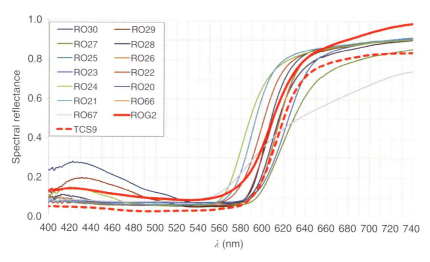

Figure 7.31 Saturated red group of oil colors including strong red (ROG2) and TCS9 as a similar CIE test color sample.

this range. The peak wavelength of the blue pumping LEDs should be selected to be at 460 nm.

7.6.2.3 Qualification and Prioritization of the Color Objects for the Optimization of Museum Lighting

For museum lighting, the colors of these 79 oil color objects play the main role while skin, hair, textile, and clothes play an important role in the optimization for film and shop lighting. In order to quantify the color rendering property of the hybrid LED lamp regarding these important color objects, their object specific CIE

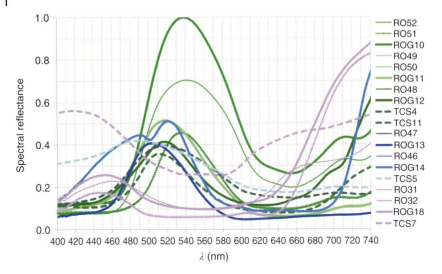

Figure 7.32 Green group of oil colors including ROG10 and not similar to CIE TCS; emerald green group including ROG12 and not similar to CIE TCS; chromium oxide green group including ROG12 similar to CIETCS4 and CIETCS11; phthalate green group including ROG13 and not similar to CIE TCS; cobalt turquoise group including ROG14 similar to CIETCS5; violet group including ROG18 similar to CIETCS7.

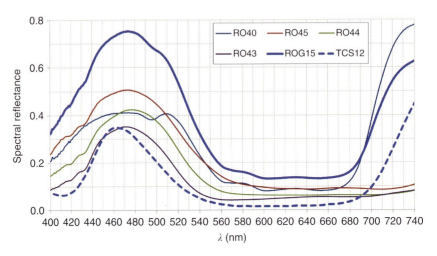

Figure 7.33 Saturated blue group of oil colors including ROG15 similar to CIETCS12.

color rendering indices (see Section 5.4) were computed for the earlier described types of skin, hair, textile, and clothes. These object specific color rendering indices of the 79 oil color objects were denoted by $RO_1 - RO_{79}$ and their average value was denoted by AVR_{O1-79} and called the oil color index. Similarly, the average value of the skin, hair, textile, and clothes color rendering indices are called skin color index, hair color index, textile color index, and clothes color index.

7.6.3
Optimization Process for the Hybrid LED Combination with High Color Quality

7.6.3.1 Role of LED Components and Primary Proposals for LED Selection

There are many different color semiconductor LEDs and different warm white PC-LEDs belonging to diverse binning groups of various LED manufacturers in the world. All of them are possible to be used as a WW-R-G-B (warm white and RGB-LEDs) LED combination. However, yellow, amber, and orange semiconductor LEDs will not be used because their temperature stability and current stability is by far not optimal (see Section 4.2). In the laboratory of the authors, many different LEDs such as royal blue ($\lambda_p \sim 440-450$ nm), blue ($\lambda_p \sim 465-470$ nm), green ($\lambda_p \sim 520-530$ nm), red ($\lambda_p \sim 630-640$ nm), deep red ($\lambda_p \sim 660-670$ nm) semiconductor LEDs, and many different types of warm white PC-LEDs ($CCT \sim 2700-3000$ K) as well as their combinations from three well-known leading LED manufacturers (called A, B, and C in this book) were studied. Each color semiconductor LED plays its own specific role to compensate for the spectral gaps in the spectrum of the warm white PC-LED.

Particularly, the royal blue LED is to complement for the royal blue spectral region in order to enhance the color quality of blue objects such as blue jeans, the oil royal blue (ROG15), or azure blue (RO45). Experimentally, the royal blue LED ($\lambda_p \sim 440$ nm) of manufacturer A achieved good color rendering for the CIE blue test color sample (TCS12). Similarly, the green LED ($\lambda_p \sim 520-530$ nm), the red LED ($\lambda_p \sim 630$ nm), and the deep red LED ($\lambda_p \sim 660$ nm) are intended to fill the spectral gaps in the green, red, and deep red spectral regions in order to increase the color rendering indices for the green, red, or deep red objects, respectively. Especially, the blue LEDs ($\lambda_p \sim 465-470$ nm) are intended to adjust the spectral gap between the blue spectral component and the phosphor spectral component in the original spectrum of the warm white PC-LED. In practice, this PC-LED gap plays a special, important role in the improvement of the general CRI of all hybrid LED lamps.

7.6.3.2 LED Selection for the Most Available LED Combination of the Hybrid LED Lamp

In the example described in this section, to enhance color quality, temperature stability, luminous efficacy, and feasibility of composition and operation of the hybrid LED lamps, the following LED combination was chosen:

- As semiconductor LEDs, the royal blue ($\lambda_p \sim 450$ nm), blue ($\lambda_p \sim 470$ nm), green ($\lambda_p \sim 530$ nm), red ($\lambda_p \sim 630$ nm) semiconductor LEDs of manufacturer C, and the deep red semiconductor LED ($\lambda_p \sim 660$ nm) of manufacturer B turned out to be the most appropriate choice at the time of writing. The reason was their temperature stability, radiant efficiency, and color quality reflected by their luminous efficacy, temperature stability, and the values of the object specific color rendering indices.
- As warm white PC-LED ($CCT \sim 3000$ K), manufacturer C was chosen because this warm white PC-LED was the most efficient (about 105 lm W^{-1} at 350 mA and 60 °C) and this one had the best temperature dependence (under three

MacAdam ellipse sizes while the operating temperature was allowed to change from 25 to 100 °C in two cases, 350 and 700 mA). In addition, the peak wavelength of its phosphor spectral component was about 600 nm which can well replace the unstable yellow or orange semiconductor LEDs.

7.6.3.3 Optimization of the Hybrid LED Lamp for Oil Color Paintings

The idea of the control system structure as a multi-input multi-output system (MIMO system) can be applied in case of the hybrid LED lamps. In that control system structure, the central objects of the MIMO system with their inputs and its outputs (or their set signals and feedback signals) must be described.

Objects of the MIMO System The investigated color objects are the oil colors, the skin colors, the hair colors, the textile colors, and the clothes colors. The parameters for the identification of these color objects are quantified according to the standard definitions and equations of the CIE CRI. Otherwise, royal blue ($\lambda_p \sim 450$ nm), blue ($\lambda_p \sim 470$ nm), green ($\lambda_p \sim 530$ nm), and red ($\lambda_p \sim 630$ nm) semiconductor LEDs of manufacturer C, the deep red semiconductor LED ($\lambda_p \sim 660$ nm) of manufacturer B, and the warm white PC-LED ($CCT \sim 3000$ K) of manufacturer C are characterized by their spectral LED models. Thus, in the MIMO system of the hybrid LED lamp, the spectral LED models of the color semiconductor LEDs and the warm white PC-LED become the central objects of the closed loop control structure illustrated in Figure 7.34 in order to satisfy the color quality demands for the color objects.

Inputs of the MIMO System The main inputs of the MIMO system are the forward currents for all LEDs and the subordinate input is the operating temperature of the entire hybrid LED lamp. In the optimization process for the hybrid LED lamp, this subordinate input is just to reflect the hot binning or the cold binning condition. In Section 4.4, it was shown that hot binning is better than cold binning concerning the selection of LEDs and the construction of LED luminaires. Therefore, the operating temperature in the optimization process was always chosen to be 80 °C.

As it is well known, the higher the forward current of the warm white PC-LED, the lower its luminous efficacy. Consequently, the forward current of 700 mA or lower (about 500 mA) should be selected as a compromise between output luminous flux and luminous efficacy so that the luminous efficacy of the finished hybrid LED lamp should be always around 90 lm W^{-1} as the acceptable limit compared to other white PC-LEDs with a high CRI. Then, in the condition of 80 °C for the operating temperature and 700 mA for the warm white PC-LED's forward current, the optimization process shall be carried out to achieve an appropriate control for the forward currents of the royal blue, blue, green, red, and deep red semiconductor LEDs.

Outputs (or the Set Signals and the Feedback Signals) of the MIMO System Three main set signals including the set signal for correlated color temperature (CCT_{set}),

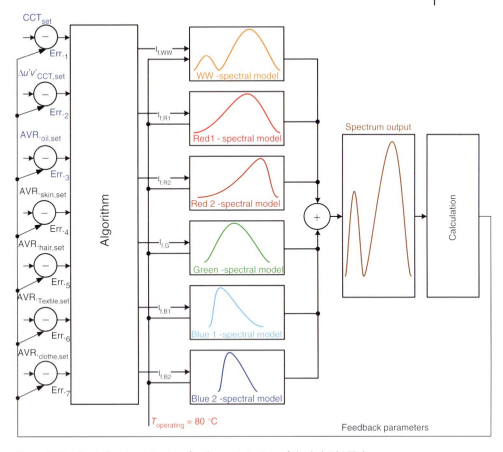

Figure 7.34 Control system structure for the optimization of the hybrid LED lamp.

whiteness ($\Delta u'v'_{CCT,set}$) and the average oil CRI ($AVR_{oil,set}$) and four subordinate set signals including the set signals for the (average) skin color (rendering) index ($AVR_{skin,set}$), the (average) hair color (rendering) index ($AVR_{hair,set}$), the (average) textile color (rendering) index ($AVR_{textile,set}$), and the (average) clothes color (rendering) index ($AVR_{clothe,set}$) had to be considered by the algorithm in the optimization process.

If the subordinate parameters such as the skin color index, hair color index, textile color index, clothes color index are required only to fulfill the acceptance limit (i.e., to be higher than 80) then the correlated color temperature must be kept at specific constant values such as 3000, 4000, 5000, or 6500, and the whiteness must be kept at its appropriate level (i.e., less than $\Delta u'v' = 0.001$) and the oil color index has to be optimized to reach its maximal value. Otherwise, a signal bus for all feedback signals determined from the output spectrum of the hybrid LED lamp is necessary to calculate the offsets between the set values and the feedback values (denoted by $Err._1 - Err._7$). Finally, based on these errors, the controller of the

MIMO system in the control system structure performs its algorithm to achieve the specified optimal values, see Figure 7.34.

7.6.3.4 Optimized Spectral Power Distributions of the Hybrid LED-Lamps

Figure 7.35 shows four optimized spectra for museum lighting, for the correlated color temperatures 3000, 4000, 5000, and 6500 K, respectively. In case of 3000 K, both the red LED and the deep red LED play a very important role whereas the green LED and the blue LED are just included to support color balance and to reduce the spectral gap between the blue spectral component and the phosphor spectral component. The royal blue LED has nearly no influence on this optimization because its role is substituted by the royal blue being the pumping light source in the warm white PC-LED.

On the contrary, for 5000 and 6500 K, the royal blue LED yields an important contribution to the final, mixed spectrum because the spectral amount of royal blue which is built in the warm white PC-LED is not intensive enough for the entire color mixing of this hybrid LED lamp. The green LED and the deep red LED are to support color balance and to increase the CRI for the whole spectrum of the hybrid LED lamp. In contrast, the red LED plays only a minor role in these cases because the wide phosphor spectrum of the warm white PC-LED replaces its role. Finally, in case of 4000 K, the role of the royal blue LED and the red LED become more similar and the weight of the deep red LED and especially the green LED become much more important than the others for the entire color mixing in the hybrid LED lamp because these LEDs must enhance both the right and the left spectral components of the original spectrum of the warm white PC-LED in order to ensure the correlated color temperature of 4000 K. Note that the measure of white point quality is always constrained to be less than $\Delta u'v' = 0.001$ and the optimization target is the maximization of the oil color index.

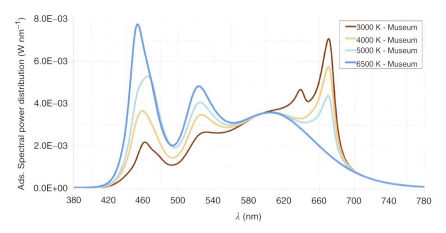

Figure 7.35 Optimized absolute spectral power distributions (SPDs) for 3000, 4000, 5000, and 6500 K, for museum lighting.

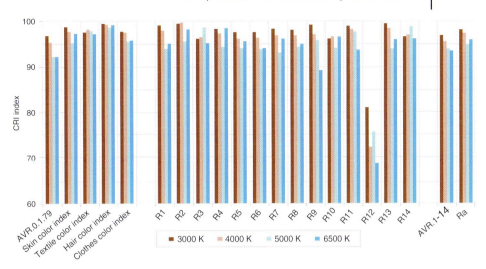

Figure 7.36 CRI values of the four optimized spectra.

Figure 7.36 shows the different color rendering indices for the above-mentioned *optimized* cases (3000, 4000, 5000, and 6500 K) for museum lighting. As can be seen from Figure 7.36, the average value of the object specific oil color rendering indices ($AVR_{\cdot O,1-79}$) as the main color quality parameter is always higher than 90 in all four cases. And the highest oil color index is achieved in case of 3000 K ($AVR_{\cdot O,1-79} = 97$).

In addition, the subordinate color quality parameters such as the skin, hair, textile, and clothes color indices are also always higher than 95 and their highest values are 98, 97, 99, and 97 in case of 3000 K, respectively. Likewise, the average value of the 14 CIE color rendering indices and the general CIE CRI of all cases are also higher than 93 and they reach their highest value in case of 3000 K. However, the twelfth (blue) CRI (R_{12}) is equal only to 82, 73, 76, and 68 for the case of 3000, 4000, 5000, and 6500 K, respectively, because of the use of the royal blue LED with the peak wavelength of only 450 nm. This value will be higher if a good royal blue LED with a peak wavelength of about 440 nm is used in the hybrid LED lamp while it still has to ensure the requirements of other lighting quality parameters such as current stability, temperature stability, luminous efficacy, and feasibility of construction and operation as a real hybrid LED lamp. Moreover, the other special CIE color rendering indices are higher than 90 and their highest values are almost as high as those in case of 3000 K except for R_3 and R_{14} (these two *CRI*s reach the highest value in case of 5000 K).

An important new finding is that, although the optimization was originally intended for museum lighting only, these optimization results are also usable for film lighting, shop lighting, or fashion show lighting and a differentiation is not necessary.

It can be stated that the optimized spectra satisfy the high color quality demand of oil paintings in museum lighting and, although the same LED system with a similar combination of the color semiconductor LEDs and the warm white PC-LED was applied, the warm white spectrum with $CCT = 3000$ K achieved the highest levels of color quality parameters compared with the other optimized spectra. Therefore, the warm white SSL system (3000 K) should be used for high quality solid state museum lighting applications. Furthermore, in all the earlier optimized cases, the correlated color temperature and the whiteness were always constrained to equal the predefined values (3000, 4000, 5000, and 6500 K) and to be less than $\Delta u'v' = 0.001$, respectively. Their luminous efficiencies always achieved about 90 lm W^{-1}.

7.6.4
Stabilization of the Lighting Quality Aspects of the Hybrid LED Lamp

Section 7.6.3 described the optimization of important lighting quality aspects of the hybrid LED lamp. The chosen color objects were relevant to a museum lighting application with oil color paintings and the optimization target was the average oil CRI for the 79 dedicated oil color objects with 3000, 4000, 5000, and 6500 K under the hot binning condition (80 °C). In a museum, the ambient temperature can be kept constant because of the severe requirements of temperature and humidity for painting conservation. However, the operating temperature of the hybrid LED lamp in other applications is not always constant but changes depending on different factors such as weather, seasons, daytime or nighttime, regions, or other operating conditions. When there is a change in the operating temperature then the previous color mixing rate will not be correct any longer. This will cause a change of all lighting quality parameters of the hybrid LED lamp such as its chromaticity, correlated color temperature, whiteness, color rendering indices, luminous flux, and luminous efficacy. Therefore, in this section, a control system structure similar to the one described in Figure 7.34 is used to optimize and stabilize the object specific color rendering indices of the warm white hybrid LED lamp (3000 K) applied to the objects of the general shop lighting application.

The hybrid LED lamp is established from the various semiconductor LEDs and the different warm white PC-LEDs of three manufacturers (A, B, and C) as well as their combinations with the hot binning condition (80 °C). Then, the optimized spectra were investigated at different operating temperatures and the change of lighting quality parameters are pointed out. Now, an important new item is the operating temperature: it must be added as a new feedback signal to the controller in the control system structure and a new control algorithm is used to stabilize the correlated color temperature and the whiteness of the hybrid LED lamp while its color rendering indices, luminous flux, and luminous efficacy is kept at stable and high levels. In this section, the general shop lighting applications are used as an example for the stabilization of the important lighting quality aspects of the hybrid LED lamp. The application selected for this example is a (conventional) shop with red color objects and the optimized target is the average value of the object specific color rendering indices of these red objects (in other words, the red color index) and the stabilized main lighting quality parameter is the correlated color temperature of the hybrid LED lamp.

7.6.4.1 Control System Structure for the Stabilization of Lighting Quality

Improvement of Lighting Quality Parameters Figure 7.37 shows the control system structure to stabilize the correlated color temperature and the whiteness of the hybrid LED lamp. The control system keeps color rendering indices, luminous flux, and luminous efficacy at high and stable levels when the operating temperature changes.

Limit of the Operating Temperature In the real operation of LED lamps, the available operating temperature of the hybrid LED lamp is usually in the range from about 25 to about 80 °C according to the investigations of the authors. In modern indoor lighting applications, the temperature in the rooms such as in offices, shops, museums, or churches is often about 25 °C because there are often air-conditioning systems in those rooms. In a high quality hybrid LED lamp, the heat sink is designed so as to transport the thermal power from the inside to the outside efficiently and therefore, its operating temperature never exceeds 80 °C.

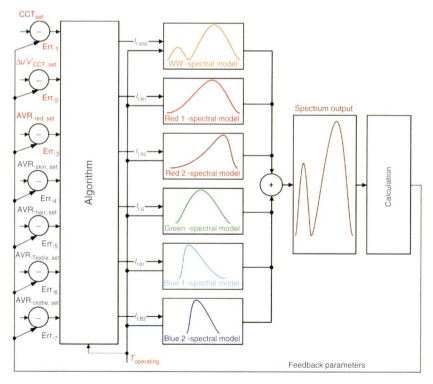

Figure 7.37 Control system to stabilize and improve the important lighting quality parameters of a hybrid LED lamp.

Control System Structure Basically, this control system structure (Figure 7.37) is similar to the one shown in Figure 7.34. A difference is that *operating temperature* becomes a *new feedback signal* brought into the controller, and the average value of the 79 object specific color rendering indices of the oil color objects ($AVR_{\text{oil,set}}$) is replaced by the one of the 42 object specific color rendering indices of red objects in general shop lighting applications ($AVR_{\text{red,set}}$) illustrated in Figure 5.24. In addition, the seven previous feedback signals are still directed into the feedback bus in order to calculate the seven error measures ($Err._1 - Err._7$) for the controller similar to the control system in Figure 7.34. But the operating temperature feedback signal is brought directly into the controller so that the controller is able to control the currents.

Objects of the MIMO System in the Control System For the MIMO system described in Figure 7.37, the forward current of the main component (the warm white PC-LED at 3000 K of manufacturer C) is 500 mA in order to enhance the total luminous efficacy of the hybrid LED lamp from 87 lm W^{-1} to greater than 90 lm W^{-1} (according to Table 7.5) and to achieve a longer lifetime of the LEDs (see Section 4.10). Consequently, the color mixing rate must be established for the case of 500 mA by the new optimization before stabilization. In addition, the total luminous flux must deliver 3000 lm corresponding to the conventional luminous flux value of traditional luminaries used in shop lighting (e.g., a conventional down light with 35 W discharge lamps with a luminous efficacy of 100 lm W^{-1}). In a real implementation, the LED shall be arranged on the circuit board so as to achieve optical homogeneity.

7.6.4.2 Results of Optimization and Stabilization

Figure 7.38 illustrates the control of the emission spectra of the individual LED components in order to stabilize the lighting quality parameters of the hybrid LED

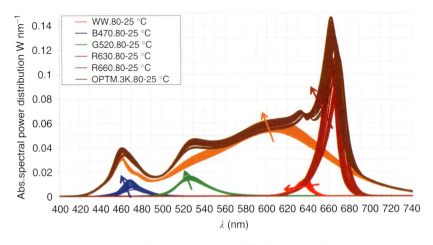

Figure 7.38 Control of the LED components to stabilize lighting quality parameters.

lamp when the operating temperature is reduced from 80 down to 25 °C. Note that 80 °C corresponds to the temperature at the hot binning condition for the original optimization in Section 7.6.3. Under uncontrolled conditions, this temperature is not always constant (80 °C) like in the binning condition and this temperature change may cause undesired variations of correlated color temperature, color rendering indices, and the white point of the LED luminaire.

Therefore, based on the control system structure described above, a current control algorithm was established for each LED component within the chosen temperature interval (in 5 °C steps). Consequently, by means of the adjustment of the currents of the individual LED components, each spectral component of the hybrid LED lamp was changed by the current control algorithm appropriately in order to stabilize its correlated color temperature (at 3000 K) and its whiteness (to be less than $\Delta u'v' = 0.001$). Its color rendering indices, luminous flux, and luminous efficacy were always kept at a high and stable level.

Table 7.6 describes the lighting quality parameters of the hybrid LED lamp at different operating temperatures (from 80 down to 25 °C) achieved by the current control algorithm. The results show that the correlated color temperature and whiteness are kept constant at 3000 K and with less than $\Delta u'v' = 0.001$. There is nearly no change with temperature for the main color rendering indices such as the red object specific color index, the CIE CRI R_9 for the saturated red test color sample (TCS9), the general CRI (R_a), the average value of the 14 CIE TCSs (AVR_{1-14}), the skin color index, the hair color index, the textile color index, or the clothes color index. Moreover, some red object specific color rendering indices such as red rose, red salami, and red tomato also remain constant and stable at a high level.

Table 7.6 Lighting quality parameters of the hybrid LED lamp after applying the control algorithm.

$T_{operating}$ (°C)	80	75	70	65	60	55	50	45	40	35	30	25
$R_{Red\ rose}$	99	99	99	99	99	99	99	99	99	99	99	99
R_{Salami}	98	98	98	98	98	98	98	98	97	97	97	97
R_{Tomato}	99	99	99	99	99	99	99	99	99	100	100	100
R_9	97	97	97	97	97	97	97	97	97	97	97	97
AVR_{1-14}	97	97	97	96	96	96	96	96	96	96	96	96
R_a	98	98	98	98	98	98	98	98	98	97	97	97
$R_{Red\ color\ index}$	98	98	98	98	98	98	98	98	98	98	98	98
$R_{Skin\ color\ index}$	99	99	99	99	99	99	99	99	99	99	99	99
$R_{Textile\ color\ index}$	98	98	98	98	98	98	98	98	98	97	97	97
$R_{Hair\ color\ index}$	100	100	100	100	100	100	100	100	100	100	100	100
$R_{Clothes\ color\ index}$	98	98	98	98	98	98	98	98	98	98	97	97
Φ_v (lm)	3098	3141	3186	3233	3281	3327	3372	3415	3458	3504	3558	3504
η_v (lm W^{-1})	91	92	93	94	96	97	99	99	100	100	101	104
CCT (K)	3000	3000	3000	3000	3000	3000	3000	3000	3000	3000	3000	3000
$\Delta u'v'_{CCT}$	10^{-4}	10^{-4}	10^{-4}	10^{-4}	10^{-4}	10^{-4}	10^{-4}	10^{-4}	10^{-4}	10^{-4}	10^{-4}	10^{-4}

The energy consumption aspect can also be considered. When the real operating temperature goes below 80 °C then both the luminous flux and the luminous efficacy of the hybrid LED lamp increase. Table 7.6 shows that, when the operating temperature is decreased from 80 down to 25 °C, the luminous flux and the luminous efficacy of the LED lamp increase from 3098 to 3504 lm and from 91 to 104 lm W^{-1}, respectively. Obviously, these results confirm that the current control algorithm described above fulfills very well its purposes including the stabilization of the lighting quality parameters and the improvement of the energy consumption parameters of the hybrid LED lamp when operating temperature changes.

References

1. ISO/CIE ISO 8995-1:2002(E)/CIE S 008/E:2001. (2001) *Joint ISO/CIE Standard, Lighting of Work Places Part 1: Indoor*, ISO/CIE.
2. DIN EN 12464-1. (2011) *Ausgabe: 2011-08: Licht und Beleuchtung – Beleuchtung von Arbeitsstätten – Teil 1: Arbeitsstätten in Innen-räumen; Deutsche Fassung EN 12464-1:2011*, Beuth Verlag GmbH, Berlin.
3. Völker, S. (1999) Eignung von Methoden zur Ermittlung eines notwendigen Beleuchtungsniveaus. PhD thesis. Technische Universität Ilmenau.
4. Energy Star (2012) Energy Star Program Requirements for Luminaires. Partner Commitments, Eligibility Criteria Version 1.2
5. Vinh, T.Q. and Khanh, T.Q. (2011) Gefährliche Mischung – Wirkung von Strom und Temperatur auf die LED-Lebensdauer. Zeitschrift Licht, Heft 11-12, pp. 76–80.
6. Vinh, T.Q. and Khanh, T.Q. (2012) *Eine Analyse weißer LEDs – Lichtausbeute, Farbwiedergabe und Farbtemperatur*, Zeitschrift Licht, Heft 11-12, Pflaum Verlag, München, pp. 70–75.
7. Krause, N., Bodrogi, P., and Khanh, T.Q. (2012) Bewertung der Farbwiedergabe – Reflexionsspektren von Objekten unter verschieden weißen Lichtquellen. Zeitschrift Licht, Heft 5, pp. 62–69.
8. Pepler, W. and Khanh, T.Q. (2013) Museumsbeleuchtung – Lichtquellen, Reflexionsspektren, optische Objektschädigung, Teil 1. Zeitschrift Licht, Heft 1–2, pp. 70–77.

8
Optimization and Characterization of LED Luminaires for Outdoor Lighting

Tran Quoc Khanh, Quang Trinh Vinh, and Hristo Ganev

8.1
Introduction

Until 2006, LED components were used rather for signaling purposes (traffic lamps or signal lamps for household equipments like radios or TV receivers) or for automotive lighting applications such as rear lamps. With the dynamic development of white LEDs since 2007 with moderate color quality (CRI values in the order 60–78), white LED technology has found its way first to outdoor lighting (park lighting, street lighting, facade lighting) because of the following advantages:

- The white LED components can be switched on and off with a rise time and fall time in the order of less than 10 μs. It can also be efficiently dimmed from 100% down to 0% of the maximal light value (see Section 4.12) so that intelligent and smart street lighting systems can be conceived with LEDs as light sources. With smart lighting systems the users hope to see the highest potentials for dynamic control with abundant possible illuminating effects, flexible change, optimal quality, and environmental protection, besides the advantages of energy saving (e.g., luminous efficacy).
- The luminous flux of the white LEDs as components or in LED modules depends on temperature. At lower temperatures, luminous flux can be increased (see Section 4.3) and the lifetime of the LED luminaire can be longer (see Section 4.10). These positive aspects must be mentioned because street lighting systems are active mostly during the night time hours with relatively lower air temperatures.
- With the lifetime of conventional light sources in the range of 3–5 years for high pressure mercury and high pressure sodium lamps, maintenance costs for changing the lamps are too high. On the basis of the expected lifetime of the LED components and LED electronics of more than 50 000 h or approximately 12 years these costs can be substantially reduced. The conditions for this

LED Lighting: Technology and Perception, First Edition.
Edited by Tran Quoc Khanh, Peter Bodrogi, Quang Trinh Vinh and Holger Winkler.
© 2015 Wiley-VCH Verlag GmbH & Co. KGaA. Published 2015 by Wiley-VCH Verlag GmbH & Co. KGaA.

assumption are an optimal design of the LED luminaire taking all characteristics of all luminaire unit groups into account.
- With the dynamic development of the luminous efficacy of white LEDs within a short time period to the values of around 150 lm W^{-1} measured under realistic operating conditions (500–700 mA, 85 °C), the luminous efficacy of LEDs is higher than the one of the best conventional light sources (see Figure 8.1) enabling the development of efficient LED luminaires.
- In comparison to conventional light sources, especially to the low and high pressure sodium lamps, white LEDs for street lighting with CRI values in the range of 65–85 do have better or comparable color rendering quality so that objects in the street lighting domain can be well detected and identified by their color and their color contrast (see Table 8.1 and Figure 8.2).

With the positive potentials of white LEDs it is possible to have long-lasting, energy efficient and reliable LED street luminaires with reasonable color quality for general street lighting. This requires knowledge on the LED luminaire structure and the methodologies of luminaire design process. These two aspects are the subjects of consideration and analysis in the next sections.

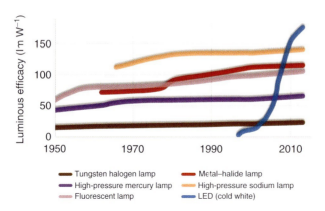

Figure 8.1 Development of the luminous efficacy of different light sources.

Table 8.1 Typical color rendering indices for different light sources used in street lighting.

No.	Types of lamps	General color rendering index R_a
1.	High pressure mercury-vapor lamps	60–70
2.	Fluorescent lamps	70–80
3.	Compact fluorescent lamps	70–85
4.	Halogen metal–halide lamps	65–80
5.	High pressure sodium-vapor lamps	20–25
6.	White LEDs	65–80

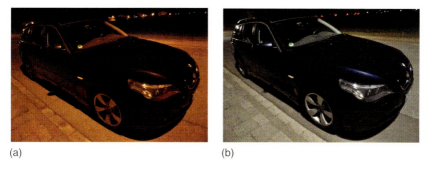

(a)   (b)

Figure 8.2 A blue car under the light of a high pressure sodium vapor lamp (a) and a white LED (b). (Photo Source: Technische Universität Darmstadt.)

8.2
Construction Principles of LED Luminaire Units

Generally, a modern LED street luminaire consists of the following functional units (see Figure 8.3):

8.2.1
Mechanical Unit

The housing should be esthetic and contain a high thermal conduction and convection. The LEDs and their PCBs (printed circuit boards) are thermally connected with the luminaire housing over the heat sink systems and adhesive. The housing will give then the thermal energy via convection to the ambient air. The surface of the housing on the outside should have a high reflectance in order to minimize the absorption of radiation from the environment (sun and sky radiation) in summer time (see Figure 8.4). In almost all countries of Asia, Europe, and America, the global sun irradiance is about $900-1000\,\text{W}\,\text{m}^{-2}$ at noon in the summer. With a housing surface of about $0.4\,\text{m} \times 0.4\,\text{m}$ and with a dark color of

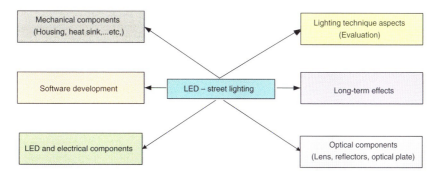

Figure 8.3 Different functional units of a modern LED street luminaire.

Figure 8.4 Housing of an LED luminaire with flat and light surface. (Reproduced with permission from the company WE-EF (Bispingen, Germany).)

the housing (dark grey) with an absorbance of 75%, the absorbed thermal radiation can have a power of $1000\,\text{W}\,\text{m}^{-2} \times 0.4 \times 0.4\,\text{m}^2 \times 0.75 = 120\,\text{W}$. This thermal power from the sun has a high influence on the energy flow of almost all street luminaires which are designed with an electrical power of 40–100 W. This surplus of thermal energy can heat up the housing of the luminaire to a temperature of 65–70 °C. Such a high temperature negatively influences the luminous efficacy and lifetime of the LEDs and electronic components (e.g., integrated circuit components, liquid capacitors) inside the housing. The outside of the housing should be formed for a maximal convection but relatively flat in order to avoid the deposition of different dirt substances on the housing leading to a possible reduction of convection.

8.2.2
Electronic Unit

The electronic unit contains the voltage supplying circuit, control unit (e.g., microcontroller), driving electronics and PCB with the soldered white LEDs. All these groups are connected over cables and plugs and with the optional sensors like temperature sensors, movement detection sensors or light sensors. The software of the microcontroller overtakes the operation, regulation, and even monitoring function for the whole luminaire.

In Figure 8.5, the scheme of the electronic unit of a typical LED street luminaire is illustrated. The supply voltage of 230 V/AC (110 V/AC in the United States) can be directly converted inside a box in the pole foot into a DC voltage of 12, 24, or 48 V, and, in this manner, it should be the best solution because the luminaire on the pole can be operated with a low DC voltage. The alternative option is to convert the AC voltage inside the luminaire mounted on the pole head into a defined low DC voltage (VE).

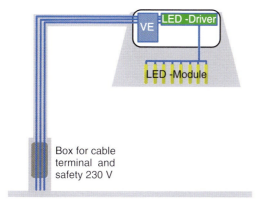

Figure 8.5 Electronic unit arrangement inside an LED street luminaire.

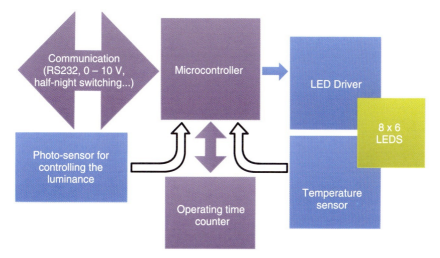

Figure 8.6 Scheme of the electronic group of a modern LED street luminaire.

In Figure 8.6, the scheme of the electronic group of a modern LED street luminaire is shown. The central component is the microcontroller which has the following tasks:

1) *Communication and Control* : For the communication and controlling tasks different analog and digital interfaces can be used: RS232, an analog voltage between 1 and 10 V, power line or DALI interface and a circuit for the activation of the dimming triggering signal for half-night-dimming (dimming after midnight time if traffic intensity is substantially reduced). With these interfaces, the level of street illumination, the luminous intensity distribution (LID) or, optionally (more for the future), the spectral distribution of the light sources of the LED luminaire can be changed reasonably depending on traffic frequency, weather (rain, fog, snow), and time during the night. Looking into

the future, a time counter built in in the electronics can be used in order to monitor and control the compensation of luminous flux depreciation because of the aging effect of all components inside the luminaire (see also Section 8.5 on maintenance factor (MF)). With a movement sensor based on the ultrasonic principle or by the aid of an infrared camera, the illumination level on those parts of the street that are currently used by traffic participants can be increased up to the level required by international standards for traffic lighting. For the time period in which the road is not in use, however, the illumination level can be reduced by about 18–20% and, possibly, the pedestrian zone can be illuminated with more light in order to prevent crime. If a temperature sensor is built on the electronic board of the LEDs (Figure 8.7), the temperature profile can be scanned and stored in a memory component (e.g., EEPROM). When the maximal allowable temperature is reached, the LED current or the duty cycle of the pulse-width modulated LEDs can be lowered. This can be done in many southern countries in summer evenings when the air temperature and the luminaire housing temperature is high but it is already dark so that the street should be illuminated and the LED luminaires must be switched on and, consequently, the LEDs heat up beyond their predefined maximal temperature. After midnight, as the ambient temperature decreases, the maximal current of the LEDs can be adjusted if necessary.

2) *Electric Supply and Dimming for the LED Components* : The microcontroller sends out a control signal that adjusts the current (i.e., the current amplitude) and the PWM duty cycle. The driving electronics adjusts correspondingly the current or the dimming factor for the LED group. Because of the low voltages of the LEDs (approximately 3 V and lower), several LEDs can be operated from the same driving electronic unit and built on the same MCPCB (metal core printed circuit board). The advantage of this topology is that in case of failure of one LED, only one MCPCB must be replaced and all other MCPCBs

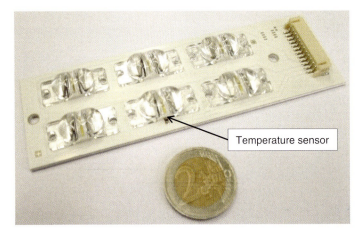

Figure 8.7 A temperature sensor on the LED board. (Photo Source: Technische Universität Darmstadt.)

can continue operating. The microcontroller controls and dims each MCPCB separately. The converting electronics transforms the voltage from 230 V/AC to 24 V or 48 V/DC and must withstand a voltage strength of min. 4 kV.

8.2.3
Optical System

Because of the emission characteristics of the blue LED chip and of the phosphor mixture, the luminous intensity of the white LED corresponds to the angular LID of a Lambertian diffuse radiator (see Figure 8.8). Therefore, an optical system is necessary to rearrange the angular distribution of the light emission from the luminaire on the pole head above the street to the road (main traffic area) and to the pedestrian zone. At the beginning of LED street lighting technology, some developers formed the optics so that no light could reach the house façades with the explanation that this illumination method should increase energy efficiency. Meanwhile, a small and well-established part of luminaire light should be

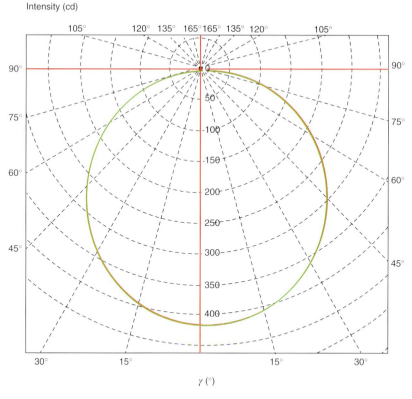

Figure 8.8 Luminous intensity of a diffusely emitting white LED. (Image source: Technische Universität Darmstadt.)

intended to illuminate the house façades to focus the attention of the citizens and the tourists on the architecture and the beauty of a city in the night.

For those streets that have to comply with standard *illuminance* criteria (e.g., the S class according to the standard EN 13201 in Europe), a homogenous illumination on the road and on the pedestrian zone should be achieved. For streets with *luminance* criteria (ME class), it is not only the average luminance on the road, but also the general uniformity (U_0) and the uniformity along the road center line (U_1) that are very important. Besides these factors, the glare index *TI* (threshold increment according to EN 13201) and the illuminance (in lx) ratio between the pedestrian zone and the road have to be taken into account. From the point of view of luminaire developers, the fulfillment of the uniformity U_1 is a difficult issue requiring more effort during the optical development. The forming of the luminaire's LID for LED luminaires with a pole height of lower than 5 m without glaring effect is also a complicated task and is described in more detail in the next sections.

In the era of conventional light sources in the past several decades, luminous flux has been emitted into all directions (a solid angle of 4π sr in case of discharge lamps) and, therefore, reflectors with aluminum or silver coating have been used. Because of the fact that the LID of white LEDs is emanating only into a half-spherical space (a solid angle of 2π sr, see Figure 8.8), glass and plastic lenses can be considered. Meanwhile, almost all plastic lenses are designed as free-form optics with an optical efficiency of about 85–90% (see Figure 8.9).

Because of the discussions on the long-term stability of all components of LED luminaires for street lighting, different optical concepts have been formulated including the application of glass lenses or hard coat reflectors. Plastic lenses can withstand temperatures that are in the range of 90–130 °C and are mounted directly on the hot LEDs. Consequently, lens materials like PMMA or Polycarbonate can exhibit yellowing or opaque aging effects. The advantages of plastic lenses are their well-controllable manufacturing process, low cost, and light weight. In contrast, glass has a higher weight and the manufacturing of free-form glass optics must be optimized. An alternative option is reflector optics allowing the design of a high efficient optics with LED components or LED arrays.

Because each high power white LED has only a luminous flux of 100–250 lm depending on the operating current, the design of an LED luminaire for illuminating a street with several thousand lumens should consist of many LEDs which can be divided into different groups in several different MCPCBs. Previously, the

Figure 8.9 Reflector (left) and lens optics for LED street luminaires.

Figure 8.10 Chain of the components that determine the luminaire's luminous efficacy.

optical concept of the LED street luminaire was that each white LED with its own optics can illuminate the whole street with a characteristic and predefined LID curve and all LIDs of all LEDs in the luminaire can be summed up into a total LID of the same shape. The advantage of this concept is that, in case of defect lenses, only the absolute luminance on the road will be lowered although the uniformity of illumination remains constant.

The chain of the components that determine the luminaire's luminous efficacy can be seen in Figure 8.10.

The electrical efficiency of modern LED electronics is in the range of 90–94%. In 2014, cold white LEDs have had a luminous efficacy of about 150 lm W^{-1} at 350 mA and 85 °C while the optical efficiency of almost all primary optics has been about 92%. The protection glass of the luminaire housing has a Fresnel reflectance of 8% or a transmittance of max. 92%. From these assumed values, the maximum luminous efficacy of the whole LED luminaire can be estimated as follows:

$$\eta_{\text{Luminaire}} = 0.92 \cdot 150 \text{ lm W}^{-1} \cdot 0.92 \cdot 0.92 = 116 \text{ lm W}^{-1}$$

At the time of writing, the efficiency of the electrical and optical systems has already approached the technologically best possible values. If the luminous efficacy of the luminaires remains their decisive factor and if this can be improved in the future then the key factors of development are therefore the luminous efficacy *of the white LEDs* and their *thermal management*.

8.3
Systematic Approach of LED Luminaire Design for Street Lighting

8.3.1
General Aspects

A design process for LED street luminaires consists of several steps. Some of these steps can be conducted parallel and some of them must be performed stepwise whereas the results of one step can influence the logic, the organization and the results of other procedures. These steps are systematized in the flowchart of Figure 8.11.

8.3.2
Definition of Specifications – Collection of Ideas

An LED street luminaire is a combination of mechanical, thermal, electronic, semiconductor, and optical components. At the end of the design process, a

452 | 8 *Optimization and Characterization of LED Luminaires for Outdoor Lighting*

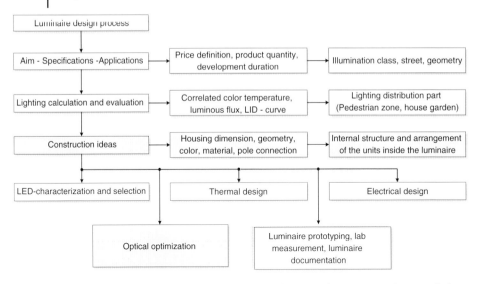

Figure 8.11 The systematic approach of the LED luminaire design process for street lighting.

luminaire is a lighting product and it is used for illumination purposes. So, its most important specifications are related to its lighting engineering characteristics.

At the beginning of the process, the designers and product managers should define what kind of streets the luminaires have to be developed for. It depends on the type, class and geometry of the streets, on the one hand. On the other hand, the following questions have to be answered:

1) Are the luminaires intended for a new street and with new pole distances which can be defined in the planning period of the street (see Figure 8.12 and Table 8.2)? In this case, the housing of the luminaire can be a new one

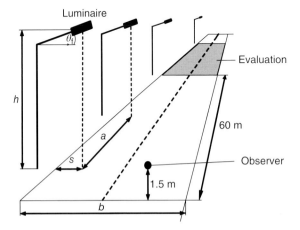

Figure 8.12 A typical street geometry, see Table 8.2.

8.3 Systematic Approach of LED Luminaire Design for Street Lighting

Table 8.2 Parameters of street geometry

Symbols	Meaning
b	Road width
h	Pole height
s	Lighting point overhang
θ_f	Tilt angle of the luminaire
a	Pole distance

intended for a new and modern street and the LID can be optimized for the greatest possible pole distance for maximal energy saving for the whole street.
2) Or is the luminaire rather intended for typical old streets with well-defined pole heights and distances and with a known street geometry (e.g., where the position of the poles cannot be shifted)?

For example, for the development of luminaires to be used in the central regions of Germany, typical values of street geometry were measured in different cities and typical streets, see Figure 8.13.

With different average street geometries, street luminaire developers can define the required illumination class of the street as it is categorized according to the different applications of street luminaires. Different ME class or S class streets usually exhibit different geometries (see Figures 8.14 and 8.15). The luminaires can therefore consist of different LED units with the same LID and the transition from one street class to the next street class can be achieved by mounting more or less LED units. The advantage of this way of thinking is to be able to build different

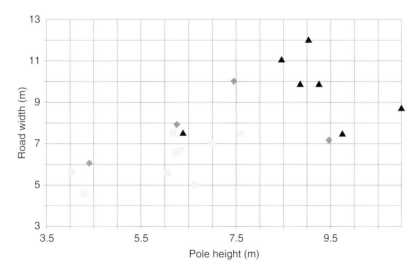

Figure 8.13 Road width versus pole height of typical streets in central Germany. (Data Source: Technische Universität Darmstadt.)

Figure 8.14 An ME2-class main street in Darmstadt, Germany, illuminated partly by cool white LEDs at 5700 K and partly by yellowish-orange high pressure sodium luminaires. (Photo Source: Technische Universität Darmstadt.)

Figure 8.15 An S4-class street in Darmstadt, Germany, illuminated partly by white high pressure mercury luminaires at 4000 K and partly by yellowish-orange high pressure sodium luminaires. (Photo Source: Technische Universität Darmstadt.)

luminaires consisting of numerous LED modules of the same LID and therefore with different luminous fluxes of the luminaire for different street categories.

A marketing study prior to luminaire development provides valuable information on the price, design features, technological development tendencies, and user acceptance for the future luminaires to be designed. Development engineers can evaluate the intended application of the luminaire like the envisioned temperature and humidity range, rain intensity, wind strength, quality of AC voltage and the possibilities of mounting the luminaire onto the pole.

An important decision for the user acceptance of the envisioned luminaire is the definition of its correlated color temperature (CCT). It is also an intercultural or interregional question depending on the culture or country of the intended users. It is well known that different countries, regions or cultures do have a different CCT preference (e.g., either warm white in the range of 2700–3200 K or cool white between 5000 and 6500 K). For the engineers, the decision for a white light in warm white or in cool white is an important analytical aspect. A cool white LED has about 15–20% more luminous flux than a similar warm white LED so that a luminaire with the same luminous flux needs more warm white LEDs, that is, there is a higher cost and much more technical effort. Or, a luminaire with warm white LEDs requires a higher LED current with the adverse consequence on the lifetime of the warm white LEDs or on the higher LED voltage and higher cost of voltage supply units. According to Figure 7.10, a warm white LED exhibits a higher junction temperature than a cold white LED at the same board temperature and this leads to a shorter lifetime of the LED components. As of today, a serious attention of product managers is required to be paid to the selection of and the decision for warm white LEDs.

The most important decision from the lighting engineer's point of view is the definition of the LID curve which is decisive for the uniform illumination of the street. The following questions must be answered:

1) How does the LID curve influence uniformity and glare in the context of lighting quality?
2) How can an optimal LID curve be conceived or calculated in order to fulfill the minimal requirements on uniformity and on glare avoidance with the minimum necessary luminous flux of the luminaire for a certain type of street geometry?

The process of the selection of the best LID curve is time consuming but investing this time is meaningful. The steps of selecting the most appropriate LID curve are shown in Figure 8.16. Once the best LID curve has been selected, development engineers can use the shape of this LID curve to design and form the optical system (i.e., the lens, the micro-optics or the reflectors) and determine the absolute luminous flux of the luminaire. From the latter value, they will be able to calculate the luminous flux of all white LEDs and then, in turn, when the luminous efficacy of the LED components is known, the necessary number of white LEDs at a predefined LED current can be computed.

In Figure 8.17, different LID-curves for different street types and street geometries are shown. As can be seen from Figure 8.17, the left diagram has a maximal luminous intensity at $\gamma = 50°$ and the red curve has a higher luminous intensity in the range between 15° and 45° to achieve a better uniformity on an ME class street. The blue curve exhibits weaker angular intensity maxima so that house façades on the luminaire's side get more illumination. It can also be recognized that the right diagram has a red curve with a strong luminous intensity from 15° until 75° which can be used for an ME class type street with a longer pole distance. But in this case (right diagram), there is a high glare potential because of the strong luminous intensity even at 72°–75°. The central diagram has a very low luminous

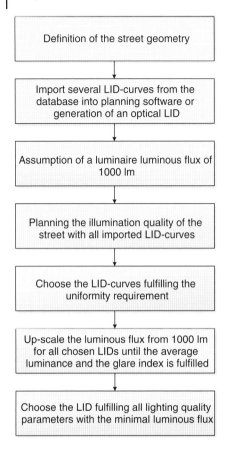

Figure 8.16 Selection of the best luminous intensity distribution.

intensity at $\gamma = 0°-30°$ with a maximum at 67° and comprises a typical LID curve for S class streets in Germany and Europe.

To find out the best construction parameters, besides material selection for the housing for the best thermal conduction and convection and the housing color for the best reflection of the radiation of the sun (see Section 8.2), construction methodology and material combination for a durable and close-fitting luminaire are very important. Users and luminaire developers should know that the lifetime intention of an LED luminaire or its LED units of over 50 000 h can be fulfilled only if all luminaire components are stable over the entire burning time. Therefore, a leak-proof and solid luminaire construction is essential.

8.3.3
LED Characterization and Selection

The process of LED characterization and selection comprises the aspects shown in Figure 8.18.

8.3 Systematic Approach of LED Luminaire Design for Street Lighting | 457

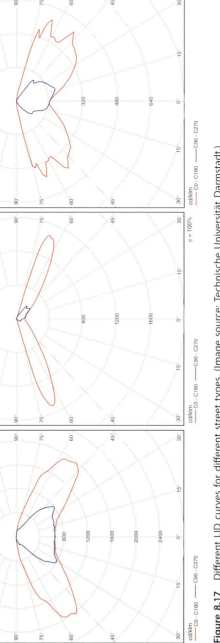

Figure 8.17 Different LID curves for different street types. (Image source: Technische Universität Darmstadt.)

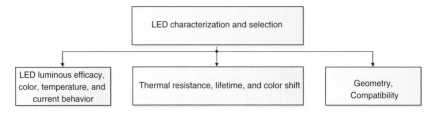

Figure 8.18 Aspects of LED selection for street lighting.

Numerous aspects and criteria of LED selection have already been analyzed in Section 7.3. But from the point of view of street lighting, some further aspects should be mentioned:

- As a compromise between warm white and cold white LEDs, a color temperature of about 4000–4500 K can be used for a broad range of applications in different regions and countries.
- The data of the white LEDs to be used can be received directly from the LED manufacturers who measure these LEDs under the *hot* binning condition, that is, 85 °C, 350 or 700 mA.
- LEDs with the smallest possible thermal resistance can be used for a long LED luminaire lifetime.

8.3.4
Thermal and Electronic Dimensioning

Thermal dimensioning has the task of minimizing the junction temperature of the LEDs for a well-known thermal LED power and at a predefined ambient temperature. This requirement can be fulfilled by the optimization of all components of the thermal transfer chain (PCB, adhesive selection, adhesive processing, soldering process, heat sink material, and form). Electronic dimensioning comprises the optimization of the voltage supply unit, the control unit, the driving unit and the sensor systems. The scheme of thermal and electronic optimization is illustrated in Figure 8.19.

In Figure 8.20, the temperature profile at the internal side of the housing of an LED street luminaire in Offenbach (in the central region in Germany) in summer

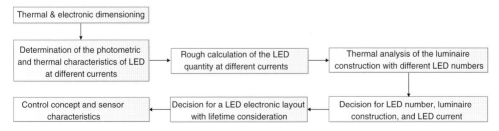

Figure 8.19 Thermal and electronic dimensioning of LED street luminaires.

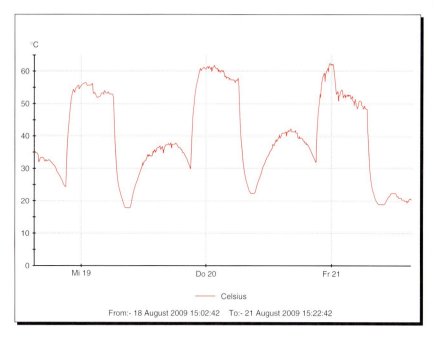

Figure 8.20 Temperature profile at the internal side of a luminaire's housing during three consecutive summer days in Offenbach, Germany, in 2009. (Data Source: Dr. K. Petry – Offenbach.)

2009 is illustrated. On the basis of this knowledge, it becomes possible to calculate the thermal and electronic values of the luminaire and this calculation is shown here as an example, at a temperature of the housing of 62 °C and assuming

- a thermal resistance of 12 K W^{-1} between the LED junction and the side beneath the MCPCB;
- a thermal resistance of the heat sink of 12 K W^{-1};
- a thermal resistance of the interface heat sink/housing and MCPCB/heat sink (with a thin and uniform adhesive layer) of 11 K W^{-1}.

A serially connected thermal resistance of 35 K W^{-1} can be estimated between the junction and the luminaire housing. At the *hot binning* condition of 350 mA, 0.285 V and 85 °C, a good neutral white LED can have, for example, 140 lm W^{-1} with an optical power of 0.45 W and a thermal power of 0.55 W and, consequently, the junction temperature can be calculated as follows:

$$T_j = 62\ °C + 0.55\ W \cdot 35\ K\ W^{-1} = 81.2\ °C$$

This (81.2 °C) is a moderate temperature. But if the luminaire company decides a current of 700 mA and a voltage of 3.2 V then a thermal power of about 1.5 W can be expected and thus the junction temperature equals

$$T_j = 62\ °C + 1.5\ W \cdot 35\ K\ W^{-1} = 114\ °C$$

This rather high temperature (114 °C) reduces the lifetime of the LEDs and of the electronic components. To improve this situation, the following measures can be adopted:

- Automatic dimming of LED current if the maximal allowable temperature is exceeded;
- Reduction of thermal resistance by a better thermally conductive housing material, by a well-formed aluminum heat sink or ceramic (see Table 3.7 and Figure 3.69);
- Selection of suitable materials for the PCB (see Table 3.7);
- Avoiding a high LED current between 700 and 1000 mA. From Figure 7.10, it is obvious that a white LED with a color temperature of 5000 K, a board temperature of 60 °C driven by 500 mA has a junction temperature of 74 °C while at 700 mA, it has a junction temperature of about 81 °C. This difference of 7 K means a remarkable lifetime reduction. The reason for increasing the LED current is to reduce the number of LEDs but it is false as it results in a shorter lifetime of the whole luminaire and hence this should be avoided.

8.4
Degradation Behavior of LED Street Luminaires

At the Laboratory of Lighting Technology of the Technische Universität Darmstadt, a research has been conducted with the aim of identifying the degradation mechanisms inside the LED luminaires and making recommendations on how these mechanisms can be prevented [1].

The first aging test of LED luminaires started in March 2010. Eight LED luminaires underwent a field test in Hofheim (near Frankfurt am Main in Germany). The luminaires were switched on, switched off and exposed to the same weather conditions like in normal street lighting. Their burning time was around 4000 h *per year*. Figure 8.21 shows the test field and the installed luminaires.

The main goal of this test was to see how the luminaires changed their performance depending on their aging time. The luminaires were measured electrically and photometrically every 1000 h with the aim of pointing out how fast their luminous flux decayed and which mechanism played a major role for such a behavior. Figure 8.22 shows the aging behavior of the tested luminaires during the test of 7000 h.

As can be seen from Figure 8.22, the eight luminaires performed very differently (as expected). The majority of the eight luminaires kept their luminous flux within the range of about 10% change. If measurement uncertainties are also considered (goniometer Type C) then this is quite a good performance for LED luminaires at the early stage of LED street lighting technology. But there are two luminaires (number 04 and 05) for which the drop of luminous flux is unexpectedly high. It is very interesting to examine the reasons and try to explain the mechanisms that lead to such an accelerated decay of luminous flux.

8.4 *Degradation Behavior of LED Street Luminaires* | 461

Figure 8.21 LED luminaires in the test field in Hofheim (near Frankfurt am Main, Germany).

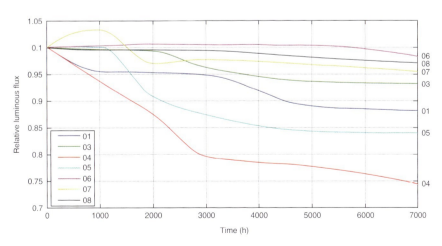

Figure 8.22 Luminous flux decay of the tested LED luminaires during the testing of 7000 h.

In luminaire number 04 (see the red curve in Figure 8.22), the luminous flux decay went along with a change in CCT from 6300 K at the beginning to 7750 K at the point of 5000 h. Such a color shift can be explained by a destruction of the phosphor caused by the overheating of the LED. In order to verify this hypothesis, measurements were conducted resulting in a very high board temperature of the LED, that is, 104 °C, which is a clear evidence for overheating following the bad thermal design of the luminaire.

The case of luminaire number 05 was much more interesting. Here the luminous flux drop was not so catastrophic and it seemed to slow down after about 3000 h (see the cyan curve in Figure 8.22). Its CCT shift was relatively low (around 200 K). It was decided to open and inspect the luminaire. By doing so, the reason behind why the luminous flux decreased was found quickly. Figure 8.23 shows one of the LEDs of luminaire No. 05 before and after cleaning.

Cleaning resulted in almost 10% increase of luminous flux for this particular LED (Figure 8.23). If the same procedure had been applied to all LEDs in the luminaire then its luminous flux would have increased by about 8–10%. But it was more important to find out where the dirt came from. This question could be answered after looking at the MCPCB and especially at the mounting holes. An emulsion consisting of solid particles and liquid could be found, see Figure 8.24.

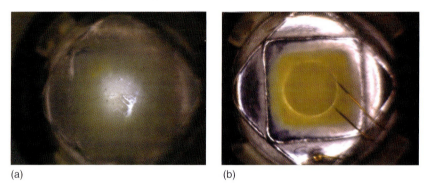

Figure 8.23 One of the LEDs of luminaire No. 5 before and after cleaning.

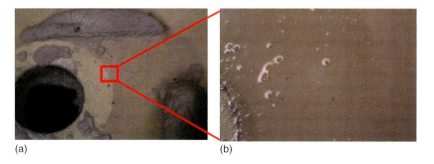

Figure 8.24 Emulsion settled around one of the mounting holes of the MCPCB (a). Zoom on the liquid part (b).

It could be found that this emulsion (Figure 8.24) contained debris from the thermal paste used to ensure a better thermal contact between the LED board and the heat sink. The conclusion from this test is that there may be different mechanisms that can lead to a premature aging of an LED luminaire.

8.5 Maintenance Factor for LED Luminaires

Since the beginning of the LED era, the greatest advantage of LED components and LED luminaires has been expected to be their luminous efficacy which is estimated to increase up to $200-220\,\mathrm{lm\,W^{-1}}$ in the next future and the long lifetime of LED components. This has led to the expectation of long maintenance periods and low personnel costs, and this seemed to guarantee the return of investment costs. Meanwhile, the development of LED products according to their luminous efficacy, luminous flux and color quality and the related smart lighting solutions have achieved a very positive progress so that the focus of research, technological optimization, and application methodology has been directed to measures to improve the reliability and the lifetime of the LED components and the LED luminaires. In this context, the so-called maintenance factor (MF) of LED lighting systems plays an extraordinary role. This section deals with the basics of maintenance for general light sources and street luminaires and the interesting subject of MF for LED lighting systems.

8.5.1 Basic Aspects of Maintenance Factor

The aim of every street lighting installation is to enable traffic participants to perceive an appropriate brightness level on the road and in its surroundings to be able to detect the objects on and besides the road, to identify persons and vehicles under different viewing conditions, and, finally, to have a secure feeling when walking or driving in the night. This safety and security can be fulfilled by means of the lighting requirements in terms of standard luminance/illuminance values on the road, glare prevention and uniformity of illumination from the beginning of any new lighting installations until the end of a predefined time period. In other words, lighting installations must fulfill the specifications of national and international standards for street lighting and take the depreciation of luminous flux and luminous intensity during the time period of usage into account.

The degradation of the luminaire and of all of its components is grounded in the following factors:

1) Light loss of the light sources because of different physical effects like a change of lamp pressure, glass coating, glass structure, and electrodes;
2) Change and depreciation of the electronic components (ballast, cable, capacitors);
3) Degradation, yellowing, and dulling of optical components (glass, reflector layers, lens transmittance reduction);

4) Deposition of dirt, dust, and moisture on the optically active components;
5) Failure of electronic, optical, and lighting components.

The degree of these degradations and failures depends on the ambient conditions (humidity, temperature, vibration), on the construction of the luminaire (tight, open, sealing materials, housing materials like plastic or aluminum housing body) and on the users (careful handling, cleaning method, and period). Because the fulfillment of the international lighting standards must always be guaranteed, the degree of depreciation must be known and taken into consideration already during the planning period. In CIE publication No. 154 [2], the term *maintenance factor* is defined as the "ratio of the average luminance/illuminance on the working plane after a certain period of use of a lighting installation to the average luminance/illuminance obtained under the same conditions for the installation considered conventionally as new."

Mathematically, the MF can be expressed by Eq. (8.1).

$$MF = \frac{E_m}{E_{in}} \quad (8.1)$$

where E_m = maintained luminance/illuminance, E_{in} = initial luminance/illuminance.

In this definition, the luminous quantity used is either the luminance or the illuminance in the working area (e.g., on the road surface). It is subject to change if the luminous flux of the luminaire or the LID or both of them change. In the current worldwide discussions, only the depreciation of luminous flux is the subject of analysis although LID curve changes or color shifts could also be regarded as important criteria. In this special context, the MF formula can be rewritten; see Eq. (8.2).

$$MF = \frac{\phi_m}{\phi_{in}} \quad (8.2)$$

where ϕ_m = maintained luminous flux of the luminaire, ϕ_{in} = initial luminous flux of the luminaire.

The depreciation behavior of a luminaire is illustrated in Figure 8.25.

Taking a look at Figure 8.25, the following aspects can be formulated:

1) If ϕ_m or E_m represents the values required by lighting standards that have to be fulfilled then the initial values ϕ_{in} or E_{in} must be higher by the factor $(1/MF)$ in order to take the depreciation effect into account;
2) The time t_p in Figure 8.25 is the time after which the conventional light source has to be replaced and the luminaire must be cleaned. In the LED era, t_p has been expected to be long enough and in the range of 50 000 and 100 000 h.
3) During the whole time period t_p, the actual luminous flux is always higher than the necessary luminous flux ϕ_m in order to fulfill the lighting standards, and the surplus of the consumed electrical energy must be paid by the street management. The higher the MF value is, the lower this surplus of electrical energy. Therefore luminaires with relatively high MF values are used. The ideal MF value equals 1.0.

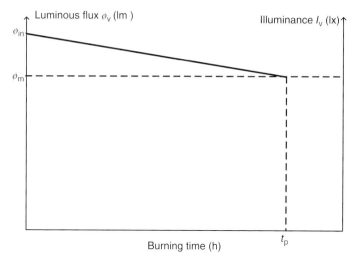

Figure 8.25 Definition and illustration of maintenance factor.

According to the CIE publication No. 154:2003 [2], the *MF* can be calculated as shown in Eq. (8.3).

$$MF = (LLMF \cdot LSF) \cdot LMF \cdot (SMF) \tag{8.3}$$

In Eq. (8.3), symbols have the following meaning:

- *MF*: Maintenance Factor
- *LLMF*: Lamp Lumen Maintenance Factor
- *LSF*: Lamp Survival Factor
- *LMF*: Luminaire Maintenance Factor
- *SMF*: Surface Maintenance Factor

The *LLMF* considers the depreciation of the light sources. In Table 8.3, the *LLMF* values of different conventional lamp types are listed. It can be seen that the high pressure sodium lamps and the tubular fluorescent lamps (tri-phosphor) have a higher *LLMF* factor in comparison to the other lamp types.

According to CIE publication 154:2003, "lamp survival is the probability of lamps continuing to operate for a given time. The survival rate depends on

Table 8.3 LLMF values of conventional lamps [2].

Lamp type	Burningtime × 10³ h				
	4	6	8	10	12
High pressure sodium	0.98	0.97	0.94	0.91	0.90
Metal halide	0.82	0.78	0.76	0.74	0.73
Fluorescentlamp (tri-phosphor)	0.95	0.94	0.93	0.92	0.91
Compact fluorescent at 25°C	0.91	0.88	0.86	0.85	0.84

Table 8.4 LSF values of conventional lamps [?]

Lamp type	Burningtime × 10³ h				
	4	6	8	10	12
High pressure sodium	0.98	0.96	0.94	0.92	0.89
Metal halide	0.98	0.97	0.94	0.92	0.88
Fluorescent lamp (tri-phosphor)	0.99	0.99	0.99	0.98	0.96
Compact fluorescent at 25 °C	0.98	0.94	0.90	0.78	0.50

the lamp type and, particularly, in the case of discharge lamps, the wattage, frequency of switching, and the ballasting system. Failed lamps cause reduction in illuminance and uniformity."

In Table 8.4, the LSF values of conventional lamp types are listed. After 3 years or 12 000 h, fluorescent lamps have the highest LSF values of 0.96 and one half (50%) of the compact fluorescent lamps will have a failure case.

The SMF value takes the aging effect of the surface materials into consideration. In case of street lighting systems, it is the depreciation of the road surface (a very difficult topic).

The LMF takes the amount of dirt on the lamps and the luminaires into account. In general, this factor causes the most serious loss of light. According to CIE publication 154:2003 [2], "the amount of light loss depends on the nature and density of airborne dirt, luminaire design and lamp type. Dirt accumulation on reflecting surfaces can be minimized by sealing the lamp compartment against entry of dust and moisture. Significant benefits can be obtained with the luminaire optical compartment sealed to at least IP5-protection."

Because the lifetime expectation of LED components of more than 50 000 h is high and the silicone in the phosphor mixture is sensitive to dirt, chemical substances, gas, and moisture, sealing quality must be high, corresponding to the IP6x-rating (IP66, IP67, or at least IP65). The LMF values depend not only on the IP rating but also on *pollution categories* defined by CIE Publ. 154:2003 [2] as follows:

Low: No nearby smoke or dust generating activities and a low ambient contaminant level, light traffic, generally limited to residential or rural areas. The ambient particulate level is no more than 150 µg m^{-3}.

Medium: Moderate smoke or dust generating activities nearby, moderate to heavy traffic, and the ambient particulate level is no more than 600 µg m^{-3}.

High: Smoke or dust plumes generated by nearby activities are commonly enveloping the luminaires.

From Tables 8.3–8.5, for an IP66 luminaire and for the calculated time of 12 000 h or 3 years with LLMF = 0.9, LSF = 0.89 and the LMF value for the low pollution category of 0.9, an MF value of 0.72 can be achieved. This rather moderate value of MF means that the initial luminous flux of the new luminaire

Table 8.5 LMF values in the different pollution categories for the IP6x-rating [2].

Pollution category	Exposure time (years)					
	1	1.5	2	2.5	3	4
High	0.91	0.90	0.88	0.85	0.83	0.80
Medium	0.92	0.91	0.89	0.88	0.87	0.86
Low	0.93	0.92	0.91	0.90	0.90	0.89

will have a factor $1/0.72 = 1.39$ higher than the required luminous flux in order to fulfill the lighting standards after 3 years, too. A more thorough analysis of these three tables and a comparison of different lamp types imply that high pressure sodium lamps have the highest MF values and all conventional lamp types deliver lower MF values requiring a high initial luminous flux and high energy consumption during the running time of the luminaire.

During the history of lighting technology with luminaires equipped with conventional lamps, luminaire manufacturers received the lamps from a small number of lamp manufacturers and the ballast sets and cable sets from electrotechnical companies so that they were responsible for the LMF values only in terms of luminaire construction and mechanical design. The depreciation of the luminous flux and the survival of the lamp was an issue related to the lamp and ballast manufacturers. With the introduction of LED technology, luminaire companies have developed their own electronic circuit boards and purchased themselves LED components and optical devices. Some of them have also developed LED electronics, optics, and designed the PCB layout so that in the LED era, luminaire manufacturers are also responsible for the MFs of the LED components, LED modules and the whole LED luminaire.

8.5.2
Basic Aspects of the Maintenance Factor of LED Street Luminaires

With the introduction of white LEDs, expectations have been related to long lifetime, high luminous efficiency, and good dimming ability of the LED components so that a generally high MF has been communicated within the lighting industry worldwide and between the luminaire companies and their users. This situation needs a deep analysis of the subject with the following results:

1) Considering Figure 8.22 on the aging behavior of LED street luminaires under realistic conditions after 7000 h of aging and testing, it is clear that different LED luminaire types exhibit different degradation behaviors with the luminaires 03, 06, 07, and 08 being relatively constant at a high level (better than 95%) after 7000 h while the luminaires 01, 04, and 05 lose their luminous flux more rapidly. The luminaire No. 04 reached the 74% level after 7000 h or less than 2 years only. Therefore, different LED luminaires can be assigned different MFs.

2) Analyzing Figures 4.71 and 4.72 and describing the luminous flux depreciation of two LED types of two leading LED manufacturers at the same temperature (85 °C), it can be seen that two different LED types may exhibit remarkably different aging behaviors and that aging behavior depends on the current of the LEDs during the aging time. The consequence of this different behaviors is that different LED types have different lifetimes according to the L_{70} definition illustrated in Figures 4.73 and 4.74. From Figure 4.73, it is clear that a lifetime $L_{70}B_{50}$ of more than 50 000 h can be achieved only if the operating current of the LED is lower than 700 mA and/or the operating temperature is lower than 85 °C. Contrary to conventional lamps, the users or luminaire developers can actively influence the lifetime of LED based luminaires.

3) The analysis of the aged LED street luminaires in Figures 8.23 and 8.24 reveals that it is not only the lifetime of the LEDs that is responsible for the degree of aging and the lifetime of the entire luminaire. The construction concept and the performance of the luminaire (sealing, leak-proof internal structure, protection of humidity input) and the selection of the right materials for all components of the luminaire are also important.

4) In the past when conventional lamps were used, it was well known that the magnetic ballasts for discharge lamps were not efficient but they had a long-term stability and reliability. With the electronic ballasts for discharge lamps, the efficiency of the luminaire can be significantly improved but the reliability and the lifetime of electronic components are often critical. In the definition of MF in Eq. (8.3), the failure rate or degradation of the electronic units in the LED luminaire are not included and is the subject of further research.

5) The MF value, the performance of the whole LED luminaire and all operation parameters have to be measured by definition under the ambient temperature of 25 °C. This is meaningful for the case of conventional lamps because high pressure discharge lamps have an operating temperature of, for example, 800 °C. For LED technology, with a junction temperature between 65 and 125 °C in reality, the ambient temperature of the luminaire is not the key factor. However, the housing temperature of the outdoor luminaire irradiated by the sun is essential and this can be as high as 70 °C, see Figure 8.20. From this point of view, users need a good MF value for the luminaires operated under realistic conditions over a long time, for example, 16 000 h under natural outdoor conditions.

For the construction of an LED luminaire comparable to the mechanical construction principle of a conventional luminaire, the LLMF values of CIE publication 154:2003 can be used, see Table 8.3. As seen in Table 8.4, the LSF value of conventional discharge lamps is too low indicating a high failure rate. In the publications and conference contributions of LED semiconductor industry, the opinion has been communicated that LED components have a very low failure rate and they depreciate gradually. In current discussions, an abrupt or catastrophic failure rate (C_z) of max. 2% was specified so that the LSF value of 0.98 can be used.

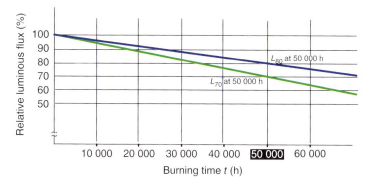

Figure 8.26 Illustration of the simplified luminous flux depreciation [3].

In the LED luminaire industry, the term *utility lifetime* (L) is defined as the time at which the luminous flux of the LED luminaire, the LED components or the LED modules are reduced down to $x\%$ of the initial value. The value $x\% = 70\%$ is often used in the LED manufacturing industry and is denoted by L_{70}. In European (including German) luminaire industry, it has been recommended to use the term L_{80}, which means $x\% = 80\%$ reduction of the initial luminous flux, see Figure 8.26. A preferred lifetime is often intended to be 50 000 h.

With LLMF = 0.8 for L_{80} of 50 000 h, with LSF = 0.98 and with an IP66 rating for the luminaire applied in a region with low pollution (LMF = 0.89), an MF = 0.7 for the utility time of 50 000 h can be calculated. This value is comparable to the MF value of 0.72 for conventional lamps (for 3 years, see Section 8.5.1), even for 50 000 h. However, the assumption for L_{80} with 50 000 h is very optimistic according to the experience of the authors of this book.

In the development departments of LED luminaire industry, engineers utilize the experience that the depreciation of luminous flux can be compensated for if the current (or duty cycle of the PWM circuit) is increased delivering the same surplus of luminous flux which has been reduced by the aging effect of the LED, see Figure 8.27. By doing so, initial luminous flux and maintained luminous flux are equal so that the LLMF value for the luminaire is not only 0.8 or 0.7, but can also be 1.0 in an ideal case.

This idea is to increase the MF value of the LED luminaire based on the possibility of dimming the LED as a function of burning time which was not possible for the operation of conventional lamps. With an ideal LLMF value of 1.0 for 50 000 h, with LSF = 0.98 and with an IP66 rating for the luminaire applied in a region with low pollution (LMF = 0.89), an MF of 0.877 for the utility time of 50 000 h can be achieved.

In reality, however, this has not been very simple because the compensation curve (the red curve in Figure 8.27) is appropriately the inverse function of the depreciation curve of the aging LED and this must be known as a dataset to be programmed into the microcontroller. From Figures 4.71 to 4.74, it is obvious that the depreciation curves are not linear and depend on temperature and LED current.

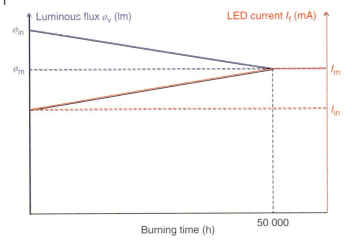

Figure 8.27 Principle of the luminous flux constancy control.

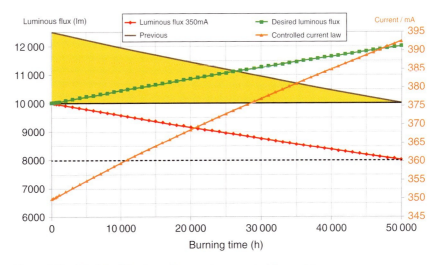

Figure 8.28 Principle of luminous flux constancy control (example).

Figure 8.28 explains the principle of the luminous flux constancy circuit in more detail. The luminous flux of 10 000 lm is necessary to illuminate the street according to the lighting standards. If additional measures or compensations are not activated then the luminous flux would be reduced to 8000 lm after 50 000 h according to the L_{80} definition. If the method used for conventional lamps is considered, then the initial luminous flux should be 12 500 lm for an often used MF value of 0.8.

If the luminous flux of 10 000 lm can be kept constant over 50 000 h, then the necessary number of white LEDs is driven by an initial current of 350 mA and this can be increased stepwise to compensate for the reduction of luminous

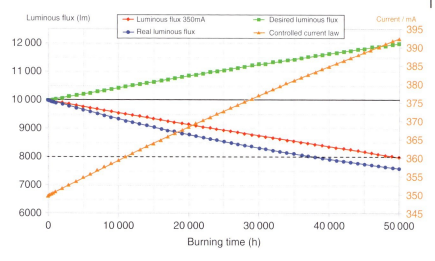

Figure 8.29 Real result of luminous flux constancy control (example).

flux and achieve the value of about 392 mA at 50 000 h. But this solution is not problem-free. The deprecation curve (i.e., the aging curve) to be compensated for was determined for a certain current (e.g., 350 mA) at a defined temperature (e.g., 85 °C). If the current of the LEDs is increased stepwise from 350 to 393 mA then the current density is higher and the aging behavior is accelerated correspondingly. Consequently, the assumed aging curve will not be valid. With a higher current at 370 or at 385 mA, the white LED is not so efficient in comparison to 350 mA and therefore a higher thermal power is produced heating up the LED and lowering its luminous flux. As a result, the real luminous flux is lower than 10 000 lm after 37 000 h; see the blue curve in Figure 8.29.

8.6
Planning and Realization Principles for New LED Installations

At the last stage of LED luminaire development – the fabrication and application process – the LED lighting product is applied in practice, in outdoor or indoor installations. It should contribute to the worldwide aim of enhancing lighting and color quality of illumination and reducing energy consumption in order to protect our environment. Generally, three main questions arise:

1) How can the users (city managements, facility managements, planning offices) evaluate the degree and absolute amount of energy saving by using a new LED lighting system in comparison to an existing conventional lighting system?
2) How can the users determine a correct energy efficiency level together with an appropriate lighting quality in an actual lighting project (illuminating a

street, a building, a museum, or a school) from a variety of LED manufacturers worldwide in the global market?
3) How can the users and LED luminaire manufacturers ensure that the lighting quality of the LED luminaire is accepted by average citizens as well as by professional users?

At the competition "Municipalities in New Light" in Germany from 2010 to 2013, 10 selected demonstration projects concerning the application of LED technology were initiated and realized [4]. The primary aim of these projects was to collect and examine practical results and experience on the transition from conventional lamp technology to LEDs as well as on the added ecological and economic value of LEDs in real applications. Within the framework of this competition, all projects were evaluated regardless of their differences in their individual aims and tasks. In this section, results from the Technische Universität Darmstadt and its associated research partners are presented with all stages including planning, selection, purchase, and installation of the luminaires and documentation. The majority of the projects were related to outdoor lighting including road lighting (see Figure 8.30) and this is accentuated in this section.

It has to be emphasized that the following description is extensive and scientific and the aim is to show the complete approach to the realization of these projects. But not all steps have to be applied in every new LED lighting system.

Figure 8.30 Residential access street with LED lighting in Rietberg. (Reproduced with permission from VDI Technologiezentrum GmbH.)

8.6.1
Technical Approach

Although the lighting systems of the ten towns show different historical developments as well as technical characteristics and are situated in different regions with specific infrastructures, the general course of every project can be divided into the following four phases:

1) *Coordination*: Timetables, processes, existing states, working plans, time, costs, and places of measurements.
2) *Measurement of the old lighting system*:
   a. On-site: Measurement of the old system with maintained luminaires and new lamps (preaged for 100 h), carrying out the public opinion poll;
   b. In the laboratory: Measurement of the old luminaires (polar curve of luminous intensity, spectrum, etc.), simulation.
3) *Measurement of the (new) LED system*
   a. On-site: Measurement of the LED system, public opinion poll;
   b. In the laboratory: Measurement of the LED luminaires (polar curve of luminous intensity, spectrum, etc.), simulation.
4) *Documentation by the scientific partners and public opinion poll*
   a. Recording and processing all relevant measurement data regarding communication with the citizens.

The earlier-mentioned four phases are characterized by the following technical and administrative tasks and focal points:

8.6.1.1 Phase 1: Coordination
In general, the first project phase contains the coordination of all stakeholders within the framework of the project. This comprises the municipal administration, the local research partner, the central research institution as well as further participating partners (i.e., expert planners, producers of LED luminaires). The core aspects are the coordination of the timetable, of the extent of the replacement, the determination of all applicable lighting standards as well as the measurement profile for the validation.

8.6.1.2 Phase 2: Measurement and Evaluation of the Old System
The validation of the old system represents the basis for the subsequent comparison with the new LED lighting installation. The points to be tested are determined according to the international and national street lighting standards and the measurement profile is prepared (see Figure 8.31). For the sake of the measurements of the lighting installation, the road was blocked during the measurement period and tree branches overhanging the road were cut. The street surface was cleaned. Usually, the measurement of the old system is carried out with clean luminaires and new lamps (which had been preaged for 100 h) only in order to result in a valid comparison.

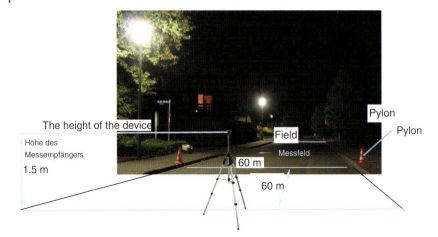

Figure 8.31 Measurement of pixel-resolved luminance distribution on the road. (Photo Source: Technische Universität Darmstadt.)

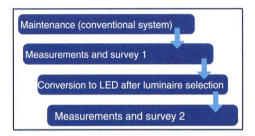

Figure 8.32 Scheme of the workflow in the project phases 2 and 3.

The measurements of the light source parts and electrical parts of the luminaire were carefully organized. In all 10 projects, the old luminaires were cleaned and sent together with the new lamps to the Laboratory of Lighting Technology of the Technische Universität Darmstadt to measure their lighting parameters and electrical parameters. The datasets obtained serve as introductory data for future lighting planning and development programs and are intended to help simulate LIDs and verify them afterward by measurement values about real lighting situations (see Figure 8.32).

8.6.1.3 Phase 3: Lighting Measurement and Evaluation of the LED System

The measurement process of the LED system in the project's third phase corresponds broadly to the measurement system used in the second phase, see Figures 8.31 and 8.32. An additional aspect, which is crucial for the whole project, is the selection of an appropriate luminaire manufacturer for the individual demands.

In addition to the energetic optimization of the lighting system, one of the key lighting objectives of the replacement with LEDs is to comply with all electrical

8.6 Planning and Realization Principles for New LED Installations

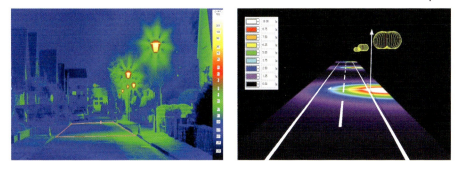

Figure 8.33 Luminance measurement on the road and planning simulation with the old lighting system. (Image source: Technische Universität Darmstadt.)

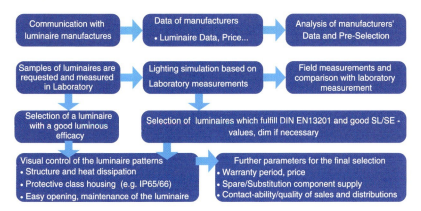

Figure 8.34 Schematic presentation of the process of qualification and selection of the appropriate luminaire.

and lighting standards. In case of road lighting, this standard is DIN EN 13201 [5] whose content is similar to the core content of other international street lighting standards (e.g., in China, United States, and Japan). In order to assess the suitability of an LED luminaire for the given lighting system in advance, it is imperative to carry out lighting simulations (see Figures 8.33 and 8.35). The corresponding luminaire data can be obtained in advance from qualified LED luminaire manufacturers.

After the replacement, simulation, and measurement results must prove that the lighting standards are met and the measurement data can be harmonized with simulation results [6, 7].

Consequently, the project's third phase comprises two consecutive steps which are explained in more detail here:

- Qualification of the LED luminaires and selection of the luminaire (see Figure 8.34);
- Installation and measurement of LED lighting.

Figure 8.35 Luminance measurement on the road and planning simulation with the new LED lighting system. (Photo Source: TU Darmstadt.)

8.6.2
Qualification of the LED Luminaires and Selection of the Luminaires

The qualification of appropriate LED luminaires is a key step in the third phase of the project. There are a large number of luminaire manufacturers on the market. The amount of available information they are ready to provide on the lighting and electric characteristics of the LED luminaires varies considerably. For municipalities (and especially planning agencies) it is a huge challenge to stick to an *objective* overview. Important factors for the selection of appropriate LED luminaires include:

- Overview of qualified luminaire manufacturers on the market and their respective product range;
- Obtaining exact data on the old lighting system which has to be replaced (road geometry, structure of residents, electricity supply system, traffic situation in daytime and nighttime);
- Overview of the expenses and their development.

For any individual lighting system which has to be newly designed, the process of identifying an appropriate partner to obtain the luminaire from can be organized as follows:

- *Communication with the luminaire manufacturers:* In this process, the aspects like the origin of the luminaire companies, distribution and repair network of the company as well as the accuracy of their data presentation should be taken into account because the lifespan of the lighting system is estimated to be 20 years.
- *Gathering and analyzing the manufacturer's data, process of elimination:* On the basis of the catalogue, the appropriate LED luminaires are preselected and others are eliminated. For this selection, the manufacturers are asked to provide the available planning data. The computer then calculates the lighting parameters for lighting the objects according to street lighting standards, the available lighting data and on the basis of the geometry of the road that has to be lit. The results of the simulation are the number and order of luminaires as well as the brightness level, the glare index, and the uniformity of illumination. Then, the required electrical power load of all luminaires can be calculated. Samples

of those luminaire types can be obtained which are most appropriate for the lighting installation and with an appropriate power level in order to be able to evaluate them visually and carry out an independent evaluation measurement in a laboratory.
- *Luminaire measurement in the laboratory:* The measurement and characterization of the luminaires by an independent laboratory serves to compare the measured values with the manufacturer's data and for the exact planning of the intended illumination.

In compliance with the lighting standards, energy efficiency is calculated by relating the necessary electrical power of the luminaire to the average illuminance E (for roads of category S) or to the average luminance L (for roads of categories ME) and to the illuminated driving lane area A. It is important that the analyzed type of luminaire fulfills the requirements set by international street lighting standards regarding the specific road geometry. These two efficiency criteria can be calculated by Eqs. (8.4) and (8.5).

$$SE = \frac{P_{\text{Luminaire}}}{E_{\text{Road}} \cdot A_{\text{Road}}} \tag{8.4}$$

$$SL = \frac{P_{\text{Luminaire}}}{L_{\text{Road}} \cdot A_{\text{Road}}} \tag{8.5}$$

where

SL	=	Energy criterion luminance
SE	=	Energy criterion illuminance
$P_{\text{Luminaire}}$	=	System power of lamp and ballast per light point
L_{Road}	=	Maintenance value of the average luminance
E_{Road}	=	Maintenance value of the average illuminance
A_{Road}	=	Effective driving lane area (distance between the poles' times and road width)

From all potential LED luminaire candidates fulfilling the lighting standards for the intended street, the luminaire types with the smallest SE and SL values should be kept in the final selection circle.

In addition to the lighting and energetic evaluations, there are some crucial contractor aspects such as guarantee time, price, and other contract conditions to be considered when choosing the luminaire from a certain manufacturer.

8.6.3
Installation and Measurement of the LED System

After the installation of the selected LED luminaires, they should be preaged for approximately 100 h or 10 nights before the measurements are carried out. The measurement itself takes place as described for the old lighting system.

8.6.3.1 Phase 4: Documentation by the Scientific Partner and Public Opinion Poll

The documentation contains the technical description of the method as well as all relevant measurement data and decision-making processes.

The acceptance of a new lighting system has a decisive influence on the success of the project. Consequently, it makes sense to involve the users and the population early and survey their opinion, especially in the context of road lighting and architectural lighting where there is a huge number of acceptance aspects which vary depending on the project. For this reason, there is no general directive on how to design a questionnaire. From a scientific perspective, the following lighting, psychological and architectural aspects should be taken into consideration:

- Perception of brightness;
- Perception of glare;
- Visibility and perception of object contrast;
- Facial recognition;
- Light color (warm, neutral or cool white, yellowish);
- Uniformity of the illumination;
- Identification of the old and the new lighting situation;
- Change of safety and security (perceived or verifiable);
- Possibility of highlighting the illuminated buildings or squares according to their historic or cultural importance;
- Directing attention to the relationship among the illuminated objects.

One option was to interview residents and other stakeholders on-site. Another possibility to survey the inhabitants' opinion was to send questionnaires to the residents or to ask them to complete the questionnaires online. The completed questionnaires were evaluated and served as a confirmation for the selected replacement solutions. At the same time, they also serve as a decision aid for future projects.

References

1. Ganev, H. and Khanh, T.Q. (2013) Analysing the degradation mechanisms of LED luminaires. Conference Lux Europa, Poland, Krakow, Oral Presentation.
2. Commission Internationale de L'Eclairage (2003) The Maintenance of Outdoor Lighting Systems, CIE 154:2003.
3. ZVEI-Paper (2014) Planungssicherheit in der LED Beleuchtung, Zentral Verband der Elektrotechnischen Industrie.
4. VDI Technologizentrum GmbH- Düsseldorf/Germany (2013) *Handbook Municipalities in New Light- Practical experiences on LED in Public Lighting*, June 2013, Supported by the Bundesministerium für Bildung und Forschung (BMBF, Federal Ministry for Education and Research) within the project "*EvalKomm*" (13N11100).
5. European Committee for Standardization (2003) *Road Lighting, Part 2: Performance Requirements, Standard 13201–2*, European Committee for Standardization, Brussels, 16 p.
6. Kuhn, T., Schiller, C., Haferkemper, N., and Khanh, T.Q. (2009) LED-street lighting- technological, energy, lighting aspects and results of real tests on the roads. Conference Lux Europa, Istanbul, 2009.
7. Khanh, T.Q. et al. (2010) *Lichttechnische und technologische Aspekte der LED Straßenbeleuchtung*, Zeitschrift Licht, H.7-8, Pflaum-Verlag.

9
Summary

Peter Bodrogi and Tran Quoc Khanh

The aim of this book is to describe new and prospective aspects of LED technology from the point of view of human light and color perception in order to develop durable and energy-efficient LED luminaires for indoor and outdoor lighting with high color and lighting quality. According to the recent development of our scientific understanding, lighting shall be designed to improve the *combination* of the visual perceptual aspect (the imaging effect including the human eye, retinal cell structures, their receptive fields, and visual signal processing in the brain) and the light and health aspect (the nonimaging pathway including the circadian clock).

To prepare the reader for the comprehension of all concepts related to these issues in relation to LED technology, Chapter 2 (as a basic chapter) conveyed the definitions of radiant, photometric, colorimetric, and light and health (circadian) related quantities. The circadian stimulus (as a possible forthcoming aspect of LED lighting) was analyzed in Section 5.10 for the example of 34 typical white LEDs by means of a modern prediction model that takes the activation of all contributing receptor types and their interactions on the retina into account. It turned out that, for these white LEDs, the value of CS can be computed from a simple measure, the so-called ipRGC to P ratio (α), which is easy to determine (both instrumentally and computationally) at every fixed illuminance level.

In Chapter 3, the principles of LED radiation generation have been formulated. The starting point was the explanation of the physical laws governing semiconductor processes and enabling the conscious engineering of high-tech semiconductor devices that create energetically and visually highly efficient LED radiation. With the recent developments of new phosphor systems in the green wavelength ranging from 505 nm on and in the red range between 605 and 670 nm, the prerequisites for the development of white LEDs with high color quality and environment-resistant characteristics could be established.

In Section 3.5, the principles of mixing the radiation of a blue LED chip and a green–red phosphor system were illustrated. Results of this practical analysis showed that red phosphors with emission peaks between 630 and 660 nm were needed as a precondition for white LEDs with high color rendering indices. Otherwise reddish object colors cannot be rendered well enough. After the description of color LEDs, phosphor systems, and phosphor mixtures, today's LED packaging

LED Lighting: Technology and Perception, First Edition.
Edited by Tran Quoc Khanh, Peter Bodrogi, Quang Trinh Vinh and Holger Winkler.
© 2015 Wiley-VCH Verlag GmbH & Co. KGaA. Published 2015 by Wiley-VCH Verlag GmbH & Co. KGaA.

technologies were briefly introduced. It was pointed out that the use of silicone which is molded directly on the blue chip can be a reason for the quick aging process of the LEDs. Ceramic materials (e.g., AlN material) are suitable as potential heat sink materials to increase the thermal conduction of the whole LED package.

At the end of Chapter 3, a LED package was described as a result of the integration of material, optical, and electrical engineering efforts. It will undergo a thorough physical, thermal, and radiometric testing and characterization by using specific and precise measurement methods and equipment as suggested in Sections 4.1, 4.8, and 4.9. By accumulating appropriate data about the LED to be tested during these measurements, the dependence of the LED's radiant, colorimetric, and spectral properties on temperature (sensor temperature and junction temperature) and its current can be represented and modeled mathematically.

The spectral model describing the dependence of the absolute emission spectrum of a color LED or a white PC-LED on the LED's current, voltage, and a predefined board temperature is an efficient tool to control the chromaticity constancy and spectral constancy of the LED's optical radiation over a broad range of LED temperatures and currents. Electric models are important to control the LED light sources built in a complex, tunable LED luminaire to compensate for their short-term temperature changes and for their long-term aging processes, while thermal models are needed to reduce thermal load and increase lifetime and long-term stability.

In current discussions on the depreciation behavior of color and white LEDs – that is, their aging behavior, lifetime, and prognosis – correct LED aging and testing data for numerous LEDs and on a long-term basis were considered to be necessary. In such investigations, it was pointed out, the LEDs of different LED manufacturers with different packaging qualities exhibited substantially different lifetime expectations and that the different mathematical models for LED lifetime prognosis deliver dissimilar lifetime values for the same LEDs as it was shown in Section 4.10.

Chapter 5 describes recent research results on the different methods of predicting the color rendering quality of conventional light sources and white LEDs comparing the current CIE color rendering index (CRI) with the new CRI, CRI2012. In current color science, intensive discussions are conducted on the question whether the combined color preference – color rendering index of the CQS system or a classic CRI quantifying purely color fidelity – shall be the *primary* index to be used to design an illuminating system for best color quality in general lighting.

Investigations presented in Chapter 5 pointed out that the new CRI (CRI2012) and the CQS indices did not correlate well if the white LED contained sharp peaks in its spectrum (i.e., sharp local maxima with narrow bandwidths, e.g., peaks of red, green, and blue pure semiconductor LEDs without phosphors). In order to assess the quality of color rendering more easily, the users of the white LED light source need a semantic interpretation of the CRI values in order to rank the CRI values intuitively by using the categories "very good," "good" or "moderate," "low" or "bad." This semantic interpretation helps users communicate with LED luminaire manufacturers. It also fosters LED light source supplier and LED luminaire manufacturer communication.

In addition to the CRI, visual white tone quality and the classification of white LEDs into different visually categorized binning groups according to their white tone chromaticities are essential. The visually relevant white tone classification proposed in this book is very different from the well-known conventional binning system based on MacAdam ellipses. The new chromaticity binning strategy ensures, on the one hand, that the white tone of the LED does not include any disturbing chromatic shade (e.g., green or purple) and, on the other hand, the visual homogeneity of large white surfaces (e.g., white walls in a room) without disturbing white tone changes across the illuminated surfaces.

In Section 4.12, the dimming methods for LEDs including current and PWM dimming had been described. It was pointed out that a PWM circuit can cause visually disturbing flicker and stroboscopic effects under both photopic and mesopic conditions. Results of dedicated research studies presented in Sections 5.11 and 6.7 implied that a PWM frequency of as high as 400 Hz cannot entirely eliminate the stroboscopic effect (e.g., moving a hand or a computer mouse under a PWM-driven LED luminaire), which can even be observed by certain sensitive subjects with a frequency beyond 1 kHz. But, generally, a frequency of 400 Hz with a duty cycle broader than 20% is able to weaken this effect, down to the rate of about 5–10% only.

In Chapter 6, mesopic (twilight) brightness perception and visual detection performance under white LEDs as well as conventional lamp types were analyzed theoretically and through real field experiments. It turned out that, at a typical (fixed) mesopic luminance level of $0.1\ cd\ m^{-2}$, a cool white LED can provide 20% more brightness than a tungsten halogen lamp of 3200 K. Also, the conspicuity of road markings at night travels with white LEDs can be 63% higher than with a yellow light of a high pressure sodium lamp. For object identification, however, no difference was found between the diverse lamp spectra.

The detection of hazardous objects (which tend to appear at or around the visual threshold, that is, with low contrasts and off the visual axis, usually in the eccentricity angle range between 2.2° and 20°) during nighttime driving conditions (which take place in the mesopic range at typical luminance levels between 0.1 and $1.5\ cd\ m^{-2}$) is the most important visual task to optimize traffic lighting. The advantages of using cool white LEDs was pointed out in this respect.

Until now, luminance, illuminance or other $V(\lambda)$-weighted luminous quantities have been used for glare characterization and evaluation. The experiments described in Section 6.6 for discomfort and disability glare under mesopic conditions showed that the glare effect cannot be modeled and evaluated by the use of such luminous quantities like luminance and a more advanced modeling including other retinal mechanisms is needed.

The focus of Chapters 7 and 8 was on the methodology of how the development of a LED luminaire for indoor (Chapter 7) and outdoor (Chapter 8) lighting can be carried out systematically in order to take all technological and perceptual aspects of LEDs into account. The design of a LED indoor luminaire of very high color quality (with a CRI better than 93), an excellent white point, variable correlated color temperatures ranging between 2700 and 6500 K and with a high luminous efficacy is possible via the hybrid-system conception only.

Latter conception combines white phosphor-converted LEDs and color LEDs, which have to be characterized and stabilized, in turn, with a suitable spectral LED model. This hybrid-system conception shall constitute the future framework for the design and optimization of modern LED lighting systems for interior lighting (the flexibility of such spectra opens up new horizons to emphasize different aspects of color quality for different collections of colored objects of various prevailing colors) and also for exterior lighting including street and automotive lighting. In exterior lighting, the flexibility of the modern LED lighting system can also be exploited (by always satisfying high energy efficiency demand) in terms of utilizing a luminaire of appropriate spectral and spatial radiation characteristics depending on weather, the topology of the illuminated user area (highway, residential street, parking lot, or park), the colored objects to be illuminated (e.g., cars, buildings, vegetation) and regional and cultural peculiarities (including a design for Asian, European, North American, or Latin American countries).

At the last stage of the LED luminaire's development, fabrication, and application process, the LED lighting product will be applied in a practical field as a part of an outdoor or indoor installation. It shall contribute to the worldwide aim to enhance the visual quality aspects of illumination (including visual performance, circadian, and color quality aspects) and to reduce energy consumption in order to protect the environment. To comply with this ambitious objective, Section 8.6 showed the necessary steps of planning new LED installations from the engineer's point of view and the evaluation of an old installation with conventional light sources in comparison with a recently installed LED lighting system. This description was based on the authors' experience. It was obtained in the course of LED illumination projects in Europe.

Index

a

aging behavior
– chromaticity difference
– – chromaticity shifts   209
– – electrical and thermal behavior   209
– – luminous efficacy   210, 211
– – luminous flux degradation   212–214
– – thermal network analysis   211
– degradation and failure mechanism   202–204
– issues   202
– spectral distribution
– – blue chip radiation   205, 207
– – phosphor system   207, 208
– test and measurement condition   206
– test and measurement specification   206
alkaline earth ortho-silicates   89, 90, 92
alkaline earth oxy-ortho-silicates   92, 93
alkaline earth sulfides
– disadvantages   87
– hydrogen sulfide gas   87
– luminescent properties   88, 90
– quantum efficiency   86, 87
– red-emitting nitride   86
aluminum garnets
– cold white light   83
– critical distance   84
– optical bandgap   84, 85
– physical chemical and spectroscopic properties 83
– STOKES shift   83
– temperature dependence   82
– trivalent terbium absorption   85
Amber-red semiconductor   64
ANSI binning standard   299, 300
atomic force microscopy (AFM)   163
Auger recombination   58

b

background luminance   386
bandgap energy   64
bead string artifact
– age groups recognition   394
– cumulative frequency   393
– experimental setup characteristics   391
– foveal observation   392
– key question   390
– LED rear lamps   391
– peripheral observation   392
– visual experiment result   392
border function (BF)   215, 216

c

center luminescence   79
ceramic material   480
characteristic luminescence   79
charge-coupled device (CCD) sensor   135
chip-on-board (COB) technology   128
chip-packaging technology
– aspects   118, 119
– efficiency improvement
– – extraction efficiency   121, 122
– – radiative recombination   122, 123
– – refraction law   120
– phosphor molding and positioning
– – conformal coating configuration   123, 125, 126
– – gob-top configuration   123, 124
– – in-cup configuration   123
– structure of   119, 120
– substrate technology
– – COB   128
– – packaging material   129
– – single/multi-chip configuration   127, 128
– – single/multi-chip LED   125

LED Lighting: Technology and Perception, First Edition.
Edited by Tran Quoc Khanh, Peter Bodrogi, Quang Trinh Vinh and Holger Winkler.
© 2015 Wiley-VCH Verlag GmbH & Co. KGaA. Published 2015 by Wiley-VCH Verlag GmbH & Co. KGaA.

chip-packaging technology (*contd.*)
– – thermal conductivity   126, 129
– – thermal energy density   128
chromatic adaptation   27, 29
chromatic lightness model   290, 291
chromaticity binning methods
– ANSI binning standard   299
– blackbody/daylight locus   299
– LED luminaire   298
– semantic binning strategy
– – computational method   302
– – constant $\Delta u'v'$ contours   304
– – customers and manufacturers   301
– – sample set evaluation   307, 308
– – semantic contours   302
– – target white tone chromaticity   302
– – vs. visual binning experiment   304
– semiconductor and phosphor materials   298
CIE model
– detection task   344
– iterative computational method   345
– luminance values   345
– photopic luminance values   346
– S/P ratios   347
– threshold contrast   344
Circadian stimulus (CS)   479
cold binning point   157, 158
color fidelity   236
color preference assessment
– CQS method   279. *See also* color quality scale (CQS)
– R_a and $R_{a,2012}$ 273
– subjective factors   273
color quality   481, 482
color quality scale (CQS)
– CIELAB   279
– color preference scale Q_p 279
– components
– – block diagram   278
– – CCT factor   278
– – chroma increment factor   276
– – chromatic adaptation   276
– – CIE tristimulus values   275
– – CIELAB chroma difference   276
– – CIELAB color difference   276
– – CIELAB values   276
– – CQS Qa and CQS Qi   278
– – reference light source   275
– – root mean square   277
– – scaling and re-scaling   277
– – test colors   275, 276
– vs. CRI CRI2012
– – CIELAB chroma values   282, 283

– – color preference optimization   282
– – correlation   281, 284, 285
– – implication   281
– – multi-LEDs   282
– – semantic interpretation scale   283
– – test colors   283
– – white LED light sources   280, 281, 284
– emission spectra   280
– gamut area scale Q_g 279
– Qa and Qi indices   279
color rendering index (CRI)   117, 118, 153, 416, 480, 481
– CIE
– – color difference   245
– – computational method   243
– – correlated color temperature   243
– – interpretation   247
– – nonuniform color space   246
– – Ri and Ra   245
– – test color samples   243, 246
– – U^*, V^*, W^* color space   243
– – visual experiments   246
– – von Kries transformation   245
– color fidelity   242
– CRI2012 method
– – components   248
– – computation method   248, 249
– – nonlinear scaling   251, 252
– – principles   247
– – real test colors   250
– – root mean square   251, 252
– – spectral gaming   251
– – test color samples   249
– – test light source   250, 251
– sample set
– – bar charts   270–272
– – CIE CRI R_a limits   268
– – grouping scheme   268
– – object reflectances   268, 269
– – semantic interpretation scale (*R*)   269
– – white LED light sources   266
– semantic interpretation   252. *See also* semantic interpretation
– spectral reflectance
– – colored objects   262
– – general color rendering   261
– – reddish objects   264, 265
– – skin tones   262, 263
– – test color sample   265

color semiconductor
– BOLTZMANN distribution   67
– direct recombination   56
– external quantum efficiency   62
– forward current   71
– hetero-junction structure   53
– homo-junction structure   52, 53
– indirect recombination   57
– injection efficiency   60
– injection luminescence   50, 52
– internal quantum efficiency   60
– light extraction efficiency   60, 61
– luminous efficacy   62
– matter waves/de BROGLIE waves   68
– photon emission   physical mechanism 68, 69
– quantum well structure   54, 55
– radiant efficiency   62
– radiative and non-radiative recombination 57–59
– semiconductor material system
– – Amber-red semiconductor   64
– – bandgap energy   64, 65
– – green efficiency gap   66, 67
– – InGaN   63
– – parameters and growth techniques   64
– – UV-blue-green semiconductor   66
– – visible/invisible wavelength   64, 65
– spectral model
– – accuracy   177
– – combined approach   171, 173
– – description   174
– – mathematical approach   168–171
– – MIMO system   168
– – multi GAUSSIAN model   174, 175, 177
– spectral power distribution   70
– temperature and current dependence
– – chromaticity difference   147–149
– – radiant flux and radiant efficiency   145, 146
– – relative spectral power distribution   143, 144
colorimetry and color science
– blackbody radiators   40, 41
– CAM02-UCS color difference   38
– CAM02-UCS color space   38
– CAM02-UCS uniform color space   38
– chromatic adaptation   27, 29
– chromaticity points   39
– CIECAM02 color appearance
– – vs. CIELAB   35
– – color stimulus   32, 33
– – compression   35
– – computational method   33
– – F_c and N_c values   33
– CIELAB color difference   37
– CIELAB color space
– – a^*-b^* diagram   31, 32
– – function f 31
– – L^* a^*, b^* value   31
– – output quantity   31
– color matching function and tristimulus values
– – chromaticity coordinates   26
– – color perception   26
– – cone types   24
– – self-luminous stimuli   25
– – spectral locus   26
– – spectral radiance distribution   24
– color perception   23
– color stimulus   23
– correlated color temperature   40
– definitions   24
– electromagnetic radiation   23
– MacAdam ellipses   36, 37
– perceived attributes   27
– psychophysical process   22
– u', v' chromaticity diagram   36
constant current reduction (CCR) method   223
CRI   242. See also color rendering indices (CRI)
cross-sectional transmission electron microscopy (TEM)   163
cyan gap   85

d

depreciation behavior   480
dimming methods   481
– CCR method   223
– design of   224
– experimental setup   224
– history   222
– light and color regulation   223
– PWM method   223
– red LED   227, 228
– white LED
– – automotive lighting   225
– – correlated color temperature   225, 227
– – indoor and outdoor lighting   225
– – linearity test   226
– – luminous efficacy   226

e

electric model   480
– AFM/TFM   163
– diffusion-recombination current model   163

electric model (*contd.*)
– limited operating range
– – forward voltage *vs* forward current   166
– – forward voltage *vs* junction temperature   165
– – mathematical description   164, 165
– – parameters   167
– – three-dimensional representation   166, 167
– MISO system   161
– small signal scattering (S) parameters   163, 164
– theoretical approach
– – built-in voltage/diffusion voltage   161
– – depletion region, width   161
– – effective state density   161, 162
– – forward bias condition   162
– – free electron and free hole concentration   161
– – reverse bias condition   162
– – SHOCKLEY equation   160
– – threshold voltage   162, 163
– tunneling current model   163
– V-I characteristics   163
equivalent luminance   288
Ergonomic Lighting Indicator (ELI)   233
external quantum efficiency   62

g

glare illuminance   386
glare source
– adaptation luminance ranges   369
– automotive lighting applications   370
– disability
– – age parameters factor   374
– – application   381
– – background luminance   371
– – CIE equations   373
– – formulae   374
– – object luminance   372
– – observer age factor K   373
– – stray light   371
– – veiling luminance function   374
– discomfort
– – de Boer rating scale   375
– – filter wheels   377
– – LED fixation   377
– – parameters equation   375
– – spectral radiance   377
– – spectral sensitivity function   376, 378, 379
– discomfort and disability experiment   387
– fitting tanh functions   381
– interview   370

– LED street lighting installation
– – luminance image   383
– – ME-street classes   384
– – stray light effect   388
– – threshold increment   383
– – TI value   385
– perception   371
– quasi-monochromatic radiation   379
green-red phosphor system   479
– chromaticity   111, 112
– colorimetric characteristics
– – CRI   117, 118
– – G2/R1 system   115, 116
– – G2/R2 system   115, 116
– – light and color quality   115, 117
– phosphor-converting system
– – emission characteristics   110
– – excitation spectra   110, 111
– – factors   109, 110
– – photoluminescence properties   109
– WLED
– – green proportion   114, 115
– – PLANCKIAN locus   113, 114

h

Helmholtz-Kohlrausch effect   285
high-tech semiconductor devices   479
hot binning point   158, 160
human visual system
– accommodation   10
– circadian effect
– – experimental data   47
– – human activity   44
– – ipRGC   44
– – melatonin and cortisol level   45, 46
– – nonvisual photosensitive pathway   45
– colorimetry and color science   22. *See also* colorimetry and color science
– cone mosaic and spectral sensitivity
– – chromatic signals   13
– – ipRGC mechanism   14
– – L, M, and S signals   12
– – photometry   13
– – rod-free inner fovea   13
– – small-field tritanopia   12
– human eye   8
– perception   8
– pupil diameter   9
– radiometry and photometry
– – definition   17
– – electric light sources   21, 22
– – irradiance and illuminance   19
– – photopic $V(\lambda)$ function   17, 18
– – radiance and luminance   20, 21

– – radiant and luminous flux   18, 19
– – radiant and luminous intensity   19, 20
– – scotopic $V'(\lambda)$ function   17, 18
– receptive fields and spatial vision
– – achromatic contrast sensitivity   15, 16
– – chromatic contrast sensitivity   16
– – definition   14
– – double opponent   15
– – stimulation   15
– retina   10
– spectral and colorimetric quantities
– – centroid wavelength   43
– – colorimetric purity   43
– – dominant wavelength   43
– – peak wavelength   41
– – spectral bandwidth   42
– – typical spectral radiance   41, 42
– visual information   7

i
imaging effect   479
Indium Gallium Nitride (InGaN)   63
injection efficiency   60
injection luminescence   50, 52
internal quantum efficiency   60
intrinsically photoreceptive retinal ganglion cell (ipRGC)   44

l
lamp lumen maintenance factor (LLMF)   465
lamp survival factor (LSF)   466
Landolt rings   357
lateral geniculate nucleus (LGN)   7
LED outdoor luminaire design
– advantages   443
– color rendering index   444
– LED installations
– – administrative tasks   473
– – communication luminaire manufacturers   476
– – data manufacturer   476
– – lighting system   476
– – luminance measurement   475, 477
– – main questions   471
– – pixel-resolved luminance distribution   474
– – process of qualification   475
– – psychological and architectural aspects   478
– – technical approach   473
– – workflow scheme   474
– luminous efficacy   444
– maintenance factor
– – ambient temperature   468

– – components   463
– – definition and illustration   465
– – depreciation behavior   464
– – formula   464
– – LED types   468
– – LLMF values   465
– – LMF values   466, 467
– – LSF values   466
– – luminaire performance   468
– – luminous flux constancy control   470, 471
– – luminous flux depreciation   469
– – luminous quantity   464
– – MF-value   464
– – pollution categories   466
– – street luminaires   467
– street lighting
– – characteristics   452
– – characterization and selection   456
– – communication and control task   447
– – electronic group scheme   447
– – electronic unit   446
– – flat and light surface   446
– – functional units   445
– – geometry   453
– – LED board temperature sensor   448
– – LED-components   448
– – LID curve   455, 457
– – luminaire number   462
– – luminous efficacy   451
– – luminous flux   461
– – luminous intensity   449
– – luminous intensity distribution   456
– – ME2-class   454
– – mechanical unit   445
– – mounting holes   462
– – numerous aspects   458
– – optical system   449
– – reflector and lens optics   450
– – road width *vs.* pole height   453
– – S4-class   454
– – systematic approach   452
– – temperature equals   459
– – temperature measures   460
– – temperature profile   459
– – test field   461
– – thermal and electronic dimensioning   458
– – thermal resistance   459
LED package   480
LED-indoor luminaire design
– applications   399
– color rendering index   401, 402
– – LED types   422, 425

LED-indoor luminaire design (*contd.*)
– – luminous efficacy   424–426
– – relative emission spectra   422–425
– components   400
– components and units
– – angular distribution   411
– – color shift   412
– – forward voltage   413
– – geometry   410
– – optimal current   413
– – thermal resistance   413
– – warm/cold white   412
– control and regulation electronics   406
– diffuse plastic plate   404
– functional units and tasks   403
– high quality
– – color rendering index   418
– – film and TV lighting   416
– – HMI discharge lamp   417
– – long tube fluorescent lamps   417
– – museum and gallery lighting   414
– – museum tasks   415
– – requirements   420
– – shop lighting   419
– methods   403
– optical system
– – advantages/disadvantage   405
– – dark-light pattern   406
– – optimization   408
– optimization and stabilization
– – control system structure   440
– – CRI values   437
– – hybrid LED lamp structure   435
– – hybrid LED lamps   427, 433, 438
– – hybrid LED luminaire   428
– – LED components   433, 440
– – lighting quality parameters   427, 439, 441
– – MIMO input system   434
– – MIMO object system   434
– – MIMO output system   434
– – MIMO system   440
– – museum lighting   431, 436
– – operating temperature   439
– – spectral reflectance   428. *See also* spectral reflectance curves
– packaging systems   403
– parameter values   400
– primary tasks   400
– secondary lenses   404
– structure   404
– thermal management   407
– – direct and indirect light contributions   409
– – LID requirements   408

– – optimization   407
     pixel-resolved angular   410
lifetime extrapolation
– ARRHENIUS behavior   217
– border function   215, 216
– concave/convex function   218, 219
– exponential function   218
– extrapolated lifetime   221
– quadratic function   220, 221
– root function   218
– TM 21-method   215, 216
– vector acceleration   217
light emitting diodes (LED)   389
light extraction efficiency   60, 61
lighting engineering
– colorimetry and color science   1
– definition   1
– human visual system   4
– indoor/outdoor scene   3
– interdisciplinary workflow   3
– light production techniques   1
– optimization process   4
– photopic perceptual aspects   5
– principles   2
luminaire maintenance factor (LMF)   466
luminous efficacy   62
luminous flux
– calibration reference   140
– geometry measurement   139, 140
– measurement errors   141
– relative spectral responsibility   141
luminous intensity distribution (LID)   408, 464

m

maintenance factor   464
measurement methods
– cooling measurement   191
– thermal map decoding
– – EUCLIDEAN algorithm   195, 196
– – structure function   194, 195
– – thermal power and calibration factor   193, 194
– – thermal   electrical and optical properties, 192
– WSM *vs* ETM   190
mesopic brightness perception   481
mesopic models
– glare source   368. *See also* glare source
– LED light sources
– – brightness perception   348
– – H7 reference luminance levels   350
– – inter-observer variability   350, 352, 353
– – LED and Xenon brightness   350

– – photopic luminance values   347
– – spectrum   349
– PWM
– – bead string artifact   389, 390. *See also* bead string artifact
– – DC operation   389
– road markings
– – automotive headlamps   364
– – contrast effective radiance   366
– – luminance based contrast effective radiance   367
– – quality control   363
– – photopic luminance   366
– – retro-reflective   365
– – spectral characteristics   365
– – street lighting spectral radiance   364
– visual acuity
– – colorimetric properties   360
– – Landolt ring   358
– – letter contrast acuity   360, 361
– – Pelli Robson test chart   359, 361
– – Snellen chart   358
– visual performance
– – brightness impressions   337
– – brightness models   339
– – chromatic signals   341
– – CIE   344. *See also* CIE model
– – contrast perception   341
– – detection   338
– – HPS lamp   356
– – LED light sources   355
– – LMSR signals   339
– – luminance values   353
– – object recognition   338, 341
– – photopic luminance values   355
– – reaction time   341
– – S/P ratios   356
– – spectral sensitivity curves   343
– – spectral sensitivity functions   342
– – threshold detection   339
– – visual search process   338
– – X model   339
mesopic visual performance   387
multi-input multi-output (MIMO) system   168, 434

n
nitride phosphors
– backlighting units (BLUs)   93
– CASN
– – crystal structure   94
– – excitation and emission spectrum   94, 95
– – temperature dependence   96
– 2-5-8-nitrides

– – chromaticity co-ordinates   99
– – emission and excitation characteristics   97, 98
– – photoluminescence properties   97
– – temperature dependence   100
– 1-2-2-2 oxy-nitrides   99–101, 103
– β-SiAlON   103–105
non-imaging pathway   479

o
optical behavior   197. *See also* thermal model

p
PELTIER cooler system   192
phonons   58
phosphor system
– alkaline earth ortho-silicates   89, 90, 92
– alkaline earth oxy-ortho-silicates   92, 93
– alkaline earth sulfides
– – disadvantages   87
– – hydrogen sulfide gas   87
– – luminescent properties   88, 90
– – quantum efficiency   86, 87
– – red-emitting nitride   86
– aluminum garnets
– – cold white light   83
– – critical distance   84
– – optical bandgap   84, 85
– – physical   chemical and spectroscopic properties, 83
– – STOKES shift   83
– – temperature dependence   82
– – trivalent terbium absorption   85
– disadvantages   72
– divalent europium   78
– dopants   73
– emission lifetime   75, 76
– JUDD-OFELT theory   74
– luminescence mechanism
– – characteristic/center luminescence   79
– – configuration co-ordinate diagram   79–81
– – electron/charge carrier hopping   82
– – photoionization process   81
– nitride phosphors   93. *See also* nitride phosphors
– operation of   73, 74
– organic dyes   72
– phosphor coating method   104, 105
– phosphor concentration/thickness   107–109
– plasma discharge lamp   72
– quantum efficiency (QE)   74
– rare-earth ions (RE)   73

phosphor system (contd.)
- saturation effects 76, 77
- trivalent cerium 77, 78
- volumetric dispensing method 106, 107
photopic perceptual aspects
- color preference assessment
- - CQS method 279. *See also* color quality scale (CQS)
- - subjective factors 273
- - R_a and $R_{a,2012}$ 274
- color quality
- - color appearance 236
- - color fidelity 236
- - color gamut 237
- - color harmony 238
- - computational approach 238
- - correlation analysis 239
- - CQS 239
- - light source 237
- - memory color rendering 238
- - visual clarity 237
- CRI 241. *See also* color rendering indices (CRI)
- lighting quality
- - aesthetic/emotional assessments 234
- - alternative spectral weighting function 242
- - classic V(λ) based scheme 240
- - cognitive level 233
- - ELI 233
- - interior lighting 236, 239
- - ipRGC mechanism 242
- - light and health aspects 234
- - luminance levels 240
- - spatial brightness 235
- - spectral optimization 240
- - spectral radiant intensity 236
- - visual performance 240
- - V(λ) function 239
- - VL and CQ measurement 241
- White LED light sources 275. *See also* white LED light sources
pulse width modulation (PWM) method 223, 389

r

radiant efficiency 62
radiometry, photometry and colorimetry
- CCD sensor 135
- image resolution 134, 135
- luminous flux
- - calibration reference 140
- - geometry measurement 139, 140
- - measurement errors 141
- - relative spectral responsibility 141
- near-field goniophotometer 138, 139
- physical measurements 133, 134
- relative luminance deviation 138
- spectral radiant flux
- - calibration 140, 142, 143
- - spectral mismatch correction factor 142
- - spectroradiometer arrangements 140, 142
- stochastic errors 138
- systematic errors 136–138
Rea model
- CS *vs.* CCT 318
- definition 316
- ipRGC to photopic ratio 319, 320
- lighting practitioner 319
- nocturnal melatonin suppression 316
- spectral power distribution 319
- spectral sensitivity 316
- workflow model 316

s

semantic binning method 305
semantic contours 302
semantic interpretation
- CAM02-UCS color differences 256
- CIE CRI method
- - $R_{a,2012}/R_{i,2012}$ and R_a 260
- - $R(R_a)$ function 258, 259
- - advantage 260
- - non-uniformity results 258
- - perceived color difference 260
- - polynomial coefficients 259
- - R(Ra) function 260
- - test color samples 258
- color rendering index 253
- color stimuli 253
- CRI2012 method 257
- experimental method
- - color stimuli 255
- - reference light sources 254
- - three-chamber viewing booth 254
single-input single-output system (SISO system) 185
spectral model 480
- color semiconductor
- - accuracy 177
- - combined approach 171, 173
- - description 174
- - mathematical approach 168–171
- - MIMO system 168
- - multi GAUSSIAN model 174, 175, 177
- white PC-LED
- - description 178, 179

– – fourth-order GAUSSIAN function   178–180
– – ninth-order Gaussian function   180, 181
spectral power distributions (SPDs)   294
spectral radiant flux
– spectral calibration   140, 142, 143
– spectral mismatch correction factor   142
– spectroradiometer arrangements   140, 142
spectral reflectance curves
– clothes   430
– green group oil colors   432
– human hair   429
– human skin   429
– saturated blue group oil colors   432
– saturated red group oil colors   431
– seventy nine oil colors   431
– skin types   429
street geometry   453
surface maintenance factor (SMF)   466

t

thermal model
– equivalent thermal circuit   182, 183
– EUCLIDEAN algorithm   200, 201
– one-dimensional thermal model
– – first order equivalent circuit   185, 186
– – nth order equivalent circuit   188, 189
– – second order equivalent circuit   187, 188
– – SISO system   185
– – transfer function   189
– – weighting function   189
– radiant efficiency   197, 198
– structure function method   198
– structure of   182
– temperature stability   197
– thermal resistance   183, 184

w

Ware-Cowan brightness model
– L_{eq} values   288
– CCT and CRI descriptors   289
– equivalent luminance   288
– L_{eq} values   289
– quasi-monochromatic stimuli   289
white LED light sources
– automotive lighting   225
– brightness and lightness impression   285
– CCT   294
– chromatic lightness model   290–292
– chromaticity binning   298. See also chromaticity binning method
– CIE brightness model   287, 288
– circadian stimulus(CS)
– – vs. CCT   318

– – vs. color rendering index   320
– – Rea model   317. See also rea model
– contour reproduction   295
– correlated color temperature   225, 227
– first color quality experiment
– – chromatic lightness task   310, 311
– – color preference task   310, 312
– – pc-LEDs and RGB LEDs   310
– – spectral power distributions   310
– flicker and stroboscopic effect
– – definition   322
– – age histogram   327, 329
– – average assessment   329
– – Bullough study   323, 325
– – cumulative histogram   329
– – desktop   327, 328
– – dimming methods   321
– – experimental setup   325
– – integral number   328
– – laptop   328
– – mean assessments   330
– – paper and hatch   329
– – perception rating   331
– – research study   323
– Helmholtz-Kohlrausch effect   285
– human visual system   285
– indoor and outdoor lighting   225
– linearity test   226
– luminous efficacy   226
– perceived whiteness   292
– sample set analysis   296
– second color quality experiment
– – brightness   314
– – color harmony   314
– – color preference   314
– – colorimetric properties   313
– – high-quality pc-LEDs   313
– – spectral power distributions   313
– – visual color rendering property   314
– SPDs   295
– spectral design   286
– target chromaticity limit   293
– target white tone chromaticity   292
– Ware-Cowan brightness model
– – L_{eq} values   288
– – CCT and CRI descriptors   289
– – equivalent luminance   288
– – L_{eq} values   289
– – quasi-monochromatic stimuli   289
– white tone preference   292
white PC-LED
– chromaticity difference   156, 157
– CRI vs luminous efficacy   153

white PC-LED (*contd.*)
– luminous flux and luminous efficacy   154, 155
– spectral components   150, 151, 157
– spectral model
– – description   178, 179
– – fourth-order GAUSSIAN function   178–180
– – ninth-order Gaussian function   180, 181
– white point quality   153, 154
white phosphor-converted LEDs (WLED)
– green proportion   114, 115
– PLANCKIAN locus   113, 114

x

X model/unified system of photometry (USP)   339